# HARCOURT BRACE JOVANOVICH COLLEGE OUTLINE SERIES

# PROBABILITY AND STATISTICS

## Dr. Ronald I. Rothenberg

*Queens College, City University of New York*

HBJ

Books for Professionals
Harcourt Brace Jovanovich, Publishers
San Diego   New York   London

Permissions Department
Harcourt Brace Jovanovich, Publishers
8th Floor
Orlando, Florida 32887

Printed in the United States of America

Library of Congress Cataloging-in-Publication Data

Rothenberg, Ronald I.
    Probability and statistics / Ronald I. Rothenberg.
        p.   cm.—(Harcourt Brace Jovanovich college outline series)
    Includes bibliographical references and index.
    ISBN 0-15-601676-1
    1. Probabilities.   2. Mathematical statistics.   I. Title.
    II. Series.
    QA273R855   1991
    519.2—dc20                                          91-29352

ISBN 0-15-601676-1

First edition

A  B  C  D  E

# PREFACE

This Outline is intended to help those students who have had enough calculus to deal with a higher level treatment of probability and statistics—whether or not they are taking a formal course. The topics covered—ranging from the basic concepts of probability to multivariate probability distributions—are those usually taught in the first semester of a calculus-based course on probability and statistics, so this Outline can be used as a supplement for such a course. But, because of its detailed treatment and continuity, this book can also be used for independent study or even as a primary textbook.

In this Outline you'll find a development of all the basic concepts: the axioms of probability, conditional probability, the binomial and normal distributions, the expected value, descriptive statistics, multivariate probability distributions—the whole ball of wax. And all of these concepts are explained as plainly and simply as the subject matter allows.

There is a dual emphasis in this Outline, namely on theory and on real-world applications. The approach is geared to problem solving. In fact, there are hundreds of solved examples and problems that deal with theory, applications to practical situations, and computational techniques. In addition, there are over a hundred supplementary problems (with hints and answers) that let you try your hand at solving problems on your own.

Because the ideas and techniques of probability and statistics are used in a variety of disciplines, this Outline will be useful to students of the natural sciences, engineering, and the social sciences, as well as to students of mathematics. In addition, this book provides a fine review for the Course 110 examination of the Society of Actuaries. Finally, those readers interested in the application of probability concepts to gambling will find much to help them in these pages.

A note on format is in order. The symbol for summation of terms for $x$ going from 0 to $\infty$ by increments of 1 is indicated within a text discussion as $\sum_{x=0}^{\infty}$, while within a displayed equation the symbolic representation $\sum_{x=0}^{\infty}$ is used. Likewise, for other terms involving the $\sum$. Similarly, the symbol for the limit as $x$ approaches a number $a$ is written within a text discussion as $\lim_{x \to a}$ while within an equation it would be written as $\lim_{x \to a}$. The same practice is followed for other expressions involving limits.

I am pleased to acknowledge the suggestions, moral support, and inspiration provided by Executive Editor Emily Thompson of Harcourt Brace Jovanovich. Ms. Thompson was the individual who arranged for me to write this book. Also, I must mention how much I appreciate the production work of Spectrum Publisher Services, and their fine editorial liaison, Elise Oranges.

Finally, this book is dedicated to the memory of T. Freeman Cope. Professor Cope was the Chair of the Mathematics Department at Queens College who hired me, and helped pave the way for my rewarding career as a college teacher of mathematics.

*Flushing, New York*                                                                 *Ronald I. Rothenberg*

# CONTENTS

# 1 INTRODUCTION TO PROBABILITY

## THIS CHAPTER IS ABOUT

- ☑ **Introductory Ideas**
- ☑ **Elementary Concepts**
- ☑ **Events**
- ☑ **Set Theory and Venn Diagrams**
- ☑ **Axioms of Probability**

## 1-1. Introductory Ideas

**Probability** is the study of random or nondeterministic situations. In such situations, the observer doesn't know, for sure, what will happen. Probability occurs in many aspects of modern life, including gambling situations (such as roulette, card games, or dice rolling), determination of premiums by insurance companies, and inheritance of genetic characteristics such as eye color, height, or hair type.

Historically, the mathematical analysis of probability started back in the seventeenth century when a French gambler, the Chevalier de Méré, consulted the mathematicians Blaise Pascal and Pierre Fermat about some puzzling aspects of a game of chance involving dice tossing.

## 1-2. Elementary Concepts

Suppose we toss a fair die $N$ times, and we let $s$ denote the number of times a one (or ace) occurs. (A fair die is one in which the mass is uniformly distributed, and where all six faces have identical areas. Thus, each of the six faces can occur with equal likelihood in the toss of the die.)

It has been experimentally observed that the ratio $s/N$, called the relative frequency (denoted by $\hat{f}$), approaches the definite number $1/6 = .1666\ldots$, as $N$ gets larger and larger. This is shown in Table 1-1, which records the results for 60,000 simulated tosses of a die, for the number and relative frequency of aces at various points in the tossing.

*note:* The simulation was done by means of a simple computer program, written in BASIC, which employed a function that generated random numbers. The data in Table 1-1 are for a typical simulated run of the computer. In each of the runs conducted, the relative frequency appeared to approach 1/6 as $N$ got larger and larger.

As you can see, initially the relative frequency $\hat{f}$ varied a lot, but as $N$ got larger, $\hat{f}$ became very stable in value. In the last 24,000 tosses, $\hat{f}$ changed by only .0012.

To generalize, suppose we have an experiment that can be repeated over and over again. Suppose that during $N$ repetitions of the experiment a particular event $E$ occurs $s$ times. Thus, the relative frequency of $E$ is given by $\hat{f} = s/N$. If $\hat{f}$ approaches a definite number as $N$ gets larger and larger, that number is called

**TABLE 1-1: Relative Frequency of Ace During 60,000 Simulated Tosses of a Die**

| Number of tosses, $N$ | Number of aces, $s$ | Relative frequency of ace, $\hat{f} = s/N$ |
|---|---|---|
| 100 | 22 | .2200 |
| 600 | 105 | .1750 |
| 6,000 | 1,032 | .1720 |
| 18,000 | 3,107 | .1726 |
| 36,000 | 6,049 | .1680 |
| 60,000 | 10,008 | .1668 |

the **probability of the event** $E$, and it is denoted by $P(E)$. For the above die tossing example, we have that $P(\text{Ace}) = 1/6$. This is read: "the probability of Ace is 1/6."

The above description of the probability of an event is somewhat unsatisfactory because the phrases "approaches a definite number" and "as $N$ gets larger and larger" have not been precisely defined. But, at any rate, one should have an intuitive understanding that the probability of an event is an idealized relative frequency of the event. (Sometimes, one substitutes the expression "proportion of occurrence" for "relative frequency.")

A (probability) experiment is one for which we cannot state what a particular outcome will be when the experiment is performed, but for which we are able to describe the set of all possible outcomes of the experiment.

**Definition 1.1:** A **(probability) experiment** $\mathscr{E}$ is a specific set of actions, the results of which cannot be predicted with certainty.

**Definition 1.2:** A set $S$ is a **sample space** for an experiment $\mathscr{E}$ if each element of $S$ represents a unique outcome of $\mathscr{E}$, and if each outcome of $\mathscr{E}$ is represented by a unique element of $S$.

*note:* Often, more than one sample space is possible for a given experiment. The appropriateness of a particular sample space depends upon what the experimenter chooses to record or observe. Refer to Example 1-1, which follows, for an illustration of this idea.

---

**EXAMPLE 1-1:** Suppose we perform the experiment of tossing a coin twice. Indicate an appropriate sample space if (**a**) we wish to record what happens on each toss, or (**b**) we wish to record the number of heads obtained.

*Solution*

(**a**) An appropriate sample space is $S = \{TT, TH, HT, HH\}$, where, for example, the element $TH$ refers to the outcome tail on first toss followed by head on second toss.

(**b**) An appropriate sample space is $\hat{S} = \{0, 1, 2\}$, where, for example, 1 means one head was obtained. We use $\hat{S}$ here to distinguish between this sample space and the previous one. Note that the outcome 1 of $\hat{S}$ corresponds to the two outcomes $TH$ and $HT$ of $S$.

---

Both $S$ and $\hat{S}$ of Example 1-1 represent outcomes for an experiment consisting of tossing a coin twice. Clearly, $S$ provides more information than $\hat{S}$. If we know which element of $S$ occurs, we can tell which element of $\hat{S}$ occurs; the reverse is not true, however. When in doubt as to which sample space to choose for an experiment, it is a good rule of thumb to use one whose elements cannot be further *subdivided*. In Example 1-1, sample space $S$ cannot be further subdivided.

Sample spaces are categorized according to the number of elements they contain. A sample space is said to be *finite* if it has a finite number of elements. A sample space is said to be *discrete* if the elements can be placed in a one-to-one correspondence with the positive whole numbers (integers), or if it is finite. In this book, we shall deal with finite sample spaces, nonfinite but discrete sample spaces, and nondiscrete sample spaces.

---

**EXAMPLE 1-2:** Describe a sample space for the experiment of tossing a pair of dice. Here, it is convenient to distinguish between the dice; for example, they could be painted with two different colors, say white and pink.

*Solution:* Suppose we wish to record what happens on each die, where, say, the first die is the white die. Thus, the appropriate sample space (the most subdivided sample space) contains 36 elements, and it is given by

$$S = \{11, 12, \ldots, 16; 21, 22, \ldots, 26; \ldots; 61, 62, \ldots, 66\}$$

For example, 26 means 2 on the first die and 6 on the second die.

---

In the next example, we consider a sample space that is discrete, but nonfinite.

---

**EXAMPLE 1-3:** Consider the experiment of tossing a coin until a head occurs. Write down a sample space such that each element contains a sequence of tosses for which only the last toss is a head.

*Solution:* An appropriate sample space is the following discrete sample space, which contains a countably infinite number of elements.

$$S = \{H, TH, TTH, TTTH, TTTTH, \ldots\}$$

Here, for example, element $TTH$ indicates that the first head occurs on the third toss.

---

In some sample spaces, the number of elements is neither finite nor countably infinite; that is, such sample spaces are not discrete. Such is the case, for example, when one records the time it takes for an activated light bulb (i.e., one that is "turned on") to fail. It is reasonable to suppose that the collection of all such possible times (called life times or life lengths) is not discrete.

## 1-3. Events

In elementary discussions of probability, we first decide on a sample space corresponding to a probability experiment. It is fruitful to describe events of an experiment as subsets of the sample space.

**Definition 1.3:** An **event** is a subset of a sample space. A **simple event** is a subset containing a single element. In many discussions, we will liberally interchange the words "event" and "subset."

---

**EXAMPLE 1-4:** Consider the experiment of tossing a coin twice, with sample space $S = \{TT, TH, HT, HH\}$, as in Example 1-1(**a**). Represent the following verbally described events as subsets of $S$: (**a**) exactly one head; (**b**) head on the second toss; (**c**) at least one head; (**d**) head on both tosses.

*Solution*

(a) $A = \{TH, HT\}$. The elements of $A$ denote the outcomes for which exactly one head occurs. Another verbal description of this event is "one head and one tail." This illustrates the fact that usually more than one verbal description for an event is possible.

(b) $B = \{TH, HH\}$.

(c) $C = \{TH, HT, HH\}$. Another verbal description of this event is "either one or two heads."

(d) Using the letter $G$ here, we have $G = \{HH\}$. Here, we have a simple event.

*note:* In the rigorous language of mathematical probability and statistics, the term *event* is actually reserved for subsets belonging to a special class of subsets of a sample space. However, if the sample space is discrete, then all subsets are events. We will speak about the situation with regard to nondiscrete sample spaces when it is appropriate.

Figure 1-1 depicts the sample space $S$, and subsets $A$ and $B$ of Example 1-4. The elements are indicated as labeled dots, and subsets $A$ and $B$ (also known as events $A$ and $B$) are indicated by closed curves enclosing dots.

Figure 1-1 is an example of a Venn diagram (after the English logician John Venn, 1834–1883). In Venn diagrams, which are useful in discussions involving set theory, a rectangle is drawn to represent the relevant universal set. In probability discussions, the universal set is a sample space $S$. Dots (or points) within the rectangle represent elements, and subsets (events) are represented by regions within the rectangle enclosed by closed curves. Often, the individual dots are not shown. (By the way, we frequently will use the words "subset" and "set" interchangeably.)

Since events are equivalent to sets, we can obtain new events by employing the set operations of union ($\cup$), intersection ($\cap$), and complement ($'$).

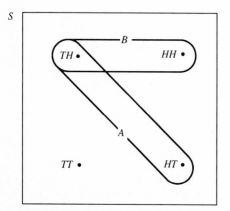

**Figure 1-1.** A Venn diagram for Example 1-4.

## 1-4. Set Theory and Venn Diagrams

### A. Elementary set theory

A **set** is simply a well-defined collection of objects (things) of any type. The objects making up the set are called its **elements** or **sample points**. We often write $a \in A$ to indicate that $a$ is an element of set $A$. Also, in this case, we sometimes say that $a$ belongs to set $A$. We often write $b \notin B$ to indicate that $b$ is not an element of set $B$.

There are several ways to specify elements of a set. In one way, we list all the elements and enclose them in braces $\{\ \}$; this is the roster method of defining a set. For example, in Example 1-1(a), we had $S = \{TT, TH, HT, HH\}$ as an illustration of the roster method. Another way to specify elements in a set is by including a statement, which gives defining properties for the elements of the set, within the enclosing braces. Thus, for example, in Example 1-2 (for tossing a pair of dice), we can describe the sample space as

$$S = \{ij \mid i = 1, 2, 3, 4, 5, 6; j = 1, 2, 3, 4, 5, 6\}$$

Here, the vertical bar $\mid$ is read as "such that" or "for which." Some authors use : (colon) for the "such that" symbol.

### B. Subsets, continued

A set may have no elements. The set with no elements is called the **empty set** or **null set**, and is denoted by the symbol $\varnothing$ or $\{\ \}$; in the latter symbol, we have braces with no elements between them. Two sets $A$ and $B$ are equal, written $A = B$, if $A$ and $B$ have exactly the same elements. That is, $A = B$ if each element of $A$ is an element of $B$, and each element of $B$ is an element of $A$. If one of the two sets has an element which is not in the other, then the sets are unequal; we indicate this by writing $A \neq B$.

We say that $A$ is a subset of $B$, written $A \subset B$ or $B \supset A$, if each element of $A$ is also an element of $B$. We agree that the null set is a subset of every other set. If $A \subset B$, we also say that $A$ is contained in $B$, or that $B$ contains $A$. We write $A \not\subset B$ to indicate that $A$ is not a subset of $B$.

---

**EXAMPLE 1-5:** Consider the experiment of tossing a coin twice, with the associated sample space $S = \{TT, TH, HT, HH\}$. In Example 1-4, we discussed the following events of this sample space:

| Event as subset of $S$ | Verbal description of event |
|---|---|
| $A = \{TH, HT\}$ | Exactly one head |
| $B = \{TH, HH\}$ | Head on the second toss |
| $C = \{TH, HT, HH\}$ | At least one head |

Determine which of these events are subsets of other events.

*Solution:* Clearly, $A \subset C$, and $B \subset C$, but $A \not\subset B$ since element $HT$ is in $A$, but $HT$ is not in $B$. Also $A \subset A$, $B \subset B$, and $C \subset C$ since any set is a subset of itself.

---

We note that, in general, two sets $A$ and $B$ are equal if $A \subset B$ and $B \subset A$, that is if $A$ is a subset of $B$ and $B$ is a subset of $A$.

When $A \subset B$, but $A \neq B$, then $A$ is said to be a proper subset of $B$. In Example 1-5, both $A$ and $B$ were proper subsets of $C$. [Note that some authors use the symbol $\subseteq$ for subset, and the symbol $\subset$ to mean proper subset.] In general, we can say that $A$ is a proper subset of $B$ if every element of $A$ is an element of $B$, but at least one element of $B$ does not belong to $A$.

Remember that in our probability discussions, we indicated that events are equivalent to subsets of some sample space $S$. Also, we say that an event $E$ "occurs" if a performance of the experiment results in an outcome symbolized by an element of the subset representing $E$. Thus, referring to Example 1-5, we say that the event $A$, "exactly one head," occurs if either the outcome symbolized by $HT$ ("head followed by tail") occurs, or the outcome symbolized by $TH$ occurs. [In the future, we will usually merely say that "$HT$ occurs" instead of saying "the outcome symbolized by $HT$ occurs."]

Suppose that set $E$ is a subset of set $F$, that is $E \subset F$. Thus, when event $E$ occurs, it is also true that event $F$ occurs. Referring to Example 1-5, we see that the occurrence of "exactly one head" implies the occurrence of "at least one head."

## C. Complement of a set; intersection and union of sets

In most discussions of sets, all the sets under discussion can be regarded as subsets of some fixed set, known as a universal set. For our discussions, the universal set will usually be a sample space $S$ for a probability experiment. Sets $A$, $B$, etc., below are understood to be subsets of $S$.

**Definition 1.4 (Complement of a Set):** The **complement** of a set $A$ with respect to $S$ (or, simply, the complement of $A$), denoted by $A'$ (some use $A^c$ or $\bar{A}$), is the set of all those elements of $S$ that do not belong to $A$.

**Definition 1.5 (Intersection of Sets):** The **intersection** of two sets $A$ and $B$, denoted by $A \cap B$, is the set of all those elements which belong to both $A$ and $B$. In symbols (where "$x \in$" means "$x$ is an element of"), we say

$$A \cap B = \{x \mid x \in A \text{ and } x \in B\}$$

This is read "$A \cap B$ is the set of elements such that each element is an element of $A$ and of $B$."

Given sets $A_1, A_2, \ldots, A_n$, each of which is a subset of sample space $S$. Then, $A_1 \cap A_2 \cap \ldots \cap A_n$ [read as the "intersection of $A_1, A_2, \ldots, A_n$,"] is the set of elements such that each element is an element of $A_1$, and of $A_2, \ldots$, and of $A_n$.

**Definition 1.6 (Union of Sets):** The **union** of two sets $A$ and $B$, denoted by $A \cup B$, is the set of all those elements that are either in $A$ alone, or in $B$ alone, or in both $A$ and $B$. Often one says that $A \cup B$ is the set of all elements that belong to $A$ or to $B$ where "or" is used in the inclusive sense of "and/or." Still another valid statement is "$A \cup B$ is the set of those elements that belong to at least one of the sets $A$ and $B$."

In symbols, where the word "or" means "and/or,"

$$A \cup B = \{x \mid x \in A \text{ or } x \in B\}$$

Given sets $A_1, A_2, \ldots, A_n$, each of which is a subset of sample space $S$. Then, $A_1 \cup A_2 \cup \ldots \cup A_n$ [read as the "union of $A_1, A_2, \ldots, A_n$,"] is the set of those elements such that each element belongs to at least one of $A_1, A_2, \ldots, A_n$.

Since the events of a sample space $S$ are equivalent to subsets of $S$, we can obtain new events by employing the set operations of complement, intersection, and union.

$A'$, the complement of $A$, is the event that occurs if event $A$ does not occur. We read $A'$ as "not $A$." $A \cap B$ is the event that occurs if both $A$ and $B$ occur. We read $A \cap B$ as "$A$ and $B$." $A \cup B$ is the event that occurs if $A$ alone occurs, $B$ alone occurs, or if both $A$ and $B$ occur (or, equivalently, if at least one of $A$ and $B$ occurs). We read $A \cup B$ as "$A$ or $B$," where we interpret the "or" as meaning and/or.

---

**EXAMPLE 1-6:** Consider the experiment of tossing a coin twice, with the associated sample space $S = \{TT, TH, HT, HH\}$, as discussed in Examples 1-1, 1-4, and 1-5. For the events $A$, $B$, and $C$ of Examples 1-4 and 1-5, determine the following events: (a) $A'$, $B'$, and $C'$; (b) $A \cup B$; (c) $A \cap B$; (d) $A \cap C'$. Recall that, as subsets, $A = \{TH, HT\}$, $B = \{TH, HH\}$, and $C = \{TH, HT, HH\}$.

*Solution*

(a) $A' = \{TT, HH\}$. In words, this is the event in which "the same result occurs on both tosses." $B' = \{TT, HT\}$. In words, this is the event "tail on the second toss." $C' = \{TT\}$, the simple event of "getting tail on both tosses."
(b) $A \cup B = \{TH, HT, HH\}$. Observe that $C = A \cup B$.
(c) $A \cap B = \{TH\}$.
(d) $A \cap C' = \varnothing$, the empty set. Sets $A$ and $C'$ have no common elements.

---

*notes*

(a) Observe that, in general, $S' = \varnothing$ and $\varnothing' = S$.
(b) Observe that we distinguish between an element $e$ of a sample space and the simple event $\{e\}$, which is a subset of the sample space. Remember

that an element by itself is not a set. Elements belong to sets. Thus, for example, element $e$ belongs to set $\{e\}$.

## D. Mutually exclusive events

Two events $A$ and $B$, which are subsets of the same sample space, are said to be mutually exclusive or disjoint if they have no elements (outcomes) in common.

**Definition 1.7 (Mutually Exclusive Events):** The pair consisting of $A$ and $B$ is mutually exclusive (disjoint) if $A \cap B = \varnothing$. [In this case, we often say that $A$ and $B$ are mutually exclusive.]

    The collection $A_1, A_2, \ldots, A_n$ is mutually exclusive if $A_i \cap A_j = \varnothing$ for $i \neq j$. [In this case, we often say that $A_1, A_2, \ldots, A_n$ *are* mutually exclusive.]

    In a verbal sense, we say that events $A$ and $B$ are mutually exclusive if they cannot occur simultaneously.

**EXAMPLE 1-7:** Which pairs of events of the following events of Example 1-6 are mutually exclusive? $A = \{TH, HT\}$, $B = \{TH, HH\}$, $C' = \{TT\}$, $A' = \{TT, HH\}$, $A \cap B = \{TH\}$.

*Solution:* $A$ and $C'$ are mutually exclusive, as are $A$ and $A'$. $B$ and $C'$ are mutually exclusive, as are $C'$ and $A \cap B$. $A'$ and $A \cap B$ are mutually exclusive.

    It is useful at this point to note that $A \cup B = B \cup A$, and $A \cap B = B \cap A$. Thus, order doesn't matter when we are dealing with either union or intersection.

## E. Venn diagrams

We briefly discussed Venn diagrams in Section 1-3. Venn diagrams are used for the purpose of representing sets and set operations (such as complement, intersection, and union) pictorially. A rectangle is drawn which represents the universal set, which, for our discussions, will usually be a sample space $S$. The points within the rectangle represent elements of $S$. On occasion, we shall specifically indicate the elements as labeled dots within the rectangle. This was done in Figure 1-1 with the elements of $S = \{TT, TH, HT, HH\}$. In certain situations, we shall not label the elements within $S$. One such situation is when no clarification in the discussion is gained by such labeling. Another situation when elements may not be labeled is when the sample space contains infinitely many elements (either countably infinite—that is, discrete and infinite—or uncountably infinite).

    A typical subset of $S$ is represented by the region within a closed curve; for instance, a circle. [Note, however, that straight line segments are acceptable, also—see Figure 1-1, for example.] The Venn diagrams of Figures 1-2a through 1-2d illustrate some of the ideas discussed in Sections 1-4C and 1-4D. In these diagrams, elements are not individually labeled. In Figure 1-2a, a typical subset (event) $A$ is represented by the region within the circle, and $A'$, the complement of $A$, is represented by the shaded region within the rectangle which lies outside the circle. Figures 1-2b and 1-2c apply to the case where the sets might have elements in common; the elements common to $A$ and $B$, if such exist, would be in the region $R_1$. Figure 1-2d is applicable if $A$ and $B$ are disjoint.

    Notice in Figures 1-2b and 1-2c that the sample space set $S$ is broken up into four disjoint regions (sets), indicated by $R_1$, $R_2$, $R_3$, and $R_4$. A subdivision of $S$ of this type is known as a partition of $S$.

**Definition 1.8 (Partition):** A partition of a set $E$ is a collection of subsets $E_1$, $E_2, \ldots, E_m$ of $E$ with the following properties:

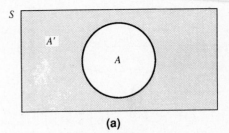

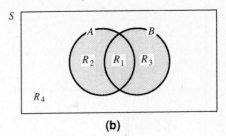

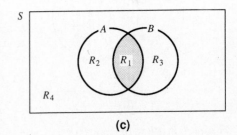

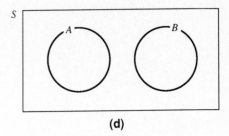

**Figure 1-2.** Simple Venn diagrams. (The $R_i$'s indicate disjoint regions.) **(a)** Set $A$ and complement set $A'$. **(b)** $A \cup B$ shown shaded. **(c)** $A \cap B$ shown shaded. **(d)** $A$ and $B$ are mutually exclusive (disjoint).

(a) $E_i \cap E_j = \varnothing$ for $i \neq j$. That is, the collection $E_1, E_2, \ldots, E_m$ is mutually exclusive. (See Definition 1.7.)

(b) $E = E_1 \cup E_2 \cup \ldots \cup E_m$. That is, $E$ equals the union of the $E_i$'s.

Thus, in Figures 1-2b and 1-2c, the collection $R_1, R_2, R_3, R_4$ constitutes a partition of $S$. We will make use of such $R_i$ sets (known as disjoint region sets) for proving and verifying laws of sets and probability, and also for practical problems.

---

**EXAMPLE 1-8:** In Figures 1-2b and 1-2c, express each disjoint region set $R_i$ as the intersection set of $A$ or $A'$ with $B$ or $B'$.

*Solution:* $R_1$ denotes the set of elements which are both in $A$ and $B$. Thus, $R_1 = A \cap B$. Also, $R_2 = A \cap B'$, $R_3 = A' \cap B$, and $R_4 = A' \cap B'$.

---

**EXAMPLE 1-9:** (a) Suppose a high school club contains 12 students whom we shall consecutively label as $1, 2, 3, \ldots, 10, 11, 12$. Suppose students 1 through 4, inclusive, are studying French, while students 3 through 9, inclusive, are studying Spanish, and students 10, 11, and 12 are studying neither language. If a probability experiment is to choose one of the students, then a reasonable sample space is $S = \{1, 2, 3, \ldots, 11, 12\}$. Suppose we let $A$ and $B$ denote the sets of students studying French and Spanish, respectively. Determine the elements in the four $R_i$ subsets, using the format of Figures 1-2b and 1-2c. (b) Repeat for a club with 12 students labeled $1, 2, \ldots, 11, 12$, if students 1, 2, 3, 4 are studying French, and students 5, 6, 7, 8, 9 are studying Spanish, with the rest studying neither language.

*Solution*

(a) Refer to Figure 1-3a. Here, $A = \{1, 2, 3, 4\}$, and $B = \{3, 4, 5, 6, 7, 8, 9\}$. Thus, $R_1 = A \cap B = \{3, 4\}$; $R_2 = A \cap B' = \{1, 2, 3, 4\} \cap \{1, 2, 10, 11, 12\} = \{1, 2\}$. Likewise, $R_3 = A' \cap B = \{5, 6, 7, 8, 9\}$, and $R_4 = A' \cap B' = \{10, 11, 12\}$.

*note:* We see that the four $R_i$'s just found are clearly disjoint; that is, no two of them have any elements in common. Also, we now have a complete subdivision of the students in terms of language study. For example, we know that students 5 through 9, inclusive, are taking Spanish but not French— see $R_3 = A' \cap B$.

(b) Here, $A$ and $B$ are disjoint. That is, the region designated as $R_1$ [again, $R_1 = A \cap B$] is shown without any labeled dots in it. Also, $R_2 = A$, $R_3 = B$, and $R_4 = \{10, 11, 12\}$. See Figure 1-3b.

Thus, we see that a Venn diagram representation with four labeled $R_i$ regions is useful even for special cases. Here, we have the special case where $R_1 = \varnothing$, but, in similar fashion we could also handle other special cases, as with other empty $R_i$'s.

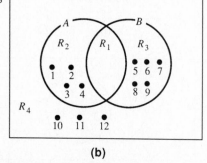

**Figure 1-3.** Venn diagrams for Example 1-9. In (b), $R_1 = \varnothing$.

## 1-5. Axioms of Probability

We shall denote the probability of an event [subset] $A$ by the symbol $P(A)$. As stated previously, the elements of a sample space $S$ symbolize the possible outcomes of interest.

### A. Relative frequency as a basis for probability

We can consider the probability of an event to be an idealized relative frequency of the event; we discussed this briefly in Section 1-2.

Consider the experiment of tossing a coin twice, and the most subdivided sample space $S = \{TT, TH, HT, HH\}$. Suppose we focus on the events $A = \{TH, HT\}$, $B = \{TH, HH\}$, $C = \{TH, HT, HH\}$, and $G = \{HH\}$. These were considered in Examples 1-4 and 1-5, for instance. Remember we say that an event $E$ occurs if a performance of the experiment results in an outcome symbolized by an element of the subset representing $E$. Thus, if tossing a coin twice results in $TH$ (that is, tail followed by head), then each of the events $A$, $B$, and $C$ will have occurred.

Also, event $S$ will have occurred; of course, event $S$ will occur regardless of the outcome of an experiment, and this is true for any experiment. [Note that $S$ is an event since it is a subset of $S$. Recall that any set is a subset of itself.]

Now, suppose we repeat the double tossing of a coin 1000 times (thus, the total frequency is $n = 1000$), and we record the frequency ($f$) and relative frequency ($\hat{f} = f/n$) for each of the outcomes. See Table 1-2.

**TABLE 1-2: Relative Frequencies in 1000 Double Tosses of a Coin**

| Outcome $\mapsto$ | $TT$ | $TH$ | $HT$ | $HH$ |
|---|---|---|---|---|
| Frequency, $f$ | 220 | 260 | 265 | 255 |
| Relative freq., $\hat{f}$ | .220 | .260 | .265 | .255 |

We see that the relative frequencies of events $A$, $B$, $C$, and $G$ are as follows:

$$\hat{f}(A) = \frac{260 + 265}{1000} = .525 \tag{1}$$

$$\hat{f}(B) = \frac{260 + 255}{1000} = .515 \tag{2}$$

$$\hat{f}(C) = \frac{780}{1000} = .780 \tag{3}$$

$$\hat{f}(G) = \frac{255}{1000} = .255 \tag{4}$$

Also,

$$\hat{f}(S) = \frac{1000}{1000} = 1 \tag{5}$$

There are several obvious facts concerning the relative frequencies of events of any sample space.

Fact 1: For any event $A$, $\hat{f}(A)$ is between 0 and 1 inclusive.

Fact 2: The relative frequency of event $S$ is 1. That is, $\hat{f}(S) = 1$. [For any performance of the experiment, a particular outcome symbolized by an element of $S$ will occur.]

Fact 3: If an event $J$ can be expressed as a union of two or more disjoint (mutually exclusive) events $A_1, A_2, A_3, \ldots$, then

$$\hat{f}(J) = \hat{f}(A_1) + \hat{f}(A_2) + \hat{f}(A_3) + \ldots$$

---

**EXAMPLE 1-10:** (a) Illustrate Fact 3 with respect to events $A$, $G$, and $C$ above [for the sample space $S = \{TT, TH, HT, HH\}$]. (b) Let $E_1, E_2, E_3, E_4$ denote the simple events $\{TT\}$, $\{TH\}$, $\{HT\}$, $\{HH\}$. [Note that $E_4$ is identical to $G$.]

Illustrate Fact 3 with respect to event $C$ and the disjoint simple events whose union is $C$.

### Solution

(a) We see that $C = A \cup G$, and that $A$ and $G$ are disjoint. In words, it should be clear that one can get at least one head by getting either exactly one head or two heads; the disjointness of the latter two events follows since it is not possible to simultaneously obtain both one head and two heads during the double tossing of a coin. Thus, it should be clear that $\hat{f}(C)$ should equal $\hat{f}(A) + \hat{f}(G)$. This is indeed the case since $.780 = .525 + .255$.

(b) Here, we see that

$$C = \{TH\} \cup \{HT\} \cup \{HH\} \tag{1}$$

that is,

$$C = E_2 \cup E_3 \cup E_4 \tag{2}$$

where it's clear that $E_2, E_3$, and $E_4$ are disjoint. Thus, it is clear that we should have

$$\hat{f}(C) = \hat{f}(E_2) + \hat{f}(E_3) + \hat{f}(E_4) \tag{3}$$

Equation (3) clearly holds here since $.780 = .260 + .265 + .255$.

---

## B. The probability axioms

**Definition 1.9:** The symbol $P(A)$ indicates the probability of an event [subset] $A$ of some sample space $S$. Here, the probability $P(A)$ is a real number.

The axioms [equivalently, postulates] below are very similar in form to the three "Facts" of Section 1-5A that held for relative frequencies. Thus, we are, in a sense, regarding a probability as being a sort of idealized relative frequency.

**Axiom 1.1:** For any event $A$ of a sample space $S$, $0 \le P(A) \le 1$.

**Axiom 1.2:** $P(S) = 1$.

**Axiom 1.3:** If an event $J$ can be expressed as a union of two or more disjoint (mutually exclusive) events $A_1, A_2, A_3, \ldots$, then

$$P(J) = P(A_1) + P(A_2) + P(A_3) + \ldots \tag{a}$$

Equivalently,

$$P(A_1 \cup A_2 \cup A_3 \cup \ldots) = P(A_1) + P(A_2) + P(A_3) + \ldots \tag{b}$$

*notes*

(a) An axiom (postulate) is a statement that we accept without proof. We shall build up the structure of the theory of probability by deducing theorems from axioms and/or previously proved theorems.

(b) The first axiom above should have been written merely as $0 \le P(A)$. In Problem 1-9 [of the Solved Problems section], we will deduce that $P(A) \le 1$.

(c) The axioms of probability as stated above apply when the sample space is discrete [that is, when $S$ has either a finite number of elements or a countably infinite number of elements]. In Axiom 1.3, we allow for the countably infinite possibility by writing "$\cup \ldots$" on the left side, and "$+ \ldots$" on the right side. The symbolism "$+ \ldots$" indicates that we should compute the sum of an infinite series if the sequence $A_1, A_2, A_3, \ldots$ has a countably infinite number of terms.

**Axiom 1.3′:** In Axiom 1.3, if $J$ is equal to a union of a finite number [say, $k$] of disjoint events $A_1, A_2, \ldots, A_k$ [that is, $J = A_1 \cup A_2 \cup \ldots \cup A_k$], then we may rewrite Eqs. (a) and (b) of Axiom 1.3 as follows:

$$P(J) = P(A_1) + P(A_2) + \ldots + P(A_k) \tag{a′}$$

$$P(A_1 \cup A_2 \cup \ldots \cup A_k) = P(A_1) + P(A_2) + \ldots + P(A_k) \tag{b′}$$

*note:* Axiom 1.3 or 1.3′ will occasionally be referred to as the Addition Rule axiom.

Now, we are in a position to prove some useful theorems, which will enable us to apply probability concepts to realistic problems.

**Theorem 1.1:** If $A$ is a subset [event] of sample space $S$, then $P(A)$ equals the sum of the probabilities corresponding to the elements belonging to $A$.

**EXAMPLE 1-11:** Prove Theorem 1.1.

*Proof:* Suppose, for simplicity, that there are a finite number of elements in $A$, and that they are labeled consecutively as $e_1, e_2, \ldots, e_a$:

$$A = \{e_1, e_2, \ldots, e_a\} \tag{1}$$

Now, we can write $A$ as the following union of simple events:

$$A = \{e_1\} \cup \{e_2\} \cup \ldots \cup \{e_a\} \tag{2}$$

[The set on the right side of Eq. (1) is equal to the union of sets on the right of Eq. (2).] The simple events $\{e_1\}, \{e_2\}, \ldots, \{e_a\}$ are mutually exclusive [to each belongs a different element].

Thus, by applying Axiom 1.3′, we have

$$P(\{e_1\} \cup \{e_2\} \cup \ldots \cup \{e_a\}) = P(\{e_1\}) + P(\{e_2\}) + \ldots + P(\{e_a\}) \tag{3}$$

Now, applying $P$ to both the left and right sides of Eq. (2) yields

$$P(A) = P(\{e_1\}) + P(\{e_2\}) + \ldots + P(\{e_a\}) \tag{4}$$

after making use of Eq. (3). Equation (4) is what we wished to prove. $\square$

*notes*

(a) If there were a countably infinite number of elements in $A$, then Eq. (4) would be replaced by $P(A) = P(\{e_1\}) + P(\{e_2\}) + \ldots$, where the summation on the right indicates the sum of an infinite series [provided it exists].

(b) Note that it is important not to write something like $P(e_i)$ instead of $P(\{e_i\})$. In Definition 1.9, we see that the probability of a subset is meaningful; the probability of an element is not meaningful. [Recall that the element $e_i$ belongs to the subset $\{e_i\}$.]

(c) Often, we will relax the symbolism, and write $P\{e_i\}$ instead of $P(\{e_i\})$.

**EXAMPLE 1-12:** A particular unbalanced coin is tossed twice. For the sample space $S = \{TT, TH, HT, HH\}$, the following probabilities are assigned, based on repeating the double toss several million times, and observing corresponding relative frequencies:

$$P(\{TT\}) = .16; \qquad P(\{TH\}) = P(\{HT\}) = .24; \qquad P(\{HH\}) = .36$$

(a) Compute the probability of getting exactly one head. (b) Repeat for the event "head on the second toss." (c) Repeat for the event "at least one head."

*Solution:* Let us label the events $A$, $B$, and $C$ as in Examples 1-4 and 1-5, for instance.

(a) Here, $A = \{TH, HT\}$, and thus by using Theorem 1.1, we obtain

$$P(A) = P(\{TH\}) + P(\{HT\}) = .48$$

(b) $B = \{TH, HH\}$, and thus $P(B) = P(\{TH\}) + P(\{HH\}) = .60$.

(c) $C = \{TH, HT, HH\}$, and thus $P(C) = P(\{TH\}) + P(\{HT\}) + P(\{HH\}) = .84$. Alternatively, $P(C) = 1 - P(\{TT\}) = 1 - .16 = .84$.

---

*note:* If the coin were balanced [fair], then we would expect that $P(\{TT\}) = P(\{TH\}) = P(\{HT\}) = P(\{HH\}) = 1/4$. Then, we would have $P(A) = 1/2 = P(B)$, and $P(C) = 3/4$.

**Theorem 1.2 (Equally Likely Simple Events Theorem):** Suppose all $n$ of the simple events associated with a finite sample space $S = \{e_1, e_2, \ldots, e_n\}$ are equally likely. That is, $P(\{e_1\}) = P(\{e_2\}) = \ldots = P(\{e_n\})$. Then,

(a)  $P(\{e_i\}) = 1/n$ for $i = 1, 2, \ldots, n$, and

(b)  $P(A) = a/n$ for any event $A$ to which $a$ elements belong. Put differently, if we let $N(A)$ and $N(S)$ stand for the number of elements in $A$ and $S$, respectively, then we can write

(b') $P(A) = N(A)/N(S)$

*note:* Often one says that the outcomes or elements are equally likely instead of saying that the simple events are equally likely. This is taken to mean the same thing.

---

**EXAMPLE 1-13:** A pair of dice is tossed [see Example 1-2]. Determine the probability that (a) a two appears on at least one die, (b) the sum of the numbers [of dots] on both dice equals six, and (c) the same number of dots is obtained on both dice.

*Solution: Preliminaries:* The most subdivided sample space is

$$S = \{11, 12, \ldots, 16; 21, 22, \ldots, 26; \ldots; 61, 62, \ldots, 66\}$$

We see that $S$ has 36 elements. It is reasonable to assume that all the elements are equally likely. Thus, by Theorem 1.2, we have $P(E) = N(E)/N(S) = N(E)/36$ for any event $E$ with $N(E)$ elements.

(a) Denoting the event as $A$, we have the following subset for $A$:

$$A = \{21, 22, 23, 24, 25, 26; 12, 32, 42, 52, 62\}$$

Since $N(A) = 11$, we have $P(A) = \frac{11}{36}$.

(b) Denoting the event as $B$, we have $B = \{15, 24, 33, 42, 51\}$, and thus,

$$P(B) = \frac{N(B)}{N(S)} = \frac{5}{36}$$

(c) Denoting the event as $C$, we have $C = \{11, 22, 33, 44, 55, 66\}$, and thus,

$$P(C) = \frac{N(C)}{N(S)} = \frac{6}{36}$$

---

**EXAMPLE 1-14:** Prove parts (a) and (b) of Theorem 1.2.

*Proof*

(a) First, we have

$$S = \{e_1, e_2, \ldots, e_n\} \tag{1}$$

Taking the probability of both sides, we get

$$P(S) = P(\{e_1, e_2, \ldots, e_n\}) \tag{2}$$

Applying Axiom 1.2 to the left side, and Theorem 1.1 to the right side, we obtain

$$1 = P(\{e_1\}) + P(\{e_2\}) + \ldots + P(\{e_n\}) \tag{3}$$

By hypothesis,

$$P(\{e_1\}) = P(\{e_2\}) = \ldots = P(\{e_n\}) \tag{4}$$

Thus, using Eq. (4) to express each $P(\{e_i\})$ as equal to $P(\{e_1\})$ for $i = 2, 3, \ldots, n$, we get

$$1 = P(\{e_1\}) + P(\{e_1\}) + \ldots + P(\{e_1\}) = n \cdot P(\{e_1\}) \tag{5}$$

Thus, from Eqs. (4) and (5), we have

$$P(\{e_i\}) = \frac{1}{n} = \frac{1}{N(S)} \qquad \text{for } i = 1, 2, \ldots, n \tag{6}$$

**(b)** Suppose, as in Example 1-11, that the $a = N(A)$ elements in $A$ are consecutively labeled as $e_1, e_2, \ldots, e_a$. Thus,

$$A = \{e_1, e_2, \ldots, e_a\} \tag{7}$$

Taking the probability of both sides, we get

$$P(A) = P(\{e_1, e_2, \ldots, e_n\}) \tag{8}$$

Applying Theorem 1.1 to the right side, we get

$$P(A) = P(\{e_1\}) + P(\{e_2\}) + \ldots + P(\{e_a\}) \tag{9}$$

Thus, from the part (a) result, we have

$$P(A) = \underbrace{\frac{1}{n} + \frac{1}{n} + \frac{1}{n} + \ldots + \frac{1}{n}}_{a \text{ terms}} = \frac{a}{n} = \frac{N(A)}{N(S)} \tag{10} \quad \square$$

## C. Some probability theorems

By making use of the three axioms of probability (Axioms 1.1, 1.2, and 1.3) in Section 1-5B, we can derive many important theorems. Often, such probability theorems are known as "rules."

**Theorem 1.3:** For the complement $A'$ of any event $A$, $P(A') = 1 - P(A)$.

**Theorem 1.4:** For the empty set $\varnothing$, $P(\varnothing) = 0$.

**Theorem 1.5:** If $A$ and $B$ are events in a sample space $S$, and $A \subset B$, then $P(A) \leq P(B)$.

**EXAMPLE 1-15:** Prove Theorems 1.3 and 1.4.

***Proof of Theorem 1.3:*** First, $S = A \cup A'$, where $A$ and $A'$ are clearly disjoint. Applying $P$ to both sides of this equation yields

$$P(S) = P(A \cup A') \tag{1}$$

Applying Axioms 1.2 and 1.3 [or 1.3′] to Eq. (1) results in

$$1 = P(A) + P(A') \tag{2}$$

Thus, $P(A') = 1 - P(A)$. $\qquad\qquad\qquad\qquad\qquad\qquad\qquad\square$

***Proof of Theorem 1.4:*** Observe that $\varnothing = S'$. Thus, using Eq. (2) above, we have

$$P(\varnothing) = P(S') = 1 - P(S) \tag{3}$$

But $P(S) = 1$ from Axiom 1.2, and thus,

$$P(\varnothing) = 1 - 1 = 0 \tag{4} \quad\square$$

**EXAMPLE 1-16:** A pair of dice is tossed. (See Examples 1-2 and 1-13.) What is the probability of getting a sum [of dots] that is at least 3?

***Solution:*** Let $A$ be the event that the "sum is at least 3." Then, the complement $A'$ is the event that the "sum is not at least 3." Now,

$$P(A') = P(\text{"sum is not at least 3"}) = P(\text{"sum is 2"}) \tag{1}$$

Thus, $P(A') = P(\{11\}) = 1/36$. Thus, from Theorem 1.3, we have

$$P(A) = 1 - P(A') = 1 - \frac{1}{36} = \frac{35}{36} \tag{2}$$

**EXAMPLE 1-17:** Prove Theorem 1.5.

***Proof:*** Refer to Figure 1-4. Recall that $A \subset B$ means that $A$ is a subset of $B$. We see that in this case, we can write

$$B = A \cup (A' \cap B) \tag{1}$$

where events $A$ and $A' \cap B$ are mutually exclusive. Thus, from Axiom 1.3 [or 1.3′], we have

$$P(B) = P(A) + P(A' \cap B) \tag{2}$$

Now, by Axiom 1.1, we have

$$P(A' \cap B) \geq 0 \tag{3}$$

since $A' \cap B$ is an event of $S$. Thus, from (2) and (3), we have

$$P(B) - P(A) \geq 0 \tag{4}$$

and hence, $P(A) \leq P(B)$.

***note:*** We could also use the more general diagram in Figure 1-2b, say, if we observe that $R_2 = A \cap B' = \varnothing$, and $R_1 = A \cap B = A$ when $A \subset B$. Then $B = R_1 \cup R_3 = A \cup (A' \cap B)$, and we have Eq. (1) above. $\qquad\square$

**EXAMPLE 1-18:** A coin, which is not necessarily fair, is tossed twice. Prove that the probability of getting two heads is less than or equal to the probability of getting head on the second toss.

***Solution:*** As subsets, the events are $\{HH\}$ and $\{TH, HH\}$, respectively, and $\{HH\}$ is a [proper] subset of $\{TH, HH\}$. Thus, by Theorem 1.5,

$$P(\{HH\}) \leq P(\{TH, HH\})$$

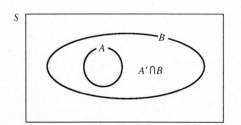

**Figure 1-4.** Diagram for $A \subset B$.

# SOLVED PROBLEMS

## Elementary Concepts

**PROBLEM 1-1**   A probability experiment is to toss a coin three times. Indicate an appropriate sample space if (**a**) one wishes to record what happens on each toss, or if (**b**) one wishes to record the total number of heads obtained.

*Solution*

(**a**) $S = \{TTT, HTT, THT, TTH, THH, HTH, HHT, HHH\}$. Here, the element $HTH$ means head on the first, tail on the second, and head on the third toss.

(**b**) $\hat{S} = \{0, 1, 2, 3\}$, where the element 1 signifies the obtaining of one head; it corresponds to the elements $HTT, THT,$ and $TTH$ of $S$.

**PROBLEM 1-2**   Given a container with three red balls, labeled 1, 2, 3, and two green balls, labeled 4 and 5. Suppose that two balls are drawn in succession, where the first ball is replaced before the second ball is drawn. [This is called drawing with replacement.] Describe a sample space if (**a**) one wishes to record the number on each ball, if (**b**) one wishes to record the color on each ball, or if (**c**) one wishes to record the total number of red balls obtained.

*Solution*

(**a**) Here, as an example, let element 23 refer to the outcome of getting 2 on the first ball and 3 on the second ball. Denote the sample space as $S_1$. Thus, we have

$$S_1 = \{11, 12, 13, 14, 15, 21, 22, 23, 24, 25, 31, 32, 33, 34, 35, 41, 42, 43, 44, 45, 51, 52, 53, 54, 55\}$$

Here, $S_1$, which contains 25 elements, is the most subdivided sample space for this experiment.

(**b**) $S_2 = \{RG, RR, GG, GR\}$. Here, for example, $RG$ refers to the outcome red on the first ball and green on the second ball.

(**c**) $S_3 = \{0, 1, 2\}$.

## Events

**PROBLEM 1-3**   For sample space $S$ of Problem 1-1, indicate the following events [described in words] as subsets of $S$: (**a**) obtaining exactly two heads; (**b**) obtaining exactly three heads; (**c**) obtaining head on the third toss; (**d**) obtaining the same result on all three tosses.

*Solution*

(**a**) $A = \{THH, HTH, HHT\}$.
(**b**) $B = \{HHH\}$.
(**c**) $C = \{TTH, THH, HTH, HHH\}$.
(**d**) $D = \{TTT, HHH\}$.

**PROBLEM 1-4**   For the sample space $S_1$ of Problem 1-2, indicate the following events as subsets of $S_1$: (**a**) obtaining the same number on both draws; (**b**) obtaining a sum of numbers equal to six; (**c**) obtaining at least one three; (**d**) obtaining exactly one red ball; (**e**) obtaining exactly two red balls; (**f**) obtaining same color on both draws.

*Solution*

(**a**) $A = \{11, 22, 33, 44, 55\}$.
(**b**) $B = \{15, 24, 33, 42, 51\}$.
(**c**) $C = \{31, 32, 33, 34, 35, 13, 23, 43, 53\}$.
(**d**) $D = \{14, 15, 24, 25, 34, 35, 41, 51, 42, 52, 43, 53\}$.

**(e)** $E = \{11, 12, 13, 21, 22, 23, 31, 32, 33\}$.
**(f)** $F = \{11, 12, 13, 21, 22, 23, 31, 32, 33, 44, 45, 54, 55\}$.

Note that $E$ is a proper subset of $F$.

## Set Theory and Venn Diagrams

**PROBLEM 1-5**   Refer to the probability experiment and sample space $S$ of Problem 1-1. **(a)** For the events [subsets] $A$, $B$, $C$, $D$ of Problem 1-3, determine which of these four events is a proper subset of another of the four events. **(b)** Express the following events as subsets of $S$:

$$A', \quad A \cap B, \quad A \cap C, \quad A \cup B, \quad B \cup C$$

*Solution*

**(a)** $B$ is a proper subset of $C$ and of $D$.
**(b)** $A' = \{TTT, HTT, THT, TTH, HHH\}$. $A \cap B = \varnothing$. That is, $A$ and $B$ are disjoint [mutually exclusive]. $A \cap C = \{THH, HTH\}$. $A \cup B = \{THH, HTH, HHT, HHH\}$. In words, $A \cup B$ is the event "at least two heads." $B \cup C = \{TTH, THH, HTH, HHH\}$. Here, $B \cup C = C$ since $B \subset C$.

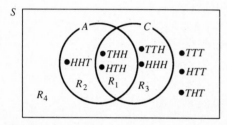

**Figure 1-5.** Venn diagram for Problem 1-6.

**PROBLEM 1-6**   Refer to Problem 1-3. **(a)** Draw a Venn diagram showing $A$ and $C$ as subsets of $S$. **(b)** Express each of the disjoint subsets $A \cap C$, $A' \cap C$, $A \cap C'$, $A' \cap C'$ as subsets of $S$.

*Solution*

**(a)** Refer to Figure 1-5.
**(b)** $A \cap C = R_1 = \{THH, HTH\}$; $A \cap C' = R_2 = \{HHT\}$; $A' \cap C = R_3 = \{TTH, HHH\}$; $A' \cap C' = R_4 = \{TTT, HTT, THT\}$. These four subsets constitute a partition of $S$.

## Axioms of Probability

**PROBLEM 1-7**   Refer to Problems 1-1, 1-3, 1-5, and 1-6. Suppose the coin that is tossed three times is unbalanced, and that the following are the probabilities of the simple events of sample space $S$: $P\{TTT\} = .064$; $P\{HTT\} = P\{THT\} = P\{TTH\} = .096$; $P\{THH\} = P\{HTH\} = P\{HHT\} = .144$; $P\{HHH\} = .216$. Recall that events $A$, $B$, $C$, and $D$ are listed in the answer of Problem 1-3. Calculate the probabilities of the following events:

$$A, \quad B, \quad C, \quad D, \quad A', \quad A \cap B, \quad A \cap C, \quad A \cup C, \quad B \cup C$$

*Solution  Preliminary Note:* For this problem, we apply Theorem 1.1, which indicates that the probability of any subset [event] is equal to the sum of the probabilities associated with the elements of the subset.

$$P(A) = P\{THH\} + P\{HTH\} + P\{HHT\} = 3(.144) = .432$$

$$P(B) = P\{HHH\} = .216.$$

$$P(C) = P\{TTH\} + P\{THH\} + P\{HTH\} + P\{HHH\} = .6$$

$$P(D) = P\{TTT\} + P\{HHH\} = .064 + .216 = .280$$

From Theorem 1.3, $P(A') = 1 - P(A) = .568$. $P(A \cap B) = P(\varnothing) = 0$, as we see from Theorem 1.4.

$$P(A \cap C) = P\{THH, HTH\} = 2(.144) = .288$$

$$P(A \cup C) = P\{THH\} + P\{HTH\} + P\{HHT\} + P\{TTH\} + P\{HHH\}$$

$$= 3(.144) + .096 + .216 = .744$$

$$P(B \cup C) = P(C) = .6$$

since $B \subset C$.

**PROBLEM 1-8**  Suppose the coin of Problems 1-1, 1-3, 1-5, and 1-6 is balanced [fair]. Thus, it is reasonable to assume that the probabilities corresponding to the simple events of sample space $S$ are identical. That is, $P\{TTT\} = P\{HTT\} = \ldots = P\{HHH\} = 1/8$. Rework Problem 1-7 for the case of a balanced coin.

**Solution**  *Preliminary Note:* For this problem, Theorem 1.2 applies. Thus, for any event $E$, $P(E) = N(E)/N(S)$.

$$P(A) = \frac{N(A)}{N(S)} = \frac{3}{8}$$

$$P(B) = P\{HHH\} = \frac{1}{8}$$

$$P(C) = \frac{N(C)}{N(S)} = \frac{4}{8}$$

$$P(D) = \frac{N(D)}{N(S)} = \frac{2}{8}$$

$$P(A') = 1 - P(A) = \frac{5}{8}$$

$$P(A \cap B) = P(\varnothing) = 0; \qquad P(A \cap C) = \frac{N(A \cap C)}{N(S)} = \frac{2}{8}$$

$$P(A \cup C) = \frac{N(A \cup C)}{N(S)} = \frac{5}{8}$$

$$P(B \cup C) = P(C) = \frac{4}{8}$$

**PROBLEM 1-9**  Make use of Theorem 1.5 to prove that $P(A) \leq 1$ for any event [subset] $A$ of a sample space $S$.

**Solution**  Since $A$ is a subset of sample space $S$, it follows that $P(A) \leq P(S)$ from Theorem 1.5. Since $P(S) = 1$ [from Axiom 1.2], we have $P(A) \leq 1$. [*Note:* This means that Axiom 1.1 should have been given merely as $0 \leq P(A)$.]

## Supplementary Problems

**PROBLEM 1-10**  An experiment and associated sample space for the tossing of a coin until a head occurs is discussed in Example 1-3. There, we saw that $S = \{H, TH, TTH, TTTH, \ldots\}$. **(a)** Indicate [as a subset] the event for which head occurs on the fifth toss of the coin. **(b)** Indicate the event for which a head occurs at or before [equivalently, not later than] the fifth toss of the coin. **(c)** Indicate the event for which a head occurs after the fifth toss of the coin.

**Answer**  **(a)** $A = \{TTTTH\}$    **(b)** $B = \{H, TH, TTH, TTTH, TTTTH\}$
**(c)** $C = \{TTTTTH, TTTTTTH, \ldots\}$. Note that $C = B'$.

**PROBLEM 1-11**  Refer to Problems 1-2 and 1-4, which deal with drawing two balls, with replacement, from a container with five numbered and colored balls. The events $A, B, C, D, E,$ and $F$ are listed in the Solution of Problem 1-4.

It is reasonable to assume that all 25 elements of sample space $S_1$ are equally likely. Make use of that assumption to compute: **(a)** $P(A)$, $P(B)$, $P(C)$, $P(E)$; **(b)** $P(C')$, $P(F')$; **(c)** $P(A \cap B)$, $P(A \cap C)$, $P(D \cap E)$; **(d)** $P(A \cup B)$, $P(A \cup C)$, $P(E \cup F)$. *Hint:* Theorem 1.2 is applicable.

*Answer* (a) $P(A) = N(A)/N(S) = 5/25$; $P(B) = 5/25$; $P(C) = 9/25$; $P(D) = 12/25$; $P(E) = 9/25$
(b) $P(C') = 1 - P(C) = 16/25$; $P(F') = 1 - P(F) = 1 - 13/25 = 12/25$    (c) $P(A \cap B) = P(\{33\}) = 1/25$;
$P(A \cap C) = P(\{33\}) = 1/25$; $P(D \cap E) = P(\emptyset) = 0$    (d) $P(A \cup B) = N(A \cup B)/N(S) = 9/25$;
$P(A \cup C) = N(A \cup C)/N(S) = 13/25$; $P(E \cup F) = P(F) = 13/25$ since $E \subset F$

**PROBLEM 1-12**    A large urn containing many balls has the balls labeled with the letter $a$, $b$, $c$, or $d$ in the following percentages: 30% for $a$, and 20%, 40%, and 10% for $b$, $c$, and $d$, respectively. (a) For the experiment of drawing a single ball, develop a sample space $S$ in which the letter on a ball is recorded. (b) Determine the probabilities corresponding to the elements of $S$. (c) What is the probability of drawing either letter $a$ or $d$? (d) What is the probability of drawing a ball that does not have the letter $b$ on it?

*Answer* (a) $S = \{a, b, c, d\}$    (b) $P\{a\} = .3$, $P\{b\} = .2$, $P\{c\} = .4$, $P\{d\} = .1$    (c) $P(\text{Either } a \text{ or } d) = P\{a, d\} = P\{a\} + P\{d\} = .4$    (d) $P(\text{not } b) = 1 - P\{b\} = .8$

**PROBLEM 1-13**    In Problem 1-2, we spoke of a container with three red balls labeled 1, 2, and 3, and two green balls, labeled 4 and 5. Suppose two balls are drawn in succession, where the second ball is drawn from the four balls that remain in the container [that is, the first ball is not returned after it is drawn]. This is known as drawing without replacement. Describe a sample space if (a) one wishes to record the number on each ball, or (b) one wishes to record the color of each ball.

*Answer* (a) $S_1 = \{12, 13, 14, 15, 21, 23, 24, 25, 31, 32, 34, 35, 41, 42, 43, 45, 51, 52, 53, 54\}$
(b) $S_2 = \{RG, RR, GG, GR\}$

**PROBLEM 1-14**    For the sample space $S_1$ of Problem 1-13, indicate the following events as subsets of $S_1$: (a) obtaining a sum of numbers equal to six; (b) obtaining exactly one ball labeled with a three [here, this is equivalent to at least one ball labeled with a three since more than one ball labeled with a three cannot occur if drawing is without replacement]; (c) obtaining exactly one red ball; (d) obtaining no red balls; (e) obtaining the same color on both draws.

*Answer* (a) $A = \{15, 24, 42, 51\}$    (b) $B = \{31, 32, 34, 35, 13, 23, 43, 53\}$    (c) $C = \{14, 15, 24, 25, 34, 35, 41, 51, 42, 52, 43, 53\}$    (d) $D = \{45, 54\}$    (e) $E = \{12, 13, 21, 23, 31, 32, 45, 54\}$

**PROBLEM 1-15**    Given the sample space $S_1$ of Problem 1-13, and the events [subsets] $A$, $B$, $C$, $D$, $E$ of the Answer of Problem 1-14, express the following events as subsets of $S_1$: $C'$, $E'$, $A \cap B$, $A \cap C$, $B \cap E$, $A \cup B$, $A \cup C$, $B \cup E$.

*Answer* $C' = \{45, 54, 12, 13, 21, 23, 31, 32\}$; $E' = \{14, 15, 24, 25, 34, 35, 41, 51, 42, 52, 43, 53\}$; $A \cap B = \emptyset$;
$A \cap C = \{15, 24, 42, 51\} = A$ since $A \subset C$; $B \cap E = \{31, 32, 13, 23\}$; $A \cup B = \{15, 24, 42, 51, 13, 23, 31, 32, 34, 35, 43, 53\}$; $A \cup C = C$ since $A \subset C$; $B \cup E = \{31, 32, 34, 35, 13, 23, 43, 53, 12, 21, 45, 54\}$

**PROBLEM 1-16**    Refer to Problems 1-13 to 1-15, which deal with drawing two balls, without replacement, from a container with five numbered and colored balls. It is reasonable to assume that all 20 elements of sample space $S_1$ are equally likely. Compute the following: (a) $P(A)$, $P(B)$, $P(C)$, $P(D)$, $P(E)$. (b) $P(C')$, $P(E')$. (c) $P(A \cap B)$, $P(A \cap C)$, $P(B \cap E)$. (d) $P(A \cup B)$, $P(A \cup C)$, $P(B \cup E)$.

*Answer* Theorem 1-2 applies. (a) $P(A) = N(A)/N(S) = 4/20$; $P(B) = 8/20$; $P(C) = 12/20$; $P(D) = 2/20$;
$P(E) = 8/20$    (b) $P(C') = 1 - P(C) = 8/20$; $P(E') = 1 - P(E) = 12/20$    (c) $P(A \cap B) = P(\emptyset) = 0$;
$P(A \cap C) = P(A) = 4/20$ since $A \subset C$; $P(B \cap E) = 4/20$    (d) $P(A \cup B) = 12/20$ [note that $P(A \cup B) = P(A) + P(B)$ since $A$ and $B$ are mutually exclusive]; $P(A \cup C) = P(C) = 12/20$ since $A \subset C$; $P(B \cup E) = 12/20$

**PROBLEM 1-17**    Suppose a probability experiment is to draw a card from a standard deck of 52 cards. [Such a deck contains four suits—clubs, diamonds, hearts, and spades—with thirteen cards running from ace (one) to ten, and then Jack, Queen, and King in each suit. The last three types, which total $3 \cdot 4 = 12$ cards, are called face cards.] Compute the probability that the card drawn is (a) an ace, (b) a face card, (c) a spade, (d) an ace of spades, (e) either an ace or a face card.

*Answer* (a) $4/52 = 1/13$    (b) $12/52 = 3/13$    (c) $13/52 = 1/4$    (d) $1/52$    (e) $(4 + 12)/52 = 16/52 = 4/13$

# 2 PROBABILITY— MORE BASIC IDEAS

## THIS CHAPTER IS ABOUT
- ☑ **Probability and Set Theory**
- ☑ **Elementary Probability Problems**

## 2-1. Probability and Set Theory

Many basic theorems of both set theory and probability can be proved with the aid of Venn diagrams. Refer to Figures 1-2 through 1-5 and Part E of Section 1-4 for illustrations of this.

### A. Using Venn diagrams for proofs and problem solving

**Theorem 2.1 (General Addition Rule for Two Events):** If $A$ and $B$ are any two events in a sample space $S$, then

$$P(A \cup B) = P(A) + P(B) - P(A \cap B)$$

---

**EXAMPLE 2-1:** Prove Theorem 2.1.

**Proof:** Consider Figure 2-1, which shows sets $A$ and $B$ intersecting in the most general way possible. There, $R_1 = A \cap B$, $R_2 = A \cap B'$, $R_3 = A' \cap B$, and $R_4 = A' \cap B'$. The collection of $R_i$'s constitutes a partition of sample space $S$; in particular, the $R_i$'s are disjoint so that we can make use of Axiom 1.3′. First, let us express the left side symbol $P(A \cup B)$ in terms of $P(R_i)$ symbols:

$$A \cup B = R_1 \cup R_2 \cup R_3 \tag{1}$$

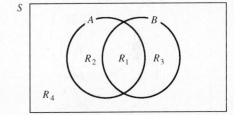

**Figure 2-1.** Sets $A$ and $B$.

Thus,

$$P(A \cup B) = P(R_1 \cup R_2 \cup R_3) \tag{2}$$

and hence,

$$P(A \cup B) = P(R_1) + P(R_2) + P(R_3) \tag{3}$$

after applying Axiom 1.3′ to the right side of Eq. (2).

Now, it remains to show that the right side of the Theorem 2.1 equation is equal to $P(R_1) + P(R_2) + P(R_3)$. Since $A = R_1 \cup R_2$ and $B = R_1 \cup R_3$, it follows that

$$P(A) = P(R_1) + P(R_2) \quad \text{and} \quad P(B) = P(R_1) + P(R_3) \tag{4, 5}$$

Also, $A \cap B = R_1$, and so

$$P(A \cap B) = P(R_1) \tag{6}$$

From Eqs. (4), (5), and (6), we have

$$P(A) + P(B) - P(A \cap B) = P(R_1) + P(R_2) + P(R_1) + P(R_3) - P(R_1)$$

$$= P(R_1) + P(R_2) + P(R_3) \tag{7}$$

We see from Eqs. (3) and (7) that both the left and right sides of Theorem 2.1 are equal to $P(R_1) + P(R_2) + P(R_3)$. Thus, we have proved Theorem 2.1.

$\square$

**EXAMPLE 2-2:** Refer to Problem 1-7, which dealt with tossing an unbalanced coin three times. Use Theorem 2.1 to compute $P(A \cup C)$ from $P(A)$, $P(C)$ and $P(A \cap C)$. Recall that, in words, $A$ was the event "exactly two heads" and $C$ was the event "head on the third toss."

*Solution:* Since $P(A) = .432$, $P(C) = .600$, and $P(A \cap C) = .288$, we have from Theorem 2.1 that

$$P(A \cup C) = .432 + .600 - .288 = .744$$

This agrees with the value for $P(A \cup C)$ obtained in Problem 1.7.

We see that the Venn diagram approach involving disjoint subsets [for example, the $R_i$'s] can be very useful for doing calculations and proofs relating to sets. In much of the work to come, we shall try to show sets intersecting in the most general way possible.

**EXAMPLE 2-3:** Draw a Venn diagram showing three subsets $A$, $B$, and $C$ contained in some sample space $S$. The diagram should allow for all possible intersections among $A$, $B$, and $C$. Draw three mutually overlapping closed curves corresponding to the sets $A$, $B$, and $C$, and label the resulting disjoint region sets as $R_i$'s. Note that there should be $2 \cdot 2 \cdot 2 = 2^3 = 8$ such $R_i$'s since each $R_i$ can be represented as

$$R_i = \left\{ \begin{array}{c} A \\ \text{or} \\ A' \end{array} \right\} \cap \left\{ \begin{array}{c} B \\ \text{or} \\ B' \end{array} \right\} \cap \left\{ \begin{array}{c} C \\ \text{or} \\ C' \end{array} \right\} \tag{1}$$

There are two possible symbols for each of the first, middle, and last terms in (1). [If we started with $n$ subsets of $S$ instead of three, then there would be $2 \cdot 2 \cdot 2 \cdot \ldots \cdot 2 = 2^n$ such disjoint $R_i$'s.]

*Solution:* The Venn diagram appears in Figure 2-2. Here, the partition of $S$ is indicated by

$$S = R_1 \cup R_2 \cup \ldots \cup R_8 \tag{2}$$

where $R_i \cap R_j = \varnothing$ for $i \neq j$, and $i = 1, 2, 3, \ldots, 8$, and $j = 1, 2, 3, \ldots, 8$.

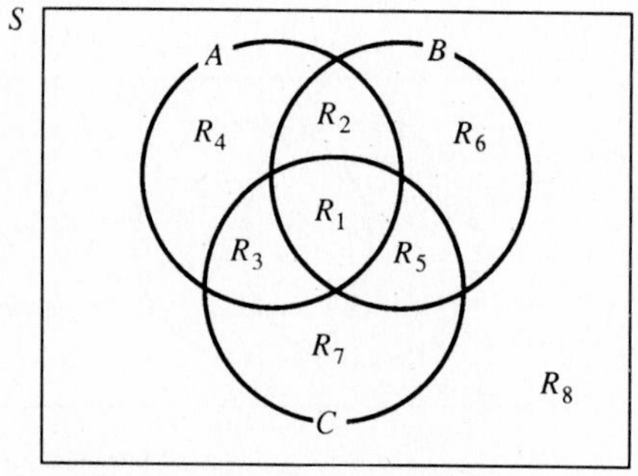

**Figure 2-2.** Sets *A*, *B*, and *C* intersecting in all possible ways.

*note:* We see, for example, that $R_1 = A \cap B \cap C$ since each element in $R_1$ is also in the subsets labeled $A$, $B$, and $C$. For $R_4$, each element in $R_4$ is in $A$, is *not* in $B$, and is *not* in $C$. Thus $R_4 = A \cap B' \cap C'$. For another example, $R_3 = A \cap B' \cap C$ since each element in $R_3$ is in both $A$ and $C$, but not in $B$.

The following theorem, which parallels Theorem 2.1, can be proved by using the method of proof of Example 2-1; it is proved in Problem 2-6.

**Theorem 2.2 (General Addition Rule for Three Events):** If $A$, $B$, and $C$ are any three events in a sample space $S$, then

$$P(A \cup B \cup C) = P(A) + P(B) + P(C)$$

$$- P(A \cap B) - P(A \cap C) - P(B \cap C) + P(A \cap B \cap C)$$

*note:* In much of our work dealing with set manipulations, we shall proceed in an intuitive fashion. That is, we shall accept many set theory results without proof. Statements of some major set theory laws are presented in Section 2-1C.

---

**EXAMPLE 2-4:** The probability that a student will attend the first meeting of a class is .75, the probability that a student will attend the second meeting is .65, and the probability that a student will attend both meetings is .60. (a) What is the probability that a student will attend at least one of the first two meetings of the class? (b) What is the probability that a student will miss both the first and second meetings of the class? (c) What is the probability that a student will attend exactly one of the first two meetings of the class?

*Solution: Preliminaries:* First, we draw a Venn diagram [Figure 2-3] in which $A$ refers to the event "a student attends the first meeting," and $B$ refers to the event "a student attends the second meeting." Then, the terms $R_1, R_2, R_3$, and $R_4$ denote mutually exclusive (disjoint) events determined from the given information. Next, we calculate the probabilities associated with $R_1, R_2, R_3$, and $R_4$.

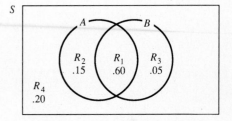

**Figure 2-3.** Venn diagram for Example 2-4.

Since we are given that $P(A) = .75$, $P(B) = .65$, and $P(A \cap B) = .60$, we can calculate the $P(R_i)$'s as follows: $P(R_1) = P(A \cap B) = .60$; $P(R_2) = P(A) - P(R_1) = .15$; $P(R_3) = P(B) - P(R_1) = .05$; $P(R_4) = P(S) - [P(R_1) + P(R_2) + P(R_3)] = .20$.

(a) The event "a student will attend at least one meeting" is equivalent to $A \cup B$. Since $A \cup B = R_1 \cup R_2 \cup R_3$, where the $R_i$'s are mutually exclusive, we apply Axiom 1.3' to obtain

$$P(A \cup B) = P(R_1 \cup R_2 \cup R_3) = P(R_1) + P(R_2) + P(R_3)$$

$$= .60 + .15 + .05 = .80$$

We can also obtain the same result by applying Theorem 2.1: $P(A \cup B) = P(A) + P(B) - P(A \cap B) = .75 + .65 - .60 = .80$.

(b) The event "a student will miss both meetings" is equivalent to $A' \cap B'$ where, for example, $A'$ means that a student will miss the first meeting. Expressing both $A'$ and $B'$ in terms of mutually exclusive $R_i$'s, we have $A' = R_3 \cup R_4$, $B' = R_2 \cup R_4$, and thus $A' \cap B' = (R_3 \cup R_4) \cap (R_2 \cup R_4) = R_4$. So, the answer is

$$P(A' \cap B') = P(R_4) = .20$$

(c) The event $E$ that "a student will attend exactly one of the first two meetings" is equivalent to "a student will attend meeting 1 or meeting 2, but not both." Thus, $E = R_2 \cup R_3$, and thus,

$$P(E) = P(R_2 \cup R_3) = P(R_2) + P(R_3) = .15 + .05 = .20$$

Here, again, we used Axiom 1.3' in expressing $P(R_2 \cup R_3)$ as $P(R_2) + P(R_3)$.

---

Refer again to Figure 2-2. In any particular case involving three subsets $A$, $B$, and $C$ of $S$, it might happen that some of the eight $R_i$ sets do not contain elements. For example, if $A$ and $B$ happen to be disjoint, then sets $R_1$ and $R_2$ would contain no elements. That is, $R_1 = R_2 = \varnothing$.

In Section 2-1B, we shall verify several of the general algebraic laws pertaining to sets by using a Venn diagram approach involving disjoint $R_i$ sets. The diagrams of Figure 2-1 (for two sets) and Figure 2-2 (for three sets) allow for intersections that are as general as possible among two or three sets, respectively. It should be understood, however, that we do allow for the possibility that some of the $R_i$'s may, in fact, be empty sets in certain situations.

Suppose a discussion involves the sets $A$ and $B$ and various complements, unions, and intersections pertaining to them. It is possible to express derived sets, such as $A \cup B'$ for example, strictly in terms of a *union* involving several or none of the $R_i$'s. We define a derived set as a set obtained by using at least one of the complement, union, and intersection operations on the given sets, as on $A$ and $B$ above.

Once we have expressed a derived set in terms of a union of $R_i$'s, we can then easily compute the probability of the derived set by making use of Axiom 1.3'.

---

**EXAMPLE 2-5:** Refer to Figure 2-1. **(a)** Express the sets $A'$, $B'$, $A' \cap B'$, and $A \cup B'$ as unions involving the $R_i$'s. **(b)** Express the probabilities of the sets of part (a) in terms of probabilities of the $R_i$'s.

***Solution***

**(a)**     $A' = R_3 \cup R_4; \; B' = R_2 \cup R_4; \; A' \cap B' = (R_3 \cup R_4) \cap (R_2 \cup R_4) = R_4$

since elements (points) that are in both $A'$ and $B'$ have to be in $R_4$.

$$A \cup B' = (R_1 \cup R_2) \cup (R_2 \cup R_4) = R_1 \cup R_2 \cup R_4$$

The last statement indicates that an element (point) which is in either $A$ or $B'$ has to be either in $R_1$ alone or $R_2$ alone or $R_4$ alone.

**(b)**
$$P(A') = P(R_3 \cup R_4) = P(R_3) + P(R_4)$$
$$P(B') = P(R_2 \cup R_4) = P(R_2) + P(R_4)$$
$$P(A' \cap B') = P(R_4)$$
$$P(A \cup B') = P(R_1) + P(R_2) + P(R_4)$$

---

## B. Generalizations concerning the disjoint $R_i$ sets

There are certain generalizations we can make concerning disjoint [mutually exclusive] $R_i$ sets. Consider a collection of subsets $A$, $B$, $C$, etc., of a sample space $S$, where each of these subsets can be expressed as a union of disjoint $R_i$'s, or equal to a single $R_i$, or equal to the empty set.

**(1)** The intersection of two sets (subsets) is equal to the union of those $R_i$'s such that each $R_i$ appears in the representations of both sets. When the two sets have no $R_i$'s in common, the intersection is the empty set $\varnothing$. [Similar statements apply to the intersection of three or more sets.]

**(2)** The union of two sets is equal to the union of those $R_i$'s that appear in the representation of the first set alone, or of the second set alone, or of both sets. [A similar statement applies to the union of three or more sets.]

**(3)** The complement of a set is equal to the union of those $R_i$'s that occur in the representation of $S$ [the sample space], but not in the representation of the given set.

***note:*** It is useful to recall that the complement of $S$ is the empty set, and the complement of the empty set is $S$.

**EXAMPLE 2-6:** Suppose that sets $F$, $G$, $H$, and $S$ [the sample space] are represented in terms of disjoint $R_i$'s as follows:

$$F = R_1 \cup R_2 \cup R_4 \cup R_7 \qquad G = R_1 \cup R_5 \cup R_6 \cup R_4 \cup R_2$$

$$H = R_3 \cup R_5 \cup R_6 \qquad S = R_1 \cup R_2 \cup R_3 \cup R_4 \cup R_5 \cup R_6 \cup R_7 \cup R_8$$

Express the following as unions of disjoint $R_i$'s: **(a)** $F \cup G$, **(b)** $F \cap G$, **(c)** $F \cup H$, **(d)** $F \cap H$, **(e)** $G'$.

*Solution*

**(a)** From generalization (2) above,

$$F \cup G = R_1 \cup R_2 \cup R_4 \cup R_7 \cup R_5 \cup R_6$$

**(b)** From generalization (1) above,

$$F \cap G = R_1 \cup R_2 \cup R_4$$

**(c)** $\qquad\qquad F \cup H = R_1 \cup R_2 \cup R_4 \cup R_7 \cup R_3 \cup R_5 \cup R_6$

**(d)** $\qquad\qquad\qquad\qquad F \cap H = \varnothing$

**(e)** From generalization (3) above,

$$G' = R_3 \cup R_7 \cup R_8$$

## C. The algebra of sets

We shall now prove some general algebraic laws pertaining to sets. The laws are said to be general since they don't depend on which sets we are considering, but hold for all sets. For example, the law $(A \cup B)' = A' \cap B'$ is true for any two sets $A$ and $B$.

　　Certain general laws such as $A \cup A = A$, $A \cap A = A$, $A \cup \varnothing = A$, and $A \cap \varnothing = \varnothing$ follow directly from the definitions we have already made (for union, intersection, empty set, and so on), and are fairly obvious in their meanings.

　　We shall now focus on proving and illustrating some basic laws relating to two or more sets. Our method will be to employ Venn diagrams, and we will express relevant sets as unions of disjoint $R_i$'s. We shall refer to Figure 2-1 for proofs involving two sets, and to Figure 2-2 if three sets are involved. The generalizations for intersection, union, and complement of Part B of Section 2-1 will be especially useful.

**EXAMPLE 2-7:** Prove that $(A \cup B)' = A' \cap B'$ for any two sets $A$ and $B$.

*Proof:* Refer to Figure 2-4, which is a reproduction of Figure 2-1.

　　Our general technique for proofs like this one will be to express each side of the set equation as a union of disjoint $R_i$'s. If the same $R_i$'s occur on both sides, then the proof is finished. Here, we have $A = R_1 \cup R_2$, $B = R_1 \cup R_3$, $A' = R_3 \cup R_4$, and $B' = R_2 \cup R_4$ from Figure 2-4. Thus,

$$A \cup B = (R_1 \cup R_2) \cup (R_1 \cup R_3) = R_1 \cup R_2 \cup R_3 \qquad (1)$$

Then,

$$(A \cup B)' = R_4 \qquad (2)$$

Also,

$$A' \cap B' = (R_3 \cup R_4) \cap (R_2 \cup R_4) \qquad (3)$$

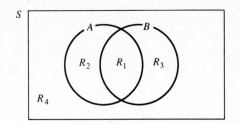

**Figure 2-4**

That is,

$$A' \cap B' = R_4 \qquad (4)$$

Since the right sides of Eqs. (2) and (4) are the same, the law is proved.

*note:* The law just proved, and the companion law $(A \cap B)' = A' \cup B'$, are called DeMorgan's set theory laws.   □

---

**EXAMPLE 2-8:** Refer to Example 2-4. Rework part (**c**) by making use of the DeMorgan law proved in Example 2-7.

*Solution:* The event $E$ whose probability we seek is equivalent to $A' \cap B'$. Now,

$$A' \cap B' = (A \cup B)' \qquad (1)$$

and thus from the DeMorgan law of Example 2-7,

$$P(A' \cap B') = P[(A \cup B)'] \qquad (2)$$

But, from Theorem 1.3, $P[(A \cup B)'] = 1 - P(A \cup B)$. Thus, we have

$$P(A' \cap B') = 1 - P(A \cup B) \qquad (3)$$

From Example 2-4(a), $P(A \cup B) = .80$, and thus

$$P(E) = P(A' \cap B') = 1 - .80 = .20 \qquad (4)$$

---

**EXAMPLE 2-9:** Prove the distributive law $A \cap (B \cup C) = (A \cap B) \cup (A \cap C)$. [Notice that this distributive law (for intersection "over" union) is similar in form to the distributive law for multiplication over addition that holds for real numbers, namely $a \cdot (b + c) = a \cdot b + a \cdot c$.]

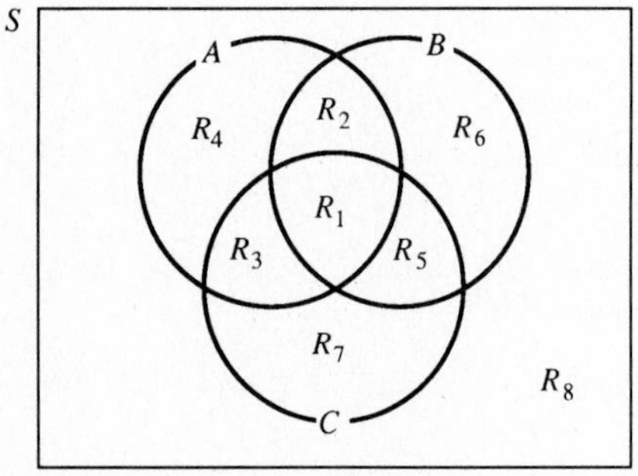

*Proof:* Refer to Figure 2-5, which is a reproduction of Figure 2-2. First, $A = R_1 \cup R_2 \cup R_3 \cup R_4$, $B = R_1 \cup R_2 \cup R_5 \cup R_6$, and $C = R_1 \cup R_3 \cup R_5 \cup R_7$. Thus,

$$B \cup C = R_1 \cup R_2 \cup R_5 \cup R_6 \cup R_3 \cup R_7 \qquad (1)$$

and thus the left side of the law can be expressed as

$$A \cap (B \cup C) = R_1 \cup R_2 \cup R_3 \qquad (2)$$

From the expressions above for $A$, $B$, and $C$,

$$A \cap B = R_1 \cup R_2 \qquad \text{and} \qquad A \cap C = R_1 \cup R_3 \qquad (3), (4)$$

Thus,

$$(A \cap B) \cup (A \cap C) = R_1 \cup R_2 \cup R_3 \qquad (5)$$

Since the right sides of Eqs. (2) and (5) are the same, the proof is complete.   □

**Figure 2-5**

In Table 2-1, we present a number of important laws obeyed by any sets $A$, $B$, and $C$ with respect to the operations of union ($\cup$), intersection ($\cap$), and complementation ($'$). The term $S$, which indicates sample space, is the universal set in probability discussions. [A discussion geared more directly to set theory might replace $S$ by $U$, which is a typical symbol for the universal set.] The nine laws presented in Table 2-1 [on occasion, a given law category is subdivided] are special cases of the laws that constitute a general mathematical system called the **Boolean algebraic system** [named after the mathematician, George Boole, 1813–1864].

TABLE 2-1: Some Major Laws of Set Theory

| Name of law | Set theory equation |
|---|---|
| Identity Laws 1a, 1b <br> 2a, 2b | $A \cup \varnothing = A; A \cap S = A$ <br> $A \cup S = S; A \cap \varnothing = \varnothing$ |
| Idempotent Laws 3a, 3b | $A \cup A = A; A \cap A = A$ |
| Commutative Laws 4a, 4b | $A \cup B = B \cup A; A \cap B = B \cap A$ |
| Complement Laws 5a, 5b <br> 6a, 6b, 6c | $A \cup A' = S; A \cap A' = \varnothing$ <br> $(A')' = A; S' = \varnothing; \varnothing' = S$ |
| Associative Laws 7a <br><br> 7b | $A \cup (B \cup C) = (A \cup B) \cup C$ [Either side can be written as $A \cup B \cup C.$] <br> $A \cap (B \cap C) = (A \cap B) \cap C$ [Either side can be written as $A \cap B \cap C.$] |
| Distributive Laws 8a <br> 8b | $A \cup (B \cap C) = (A \cup B) \cap (A \cup C)$ <br> $A \cap (B \cup C) = (A \cap B) \cup (A \cap C)$ |
| DeMorgan's Laws 9a <br> 9b | $(A \cup B)' = A' \cap B'$ <br> $(A \cap B)' = A' \cup B'$ |

# 2-2. Elementary Probability Problems

## A. Using Venn diagrams and probability theorems

**EXAMPLE 2-10:** In a certain state, 75% of all cars were equipped with seat belts, 23% were equipped with air bags, and 91% had either seat belts or air bags or both [equivalently, 91% had seat belts and/or air bags]. What is the percentage of cars in that state that are equipped with both seat belts and air bags?

*Solution: Preliminary Note:* Here, we assume that the sample space consists of all cars in the state, and that the experiment is to pick a single car, such that we will keep a record of whether a given car has seat belts or air bags or both. Thus, $P(\text{Seat Belts}) = 75/100 = .75$.

Let $A$ and $B$, respectively, denote the events that a car has seat belts and air bags. Thus, $P(A) = .75$, $P(B) = .23$, and $P(A \cup B) = .91$. Using Theorem 2.1, we obtain

$$P(A \cap B) = .75 + .23 - .91 = .07$$

Thus, 7% of the cars in the state are equipped with both seat belts and air bags.

One can show from Theorem 2.1 or Figure 2-1 that

$$P(A \cup B) \leq P(A) + P(B)$$

and from Theorem 2.2 or Figure 2-2 that

$$P(A \cup B \cup C) \leq P(A) + P(B) + P(C)$$

Theorem 2.3 below is a generalization of such inequalities.

**Theorem 2.3:** For any finite sequence of events [subsets] $A_1, A_2, \ldots, A_n$ of a sample space $S$,

$$P(A_1 \cup A_2 \cup A_3 \cup \ldots \cup A_n) \leq P(A_1) + P(A_2) + \ldots + P(A_n)$$

**EXAMPLE 2-11:** Verify that $P(A \cup B) \leq P(A) + P(B)$.

*Solution:* From Theorem 2.1, we have

$$P(A \cup B) = P(A) + P(B) - P(A \cap B) \tag{1}$$

Thus,

$$P(A) + P(B) - P(A \cup B) = P(A \cap B) \tag{2}$$

But, from Axiom 1.1 applied to the event $A \cap B$, we have $P(A \cap B) \geq 0$, and thus,

$$P(A) + P(B) - P(A \cup B) \geq 0 \tag{3}$$

and hence,

$$P(A) + P(B) \geq P(A \cup B) \tag{4}$$

*note:* From Theorem 2.3, we see that one upper bound for $P(A_1 \cup A_2 \cup A_3 \cup \ldots \cup A_n)$ is $P(A_1) + P(A_2) + \ldots + P(A_n)$. From Axiom 1.1, we see that

$$P(A_1 \cup A_2 \cup A_3 \cup \ldots \cup A_n) \leq 1$$

since $A_1 \cup A_2 \cup A_3 \cup \ldots \cup A_n$ is an event [subset] of $S$. Thus, we can say, in more precise fashion, that an upper bound for $P(A_1 \cup A_2 \cup A_3 \cup \ldots \cup A_n)$ is the minimum of $P(A_1) + P(A_2) + \ldots + P(A_n)$ and the number 1.

---

**EXAMPLE 2-12:** Surveys showed that for the workers at the AJAX Tire and Wheel Company, 70% regularly smoked tobacco and 40% drank regularly (liquor, beer, coolers, or wine). Let $A$ be the event that a worker smokes regularly, and $B$ be the event that a worker drinks regularly. Determine closest possible lower and upper bounds for **(a)** $P(A \cap B)$, and **(b)** $P(A \cup B)$. That is, for part (a), find $k_1$ and $k_2$ in

$$k_1 \leq P(A \cap B) \leq k_2$$

such that $k_1$ is as large as possible and $k_2$ is as small as possible. Note that it is clearly true that $0 \leq P(A \cap B) \leq 1$ since for any event $E$, we have the bound condition $0 \leq P(E) \leq 1$ from Axiom 1.1.

*Solution:* Observe that $A \cap B$ is the event that a worker both smokes and drinks, while $A \cup B$ is the event that a worker either is only a smoker, or is only a drinker, or both smokes and drinks.

*Method 1 for Parts (a) and (b):* First observe that $P(A) = .7$ and $P(B) = .4$ from the given information. From Axiom 1.1, we have $P(A \cup B) \leq 1$, and from Theorem 2.3 for $n = 2$, we have $P(A \cup B) \leq P(A) + P(B) \leq 1.1$. Thus, from these two inequalities, the lowest possible upper bound for $P(A \cup B)$ is given by

$$P(A \cup B) \leq 1 \tag{1}$$

For $P(A \cap B)$, we have $0 \leq P(A \cap B)$ from Axiom 1.1. But, we can find a greater lower bound than 0 for $P(A \cap B)$. Replacing $P(A \cup B)$ in Eq. (1) by its expression from Theorem 2.1 leads to

$$P(A) + P(B) - P(A \cap B) \leq 1 \tag{2}$$

Substituting $P(A) = .7$ and $P(B) = .4$ in (2) yields the following greater lower bound for $P(A \cap B)$:

$$.1 \leq P(A \cap B) \tag{3}$$

Now, $A \cap B$ is a subset of both $A$ and $B$ [see Figure 2-1, for example, where $R_1 = A \cap B$]. Thus, from Theorem 1.5,

$$P(A \cap B) \leq P(A) = .7 \qquad \text{and} \qquad P(A \cap B) \leq P(B) = .4 \tag{4, 5}$$

Thus, from (4) and (5), we have

$$P(A \cap B) \le .4 \qquad \textbf{(6)}$$

From (3) and (6), we have the closest possible bounds for $P(A \cap B)$, namely

$$.1 \le P(A \cap B) \le .4 \qquad \textbf{(7)}$$

and this is the answer to part (a). For a lower bound for $A \cup B$, we have $0 \le P(A \cup B)$, where this inequality follows from Axiom 1.1. To find a greater lower bound for $P(A \cup B)$, we apply Theorem 2.1 to (6). Thus, we get

$$P(A) + P(B) - P(A \cup B) \le .4 \qquad \textbf{(8)}$$

and thus,

$$P(A \cup B) \ge .7 + .4 - .4 = .7 \qquad \textbf{(9)}$$

From (1) and (9), we have

$$.7 \le P(A \cup B) \le 1 \qquad \textbf{(10)}$$

and this is the answer to part (b).

*Method 2 for Parts (a) and (b):* From the given information for $P(A)$ and $P(B)$, we see that "extreme case" Venn diagrams are indicated in Figure 2-6. From Figure 2-6a, we have $A \cup B = S$, and thus, from Axiom 1.2,

$$P(A \cup B) = 1 \qquad \textbf{(1')}$$

It is convenient to employ symbols for disjoint $R_i$ subsets in Figure 2-6a. From $A = R_1 \cup R_2$, and $P(A) = .7$, we have

$$P(R_1) + P(R_2) = .7 \qquad \textbf{(2')}$$

From $B = R_1 \cup R_3$, and $P(B) = .4$, we have

$$P(R_1) + P(R_3) = .4 \qquad \textbf{(3')}$$

From $S = R_1 \cup R_2 \cup R_3$, and $P(S) = 1$, we have

$$P(R_1) + P(R_2) + P(R_3) = .1 \qquad \textbf{(4')}$$

Solving Eqs. (2'), (3'), and (4') simultaneously, we obtain

$$P(R_1) = .1, \qquad P(R_2) = .6, \qquad \text{and } P(R_3) = .3 \qquad \textbf{(5')}$$

In particular, for this extreme case,

$$P(A \cap B) = P(R_1) = .1 \qquad \textbf{(6')}$$

From Figure 2-6b, $B \subset A$, which means that $B = A \cap B$, and thus

$$P(A \cap B) = P(B) = .4 \qquad \textbf{(7')}$$

Also, $A \cup B = A$ in this case, and hence

$$P(A \cup B) = P(A) = .7 \qquad \textbf{(8')}$$

Now, in general, the value for $P(A \cap B)$ is between the extreme values as given in (6') and (7'). That is,

$$.1 \le P(A \cap B) \le .4 \qquad \textbf{(9')}$$

Likewise, the value for $P(A \cup B)$ is between the extreme values as given in (1') and (8'). That is,

$$.7 \le P(A \cup B) \le 1 \qquad \textbf{(10')}$$

*note:* From Example 2-12, we can conclude that the proportion of AJAX employees who smoke and drink is definitely between 10% and 40%, inclusive, and the proportion who do at least one of smoke or drink is definitely between 70% and 100%, inclusive.

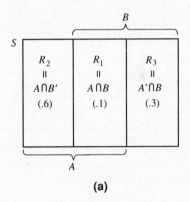

(a)

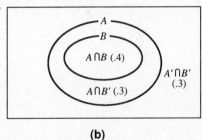

(b)

**Figure 2-6.** Extreme cases when $P(A) = .7$ and $P(B) = .4$. [Probability numbers in parentheses.] (a) $A \cup B = S$, and $R_4 = A' \cap B' = \varnothing$. (b) $B \subset A$, and thus $B = A \cap B$.

## B. Problems involving three events *A*, *B*, and *C*

**EXAMPLE 2-13:** A survey of the males who work for the BOOKOUT Publishing Company disclosed that 47% regularly smoked cigarettes, 27% regularly smoked a pipe, 18% regularly smoked cigars, 12% regularly smoked cigarettes and a pipe, 7% regularly smoked cigarettes and cigars, 5% regularly smoked a pipe and cigars, and 2% regularly smoked all three tobacco products. (**a**) What percentage of the males at the company smoked at least one of the three tobacco products? (**b**) What percentage did not smoke any of the above three tobacco products? (**c**) What is the probability a male picked at random at the BOOKOUT Publishing Company smoked only cigarettes? (**d**) What is the probability a male picked at random at the BOOKOUT Publishing Company smoked exactly one of the three tobacco products?

*Solution: Preliminaries:* Refer to Figure 2-7, which is just Figure 2-2 as applied to this problem. Remember that the $R_i$'s are disjoint, and constitute a partition of $S$. Let $A$, $B$, and $C$ refer to the events of smoking cigarettes, a pipe, and cigars, respectively. Let us fill in $P(R_i)$ values in Figure 2-7 by using the given data in the reverse of the order in which it is presented.

Given: $P(A \cap B \cap C) = .02$; $P(B \cap C) = .05$; $P(A \cap C) = .07$; $P(A \cap B) = .12$; $P(C) = .18$; $P(B) = .27$; $P(A) = .47$.

Calculated: $P(R_1) = P(A \cap B \cap C) = .02$; $P(R_5) = P(B \cap C) - P(R_1) = .05 - .02 = .03$; $P(R_3) = P(A \cap C) - P(R_1) = .07 - .02 = .05$; $P(R_2) = P(A \cap B) - P(R_1) = .12 - .02 = .10$; $P(R_7) = P(C) - [P(R_1) + P(R_3) + P(R_5)] = .18 - [.02 + .05 + .03] = .08$; $P(R_6) = P(B) - [P(R_1) + P(R_2) + P(R_5)] = .27 - [.02 + .10 + .03] = .12$; $P(R_4) = P(A) - [P(R_1) + P(R_2) + P(R_3)] = .47 - [.02 + .10 + .05] = .30$; $P(R_8) = P(S) - [P(R_1) + P(R_2) + \ldots + P(R_7)] = 1 - [.02 + .10 + .05 + .30 + .03 + .12 + .08] = 1 - .70 = .30$.

(**a**) $P(A \cup B \cup C) = P(R_1) + P(R_2) + \ldots + P(R_7) = .70$. Thus, 70% of the males smoked at least one of the three tobacco products.

(**b**) Here, the answer is $P(R_8) = .30$ or 30%, since the elements [males] in $R_8$ are neither in set $A$ [cigarette smokers], nor in set $B$ [pipe smokers], nor in set $C$ [cigar smokers].

(**c**) Here, the answer is $P(R_4) = .30$ or 30%, since elements in $R_4$ are in set $A$ [cigarette smokers], but not in set $B$ or $C$.

(**d**) Here, the answer is $P(R_4) + P(R_6) + P(R_7) = .50$ or 50%. For example, $P(R_7)$ is the probability that a male at the BOOKOUT Company smoked only cigars.

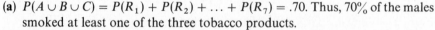

**Figure 2-7.** Venn diagram for Example 2-13. [Numbers in parentheses in $R_i$ regions are probabilities.]

## C. Relating odds to probabilities

Very often in colloquial usage and in business and gambling situations, people describe probabilities of events indirectly in terms of *odds*.

**Definition 2.1 (Fair Odds):** Let $E$ be any event. We say that *fair odds for E* or *fair odds in favor of E* are $a$ to $b$ or $a : b$ if $P(E) = a/(a + b)$, and conversely. If the fair odds for $E$ are $a$ to $b$, then the *fair odds against E* are $b$ to $a$ or $b : a$. Here, the numbers $a$ and $b$ are both nonnegative.

**EXAMPLE 2-14:** Suppose that the fair odds for event $E$ are 3 to 2. (**a**) Determine the probability, $P(E)$. (**b**) Determine fair odds against event $E$. (**c**) Determine other values for fair odds for the event $E$.

## Solution

**(a)** Here, possible values for $a$ and $b$ are $a = 3$ and $b = 2$. Thus, $P(E) = 3/(3 + 2) = 3/5$.

**(b)** Fair odds against $E$ are 2 to 3.

**(c)** Any two nonnegative numbers having the same ratio as $a/b$ may be used in place of $a$ and $b$. Thus, 3 to 2 fair odds are equivalent to 6 to 4 fair odds, or 12 to 8 fair odds, or 1.5 to 1 fair odds.

---

### notes

**(i)** Very often, the word "fair" is dropped. Thus, for example, we merely say that the odds in favor of $E$ are $a$ to $b$ if $P(E) = a/(a + b)$, and conversely.

**(ii)** Any two nonnegative numbers having the same ratio as $a/b$ may be used in place of $a$ and $b$, as in part (c) of Example 2-14. That is, if odds in favor of $E$ are $a$ to $b$, then we can also say that odds in favor of $E$ are $ka$ to $kb$, where $k$ is any positive number.

Usually, odds are expressed in terms of nonnegative integers [i.e., "whole numbers"], and very often the integers used have no common factors, as with 3 to 2 odds instead of 6 to 4 odds.

**Theorem 2.4:** Given that the probability of an event $E$ is $P(E)$. Then, odds in favor of $E$ are $P(E)$ to $P(E')$, and odds against $E$ are $P(E')$ to $P(E)$. [Recall that $P(E') = 1 - P(E)$; see Theorem 1.3.]

---

**EXAMPLE 2-15:** Prove Theorem 2.4.

***Proof:*** Let

$$P(E) = \frac{P(E)}{1} \tag{1}$$

But, since $1 = P(E) + P(E')$, we can rewrite Eq. (1) as

$$P(E) = \frac{P(E)}{P(E) + P(E')} \tag{2}$$

Thus, in the equation $P(E) = a/(a + b)$ of Definition 2.1, we can replace $a$ by $P(E)$ and $b$ by $P(E')$. That means that the odds in favor of $E$, which are $a$ to $b$, can also be expressed as $P(E)$ to $P(E')$. Also, the odds against $E$, which are $b$ to $a$, can also be expressed as $P(E')$ to $P(E)$. $\square$

**EXAMPLE 2-16:** Suppose that the odds in favor of event $A$ are 7 to 5, the odds in favor of event $B$ are 2 to 3, and the odds against event $C$ are 6 to 2. **(a)** Determine $P(A)$, $P(B)$, and $P(C)$. **(b)** Determine odds against event $A$ and odds for event $C$.

## Solution

**(a)** Using Definition 2.1, we have

$$P(A) = \frac{7}{7 + 5} = \frac{7}{12} \qquad P(B) = \frac{2}{2 + 3} = \frac{2}{5} = .40$$

For event $C$, $b = 6$ and $a = 2$. Thus,

$$P(C) = \frac{a}{a + b} = \frac{2}{2 + 6} = \frac{2}{8} = \frac{1}{4} = .25$$

**(b)** *Method 1:* Odds against $A$ are $b$ to $a$ or 5 to 7 since $P(A) = 7/(7 + 5)$. Odds for event $C$ are $a$ to $b$ or 2 to 6 since $P(C) = 2/(2 + 6)$.

*Method 2:* Using Theorem 2.4, odds against $A$ are $P(A')$ to $P(A)$, that is, 5/12 to 7/12, or 5 to 7, after multiplying both prior numbers by $k = 12$.

Odds for event $C$ are $P(C)$ to $P(C')$, that is, 1/4 to 3/4, or 1 to 3, after multiplying through both prior numbers by $k = 4$.

---

*Subjective probability* deals with personal estimates of the probabilities of various real-life events that often relate to business and gambling situations. As with the usual definition of probability, a subjective probability is a number between 0 and 1, inclusive. The subjective probability that an individual would assign to an event is often based on the person's experience with similar situations, or merely upon a so-called "gut feeling" associated with the event. Frequently, an individual will do research or hire others to do research to determine reasonable estimates for the subjective probability of an event.

Certainly, if a person feels that event $A$ is less likely to occur than event $B$, then the subjective probability for $A$ will be less than the probability for $B$.

Often, the subjective probability of an event may be determined by exposing a person to a risk-taking situation, and finding the odds at which the person would consider it fair to bet on the occurrence of the event. [The odds are then converted to a probability by using Definition 2.1.] For example, if a person feels that it is fair to bet $5000 against $3000 that a particular venture will succeed, the person would be saying, in effect, that fair odds that the venture will succeed are 5000 to 3000 or, equivalently, 5 to 3. Thus, the person's subjective probability that the venture will succeed is $5/(5 + 3) = 5/8 = .625$.

*note:* In the betting situation of the prior paragraph, the person is saying, in effect, that the most he is willing to pay to an adversary is $5000 if the business fails, provided he will receive $3000 from the adversary if the business succeeds.

---

**EXAMPLE 2-17:** An entrepreneur believes that the odds in favor of a fast-food restaurant succeeding are 3 to 1, the odds in favor of a dinner theater succeeding are 4 to 1, and the odds against success in both ventures are 2 to 3. Based on subjective probabilities deduced from these odds statements, compute **(a)** the probability that both ventures will fail; **(b)** the probability that at least one venture will succeed; and **(c)** the probability that only the fast-food restaurant will succeed.

*Solution: Preliminary Calculations*: Let $A$ and $B$, respectively, refer to the events of succeeding in the fast-food restaurant and dinner theater. Converting from odds to probabilities, we have

$$P(A) = \frac{3}{3+1} = \frac{3}{4} \qquad P(B) = \frac{4}{4+1} = \frac{4}{5} \qquad P(A \cap B) = \frac{3}{3+2} = \frac{3}{5}$$

The last probability follows from the fact that the odds in favor of both ventures succeeding are 3 to 2. Now, filling in $P(R_i)$'s in the Venn diagram of Figure 2-8, $P(R_1) = P(A \cap B) = .60$; $P(R_2) = P(A) - P(R_1) = .75 - .60 = .15$; $P(R_3) = P(B) - P(R_1) = .80 - .60 = .20$; $P(R_4) = P(S) - [P(R_1) + P(R_2) + P(R_3)] = .05$.

**(a)** $P$(both ventures fail) $= P(R_4) = .05$. [Note that $R_4 = A' \cap B'$.]
**(b)** $P$(at least one venture succeeds) $= P(A \cup B) = P(A) + P(B) - P(A \cap B) = .75 + .80 - .60 = .95$.
**(c)** $P$(only the fast-food restaurant succeeds) $= P(A \cap B') = P(R_2) = .15$.

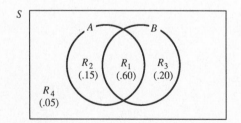

**Figure 2-8.** Venn diagram for Example 2-17.

# SOLVED PROBLEMS

## Probability and Set Theory

**PROBLEM 2-1**  Two secretaries, Ann and Bobby, work in an office. On any given day, the probability that Ann shows up for work is .9, the probability that Bobby shows up for work is .8, and the probability that at least one of the secretaries shows up for work is .92. Find the probability that, on any given day, **(a)** both secretaries show up for work, **(b)** only Bobby shows up, **(c)** neither secretary shows up, and **(d)** only one of the secretaries shows up.

*Solution* *Preliminary Calculations:* Draw a Venn diagram as shown in Figure 2-9 in which $A$ stands for "Ann shows up for work" and $B$ stands for "Bobby shows up for work."

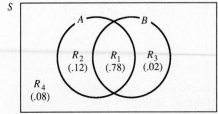

*Given:* $P(A) = .90$, $P(B) = .80$, $P(A \cup B) = .92$.

*Calculated:* $P(R_1) = P(A \cap B) = P(A) + P(B) - P(A \cup B) = .90 + .80 - .92 = .78$; $P(R_2) = P(A) - P(R_1) = .12$; $P(R_3) = P(B) - P(R_1) = .02$; $P(R_4) = 1 - [P(R_1) + P(R_2) + P(R_3)] = .08$.

**(a)** $P(\text{Both show up}) = P(A \cap B) = P(R_1) = .78$.

**(b)** $P(\text{Only Bobby shows up}) = P[(\text{not } A) \text{ and } B] = P(R_3) = .02$.

**Figure 2-9.** Venn diagram for Problem 2-1.

**(c)** $P(\text{Neither shows up}) = P[(\text{not } A) \text{ and } (\text{not } B)] = P(R_4) = .08$.

**(d)** $P(\text{Only one secretary shows up}) = P(\text{Only Ann shows up}) + P(\text{Only Bobby shows up}) = P(R_2) + P(R_3) = .14$

**PROBLEM 2-2**  Prove that for any two events $A$ and $B$ of a sample space $S$, $P(A \cup B) = P(A) + P(A' \cap B)$.

*Proof* Refer to Figure 2-9, say [but ignore the probability numbers]. We see that

$$A \cup B = (R_1 \cup R_2) \cup R_3 \tag{1}$$

where the $R_i$'s are disjoint. Thus,

$$A \cup B = A \cup (A' \cap B) \tag{2}$$

since $A = (R_1 \cup R_2)$ and $A' \cap B = R_3$. Thus, applying Axiom 1.3'

$$P(A \cup B) = P(A) + P(A' \cap B) \tag{3}$$

since $A$ and $A' \cap B$ are disjoint.

*note:* Since we can switch the roles of $A$ and $B$, we also have the following probability equation:

$$P(A \cup B) = P(B) + P(A \cap B') \tag{4} \quad \square$$

**PROBLEM 2-3**  **(a)** Given any two events $A$ and $B$ of sample space $S$. Prove that

$$P(A \text{ or } B \text{ but not both } A \text{ and } B) = P(A \cup B) - P(A \cap B)$$

Note that the event "$A$ or $B$ but not both $A$ and $B$" can also be written as "exactly one of $A$ or $B$," or "only $A$ or only $B$." **(b)** Use the result of part (a) to solve part (d) of Problem 2-1.

*Solution*

**(a)** Refer to Figure 2-9 [but ignore the probability numbers]. The event we're concerned with is $R_2 \cup R_3$, which is the same as $(A \cap B') \cup (A' \cap B)$. Now, since the $R_i$'s are disjoint,

$$P(R_2 \cup R_3) = P(R_2) + P(R_3) \tag{1}$$

Let us add on zero in the form $P(R_1) - P(R_1)$ to the right side of Eq. (1). Thus,

$$P(R_2 \cup R_3) = P(R_1) + P(R_2) + P(R_3) - P(R_1) \qquad (2)$$

That is, $P(R_2 \cup R_3) = P(R_1 \cup R_2 \cup R_3) - P(R_1)$, or

$$P(R_2 \cup R_3) = P(A \cup B) - P(A \cap B)$$

since $A \cup B = R_1 \cup R_2 \cup R_3$ and $A \cap B = R_1$.

**(b)** $P(\text{Only one secretary shows up}) = P(A \cup B) - P(A \cap B) = .92 - .78 = .14.$

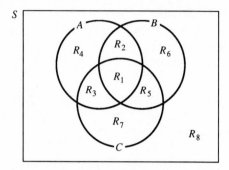

**Figure 2-10.** All possible intersections of sets *A*, *B*, and *C*.

**PROBLEM 2-4**   Refer to Figure 2-10, which duplicates Figure 2-2. Express each $R_i$ in terms of an intersection of *A* or *A'* with *B* or *B'* and with *C* or *C'*.

**Solution**   There are a total of eight of the disjoint $R_i$'s in the partition of *S* shown in Figure 2-10. Each $R_i$ is a collection of those elements [if any] which are or are not in *A* and *B* and *C*. Thus, $R_1 = A \cap B \cap C$, $R_2 = A \cap B \cap C'$, $R_3 = A \cap B' \cap C$, $R_4 = A \cap B' \cap C'$, $R_5 = A' \cap B \cap C$, $R_6 = A' \cap B \cap C'$, $R_7 = A' \cap B' \cap C$, and $R_8 = A' \cap B' \cap C'$.

For example, the elements that belong to $R_7$ don't belong to *A*, and don't belong to *B*, but do belong to *C*.

**PROBLEM 2-5**   Refer to Figure 2-10. **(a)** Express the sets $A'$, $C$, $C'$, $A' \cap C'$, and $A' \cup C$ as unions involving the $R_i$'s. **(b)** Express the probabilities of the sets of part (a) in terms of the probabilities of the $R_i$'s.

**Solution**

**(a)** $A' = R_5 \cup R_6 \cup R_7 \cup R_8, C = R_1 \cup R_3 \cup R_5 \cup R_7, C' = R_2 \cup R_4 \cup R_6 \cup R_8.$ From the unions of $R_i$'s above for $A'$ and $C'$, we obtain

$$A' \cap C' = R_6 \cup R_8$$

From the unions of $R_i$'s above for $A'$ and $C$, we obtain

$$A' \cup C = R_5 \cup R_6 \cup R_7 \cup R_8 \cup R_1 \cup R_3$$

**(b)** By applying Axiom 1.3′ to the $R_i$ unions above,

$$P(A') = P(R_5) + P(R_6) + P(R_7) + P(R_8)$$

$$P(C) = P(R_1) + P(R_3) + P(R_5) + P(R_7)$$

$$P(C') = P(R_2) + P(R_4) + P(R_6) + P(R_8)$$

$$P(A' \cap C') = P(R_6) + P(R_8)$$

$$P(A' \cup C) = P(R_5) + P(R_6) + P(R_7) + P(R_8) + P(R_1) + P(R_3)$$

**PROBLEM 2-6**   Use the method of Example 2-1 to prove Theorem 2.2, the general addition rule for three events.

**Solution**   Refer to Figure 2-10. We wish to show that the left and right sides of Theorem 2.2 are both equal to the same expression involving $P(R_i)$ terms. For the left side,

$$A \cup B \cup C = R_1 \cup R_2 \cup \ldots \cup R_7 \qquad (1)$$

and thus,

$$P(A \cup B \cup C) = P(R_1) + P(R_2) + \ldots + P(R_7) \qquad (2)$$

after applying Axiom 1.3′ to $P(R_1 \cup R_2 \cup \ldots \cup R_7)$, to which $P(A \cup B \cup C)$ is equal.

Now, let us work on the right side terms. First, $A = R_1 \cup R_2 \cup R_3 \cup R_4$, and thus

$$P(A) = P(R_1) + P(R_2) + P(R_3) + P(R_4) \qquad (3)$$

Also, $B = R_1 \cup R_2 \cup R_5 \cup R_6$, and $C = R_1 \cup R_3 \cup R_5 \cup R_7$, and thus

$$P(B) = P(R_1) + P(R_2) + P(R_5) + P(R_6) \qquad (4)$$

and

$$P(C) = P(R_1) + P(R_3) + P(R_5) + P(R_7) \tag{5}$$

Also, $A \cap B = R_1 \cup R_2$, and thus

$$P(A \cap B) = P(R_1) + P(R_2) \tag{6}$$

Likewise, $A \cap C = R_1 \cup R_3$, and $B \cap C = R_1 \cup R_5$, and thus,

$$P(A \cap C) = P(R_1) + P(R_3) \tag{7}$$

and

$$P(B \cap C) = P(R_1) + P(R_5) \tag{8}$$

Since $A \cap B \cap C = R_1$, we have

$$P(A \cap B \cap C) = P(R_1) \tag{9}$$

Now, we substitute from Eqs. (4), (5), (6), (7), (8), and (9) into the right side of the Theorem 2.2 equation to get

$$P(A) + P(B) + P(C) - P(A \cap B) - P(A \cap C) - P(B \cap C) + P(A \cap B \cap C)$$
$$= 3P(R_1) + 2P(R_2) + 2P(R_3) + 2P(R_5) + P(R_4) + P(R_6) + P(R_7)$$
$$- 3P(R_1) - P(R_2) - P(R_3) - P(R_5) + P(R_1)$$
$$= P(R_1) + P(R_2) + P(R_3) + P(R_4) + P(R_5) + P(R_6) + P(R_7) \tag{10}$$

Since the final expressions in Eqs. (2) and (10) are the same, the theorem is proved.

**PROBLEM 2-7** (a) Prove DeMorgan's law $(A \cap B)' = A' \cup B'$. (b) Derive the equation $P(A' \cup B') = 1 - P(A \cap B)$.

*Solution*

(a) Refer to Figure 2-9 [but ignore the probability numbers].

$$A = R_1 \cup R_2; \qquad B = R_1 \cup R_3 \tag{1}, (2)$$

Thus, $A \cap B = R_1$, and hence,

$$(A \cap B)' = R_2 \cup R_3 \cup R_4 \tag{3}$$

Now,

$$A' = R_3 \cup R_4 \qquad \text{and} \qquad B' = R_2 \cup R_4 \tag{4}, (5)$$

and thus,

$$A' \cup B' = R_2 \cup R_3 \cup R_4 \tag{6}$$

Since the right sides of Eqs. (3) and (6) are the same, the law is proved.
(b) From the part (a) result, we have

$$P[(A' \cup B')] = P[(A \cap B)'] \tag{1}$$

Now, $P[(A \cap B)'] = 1 - P(A \cap B)$ by Theorem 1.3, and hence

$$P(A' \cup B') = 1 - P(A \cap B) \tag{2}$$

### Elementary Probability Problems

**PROBLEM 2-8** A survey of people who went to the Johnson City Zoo in 1985 showed that 50% visited the Apes building and 40% visited the Reptiles building. Let $A$ be the event that a person visits the Apes building, and let $R$ be the event that a person visits the Reptiles building. Determine the closest possible lower and upper bounds for $P(A \cap R)$ and $P(A \cup R)$.

*Solution* Consider the two possible extreme cases depicted in Figure 2-11. In

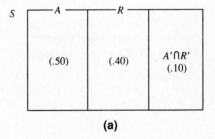

(a)

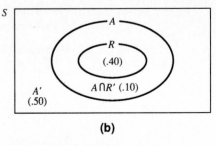

(b)

**Figure 2-11.** Extreme cases for Problem 2-8. (a) $A \cap R = \emptyset$. (b) $R$ is a proper subset of $A$.

Figure 2-11a, the sets $A$ and $R$ "fill out" $S$ as much as possible. Here, $P(A) + P(R) = .90$, and in this case, $A$ and $R$ are disjoint. That is, for the extreme case of Figure 2-11a, we have

$$P(A \cap R) = P(\varnothing) = 0 \quad \text{and} \quad P(A \cup R) = P(A) + P(R) = .90 \quad \textbf{(1), (2)}$$

In the extreme case shown in Figure 2-11b, $R$ is a proper subset of $A$. Thus, for this extreme case, we have

$$P(A \cap R) = P(R) = .40 \quad \text{and} \quad P(A \cup R) = P(A) = .50 \quad \textbf{(3), (4)}$$

Now, $P(A \cap R)$ must be between the extreme values given in **(1)** and **(3)**, and $P(A \cup R)$ must be between the extreme values given in **(2)** and **(4)**. That is,

$$0 \le P(A \cap R) \le .40 \quad \text{and} \quad .50 \le P(A \cup R) \le .90 \quad \textbf{(5), (6)}$$

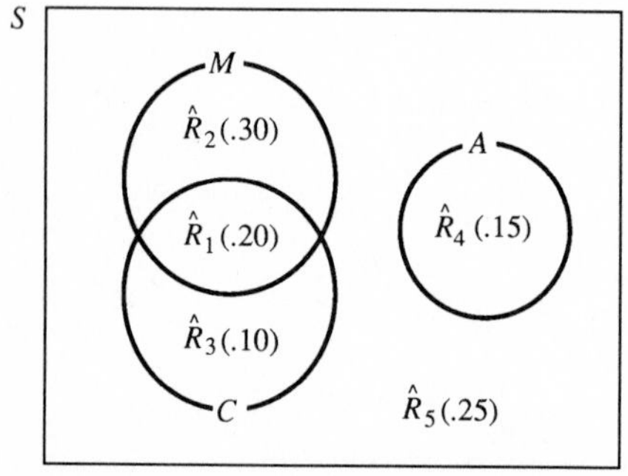

**(a)**

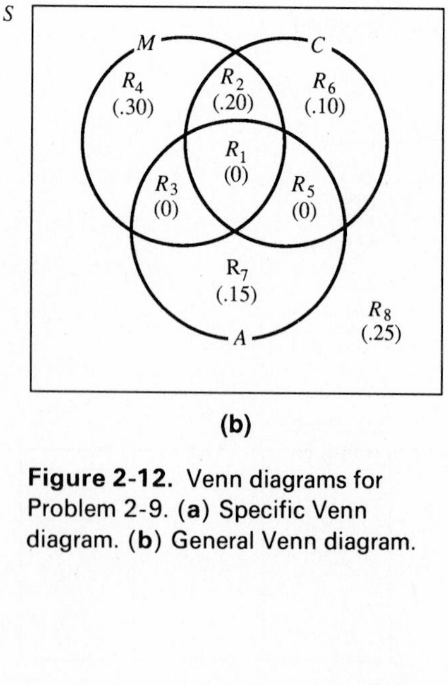

**(b)**

**Figure 2-12.** Venn diagrams for Problem 2-9. **(a)** Specific Venn diagram. **(b)** General Venn diagram.

**PROBLEM 2-9**   A survey of freshmen was conducted at Old Tech University. The letters $M$, $C$, and $A$ stand for whether a student takes courses in mathematics, computer science, or anthropology, respectively. Fifty percent of the students take courses in mathematics; that is, $P(M) = .50$. Also, $P(C) = .30$, $P(A) = .15$, $P(M \cap C) = .20$ and $M \cap A = \varnothing$ and $C \cap A = \varnothing$. The last item indicates that there are no students who take both computer science and anthropology courses. Determine the values of the following probabilities: **(a)** $P(M \cap C')$, **(b)** $P[(M \cup C)']$, **(c)** $P(M' \cap C')$, **(d)** $P(M' \cap A)$, and **(e)** $P(M \cup C \cup A)$.

*Solution   Preliminaries:* We could use either of the Venn diagrams in Figure 2-12. [The numbers denote probabilities of the indicated disjoint regions.] Note that $M \cap C \cap A = \varnothing$, since $M \cap C \cap A$ is a subset of $M \cap A$, and the latter set is empty.

Let us here use the Venn diagram of Figure 2-12a, which is specific to the problem at hand. [Problem 2-19 deals with using the general Venn diagram of Figure 2-11b for the current problem.]

**(a)** Since $M = \hat{R}_1 \cup \hat{R}_2$, and $C' = \hat{R}_2 \cup \hat{R}_4 \cup \hat{R}_5$, it follows that $M \cap C' = \hat{R}_2$, and thus, $P(M \cap C') = P(\hat{R}_2) = .30$.

**(b)** Since $C = \hat{R}_1 \cup \hat{R}_3$, it follows that $M \cup C = \hat{R}_1 \cup \hat{R}_2 \cup \hat{R}_3$. Thus, $(M \cup C)' = \hat{R}_4 \cup \hat{R}_5$, and thus, $P[(M \cup C)'] = P(\hat{R}_4) + P(\hat{R}_5) = .40$.

**(c)** One of DeMorgan's laws says the $M' \cap C' = (M \cup C)'$. Thus, $P[M' \cap C'] = P[(M \cup C)'] = .40$, after referring to part **(b)**.

**(d)** Here, $M' = \hat{R}_3 \cup \hat{R}_4 \cup \hat{R}_5$ and $A = \hat{R}_4$, and so $P(M' \cap A) = P(\hat{R}_4) = P(A) = .15$.

**(e)** Here, $M \cup C \cup A = \hat{R}_1 \cup \hat{R}_2 \cup \hat{R}_3 \cup \hat{R}_4$ and $P(M \cup C \cup A) = P(\hat{R}_1) + P(\hat{R}_2) + P(\hat{R}_3) + P(\hat{R}_4) = .75$.

**PROBLEM 2-10**   Suppose fair odds against event $A$ are 3 to 7. **(a)** Determine fair odds for event $A$. **(b)** Determine the probabilities $P(A)$ and $P(A')$.

*Solution*

**(a)** The fair odds for event $A$ are 7 to 3.
**(b)** Thus, $P(A) = 7/(7 + 3) = .7$ and $P(A') = 1 - P(A) = .3$.

**PROBLEM 2-11**   Suppose that $P(A) = .34$ and $P(B) = .72$. **(a)** Determine fair odds in favor of $A$, and against $A$. **(b)** Determine fair odds in favor of $B$, and against $B$.

*Solution*

**(a)**
$$P(A') = 1 - P(A) = 1 - .34 = .66$$

Thus, by Theorem 2.4, fair odds in favor of $A$ are $P(A)$ to $P(A')$. That is, fair odds for $A$ are .34 to .66, or 34 to 66, or 17 to 33.

**(b)**
$$P(B') = 1 - P(B) = 1 - .72 = .28$$

Thus, fair odds for $B$ are $P(B)$ to $P(B')$, that is, .72 to .28, or 72 to 28, or 18 to 7.

**PROBLEM 2-12** College student Jan estimates that the odds against getting $B$ or higher in English are 2 to 3, the odds against getting $B$ or higher in mathematics are 1 to 3, and the odds in favor of simultaneously getting $B$ or higher in both subjects are 4 to 6. Compute **(a)** the probability of Jan getting $B$ or higher in at least one of English and mathematics, **(b)** the probability of Jan getting $B$ or higher in exactly one of the two subjects, **(c)** the probability of Jan getting below $B$ in both subjects, and **(d)** the probability of Jan getting below $B$ in at least one of English and mathematics.

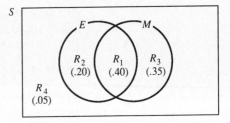

**Figure 2-13.** Venn diagram for Problem 2-12.

***Solution*** *Preliminaries:* Let $E$ be the event of Jan getting "$B$ or higher in English," and $M$ be the event of Jan getting "$B$ or higher in mathematics."

Converting from odds to probabilities, we have $P(E) = 3/5 = .60$, $P(M) = 3/4 = .75$, and $P(E \cap M) = 4/10 = .40$. Thus, we obtain the Venn diagram of Figure 2-13.

For the $P(R_i)$ values, we have:

$$P(R_1) = P(E \cap M) = .40; \qquad P(R_2) = P(E) - P(R_1) = .60 - .40 = .20;$$

$$P(R_3) = P(M) - P(R_1) = .75 - .40 = .35;$$

$$P(R_4) = 1 - [P(R_1) + P(R_2) + P(R_3)] = .05$$

**(a)** Symbolically, the event of Jan getting $B$ or higher in at least one of the two subjects is $E \cup M$. Thus,

$$P(E \cup M) = P(E) + P(M) - P(E \cap M) = .6 + .75 - .40 = .95$$

Note that $P(E \cup M)$ is also equal to $P(R_1) + P(R_2) + P(R_3)$.

**(b)** Here, the event is equivalent to $R_2 \cup R_3$, and thus the answer is

$$P(R_2 \cup R_3) = P(R_2) + P(R_3) = .55$$

**(c)** Here, the event can be represented symbolically as $E' \cap M'$, which equals $(R_3 \cup R_4) \cap (R_2 \cup R_4) = R_4$. Thus, for our answer, we have

$$P(E' \cap M') = P(R_4) = .05$$

**(d)** Here, the event can be represented symbolically as $E' \cup M'$, which equals $R_2 \cup R_3 \cup R_4$. Thus, for our answer, we have

$$P(E' \cup M') = P(R_2) + P(R_3) + P(R_4) = .60$$

## Supplementary Problems

**PROBLEM 2-13** Give an alternate proof of Theorem 2.1, the general addition rule for two events, by making use of the set equations $A \cup B = A \cup (A' \cap B)$ and $B = (A \cap B) \cup (A' \cap B)$. [*Hints:* In the first equation, $A$ and $A' \cap B$ are disjoint, and in the second equation, $A \cap B$ and $A' \cap B$ are disjoint.]

**PROBLEM 2-14** Use Theorem 2.1 to show that **(a)** $P(A \cap B) \le P(A) + P(B)$ and **(b)** $P(A) + P(B) - 1 \le P(A \cap B)$.
[*Hint:* Part **(a)** follows from $P(A \cup B) \ge 0$, and part **(b)** follows from $P(A \cup B) \le 1$.]

**PROBLEM 2-15** A survey of people who visited the FUNLAND amusement park in 1986 showed that 40% went on the parachute ride, 45% went on the roller coaster ride, and 62% went on at least one of the above two rides. What percentage of people **(a)** went on both rides, **(b)** went on at most one of the two rides, **(c)** went on exactly one of the two rides, and **(d)** didn't go on either ride?

*Answer* **(a)** 23%  **(b)** 77%  **(c)** 39%  **(d)** 38%

**PROBLEM 2-16**   Suppose that all you are given is that $P(A) = .75$, and $P(B) = .50$, where $A$ and $B$ are subsets of sample space $S$. Determine closest possible lower and upper bounds for **(a)** $P(A \cap B)$ and **(b)** $P(A \cup B)$.

*Answer* **(a)** $.25 \leq P(A \cap B) \leq .50$     **(b)** $.75 \leq P(A \cup B) \leq 1$

**PROBLEM 2-17**   Of the freshmen at Everbright College, 55% were taking an English course, 50% were taking a mathematics course, and 23% were taking a foreign language course. Also, 27% were taking an English course and a mathematics course, 11% were taking an English course and a foreign language course, and 12% were taking a mathematics course and a foreign language course. Also, 7% were taking an English course and a mathematics course and a foreign language course. What percentage of the freshmen **(a)** were taking a course in at least one of the three subject areas, **(b)** were not taking any courses in the three subject areas, **(c)** were taking a course only in mathematics, and **(d)** were taking a course in exactly one of the three subject areas?

*Answer* **(a)** 85%     **(b)** 15%     **(c)** 18%     **(d)** 49%

**PROBLEM 2-18**   Four candidates are running for Mayor of Fernville. If $A$ is twice as likely to win as $B$ [that is, $P(A) = 2P(B)$], $B$ is two-thirds as likely as $C$ to win, and $C$ is one and a half times as likely to win as $D$, what are the probabilities that **(a)** $A$ wins, and **(b)** $C$ does not win? [*Hint:* Express all probabilities in terms of $P(D)$.]

*Answer* **(a)** $P(A \text{ wins}) = P(A) = 4/11$     **(b)** $P(C \text{ doesn't win}) = P(C') = 8/11$

**PROBLEM 2-19**   Rework Problem 2-9 by using the general Venn diagram of Figure 2-12b. [*Hint:* Since, for example, $M \cap A = \varnothing$, we have $P(R_1) = P(R_3) = 0$.]

*Answer* **(a)** $P(M \cap C') = .30$     **(b)** $P[(M \cup C)'] = .40$     **(c)** $P(M' \cap C') = .40$     **(d)** $P(M' \cap A) = .15$
**(e)** $P(M \cup C \cup A) = .75$

**PROBLEM 2-20**   The manager of a store in a sporting goods chain feels that the odds are 1 to 1 in favor of her getting a holiday season bonus of $3000, and 3 to 1 against her being promoted to regional manager, and 13 to 7 in favor of her getting either the bonus or the promotion [or both]. What is the probability she will **(a)** get both the bonus and the promotion, **(b)** get exactly one of the bonus and promotion, **(c)** merely get a bonus of $3000, **(d)** get neither the bonus nor the promotion?

*Answer* **(a)** .10     **(b)** .55     **(c)** .40     **(d)** .35

# 3 CONDITIONAL PROBABILITY AND INDEPENDENCE

## THIS CHAPTER IS ABOUT

☑ **Conditional Probability**
☑ **Independent Events**
☑ **Counting Methods**
☑ **A Review of the Concepts of Mutually Exclusive and Independent Events**

## 3-1. Conditional Probability

### A. Conditional probability when elements are equally likely

Let us focus again on the experiment of tossing a pair of fair dice. [See Examples 1-2, 1-13, and 1-16.] The most subdivided sample space $S$ has 36 equally likely elements. [Or, as we said previously, $S$ has 36 equally likely simple events—refer to wording of Theorem 1.2 and to the note that follows it.] That is,

$$S = \{11, 12, \ldots, 16, 21, 22, \ldots, 26, \ldots, 61, 62, \ldots, 66\}$$

where $P(\{11\}) = P(\{12\}) = \ldots = P(\{66\}) = \frac{1}{36}$. That is,

$$P(\{ij\}) = \frac{1}{36} \quad \text{for } i = 1, 2, \ldots, 6 \quad \text{and} \quad j = 1, 2, \ldots, 6$$

---

**EXAMPLE 3-1:** Let the event $B$ be the event "a two occurs on at least one die," and $A$ be the event "sum of the numbers [of dots] on the dice equals six." Represent $B$ and $A$ as subsets of $S$, and determine $P(B)$ and $P(A)$.

**Solution:** We see that $B = \{21, 22, 23, 24, 25, 26, 12, 32, 42, 52, 62\}$ and that $A = \{15, 24, 33, 42, 51\}$. Then, apply Theorem 1.2, we have

$$P(B) = \frac{N(B)}{N(S)} = \frac{11}{36} \quad \text{and} \quad P(A) = \frac{N(A)}{N(S)} = \frac{5}{36}$$

---

Suppose we again consider the experiment of tossing a pair of fair dice as in the preceding discussion, but now suppose we wish to find the probability of obtaining "a two on at least one die" on condition that the "sum of numbers on the dice equals six." This involves the concept of conditional probability. Employing the same labels $B$ and $A$ that were given to these events in Example 3-1, it is customary to use the symbol $P(B|A)$ for this probability.

In general, the symbol $P(B \mid A)$ is read as the "conditional probability of $B$ given $A$," or the "probability that $B$ occurs on condition that [or given that] $A$ has occurred."

Let's attempt to find a logically meaningful value for $P(B \mid A)$, with $A$ and $B$ as given in Example 3-1. Since, in the probability $P(B \mid A)$, we have the condition that $A$ has occurred, it is logical to let $A$, expressed as a subset, be a reduced sample space [indicated by $\hat{S}$] for the current situation. That is, we write

$$\hat{S} = A = \{15, \underline{24}, 33, \underline{42}, 51\}$$

Our reduced sample space contains five elements; in each element, the sum of numbers [of dots] on the dice is equal to 6. Since these five elements were equally likely when regarded as elements of $S$, it seems plausible to assume that they are equally likely as elements of $\hat{S}$. Now, of the five elements of $\hat{S}$, two are favorable to event $B$, namely 24 and 42 [shown underlined above]. These are said to be "favorable to event $B$" since both elements are in the subset representation of $B$; the latter is given in Example 3-1.

Thus, we conclude that $P(B \mid A) = \frac{2}{5}$. Here, we divide the number of elements favorable to $B$ [here, 2] by the total number of elements in the reduced sample space $\hat{S}$ [here, 5]. The latter set also happens to be the set for event $A$.

Now, we will develop more systematic approaches for computing conditional probabilities. Observe that the two elements 24 and 42 discussed above are none other than the two elements in the intersection set $A \cap B$ [read as "$A$ intersect $B$" or "$A$ and $B$"]. That is, $A \cap B = \{24, 42\}$. Also, $N(A \cap B) = 2$, since there are two elements in $A \cap B$.

Observe also that the denominator term of $P(B \mid A)$ is five, which equals $N(A)$, the number of elements in $A$. Thus, we may write

$$P(B \mid A) = \frac{N(A \cap B)}{N(A)} \qquad (1)$$

The above is unchanged in value if we divide the top and bottom terms by $N(S)$, the number of elements in the original sample space [here, 36]. That is,

$$P(B \mid A) = \frac{N(A \cap B)/N(S)}{N(A)/N(S)} \qquad (2)$$

But now the top fraction is equal to $P(A \cap B)$, and the bottom fraction is $P(A)$. [For these two probabilities, we are employing Theorem 1.2 with respect to the original sample space.] Thus, Eq. (1) becomes

$$P(B \mid A) = \frac{P(A \cap B)}{P(A)} \qquad (3)$$

Though the derivation of the equation for $P(B \mid A)$ in Eq. (3) was based on a specific example, it is general and applies to any experiment for which the original sample space has equally likely elements. In addition, Eq. (3) is taken to be the definition of conditional probability for any sample space [regardless of whether the elements of the sample space are equally likely].

**Definition 3.1:** Given that $A$ and $B$ are events of sample space $S$, and $P(A) > 0$, then the conditional probability of $B$ given $A$ is defined by the equation

$$P(B \mid A) = \frac{P(A \cap B)}{P(A)}$$

*notes*

(a) The probabilities $P(A \cap B)$ and $P(A)$ in Definition 3.1 above are computed with respect to the original sample space $S$.

(b) From set theory, we have that $B \cap A = A \cap B$. That means that we can also write the equation for $P(B \mid A)$ as

$$P(B \mid A) = \frac{P(B \cap A)}{P(A)}$$

(c) Note the requirement $P(A) > 0$ in Definition 3.1. Some authors state the requirement as $P(A) \neq 0$, instead. But, remember that the only way the probability $P(A)$ can be unequal to zero is if it is positive. This is so because of Axiom 1.1, which says, in part, that $P(A) \geq 0$. That means that $P(A)$ *cannot* be negative!

---

**EXAMPLE 3-2:** Refer to Example 3-1. Determine the conditional probability of $A$ given $B$ for the events $A$ and $B$ as presented there.

*Solution:* Here, we wish to compute $P(A \mid B)$. By interchanging the symbols $A$ and $B$ in Definition 3.1, we have

$$P(A \mid B) = \frac{P(B \cap A)}{P(B)} \qquad (1)$$

Now, $B \cap A = A \cap B = \{24, 42\}$, and thus,

$$P(B \cap A) = \frac{2}{36} \qquad (2)$$

Now, $B$ and $P(B)$ are given in Example 3-1, where we have

$$P(B) = \frac{11}{36} \qquad (3)$$

Thus, substituting from Eqs. (2) and (3) into Eq. (1), we obtain

$$P(A \mid B) = \frac{2/36}{11/36} = \frac{2}{11} \qquad (4)$$

---

## B. Conditional probability when the elements of $S$ are not equally likely

The key assumption in the previous discussion, which dealt with an original sample space with equally likely elements, was to assume that the elements in the reduced sample space $\hat{S}$ were also equally likely.

Recall that we had $\hat{S} = A$ in the discussion preceding Definition 3.1. Thus, in that discussion, in effect we took each of the probabilities corresponding to the elements of $\hat{S}$ to be equal to $1/5$. That is, in effect we took

$$\hat{P}(\{15\}) = \hat{P}(\{24\}) = \hat{P}(\{33\}) = \hat{P}(\{42\}) = \hat{P}(\{51\}) = 1/5$$

Here, $\hat{P}$ symbolizes the probability corresponding to an element which belongs to the reduced sample space. Remember that such probabilities are those that apply with respect to the reduced sample space. [In the original sample space $S$, which has 36 equally likely elements, we have $P(\{15\}) = P(\{24\}) = P(\{33\}) = P(\{42\}) = P(\{51\}) = 1/36$, where $1/36$ is the probability value corresponding to any of the 36 elements of $S$.]

The key assumption to make if the original elements are not equally likely is to assume that the ratio of probabilities for any pair of elements in the reduced sample space has the same value in the original sample space as in the reduced sample space. That is,

$$\frac{P(\{e_j\})}{P(\{e_i\})} = \frac{\hat{P}(\{e_j\})}{\hat{P}(\{e_i\})} \qquad (I)$$

for any pair of elements $e_i$ and $e_j$ which belong to the reduced sample space.

---

**EXAMPLE 3-3:** Given an original sample space $S$ which is discrete. Focus on subsets $A$ and $B$ of $S$. Derive an equation for $P(B \mid A)$. Assume that (I) above applies to elements in the reduced sample space $\hat{S}$, which in this discussion is also the event $A$. Another fact that applies is that the sum of "reduced" probabilities corresponding to elements in the reduced sample space equals 1. That is

$$\sum_A \hat{P}(\{e_i\}) = 1 \qquad \text{(II)}$$

In (II), the summation is with respect to elements in subset $A$. [Observe that $A$ is the subscript of the $\sum$ symbol.]

*Solution:* Now, from Eq. (I), it follows after a little algebraic manipulation that

$$\frac{\hat{P}(\{e_j\})}{P(\{e_j\})} = \frac{\hat{P}(\{e_i\})}{P(\{e_i\})} \qquad \text{(1)}$$

for any pair of elements $e_i$ and $e_j$ in the reduced sample space $A$. That is,

$$\frac{\hat{P}(\{e_i\})}{P(\{e_i\})} = c \qquad \text{or} \qquad \hat{P}(\{e_i\}) = cP(\{e_i\}) \qquad \text{for any element } e_i \text{ of } A \quad \text{(2a), (2b)}$$

Here, $c$ is a constant. $\left(\text{Note that in Example 3-1, the } c \text{ value was } \dfrac{(1/5)}{(1/36)} = \dfrac{36}{5} = 7.2.\right)$ Substituting from Eq. (2b) into (II) above, we obtain

$$c \sum_A P(\{e_i\}) = 1 \qquad \text{(3)}$$

Now, $P(A) = \sum_A P(\{e_i\})$, and thus Eq. (3) may be written as

$$cP(A) = 1 \qquad \text{or} \qquad c = \frac{1}{P(A)} \qquad \text{(4a), (4b)}$$

Thus, substituting $c$ from Eq. (4b) into Eq. (2b) results in

$$\hat{P}(\{e_i\}) = \frac{P(\{e_i\})}{P(A)} \qquad \text{(5)}$$

Now, $P(B \mid A)$ is equal to the sum of the $\hat{P}(\{e_i\})$ probabilities for elements which are both in $A$ and in $B$, that is for elements in $A \cap B$. [See Figure 3-1.] That is,

$$P(B \mid A) = \sum_{A \cap B} \hat{P}(\{e_i\}) \qquad \text{(6)}$$

Substituting from Eq. (5) into Eq. (6) results in

$$P(B \mid A) = \frac{\sum_{A \cap B} P(\{e_i\})}{P(A)} = \frac{P(A \cap B)}{P(A)} \qquad \text{(7)}$$

From the equality of the first and last expressions of Eq. (7), we see that this equation is equivalent to the defining equation for $P(B \mid A)$ as given in Definition 3.1.

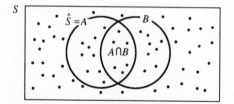

**Figure 3-1.** Diagram showing sample space $S$ and reduced sample space $\hat{S}$. Each dot ( · ) indicates an element in $S$.

If we multiply both sides of the equation of Definition 3.1 by $P(A)$, we obtain the multiplication rule theorem that follows.

**Theorem 3.1 (Multiplication Rule):** If $A$ and $B$ are any two events of a sample space $S$ and $P(A) > 0$, then

$$P(A \cap B) = P(A) \cdot P(B \mid A)$$

This reads as follows: "The probability of the event $A$ and $B$ equals $P(A)$ times the conditional probability of $B$ given $A$."

**EXAMPLE 3-4:** Consider the experiment of tossing an unbalanced coin twice. Suppose the probabilities for the four simple events of the most subdivided sample space are as follows:

$$P(\{TT\}) = .16; \qquad P(\{TH\}) = P(\{HT\}) = .24; \qquad P(\{HH\}) = .36$$

[Note that the most subdivided sample space is $S = \{TT, TH, HT, HH\}$ regardless of whether the coin is balanced—i.e., fair—or unbalanced.] **(a)** Compute the probability of getting the same result on both tosses given that at least one head is obtained. Note that this is a conditional probability. **(b)** Compute the probability of getting at least one head given that the same result is obtained on both tosses.

*Solution: Preliminaries:* Let $A$ be the event "at least one head," and let $B$ be the event "same result on both tosses." Thus, we have

$$A = \{TH, HT, HH\}, \quad B = \{TT, HH\}, \quad \text{and hence } A \cap B = \{HH\} \quad \textbf{(1), (2), (3)}$$

**(a)** Here, we wish to find $P(B \mid A)$. From Definition 3.1, we have

$$P(B \mid A) = \frac{P(A \cap B)}{P(A)} \qquad \textbf{(4)}$$

Now, from Theorem 1.1, Eqs. (1) and (3), and the given probability data, we have

$$P(A) = P(\{TH\}) + P(\{HT\}) + P(\{HH\}) = .24 + .24 + .36 = .84$$

and

$$P(A \cap B) = P(\{HH\}) = .36$$

Substituting these values into Eq. (4) yields

$$P(B \mid A) = \tfrac{.36}{.84} = \tfrac{3}{7} = .4286 \qquad \textbf{(5)}$$

**(b)** Here, we wish to find $P(A \mid B)$, and we have

$$P(A \mid B) = \frac{P(B \cap A)}{P(B)} \qquad \textbf{(6)}$$

after interchanging $A$ and $B$ in Definition 3.1. Now, $P(B \cap A) = .36$ since $B \cap A = A \cap B$. Now, from Theorem 1.1, Eq. (2), and the given probability data, we have

$$P(B) = P(\{TT\}) + P(\{HH\}) = .16 + .36 = .52 \qquad \textbf{(7)}$$

Thus, substituting into Eq. (6), we obtain

$$P(A \mid B) = \tfrac{.36}{.52} = \tfrac{9}{13} = .6923 \qquad \textbf{(8)}$$

**EXAMPLE 3-5:** Ann plans to take a mathematics course in the fall semester and a computer science course in the spring semester. She feels that the probability she will get at least a B in the mathematics course in the fall semester is 1/3. She also believes that the probability she will get at least a B in the computer science course in the spring semester given that she has received at least a B in the mathematics course in the fall semester is 5/8. Based on these subjective probability estimates, what is the probability she will get at least a B in both the mathematics and computer science courses?

*Solution:* Let $M$ stand for the event "at least a B in the mathematics course," and $C$ stand for the event "at least a B in the computer science course." We wish to compute $P(M \text{ and } C) = P(M \cap C)$. From Theorem 3.1,

$$P(M \cap C) = P(M) \cdot P(C \mid M) = \tfrac{1}{3} \cdot \tfrac{5}{8} = \tfrac{5}{24} = .208\overline{3}$$

Here, the symbol $\overline{3}$ means that the 3 is repeated ad infinitum. That is, $.208\overline{3} = .20833333\ldots$.

Theorem 3.1 can be generalized so that it applies to three or more events. For example, for three events, we have Theorem 3.2.

**Theorem 3.2 (Multiplication Rule for Three Events):** If $A$ and $B$ and $C$ are any three events of a sample space $S$ and $P(A \cap B) > 0$, then

$$P(A \cap B \cap C) = P(A) \cdot P(B \mid A) \cdot P(C \mid A \cap B)$$

The generalized theorem for $k$ events, where $k \geq 2$, is as follows:

**Theorem 3.3 (General Multiplication Rule):** If $A_1, A_2, \ldots, A_{k-1}$, and $A_k$ are any $k$ events $[k \geq 2]$ of a sample space $S$, and $P(A_1 \cap A_2 \cap \ldots \cap A_{k-1}) > 0$, then

$$P(A_1 \cap A_2 \cap \ldots \cap A_{k-1} \cap A_k)$$
$$= P(A_1) \cdot P(A_2 \mid A_1) \cdot P(A_3 \mid A_1 \cap A_2) \cdot \ldots \cdot P(A_k \mid A_1 \cap A_2 \cap \ldots \cap A_{k-1})$$

where there are $k$ terms in the product on the right.

---

**EXAMPLE 3-6:** Prove Theorem 3.2.

*Proof:* Writing $A \cap B \cap C$ as $(A \cap B) \cap C$, and using Theorem 3.1, we have

$$P(A \cap B \cap C) = P[(A \cap B) \cap C] = P(A \cap B) \cdot P(C \mid A \cap B) \qquad \textbf{(1)}$$

Here, we made use of the hypothesis that $P(A \cap B) > 0$. Now, again from Theorem 3.1, we can write

$$P(A \cap B) = P(A) \cdot P(B \mid A) \qquad \textbf{(2)}$$

provided that $P(A) > 0$. The latter is valid since $P(A \cap B) > 0$ implies $P(A) > 0$. [From Theorem 1.2, $P(A \cap B) \leq P(A)$ since $A \cap B$ is a subset of $A$. Thus, $P(A)$ is positive whenever $P(A \cap B)$ is positive.]

Substituting the expression for $P(A \cap B)$ from Eq. (2) into Eq. (1) results in

$$P(A \cap B \cap C) = P(A) \cdot P(B \mid A) \cdot P(C \mid A \cap B)$$

as was to be shown. $\qquad\qquad\qquad\qquad\qquad\qquad\qquad\qquad\qquad\qquad\qquad\qquad\qquad$ $\square$

---

The following useful theorem will be proved in the solution to Problem 3-4.

**Theorem 3.4:** Suppose $P(A) > 0$. Then $P(B' \mid A)$ is defined [equivalently, $P(B' \mid A)$ exists], and is given by $P(B' \mid A) = 1 - P(B \mid A)$.

Note the similarity of this to Theorem 1.3. The latter indicates that $P(A') = 1 - P(A)$.

---

**EXAMPLE 3-7:** A study is done on the adults [people over 21] in Newtown, USA. Suppose that 20% of the adults are college graduates. Also, 75% of the college graduates earn annual incomes of at least $25,000. **(a)** If an adult is selected at random, what is the probability that the adult is both a college graduate and an earner of at least $25,000 annually? **(b)** What is the percentage of college graduates in Newtown that earn less than $25,000 annually?

*Solution: Preliminaries:* By "at random," we mean that all possible selections of adults are equally likely from the sample space of adults in Newtown, USA. Let

*A* be the event of selecting an adult who is a college graduate, and let *B* be the event of selecting an adult who makes at least $25,000 annually.

(a) We wish to compute $P(A \cap B)$. Here, using the given information that $P(A) = .20$ and $P(B \mid A) = .75$, we have from Theorem 3.1 that

$$P(A \cap B) = P(A) \cdot P(B \mid A) = (.20)(.75) = .15$$

Thus, the probability is .15 that an adult is both a college graduate and an earner of at least $25,000 annually.

(b) Here, we wish to find $P(B' \mid A)$. From Theorem 3.4,

$$P(B' \mid A) = 1 - P(B \mid A) = 1 - .75 = .25 \qquad \text{or} \qquad 25\%$$

*notes*

(i) There is not enough information given to compute $P(B)$. The latter is equivalent to the fraction of all adults in Newtown who earn at least $25,000.

(ii) One can solve the problems posed in Example 3-7 by using common sense. Thus, in part (a), assume there are 1000 adults in Newtown. Since 20% are college graduates, this means that 200 of the adults are college graduates. Now, 75% of these 200 adults is the same as (.75)(200) or 150 adults. That is, 150 adults earn at least $25,000. The answer to part (a) is obtained by observing that 150 is 15% of 1000.

## 3-2. Independent Events

Informally and intuitively speaking, we say that two events *A* and *B* are independent if the occurrence or nonoccurrence of either of them does not affect the probability of occurrence of the other. That is, we say that events *A* and *B* of sample space *S* are independent if

- $P(A) = P(A \mid B) = P(A \mid B')$           **(1)**

and, likewise,

- $P(B) = P(B \mid A) = P(B \mid A')$          **(2)**

[For the conditional probabilities to exist, we have to assume that each of $P(B)$, $P(B')$, $P(A)$, and $P(A')$ is positive. For example, $P(A)$ must be positive in order for $P(B \mid A)$ to exist.]

Refer to Theorem 3.1, which says that $P(A \cap B) = P(A) \cdot P(B \mid A)$. Replacing $P(B \mid A)$ by $P(B)$ leads to

$$P(A \cap B) = P(A) \cdot P(B) \qquad \qquad \textbf{(3)}$$

We use Eq. (3) as the formal definition of independence. [In the derivation of Eq. (3), we assume that $P(A) > 0$ so that $P(B \mid A)$ will exist.]

### A. Definitions and theorems for independent events

In the following definition, *A* and *B* are events (subsets) of the same sample space *S*.

**Definition 3.2:** Events *A* and *B* are **independent** if and only if

$$P(A \cap B) = P(A) \cdot P(B)$$

*notes*

(a) Observe the wording "if and only if." Consider the general form *P if and only if Q*, where *P* and *Q* denote statements. [In Definition 3.2, "events

*A* and *B* are independent" has the role of *P*, and "$P(A \cap B) = P(A) \cdot P(B)$" has the role of *Q*.] In general, when *P* if and only if *Q* is valid, this means that both of the following expressions (i) and (ii) are valid:

$$\text{If } P, \quad \text{then } Q \tag{i}$$

and

$$\text{If } Q, \quad \text{then } P \tag{ii}$$

The expressions (i) and (ii) are said to be converses of one another.

(b) Using the ideas of note (a), we can replace Definition 3.2 by the pair (î) and (îî), that follows:

| | |
|---|---|
| If events *A* and *B* are independent,<br>    then $P(A \cap B) = P(A) \cdot P(B)$ | Definition 3.2, part (î) |
| If $P(A \cap B) = P(A) \cdot P(B)$, then events<br>    *A* and *B* are independent | Definition 3.2, part (îî) |

The expressions from parts (î) and (îî) are said to be converses of each other.

(c) All definitions are actually *if and only if* type expressions. We have used and will use merely the word *if* in some statements of definitions, even though the word *if* can properly be replaced by the words *if and only if*.

(d) Any time an expression of the form "if *P*, then *Q*" is valid, then by logic, the contrapositive expression "if not *Q*, then not *P*" is automatically valid. Thus, taking the contrapositive of Definition 3.2, part (î) [see note (b) above] we have the following valid expression:

| | |
|---|---|
| If $P(A \cap B) \neq P(A) \cdot P(B)$, then events<br>    *A* and *B* are not independent | Definition 3.2, part (î)—<br>Contrapositive Form |

(e) In general, statements *P* and *Q* are said to be equivalent if the expression "*P* if and only if *Q*" is valid.

(f) Some authors say that events *A* and *B* are dependent if they are not independent.

The following theorem ties together the intuitive concepts of independence with the formal definition as given in Definition 3.2.

**Theorem 3.5**

(a) Events *A* and *B* are independent if and only if $P(B \mid A) = P(B)$. [Here, assume that $P(A) > 0$ so that $P(B \mid A)$ exists.]

(b) Events *A* and *B* are independent if and only if $P(A \mid B) = P(A)$. [Here, assume that $P(B) > 0$ so that $P(A \mid B)$ exists.]

(c) Events *A* and *B* are independent if and only if $P(B \mid A') = P(B)$. [Here, assume that $P(A') > 0$ so that $P(B \mid A')$ exists.]

(d) Events *A* and *B* are independent if and only if $P(A \mid B') = P(A)$. [Here, assume that $P(B') > 0$ so that $P(A \mid B')$ exists.]

---

**EXAMPLE 3-8:** Prove Theorem 3.5, part (a).

*Solution:* Remembering the meaning of the "if and only if" wording [see note (a) following Definition 3.2], we see that we have to prove two subtheorems here.

*Proof of "if events A and B are independent, then $P(B \mid A) = P(B)$"*: By hypothesis, *A* and *B* are independent. By Definition 3.2, this means that

$$P(A \cap B) = P(A) \cdot P(B) \tag{1}$$

From Definition 3.1 [the defining equation for $P(B \mid A)$], we have

$$P(B \mid A) = \frac{P(A \cap B)}{P(A)} \qquad (2)$$

Substituting from Eq. (1) into Eq. (2) and cancelling out the $P(A)$ terms yields

$$P(B \mid A) = \frac{\cancel{P(A)} \cdot P(B)}{\cancel{P(A)}} = P(B) \qquad (3)$$

This completes the proof of the first subtheorem. Now, let us prove the other subtheorem.

*Proof of "if $P(B \mid A) = P(B)$, then events A and B are independent"*: First, by hypothesis,

$$P(B \mid A) = P(B) \qquad (1')$$

From Definition 3.1, we have

$$P(B \mid A) = \frac{P(A \cap B)}{P(A)} \qquad (2')$$

Substituting $P(B \mid A)$ from Eq. (1') into Eq. (2') and then multiplying through by $P(A)$ yields

$$P(A \cap B) = P(A) \cdot P(B) \qquad (3')$$

This indicates, in accord with Definition 3.2, that $A$ and $B$ are independent. Thus, the second subtheorem is proved.

---

The proof of Theorem 3.5, part (b) is virtually identical to the proof [just done] of Theorem 3.5, part (a). The proof of Theorem 3.5, part (c) is done in Problem 3-6.

---

**EXAMPLE 3-9:** Refer to the experiment of tossing a pair of fair dice, or, equivalently, tossing a single fair die twice in succession. Consider again the events $A$ and $B$ of Examples 3-1 and 3-2. In words, $A$ is "sum of numbers [of dots] equals six," and $B$ is "a two occurs on at least one die." Suppose $C$ is the event "an ace (one dot) occurs on the first die," and $D$ is the event "an ace occurs on the second die." (Here, if we are tossing a pair of dice, we agree to distinguish between them, and to refer to them as the first die and the second die, respectively. One way to distinguish a pair of dice is to paint them different colors, say pink for the first die and white for the second die.) **(a)** Determine if $A$ and $B$ are independent. **(b)** Determine if $C$ and $D$ are independent.

*Solution*

**(a)** *Preliminaries:* We know from Examples 3-1 and 3-2, and discussions relating to these Examples, that $P(A) = 5/36$, $P(B) = 11/36$, $A \cap B = \{24, 42\}$, and thus, $P(A \cap B) = 2/36$.

   *Method 1:* We see that $P(A \cap B) \neq P(A) \cdot P(B)$ since $P(A \cap B) = .055\overline{5}$ and $P(A) \cdot P(B) = .0424$. Thus, by Definition 3.2, it follows that $A$ and $B$ are not independent.

   *Method 2:* We know that $P(B \mid A) = P(A \cap B)/P(A) = 2/5$, and $P(B) = 11/36$. Since $P(B \mid A) \neq P(B)$, it follows from Theorem 3.5, part (a), that $A$ and $B$ are not independent.

   **note:** Here, we made use of the contrapositive of "if $A$ and $B$ are independent, then $P(B \mid A) = P(B)$," which is "if $P(B \mid A) \neq P(B)$, then $A$ and $B$ are not independent."

Remember that if a valid expression [for example, a theorem] has the form "if *P*, then *Q*," then the expression "if not *Q*, then not *P*," known as the contrapositive of the former expression, is automatically valid.

**(b)** First, observe that *C* and *D* have the following representations and are subsets of sample space *S*:

$$C = \{11, 12, 13, 14, 15, 16\}; \qquad D = \{11, 21, 31, 41, 51, 61\}$$

Observe that for each of the six elements of *C*, an ace [that is, a "1"] occurs on the first die. From the above subsets for *C* and *D*, or from common sense, $C \cap D = \{11\}$.

Now, applying Theorem 1.2 [equally likely elements theorem] to the subsets for *C*, *D*, and $C \cap D$, and recalling that $N(S) = 36$, we have

$$P(C) = \frac{6}{36} = 1/6; \qquad P(D) = \frac{6}{36} = 1/6; \qquad \text{and } P(C \cap D) = \frac{1}{36} \quad \textbf{(1), (2), (3)}$$

Since $P(C \cap D) = P(C) \cdot P(D)$, it follows from Definition 3.2 that *C* and *D* are independent events.

***note:*** It should be intuitively plausible that events *C* and *D* are independent. After all, *C* pertains to the toss of the first die, and *D* pertains to the toss of the second die, and we shouldn't expect what happens on either toss to affect what happens on the other toss.

---

The following theorem should be intuitively clear:

**Theorem 3.6:** If the two events *A* and *B* are independent, then

**(a)** the two events *A* and *B'* are also independent, and
**(b)** the two events *A'* and *B* are also independent, and
**(c)** the two events *A'* and *B'* are also independent.

---

**EXAMPLE 3-10:** Prove part (a) of Theorem 3.6.

***Proof:*** Refer to Figure 3-2, a Venn diagram which we've seen many times in Chapters 1 and 2, often with the disjoint subsets indicated by the $R_i$ labels.
Thus,

$$A = (A \cap B) \cup (A \cap B') \tag{1}$$

where $A \cap B$ and $A \cap B'$ are disjoint. Thus, by Axiom 1.3',

$$P(A) = P[(A \cap B) \cup (A \cap B')] = P(A \cap B) + P(A \cap B') \tag{2}$$

Since by hypothesis, *A* and *B* are independent, $P(A \cap B) = P(A) \cdot P(B)$, and thus Eq. (2) becomes

$$P(A) = P(A) \cdot P(B) + P(A \cap B') \tag{3}$$

Thus, solving Eq. (3) for $P(A \cap B')$, we get

$$P(A \cap B') = P(A) \cdot [1 - P(B)] = P(A) \cdot P(B') \tag{4}$$

It follows from Eq. (4) and Definition 3.2, as applied to events *A* and *B'*, that *A* and *B'* are independent.  □

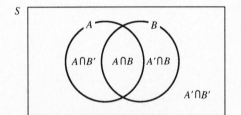

**Figure 3-2.** Venn diagram for Example 3-10. $R_1 = A \cap B$; $R_2 = A \cap B'$; $R_3 = A' \cap B$; $R_4 = A' \cap B'$.

---

The proof of part (b) of Theorem 3.6 is similar, and the proof of part (c) is given in the solution to Problem 3-10.
Let us extend the concept of independence to three events.

**Definition 3.3 (Independence of Three Events):** Three events $A$, $B$, and $C$ are independent if and only if the following four equations hold:

$$P(A \cap B) = P(A) \cdot P(B); \quad P(A \cap C) = P(A) \cdot P(C); \qquad \textbf{(ia), (ib)}$$

$$P(B \cap C) = P(B) \cdot P(C) \qquad \textbf{(ic)}$$

$$P(A \cap B \cap C) = P(A) \cdot P(B) \cdot P(C) \qquad \textbf{(ii)}$$

If just (ia), (ib), and (ic) hold, then events $A$, $B$, and $C$ are defined to be *pairwise independent*.

Definition 3.4, which follows, is a generalization of Definition 3.3. It extends the concept of independence to any collection of 2 or more events.

**Definition 3.4 (Independence of $k$ Events):** Events $A_1$, $A_2$, ..., and $A_k$ are independent if and only if the probability of the intersection of any $2, 3, \ldots$, or $k$ of these events equals the product of their respective probabilities.

Events $A_1, A_2, \ldots$, and $A_k$ are defined to be pairwise independent if for every distinct pair $A_i$ and $A_j$, $P(A_i \cap A_j) = P(A_i) \cdot P(A_j)$. This is just an extension of conditions (i) in Definition 3.3.

*notes*

(a) It is possible for three or more events to be pairwise independent without being independent. [See Problem 3-11.]
(b) It is possible for $P(A \cap B \cap C) = P(A) \cdot P(B) \cdot P(C)$ without $A$, $B$, and $C$ being pairwise independent. [See Problem 3-18.] The preceding equation is just a statement of condition (ii) of Definition 3.3.
(c) If it is given that certain events are independent, then the probability that they all will occur [which is the probability of their intersection] is equal to the product of the probabilities of the individual events. For example, if it's given that events $A$, $B$, and $C$ are independent, then it follows that $P(A$ and $B$ and $C)$, or $P(A \cap B \cap C)$, is equal to $P(A) \cdot P(B) \cdot P(C)$.

## B. Examples dealing with the independence concept

**EXAMPLE 3-11:** In a federal study of educational levels of adults in Fernwood, the 1000 adults in the city were classified according to sex [male or female], and to whether one does or does not have a high school diploma. The data are given in Table 3-1.

**TABLE 3-1: Table of Data for City of Fernwood [Example 3-11]**

|  | Has high school diploma ($H$) | No high school diploma ($H'$) |  |
|---|---|---|---|
| Male ($M$) | 360 | 120 | $N(M) = 480$ |
| Female ($M'$) | 340 | 180 | $N(M') = 520$ |
|  | $N(H) = 700$ | $N(H') = 300$ | $N(S) = 1000$ |

For example, 340 of the adults are female and have diplomas. In the above table, $N(H)$ is the symbol for the total number of adults in Fernwood who have high school diplomas, and $N(M)$ is the symbol for the total number of adults who are male. The letter $H$ indicates the event [subset] that an adult has a high school diploma, and $M$ indicates that an adult is male. (a) If an adult is picked at random, what is the probability the adult has a high school diploma? (b) Given that an adult is male, what is the probability the adult has a high school diploma?

(c) Given that an adult has a high school diploma, what is the probability the adult is male? (d) Determine if the events $M$ and $H$ are independent.

**Solution:** *Preliminaries:* It is implied that the experiment is to pick [or select, or choose, or sample] an adult at random from the 1000 adults in Fernwood. Thus, $N(S) = 1000$. By "at random," we mean that all possible selections of adults are equally likely. Thus, we may use Theorem 1.2, which says that $P(A) = N(A)/N(S)$ for any event $A$ of $S$.

(a) Since, from Table 3-1, we have $N(H) = 700$ and $N(S) = 1000$, it follows that

$$P(H) = \frac{N(H)}{N(S)} = \frac{700}{1000} = .7 \tag{1}$$

(b) Here, we wish to compute $P(H \mid M)$. From Definition 3.1, we have

$$P(H \mid M) = \frac{P(M \cap H)}{P(M)} \tag{2}$$

From Table 3-1, $N(M \cap H)$, the number of adults who are both male and have a high school diploma, is 360. Thus,

$$P(M) = \frac{N(M)}{N(S)} = \frac{480}{1000} \quad \text{and} \quad P(M \cap H) = \frac{N(M \cap H)}{N(S)} = \frac{360}{1000} \tag{3}, (4)$$

Substituting from Eqs. (3) and (4) into Eq. (2) yields

$$P(H \mid M) = \frac{360/1000}{480/1000} = \frac{N(M \cap H)}{N(M)} = \frac{360}{480} = \frac{3}{4} = .75 \tag{5}$$

(c) Here, we wish to compute $P(M \mid H)$. Thus,

$$P(M \mid H) = \frac{P(H \cap M)}{P(H)} = \frac{N(H \cap M)}{N(H)} = \frac{360}{700} = .5143 \tag{6}$$

(d) *Method 1:* From the above data, $P(M \cap H) = .360$, $P(M) = .480$, and $P(H) = .70$. Thus, $P(M) \cdot P(H) = .336$, and hence $P(M \cap H) \neq P(M) \cdot P(H)$. From Definition 3.2, we see that $M$ and $H$ are not independent events.

*Method 2:* From parts (a) and (b), we see that $P(H \mid M) = .75$, $P(H) = .70$, and thus $P(H \mid M) \neq P(H)$. Thus, from part (a) of Theorem 3.5, we see that $H$ and $M$ are not independent.

---

**EXAMPLE 3-12:** A federal study similar to that mentioned in Example 3-11 was conducted with respect to the 1400 adults in Midville, with results as indicated in Table 3-2.

**TABLE 3-2: Table of Data for City of Midville [Example 3-12]**

|  | Has high school diploma $(H)$ | No high school diploma $(H')$ |  |
| --- | --- | --- | --- |
| Male $(M)$<br>Female $(M')$ | 320<br>480 | 240<br>360 | $N(M) = 560$<br>$N(M') = 840$ |
|  | $N(H) = 800$ | $N(H') = 600$ | $N(S) = 1400$ |

Repeat parts (a), (b), (c), and (d) of Example 3-11 for the current example.

**Solution**

(a) From Table 3-2, we have $N(H) = 800$, $N(S) = 1400$, and thus

$$P(H) = \frac{N(H)}{N(S)} = \frac{800}{1400} = \frac{4}{7} = .5714$$

**(b)** Here,

$$P(M \cap H) = \frac{N(M \cap H)}{N(S)} = \frac{320}{1400} \quad \text{and} \quad P(M) = \frac{560}{1400}$$

and thus

$$P(H \mid M) = \frac{P(M \cap H)}{P(M)} = \frac{320}{560} = \frac{4}{7}$$

**(c)** Using prior results for $P(H)$ and $P(M \cap H)$ from parts (a) and (b), we obtain

$$P(M \mid H) = \frac{P(M \cap H)}{P(H)} = \frac{320}{800} = \frac{2}{5} = .400$$

**(d)** From the above data, we have $P(M \cap H) = 320/1400 = 8/35$, $P(M) = 560/1400 = 2/5$, and $P(H) = 800/1400 = 4/7$. Thus, $P(M \cap H) = P(M) \cdot P(H)$, which means that events $M$ and $H$ are independent. [See Definition 3.2.]

---

In Example 3-12, the essential features for the independence for two events are revealed for a case where the elements of a sample space are equally likely. Thus, the proportion of those with property $H$ [high school diploma] is 4/7 whether we are examining the entire adult population [see data in bottom row of Table 3-2, which indicates that $P(H) = 800/1400 = 4/7$], or the male adults [see data in first row, which indicates that $P(H \mid M) = 320/560 = 4/7$], or the female adults [see data in second row, which indicates that $P(H \mid M') = 480/840 = 4/7$].

Thus, the proportion of adults with high school diplomas is independent of whether we focus on the subset [portion] consisting of males, or on the subset consisting of females, or on the overall set [sample space] consisting of all adults. In each case, the proportion is 4/7.

Likewise, the proportion of adults with property $M$ [male] is 2/5, whether we are examining the entire adult population [see data in right-hand column, which indicates that $P(M) = 560/1400 = 2/5$], or adults with high school diplomas [see data in first column, which indicates that $P(M \mid H) = 320/800 = 2/5$], or adults without high school diplomas [see data in second column, which indicates that $P(M \mid H') = 240/600 = 2/5$].

Thus, the proportion of adults who are males is independent of whether we focus on the subset [portion] consisting of adults with high school diplomas, or on the subset consisting of adults without high school diplomas, or on the overall set [sample space] consisting of all adults. In each case, the proportion is 2/5.

## 3-3. Counting Methods

So far, in our probability experiments, we have focused on very simple sample spaces. In order to be able to handle more complex sample spaces, and to develop more systematic methods for counting things in general, we shall digress for a while and focus on counting methods [or, as they are sometimes called, combinatorial methods].

### A. Tree diagrams

A tree diagram is useful in enumerating [counting] the different ways a sequence of events can occur, where each individual event can occur in a finite number of ways. [The term "event" here refers to what one usually thinks of, namely an occurrence of some kind, and is not necessarily a probability event; the latter is defined in Definition 1.3.]

**EXAMPLE 3-13:** Suppose that Alice and Bob are playing a tennis match in which the winner is the one who wins two sets first. Use a tree diagram to determine and list all possible ways the match can be decided.

*Solution:* This type of match is often referred to as a "best two out of three sets match," although the match can end if either player wins the first two sets in a row. The relevant tree diagram is given in Figure 3-3.

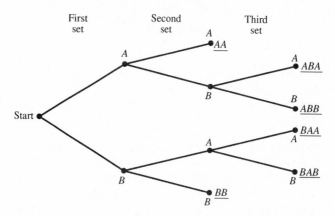

**Figure 3-3.** Tree diagram for Example 3-13. Items in list are underlined.

In the tree diagram, first we indicate the starting point on the left. Next, we draw two lines, or *branches*, called initial branches, from the starting point. The first point labeled *A*, from left to right, indicates that Alice won the first set, and the first point labeled *B* indicates that Bob won the first set. From each of these points *A* and *B*, we draw second-level branches to indicate the winners of the second set. Here, we observe two ways the match can end, namely *AA* and *BB*. Here, *AA* means that Alice won both the first set and the second set, and thus won the match. Similarly, *BB* indicates that Bob won the match by winning the first two sets.

Continuing to the right along the paths that read *AB* and *BA* thus far, we see there are four third-level branches. These branches represent the four ways the match can end if three sets are played.

There are six endpoints corresponding to the six different sequences for which the match can end. Any sequence of branches that goes from the starting point to an endpoint is called a *path*.

The six sequences for which the match can end are listed as follows: *AA*, *ABA*, *ABB*, *BAA*, *BAB*, and *BB*. [For example, *ABB* indicates that Alice won the first set, followed by Bob winning the second and third sets, and thus the match.] The six items (or sequences) in the list are shown underlined in the tree diagram, next to the six respective endpoints.

In general, in a tree diagram, one can trace backwards from an endpoint over a path of branches back to the starting point to determine, in reverse order, the sequence of intermediate results which led to the endpoint.

*note:* Some authors use "vertex" or "node" as the word for *point*, and the word "arc" in place of *branch*, in tree diagram terminology.

**EXAMPLE 3-14:** A survey is being conducted to determine whether people approve or disapprove of a certain soft drink. The interviewer wishes to file the survey forms according to the sex, age group, and opinion [approval or disapproval] of the respondents. Determine the different filing categories and the number of such categories if the age group symbols used are $Y$ [under 25], $I$ [25 to 55], and $O$ [over 55]. Here, $Y$, $I$, and $O$ stand for "young," "intermediate age," and "old," respectively. The sex groupings are $M$ [for male] and $F$ [for female], and the opinion ratings are $A$ [for approve] and $D$ [for disapprove]. Use a tree diagram to solve this problem.

*Solution:* The tree diagram of Figure 3-4 indicates the resulting twelve filing categories. These are underlined, and listed to the right of the 12 endpoints. Each path contains three branches in this Example. For example, the particular category $MIA$ indicates people who are male, intermediate in age, and who approve the soft drink.

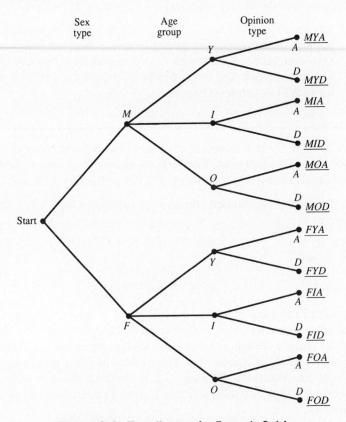

**Figure 3-4.** Tree diagram for Example 3-14.

In Example 3-14, observe that the total number of filing categories [12] is equal to the product of the number of sex types [2] times the number of age types [3] times the number of opinion types [2]. That is, $12 = 2 \cdot 3 \cdot 2$. This calculation illustrates a general theorem, known as the *fundamental counting principle* or *multiplication rule theorem*.

## B. The fundamental counting principle

This theorem counts the number of ways a combined operation consisting of several component operations can be done.

The combined operation can be thought of as consisting of component operations occurring in sequence, that is, one after the other. For example, in Example 3-14, the combined operation of forming a filing category consists of first labeling by sex, then by age group, and lastly by opinion. For general purposes, suppose the combined operation consists of $k$ component operations, labeled $1, 2, \ldots, k$.

**Theorem 3.7 (Fundamental Counting Principle):** Given a combined operation consisting of $k$ component operations. Suppose a first operation can be done in $n_1$ ways, and for each of these ways a second operation can be done in $n_2$ ways, and for each of these ways a third operation can be done in $n_3$ ways, and so on, for each of the $k$ component operations.

Then, a combined operation consisting of the $k$ component operations can be done in $n_1 \cdot n_2 \cdot n_3 \cdot \ldots \cdot n_k$ ways. [Here, we have the product of $n_1$ times $n_2$, etc., times $n_k$.]

To illustrate the notation of Theorem 3.7 with respect to Example 3-14, we have $k = 3$ and $n_1 = 2$ [sexes], $n_2 = 3$ [age groups], and $n_3 = 2$ [opinion types]. The total number of combined operations [there, the combined operations were the filing categories] is $n_1 \cdot n_2 \cdot n_3 = 2 \cdot 3 \cdot 2 = 12$.

*note:* The product $n_1 \cdot n_2 \cdot n_3 \cdot \ldots \cdot n_k$ in Theorem 3.7 can be visualized as the number of endpoints [or paths] in a tree diagram consisting of $n_1$ first-level branches, followed by $n_2$ second-level branches for each first-level branch, and so on, finally followed by $n_k$ $k$th-level branches for each $(k-1)$th-level branch.

---

**EXAMPLE 3-15:** Given a sample space $S$ and three subsets [events] $A$, $B$, and $C$ of $S$. Allowing for all possible intersections among $A$, $B$, and $C$, what is the number of disjoint subsets in a partition of $S$ determined by $A$, $B$, and $C$?

*Solution:* Previously, we labeled the disjoint subsets as $R_i$'s. Each $R_i$ can be represented as

$$R_i = \begin{Bmatrix} A \\ \text{or} \\ A' \end{Bmatrix} \cap \begin{Bmatrix} B \\ \text{or} \\ B' \end{Bmatrix} \cap \begin{Bmatrix} C \\ \text{or} \\ C' \end{Bmatrix} \tag{1}$$

as we noted in Example 2-3.

Consider a given element of sample space $S$. The first "operation" has to do with whether a given element is in set $A$, or in its complement $A'$. Thus, $n_1 = 2$. Similarly, the second "operation" has to do with whether a given element is in set $B$ or in its complement $B'$. Thus, $n_2 = 2$. In similar fashion, $n_3 = 2$, since the third operation has to do with whether a given element is in set $C$ or in its complement $C'$.

Thus, by the fundamental counting principle, there are $2 \cdot 2 \cdot 2 = 2^3 = 8$ such $R_i$'s. These $R_i$'s are shown, for example, in Figure 2-2.

---

*note:* Following the same reasoning as in Example 3-15, it's clear that the number of disjoint subsets in a partition of $S$ induced by all possible intersections among the $k$ subsets $A_1, A_2, \ldots, A_k$ is $2 \cdot 2 \cdot 2 \cdot \ldots \cdot 2 = 2^k$ subsets.

## C. Applying counting methods to probability situations

---

**EXAMPLE 3-16:** Given a miniature card deck with three spades [ace, two, and three] and two hearts [ace and two]. The three spades are labeled $AS$, $2S$, and $3S$, and the two hearts are labeled $AH$ and $2H$.

Suppose two cards are drawn in the following way. After shuffling the five-card deck, the first card is drawn. Then, the first card is returned to the deck. Finally, after another shuffling of the deck [which has five cards again], the second card is drawn. Such a process is known as drawing with replacement.

**(a)** Determine the most subdivided sample space $\mathscr{S}$ for the experiment of drawing two cards with replacement. **(b)** Let the event $S_1$ denote the event "spade on the first card." Display $S_1$ as a subset of the sample space of part (a), and compute $P(S_1)$. **(c)** Let the event $H_2$ denote the event "heart on the second card." Display $H_2$ as a subset of the sample space of part (a), and compute $P(H_2)$. **(d)** Display $S_1 \cap H_2$ as a subset of sample space $\mathscr{S}$, and compute $P(S_1 \cap H_2)$. **(e)** Determine whether the events $S_1$ and $H_2$ are independent.

***Solution:*** *Preliminary Note:* The fundamental counting principle [Theorem 3.7] is useful in this problem.

**(a)** The first card can be drawn in any of five ways as can the second since each draw is from five cards [recall that the first card is replaced]. Thus, using the fundamental counting principle, we see that the two cards can be drawn in a total of $n_1 \cdot n_2 = 5 \cdot 5 = 25$ ways. Thus, the most subdivided sample space $\mathscr{S}$ has 25 elements [that is, $N(\mathscr{S}) = 25$], some of which are indicated below:

$$\mathscr{S} = \{AS\text{-}AS, AS\text{-}2S, AS\text{-}3S, AS\text{-}AH, AS\text{-}2H; 2S\text{-}AS, \ldots, 2S\text{-}2H;$$

$$3S\text{-}AS, \ldots, 3S\text{-}2H; AH\text{-}AS, \ldots, AH\text{-}2H; 2H\text{-}AS, 2H\text{-}2S, 2H\text{-}3S,$$

$$2H\text{-}AH, 2H\text{-}2H\}$$

For example, $AS\text{-}2H$ means $AS$ [that is, ace of spades] is drawn on the first card, and $2H$ [that is, two of hearts] is drawn on the second card.

**(b)** It is reasonable to assume that all 25 elements are equally likely. Thus, Theorem 1.2 applies, with the denominator term being $N(\mathscr{S}) = 25$. To determine $P(S_1)$, we have to determine the number of elements in $S_1$, that is, for the event "spade on the first card." Then, we'll compute $P(S_1)$ from $P(S_1) = N(S_1)/N(\mathscr{S}) = N(S_1)/25$.

Since for event $S_1$, the first card must be a spade, and the second card doesn't matter [that is, it can be any of five cards], the number of elements in $S_1$ is given by $N(S_1) = n_1 \cdot n_2 = 3 \cdot 5 = 15$. Here, $n_1 = 3$ refers to the first card [there are three spades available], and $n_2 = 5$ refers to the second card. Thus,

$$P(S_1) = \frac{N(S_1)}{N(\mathscr{S})} = \frac{3 \cdot 5}{5 \cdot 5} = \frac{15}{25} = \frac{3}{5} \tag{1}$$

The subset display of the 15 elements in $S_1$ is as follows:

$$S_1 = \{AS\text{-}AS, AS\text{-}2S, AS\text{-}3S, AS\text{-}AH, AS\text{-}2H; 2S\text{-}AS, 2S\text{-}2S, 2S\text{-}3S,$$

$$2S\text{-}AH, 2S\text{-}2H; 3S\text{-}AS, 3S\text{-}2S, 3S\text{-}3S, 3S\text{-}AH, 3S\text{-}2H\}$$

**(c)** The number of elements for event $H_2$, that is for heart on the second card, is $n_1 \cdot n_2 = 5 \cdot 2 = 10$, since the first card can be any of five cards while the second card must be a heart [there are two hearts in the deck]. Thus,

$$P(H_2) = \frac{N(H_2)}{N(\mathscr{S})} = \frac{5 \cdot 2}{5 \cdot 5} = \frac{10}{25} = \frac{2}{5} \tag{2}$$

The subset display for the ten elements of $H_2$ is as follows:

$$H_2 = \{AS\text{-}AH, 2S\text{-}AH, 3S\text{-}AH, AH\text{-}AH, 2H\text{-}AH;$$

$$AS\text{-}2H, 2S\text{-}2H, 3S\text{-}2H, AH\text{-}2H, 2H\text{-}2H\}$$

**(d)** We can determine the probability of the event $S_1 \cap H_2$ in two ways. One way is to take the intersection of the sets listed in parts (b) and (c). In other words, we determine the elements common to the sets [events] $S_1$ and $H_2$. Thus,

$$S_1 \cap H_2 = \{AS\text{-}AH, AS\text{-}2H, 2S\text{-}AH, 2S\text{-}2H, 3S\text{-}AH, 3S\text{-}2H\}$$

and hence, $P(S_1 \cap H_2) = N(S_1 \cap H_2)/N(\mathscr{S}) = 6/25$.

Another approach involves merely counting the number of elements in $S_1 \cap H_2$ by using the fundamental counting principle. Thus, $N(S_1 \cap H_2)$ is given by $n_1 \cdot n_2 = 3 \cdot 2 = 6$ since the first card must be from among the three spades $[n_1 = 3]$, and the second card must be from among the two hearts $[n_2 = 2]$. Thus,

$$P(S_1 \cap H_2) = \frac{N(S_1 \cap H_2)}{N(\mathscr{S})} = \frac{3 \cdot 2}{5 \cdot 5} = \frac{6}{25} \tag{3}$$

(e) The events $S_1$ and $H_2$ are independent since from parts (b), (c), and (d), we see that $P(S_1 \cap H_2)$ is equal to $P(S_1)$ times $P(H_2)$.

---

In Example 3-16, we would expect events $S_1$ and $H_2$ to be independent since, intuitively, we feel that what happens on the first draw should have no effect on what happens on the second draw in a drawing with replacement situation.

Notice that we wrote out the relevant probabilities in "product form" after the second equals signs in Eqs. (1), (2), and (3) of Example 3-16. That is, we had

$$P(S_1) = \frac{3 \cdot 5}{5 \cdot 5} \qquad P(H_2) = \frac{5 \cdot 2}{5 \cdot 5} \qquad P(S_1 \cap H_2) = \frac{3 \cdot 2}{5 \cdot 5}$$

Such forms for expressing the probabilities in drawing with replacement and similar situations will prove useful in generating some general methods in the future development. The following example deals with a drawing without replacement situation.

---

**EXAMPLE 3-17:** Given the five-card deck of Example 3-16 [with three spades and two hearts]. Suppose that two cards are drawn from the deck in succession [or in sequence, or in order], but *without replacement* after the first draw. Thus, the second card is drawn from the four remaining cards. [Thus, we visualize the cards being drawn out one at a time, in sequence. Later, we shall often treat drawing without replacement situations somewhat differently. In particular, we shall visualize the cards as being drawn out in a group, or *batch*, and we shall *not* think of the cards as being drawn out individually, in sequence.] Rework Example 3-16 with the same parts (a) through (e). In addition, consider the following new part, which will prove to be instructive. (f) Compute $P(H_2 \mid S_1)$, which is the conditional probability of a heart on card 2 given that card 1 is a spade.

*Solution*

(a) Since the first card can be drawn in any of five ways, and the second card can be drawn in any of four ways, the fundamental counting principle tells us that the two cards can be drawn in $5 \cdot 4 = 20$ ways. Thus, the sample space $\mathscr{S}$, which is partially indicated below, has 20 elements. That is, $N(\mathscr{S}) = 20$.

$$\mathscr{S} = \{AS\text{-}2S, AS\text{-}3S, AS\text{-}AH, AS\text{-}2H; 2S\text{-}AS, 2S\text{-}3S, 2S\text{-}AH, 2S\text{-}2H;$$

$$3S\text{-}AS, \ldots, 3S\text{-}2H; AH\text{-}AS, \ldots, AH\text{-}2H; 2H\text{-}AS, 2H\text{-}2S,$$

$$2H\text{-}3S, 2H\text{-}AH\}$$

It is reasonable to assume that all 20 elements are equally likely, and so Theorem 1.2 applies. That is, for any event [subset] $A$ of $\mathscr{S}$, we have $P(A) = N(A)/N(\mathscr{S})$.

(b) Here, we have $P(S_1) = N(S_1)/N(\mathscr{S}) = N(S_1)/20$. To find the number of elements in $S_1$, which, in words, is "spade on the first card," we count those elements in the sample space for which the first card is a spade. From the sample space $\mathscr{S}$ above, we see that $N(S_1) = 12$. However, it is also instructive to count $N(S_1)$ by using the fundamental counting principle. Since the first card must be a spade, and there are three available, and the second card can be any of the four cards available on the second draw, we see that $N(S_1) = n_1 \cdot n_2 = 3 \cdot 4 = 12$. Thus,

$$P(S_1) = \frac{N(S_1)}{N(\mathscr{S})} = \frac{3 \cdot 4}{5 \cdot 4} = \frac{12}{20} = \frac{3}{5} \tag{1}$$

The subset display of the 12 elements in $S_1$ follows:

$$S_1 = \{AS\text{-}2S, AS\text{-}3S, AS\text{-}AH, AS\text{-}2H; 2S\text{-}AS, 2S\text{-}3S, 2S\text{-}AH, 2S\text{-}2H;$$

$$3S\text{-}AS, 3S\text{-}2S, 3S\text{-}AH, 3S\text{-}2H\}$$

**(c)** Here, $P(H_2) = N(H_2)/N(\mathscr{S}) = N(H_2)/20$. To find the number of elements in $H_2$, which, in words, is "heart on the second card," we count those elements in the sample space for which the second card is a heart. From the sample space $\mathscr{S}$ [part (a)], we see that $N(H_2) = 8$. The subset display of the 8 elements in $H_2$ follows; note that each element has heart on the second card:

$$H_2 = \{AS\text{-}AH, AS\text{-}2H, 2S\text{-}AH, 2S\text{-}2H, 3S\text{-}AH, 3S\text{-}2H, AH\text{-}2H, 2H\text{-}AH\}$$

Thus,

$$P(H_2) = \frac{N(H_2)}{N(\mathscr{S})} = \frac{8}{20} = \frac{2}{5} \tag{2}$$

Let us illustrate how the elements of $H_2$ can be counted by means of a tree diagram. Refer to Figure 3-5.

The eight elements of $H_2$ are listed at the right side of the tree diagram; for example, the first two elements in the list are $AS\text{-}AH$ and $AS\text{-}2H$. Thus, $N(H_2) = 8$.

Observe that the fundamental counting principle cannot *directly* be used here to count $N(H_2)$ because the number of ways to do the second operation [drawing a second card] depends on the way the first operation [drawing a first card] was done. Recall that for the application of the fundamental counting principle [Theorem 3.7] to a two-stage process, there have to be the same number of second-level branches for each first-level branch in the associated tree diagram. Such is not the case here as the tree diagram of Figure 3-5 reveals. For example, if the first card drawn is the ace of spades [$AS$], there are two second-level branches [namely, the $AH$ and $2H$ branches], while if $AH$ is the first card drawn, there is only one second-level branch [namely, the $2H$ branch].

We can indirectly employ the fundamental counting principle, however, to count $N(H_2)$. First, observe that

$$N(H_2) = N(S_1 \text{ and } H_2) + N(H_1 \text{ and } H_2) \tag{3}$$

The two terms on the right indicate that heart on the second card will occur either preceded by a spade or by a heart. [Also, observe that $(S_1 \text{ and } H_2)$ can be expressed as $(S_1 \cap H_2)$, and similarly for $(H_1 \text{ and } H_2)$.] Now, applying the fundamental counting principle to each of the individual terms $N(S_1 \text{ and } H_2)$ and $N(H_1 \text{ and } H_2)$, we have

$$N(S_1 \text{ and } H_2) = 3 \cdot 2 = 6 \quad \text{and} \quad N(H_1 \text{ and } H_2) = 2 \cdot 1 = 2 \quad \textbf{(4a), (4b)}$$

The first six elements in the list of Figure 3-5 are those pertaining to $(S_1 \text{ and } H_2)$, while the last two elements in the list are those pertaining to $(H_1 \text{ and } H_2)$. At any rate, substituting from Eqs. (4a) and (4b) into Eq. (3) yields

$$N(H_2) = 3 \cdot 2 + 2 \cdot 1 = 6 + 2 = 8 \tag{5}$$

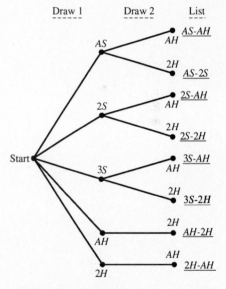

**Figure 3-5.** Tree diagram for Example 3-17, part (c).

(d) We can determine the probability of the event $(S_1 \cap H_2)$ in two ways. One way is by taking the intersection of the sets listed in parts (b) and (c). That is, we determine elements common to both sets. Thus, we obtain

$$S_1 \cap H_2 = \{AS\text{-}AH, AS\text{-}2H, 2S\text{-}AH, 2S\text{-}2H, 3S\text{-}AH, 3S\text{-}2H\}$$

Hence, $N(S_1 \cap H_2) = 6$ and $P(S_1 \cap H_2) = N(S_1 \cap H_2)/N(\mathcal{S}) = 6/20$.

Another approach for counting $N(S_1 \cap H_2)$ is to use the fundamental counting principle. Thus, counting the number of ways to draw a spade on the first card followed by a heart on the second card, we have $N(S_1 \cap H_2) = n_1 \cdot n_2 = 3 \cdot 2 = 6$. Then, since $N(\mathcal{S}) = 5 \cdot 4$, we have

$$P(S_1 \cap H_2) = \frac{N(S_1 \cap H_2)}{N(\mathcal{S})} = \frac{3 \cdot 2}{5 \cdot 4} = \frac{6}{20} = \frac{3}{10} \tag{6}$$

Here, we have a "product form" occurring after the second equals sign.

(e) The events $S_1$ and $H_2$ are not independent since from parts (b), (c), and (d), we see that

$$P(S_1 \cap H_2) \neq P(S_1) \cdot P(H_2)$$

Here, $P(S_1 \cap H_2) = 6/20 = .30$ and $P(S_1) \cdot P(H_2) = 6/25 = .24$.

(f) Since $P(H_2 \mid S_1) = P(S_1 \cap H_2)/P(S_1)$, we have the following calculation for $P(H_2 \mid S_1)$, after employing the results from parts (b) and (d):

$$P(H_2 \mid S_1) = \frac{P(S_1 \cap H_2)}{P(S_1)} = \frac{(3 \cdot 2)/(5 \cdot 4)}{(3 \cdot 4)/(5 \cdot 4)} = \frac{3 \cdot 2}{3 \cdot 4} = \frac{2}{4} = \frac{1}{2} \tag{7}$$

In Example 3-17, intuitively we would not expect the events $S_1$ and $H_2$ to be independent since what happens on the first draw should have an effect on what happens on the second draw in a drawing without replacement situation.

## 3-4. A Review of the Concepts of Mutually Exclusive and Independent Events

Recall from Definition 1.7 that two events $A$ and $B$ of sample space $S$ are mutually exclusive [disjoint] if and only if

$$A \cap B = \emptyset \tag{I}$$

From Definition 3.2, we have that two events $A$ and $B$ of sample space $S$ are independent if and only if

$$P(A \cap B) = P(A) \cdot P(B) \tag{II}$$

**EXAMPLE 3-18:** Suppose that events $A$ and $B$ are independent, and that $P(A) > 0$ and $P(B) > 0$. Prove that $A$ and $B$ cannot be mutually exclusive.

*Solution:* Since, by hypothesis, $A$ and $B$ are independent, it follows from Eq. (II) above that

$$P(A \cap B) = P(A) \cdot P(B) \tag{1}$$

Since, by hypothesis, $P(A) > 0$ and $P(B) > 0$, it follows that $P(A) \cdot P(B) > 0$. [A positive number times a positive number yields a positive number.] Substituting the latter into Eq. (1) results in

$$P(A \cap B) > 0 \qquad (2)$$

Thus, it follows that $A$ and $B$ cannot be mutually exclusive. [If $A$ and $B$ were mutually exclusive, then $A \cap B = \emptyset$, and hence $P(A \cap B) = 0$, by Theorem 1.4. The latter is contradicted by (2) above, which indicates that the number $P(A \cap B)$ is positive. Remember that a basic property of the real number system is that any real number can be only one of positive, negative, or zero.]

**EXAMPLE 3-19:** Suppose that events $A$ and $B$ are mutually exclusive, and that $P(A) > 0$ and $P(B) > 0$. Prove that $A$ and $B$ cannot be independent.

*Solution:* From the hypothesis that $A$ and $B$ are mutually exclusive, and from Eq. (I) above, we have that $A \cap B = \emptyset$, and thus

$$P(A \cap B) = 0 \qquad (1)$$

Now, let us tentatively assume that $A$ and $B$ are independent. Thus,

$$P(A \cap B) = P(A) \cdot P(B) \qquad (2)$$

from Definition 3.2—see Eq. (II) above. From Eq. (1), we see that the left side of Eq. (2) must be zero, and from the hypothesis that $P(A) > 0$ and $P(B) > 0$, we see that the right side of Eq. (2) must be positive. Thus, we have a *contradiction*. This means that our assumption that $A$ and $B$ are independent is false. That is, $A$ and $B$ cannot be independent.

**EXAMPLE 3-20:** Suppose that either $P(A)$ or $P(B)$ equals zero. Prove that $A$ and $B$ are independent events.

*Solution:* Suppose, for example, that $P(A) = 0$. Now, from Theorem 1.5,

$$P(A \cap B) \leq P(A) \qquad (1)$$

since $A \cap B$ is a subset of $A$. Thus, since $P(A) = 0$, it follows that

$$P(A \cap B) \leq 0 \qquad (2)$$

Now, since $P(A \cap B) \geq 0$ [recall that the probability of any event is non-negative, in accord with Axiom 1.1], it follows from this and Eq. (2) that

$$0 \leq P(A \cap B) \leq 0 \qquad (3)$$

Now, Eq. (3) indicates that

$$P(A \cap B) = 0 \qquad (4)$$

Now, from Definition 3.2, events $A$ and $B$ are independent since Eq. (5) below is satisfied:

$$P(A \cap B) = P(A) \cdot P(B) \qquad (5)$$

Here, both the left and right sides are equal to zero.

*notes*

(a) The demonstration in Example 3-20 would be essentially unchanged if we started off with the supposition that $P(B) = 0$.
(b) If we tried to employ part (a) of Theorem 3.5 here, that is, if we considered the equation $P(B) = P(B \mid A)$, we would be led to no conclusion. Remember that $P(B \mid A)$ doesn't exist here since $P(A) = 0$.

# SOLVED PROBLEMS

## Conditional Probability

**PROBLEM 3-1**  Suppose a pair of fair dice is tossed. (**a**) What is the probability of an even number [of dots] occurring on both dice? (**b**) What is the probability the sum of the numbers [of dots] on the dice equals eight? (**c**) What is the probability of an even number occurring on both dice given that that sum of the numbers equals eight?

*Solution Preliminaries:* We shall work relative to the most subdivided sample space $S$, which contains 36 equally likely elements. That is,

$$S = \{11, 12, \dots, 16, \dots, 61, 62, \dots, 66\}$$

where $P(\{ij\}) = 1/36$ for $i = 1, 2, \dots, 6$ and $j = 1, 2, \dots, 6$.

(**a**) Denote the event by $A$. Thus, as a subset,

$$A = \{22, 24, 26, 42, 44, 46, 62, 64, 66\}$$

and hence, $P(A) = N(A)/N(S) = 9/36 = 1/4$.

(**b**) Denote the event by $B$. Thus, as a subset,

$$B = \{26, 35, 44, 53, 62\}$$

and hence $P(B) = N(B)/N(S) = 5/36$.

(**c**) Using the symbols of parts (a) and (b), here we wish to compute $P(A \mid B)$. First, we observe that $A \cap B = \{26, 44, 62\}$, and hence $P(A \cap B) = 3/36$. Then we have, after using the value for $P(B)$ from part (b),

$$P(A \mid B) = \frac{P(A \cap B)}{P(B)} = \frac{3/36}{5/36} = \frac{3}{5}$$

**PROBLEM 3-2**  Suppose three fair coins are tossed in succession. Let $A$ be the event "at least two heads," and $B$ be the event "first coin is tails." (**a**) Determine $P(A)$, $P(B)$, and $P(A \cap B)$. (**b**) Determine $P(A \mid B)$.

*Solution*

(**a**) The most subdivided sample space is

$$S = \{HHH, HHT, HTH, HTT, THH, THT, TTH, TTT\}$$

It's reasonable to assume that all 8 elements are equally likely. As subsets, the events $A$, $B$, and $A \cap B$ are as follows:

$$A = \{HHH, HHT, HTH, THH\} \qquad B = \{THH, THT, TTH, TTT\} \qquad A \cap B = \{THH\}$$

Thus, using Theorem 1.2, we have

$$P(A) = \frac{N(A)}{N(S)} = \frac{4}{8} \quad \text{and, similarly} \quad P(B) = \frac{4}{8} \quad \text{and} \quad P(A \cap B) = \frac{1}{8}$$

(**b**)
$$P(A \mid B) = \frac{P(A \cap B)}{P(B)} = \frac{1/8}{4/8} = \frac{1}{4}$$

As expected, the added knowledge that the first coin is tails decreases the probability of getting at least two heads. This is revealed in the calculations since $P(A) = 1/2$, and $P(A \mid B) = 1/4$.

**TABLE 3-3: Table for Problem 3-3**

|  | In union less than 5 years (A) | In union at least 5 years (A') | Row totals |
|---|---|---|---|
| Answer of yes (B) | 10% | 30% | 40% |
| Answer of no (B') | 25% | 35% | 60% |
| Column totals | 35% | 65% | 100% |

**PROBLEM 3-3**  Table 3-3 classifies members of a union with respect to two characteristics: (1) the period of time a member has been in the union, and (2) the way a member answers a question pertaining to contract negotiations. The numbers in the body of Table 3-3 are percentages for each of four categories. Let $A$ be the event "in union less than 5 years," and let $B$ be the event "answer of yes to question," with $A'$ and $B'$ denoting the complements of these events. **(a)** If a union member is selected at random, what is the probability the member has been in the union at least 5 years and answered "yes" to the question? **(b)** What is the probability a union member answered "no" to the question, given that the member has been in the union for less than 5 years?

*Solution*

**(a)**  Here, we wish to find $P(A' \cap B)$. From the first row, second column entry of Table 3-3, we have $P(A' \cap B) = .30$.

**(b)**  Here, we wish to find $P(A \mid B')$. Now,

$$P(A \mid B') = \frac{P(A \cap B')}{P(B')} \tag{1}$$

Now, $B' = (A \cap B') \cup (A' \cap B')$, where $A \cap B'$ and $A' \cap B'$ are disjoint. Thus,

$$P(B') = P(A \cap B') + P(A' \cap B') \tag{2}$$

Remember that $B'$ stands for "answer of no" to the question. Thus, the entries that pertain to $A \cap B'$, $A' \cap B'$, and $B'$ are given in the second row of Table 3-3. So, using values from Table 3-3, we have $P(A \cap B') = .25$, and $P(A' \cap B') = .35$. Thus, substitution of these values into Eq. (2) and then into Eq. (1) leads to

$$P(B') = .25 + .35 = .60 \tag{3}$$

and

$$P(A \mid B') = \frac{P(A \cap B')}{P(B')} = \frac{.25}{.60} = .416\overline{6} \quad \text{or} \quad 41.6\overline{6}\% \tag{4}$$

**PROBLEM 3-4**  Prove Theorem 3.4.

*Proof*  Since $P(A) > 0$, it follows from Definition 3.1 that $P(B \mid A)$ exists. Similarly, $P(B' \mid A)$ exists. From Definition 3.1, we have

$$P(B \mid A) = \frac{P(A \cap B)}{P(A)} \quad \text{and} \quad P(B' \mid A) = \frac{P(A \cap B')}{P(A)} \tag{1a}, (1b)$$

Now, $A$ can be written as

$$A = (A \cap B) \cup (A \cap B') \tag{2}$$

where $A \cap B$ and $A \cap B'$ are disjoint. [See Example 3-10 and Figure 3-2 of that example for a review.] Thus, from Eq. (2), we have

$$P(A) = P(A \cap B) + P(A \cap B') \tag{3}$$

Solving for $P(A \cap B')$ from Eq. (3) and then substituting the result into the numerator of Eq. (1b) leads to

$$P(B' \mid A) = \frac{P(A) - P(A \cap B)}{P(A)} = 1 - \frac{P(A \cap B)}{P(A)} = 1 - P(B \mid A) \qquad \textbf{(4)}$$

after making use of Eq. (1a). $\qquad\qquad\qquad\qquad\qquad\qquad\qquad\qquad\qquad\qquad$ □

### Independent Events

**PROBLEM 3-5** Suppose both $P(A) > 0$ and $P(B) > 0$, which means that both $P(B \mid A)$ and $P(A \mid B)$ exist. Prove the following: If $P(B \mid A) = P(B)$, then $P(A \mid B) = P(A)$.

*Proof* By hypothesis, we have $P(B \mid A) = P(B)$. Thus, from Definition 3.1, it follows that

$$P(B) = \frac{P(A \cap B)}{P(A)} \qquad \textbf{(1)}$$

and thus,

$$P(A \cap B) = P(A) \cdot P(B) \qquad \textbf{(2)}$$

[In particular, this means that $A$ and $B$ are independent.] Now, from Definition 3.1,

$$P(A \mid B) = \frac{P(A \cap B)}{P(B)} \qquad \textbf{(3)}$$

[The numerator can also be written as $P(B \cap A)$.] Substituting from Eq. (2) into Eq. (3) yields

$$P(A \mid B) = \frac{P(A) \cdot \cancel{P(B)}}{\cancel{P(B)}} = P(A) \qquad \textbf{(4)}$$

which was to be proven. [We can divide through by $P(B)$ here since $P(B) \neq 0$ by hypothesis.] $\quad$ □

**PROBLEM 3-6** Prove part (c) of Theorem 3-8.

*Proof* Let us first prove "if $A$ and $B$ are independent, then $P(B \mid A') = P(B)$." We assume that $P(A') > 0$ so that $P(B \mid A')$ will exist.

Since, by hypothesis, $A$ and $B$ are independent, it follows from Definition 3.2 that

$$P(A \cap B) = P(A) \cdot P(B) \qquad \textbf{(1)}$$

Now, by Definition 3.1,

$$P(B \mid A') = \frac{P(A' \cap B)}{P(A')} \qquad \textbf{(2)}$$

Now, since $B = (A \cap B) \cup (A' \cap B)$, where $(A \cap B)$ and $(A' \cap B)$ are disjoint, it follows that

$$P(A' \cap B) = P(B) - P(A \cap B) \qquad \textbf{(3)}$$

Substitution of $P(A \cap B)$ from Eq. (1) into Eq. (3) leads to

$$P(A' \cap B) = P(B) \cdot [1 - P(A)] = P(A') \cdot P(B) \qquad \textbf{(4)}$$

Here, we used $P(A') = 1 - P(A)$. Substitution of $P(A' \cap B)$ from Eq. (4) into Eq. (2) yields

$$P(B \mid A') = \frac{\cancel{P(A')} \cdot P(B)}{\cancel{P(A')}} = P(B) \qquad \textbf{(5)}$$

This completes the proof of the first part.

Let us now prove "if $P(B \mid A') = P(B)$, then $A$ and $B$ are independent," again with the assumption that $P(A') > 0$. First, from the definition for $P(B \mid A')$, and the hypothesis that $P(B \mid A') = P(B)$, we have

$$\frac{P(A' \cap B)}{P(A')} = P(B) \qquad \textbf{(6)}$$

and thus,

$$P(A' \cap B) = P(A') \cdot P(B) \qquad (7)$$

Now, $B = (A \cap B) \cup (A' \cap B)$, where $(A \cap B)$ and $(A' \cap B)$ are disjoint. This leads to

$$P(A \cap B) = P(B) - P(A' \cap B) \qquad (8)$$

Substituting the expression for $P(A' \cap B)$ from Eq. (7) into Eq. (8) results in

$$P(A \cap B) = P(B) - P(A') \cdot P(B) = P(B) \cdot [1 - P(A')] \qquad (9)$$

and thus, since $P(A) = 1 - P(A')$,

$$P(A \cap B) = P(A) \cdot P(B) \qquad (10)$$

Thus, from Definition 3.2, it follows that $A$ and $B$ are independent. $\qquad \square$

**PROBLEM 3-7**   Refer to Problem 3-1. Determine if events $A$ and $B$ are independent.

**Solution** *Method 1:* There, $P(A) = 1/4$, $P(B) = 5/36$, and $P(A \cap B) = 1/12$. Since $P(A \cap B) \neq P(A) \cdot P(B)$, it follows from Definition 3.2 that events $A$ and $B$ are not independent.
   *Method 2:* Since $P(A \mid B) = 3/5$ and $P(A) = 1/4$, we see that $P(A \mid B) \neq P(A)$. Thus, by part **(b)** of Theorem 3.5, it follows that events $A$ and $B$ are not independent.

**PROBLEM 3-8**   Refer to Problem 3-2. Determine if events $A$ and $B$ are independent.

**Solution**   From the solution to Problem 3-2, $P(A) = 1/2$ and $P(A \mid B) = 1/4$. Thus, it follows from part **(b)** of Theorem 3.5 that events $A$ and $B$ are not independent.

**PROBLEM 3-9**   Refer to Problem 3-3. **(a)** Show that the events $A$ and $B$ are not independent. **(b)** Refer to Table 3-3 of Problem 3-3. Suppose that the entries in the "Answer of Yes" row are not changed. What changes in the first two entries in the "Answer of No" row will cause events $A$ [which is "In Union Less than 5 Years"] and $B$ [which is "Answer of Yes"] to be independent?

**Solution**

**(a)** From Table 3-3, we see that $P(A) = .35$, $P(B) = .40$, and $P(A \cap B) = .10$. Since $P(A \cap B) \neq P(A) \cdot P(B)$, it follows that events $A$ and $B$ are not independent.

**(b)** If $P(A \mid B)$ were to equal $P(A)$, then by part (b) of Theorem 3.5, events $A$ and $B$ would be independent. Now,

$$P(A \mid B) = \frac{P(A \cap B)}{P(B)} \qquad (1)$$

From the "Answer of Yes" row of Table 3-3, we see that $P(A \cap B) = .10$ and $P(B) = .40$. Thus, substitution of these values into Eq. (1) leads to

$$P(A \mid B) = \frac{.10}{.40} = .25 \qquad (2)$$

Thus, we now set $P(A)$ equal to .25 also. Then, from the equation $P(A) = P(A \cap B) + P(A \cap B')$, we substitute in $P(A \cap B) = .10$ and $P(A) = .25$, and obtain

$$P(A \cap B') = .25 - .10 = .15 \qquad (3)$$

Next, from $P(B') = P(A \cap B') + P(A' \cap B')$, and the values $P(B') = .60$ and $P(A \cap B') = .15$, we obtain

$$P(A' \cap B') = .60 - .15 = .45 \qquad (4)$$

Thus, the first two entries in the "Answer of No," or $B'$, row should be altered to 15% and 45%, respectively. Refer to Table 3-4, for which events $A$ and $B$ are now independent.

**TABLE 3-4:  Table for Problem 3-9(b); *A* and *B* are Independent**

|  | In union less than 5 years (*A*) | In union at least 5 years (*A'*) | Row totals |
|---|---|---|---|
| Answer of yes (*B*) | 10% | 30% | 40% |
| Answer of no (*B'*) | 15% | 45% | 60% |
| Column totals | 25% | 75% | 100% |

As a further check to show that events *A* and *B* are now independent, we see from the new entries in the *A* and *A'* columns that $P(B \mid A) = .10/.25$, and $P(B \mid A') = .30/.75$. Both of these fractions are equal to .40, which, as expected, is the value of $P(B)$.

**PROBLEM 3-10**    Prove part (c) of Theorem 3.6.

*Proof*  Here, by hypothesis, we have that *A* and *B* are independent, and we wish to show that, as a consequence, *A'* and *B'* are independent. Thus, from the hypothesis, it follows that Eq. (1) below is true:

$$P(A \cap B) = P(A) \cdot P(B) \tag{1}$$

Now, $B' = (A \cap B') \cup (A' \cap B')$, where $(A \cap B')$ and $(A' \cap B')$ are disjoint. [Refer, for example, to Figure 3-2; previously, these disjoint subsets were sometimes known as $R_2$ and $R_4$.] Thus,

$$P(B') = P(A \cap B') + P(A' \cap B') \tag{2}$$

From part (a) of Theorem 3.6,

$$P(A \cap B') = P(A) \cdot P(B') \tag{3}$$

Substituting the expression for $P(A \cap B')$ from Eq. (3) into Eq. (2) yields

$$P(A' \cap B') = P(B') - P(A) \cdot P(B') = P(B') \cdot [1 - P(A)] \tag{4}$$

This becomes

$$P(A' \cap B') = P(A') \cdot P(B') \tag{5}$$

thus indicating that *A'* and *B'* are independent.  □

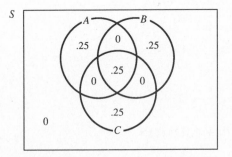

**Figure 3-6.** Venn diagram for Problem 3-11.

**PROBLEM 3-11**    Given 100 chips in a bag, where 25 are stamped only with the letter *A*, 25 are stamped only with the letter *B*, and 25 are stamped only with the letter *C*. All of the letters *A*, *B*, and *C* are stamped on each of the remaining 25 chips. The experiment is to draw a single chip from the bag. Now, let *A* be the *event* "a chip is stamped with letter *A*," and similarly for the events *B* and *C*. Show that the events *A*, *B*, and *C* are pairwise independent, but not independent. [Refer to Definition 3.3 for the definitions of independence and pairwise independence among three events.]

*Solution*  The Venn diagram of Figure 3-6 describes the situation for this problem; the numbers denote probabilities for the respective disjoint regions. From the given information, we have

$$P(A \cap B' \cap C') = P(A' \cap B \cap C') = P(A' \cap B' \cap C) = .25 \tag{1a), (1b), (1c}$$

$$P(A \cap B \cap C) = .25 \tag{2}$$

Thus, by referring to Figure 3-6, for example, we can deduce that

$$P(A) = P(B) = P(C) = .50 \tag{3a), (3b), (3c}$$

and

$$P(A \cap B) = P(A \cap C) = P(B \cap C) = .25 \tag{4a), (4b), (4c}$$

Thus, we see that

$$P(A \cap B) = P(A) \cdot P(B); \qquad P(A \cap C) = P(A) \cdot P(C); \qquad \text{and} \qquad \textbf{(5a), (5b)}$$

$$P(B \cap C) = P(B) \cdot P(C) \qquad \qquad \textbf{(5c)}$$

since both sides of each of Eqs. (5a), (5b), and (5c) equal .25. This means that the three events $A$, $B$, and $C$ are pairwise independent. However, we see that

$$P(A \cap B \cap C) \neq P(A) \cdot P(B) \cdot P(C) \qquad \qquad \textbf{(6)}$$

since the left side equals .25 [see Eq. (2)] but the right side equals $(.5)^3 = .125$. Thus, condition (ii) of Definition 3.3 is not satisfied; that means that events $A$, $B$, and $C$ are not independent even though they are pairwise independent.

## Counting Methods

**PROBLEM 3-12**  Suppose Alice and Bob are playing a tennis match in which the winner is the one who wins three sets first [this is a so-called "best three out of five sets match"]. Use a tree diagram to determine all possible ways the match can be decided.

*Solution*  The portion of the tree diagram indicating the 10 ways the match can end if Alice ($A$) wins the first set is presented in Figure 3-7. [The 10 ways the match can end are listed to the right of the 10 endpoints. The items in the list are shown underlined.] One can determine the 10 additional ways the match can end if Bob ($B$) wins the first set by interchanging the letters $A$ and $B$ in the 10 ways presented in Figure 3-7.

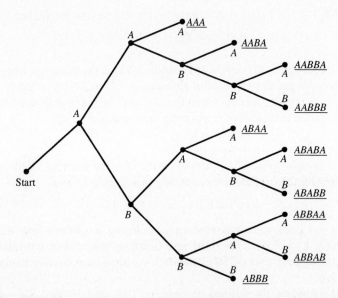

**Figure 3-7.**  Tree diagram for Problem 3-12. Items in list are underlined.

Thus, the match can end in a total of 20 ways. The 10 ways if Alice wins the first set [from Figure 3-7] are *AAA, AABA, AABBA, AABBB, ABAA, ABABA, ABABB, ABBAA, ABBAB,* and *ABBB*.

The 10 ways if Bob wins the first set are *BBB, BBAB, BBAAB, BBAAA, BABB, BABAB, BABAA, BAABB, BAABA,* and *BAAA*.

**PROBLEM 3-13**  Given a five-card deck with three spades [ace, two, and three] and two hearts [ace and two], as in Example 3-16. Suppose three cards are drawn with replacement. [This means that each of the three cards is drawn from a well-shuffled deck containing five cards.] **(a)** Determine the number of elements in the most subdivided sample space for the experiment of drawing three cards with replacement. **(b)** Compute the probability of the event "$S_1$ and $H_2$ and $S_3$." This event is taken to mean obtaining spade on card 1, heart on card 2, and spade on card 3.

### Solution

(a) Let $\mathscr{S}$ denote the sample space of all possible orders for the drawing of the three cards with replacement. Since each card can be drawn in five possible ways, this means that $N(\mathscr{S})$, which stands for the number of elements in the most subdivided sample space, is given by

$$N(\mathscr{S}) = n_1 \cdot n_2 \cdot n_3 = 5 \cdot 5 \cdot 5 = 5^3 = 125 \qquad (1)$$

Here, we used Theorem 3.7, the fundamental counting principle. It is reasonable to suppose that these 125 elements are equally likely; this means, of course, that Theorem 1.2 is applicable.

(b) Since Theorem 1.2 applies, the desired probability, $P(S_1$ and $H_2$ and $S_3)$—we could write this as $P(S_1 \cap H_2 \cap S_3)$—is given by

$$P(S_1 \text{ and } H_2 \text{ and } S_3) = \frac{N(S_1 \text{ and } H_2 \text{ and } S_3)}{N(\mathscr{S})} \qquad (2)$$

The quantity $N(S_1$ and $H_2$ and $S_3)$, which is the number of orders [elements] favorable to getting spade followed by heart followed by spade, is given by

$$N(S_1 \text{ and } H_2 \text{ and } S_3) = 3 \cdot 2 \cdot 3 = 18 \qquad (3)$$

in accord with the fundamental counting principle. Substitution from Eqs. (1) and (3) into Eq. (2) results in

$$P(S_1 \text{ and } H_2 \text{ and } S_3) = \frac{3 \cdot 2 \cdot 3}{5 \cdot 5 \cdot 5} = \frac{18}{125} = .144 \qquad (4)$$

Thus, the answer to part (b) is 18/125 or .144.

Note that $P(S_1) = 3/5$, $P(H_2) = 2/5$, and $P(S_3) = 3/5$, so that Eq. (4) can also be written as

$$P(S_1 \text{ and } H_2 \text{ and } S_3) = P(S_1) \cdot P(H_2) \cdot P(S_3) \qquad (5)$$

**PROBLEM 3-14**   Repeat parts (a) and (b) of Problem 3-13 for the situation where all is the same except that the three cards are drawn without replacement. Here, if we visualize the cards being drawn one at a time, in sequence, then the second card is drawn from a deck containing four cards, and the third card is drawn from a deck containing three cards.

### Solution

(a) Let $\mathscr{S}$ denote the sample space of all possible orders for the drawing of three cards without replacement. Applying the fundamental counting principle, we have

$$N(\mathscr{S}) = n_1 \cdot n_2 \cdot n_3 = 5 \cdot 4 \cdot 3 = 60 \qquad (1)$$

Note that $n_2 = 4$ since, when the second card is drawn, it is drawn from a deck containing four cards; similarly, $n_3 = 3$ since the third card is drawn from a deck containing three cards. It is reasonable to suppose that all 60 elements [or orders] are equally likely; thus, we may apply Theorem 1.2.

(b) Since Theorem 1.2 applies, the desired probability, $P(S_1$ and $H_2$ and $S_3)$—again, we could write this as $P(S_1 \cap H_2 \cap S_3)$—is given by

$$P(S_1 \text{ and } H_2 \text{ and } S_3) = \frac{N(S_1 \text{ and } H_2 \text{ and } S_3)}{N(\mathscr{S})} \qquad (2)$$

Here, the quantity $N(S_1$ and $H_2$ and $S_3)$, which is the number of orders [elements] favorable to getting spade followed by heart followed by spade, is given by

$$N(S_1 \text{ and } H_2 \text{ and } S_3) = 3 \cdot 2 \cdot 2 = 12 \qquad (3)$$

in accord with the fundamental counting principle. Note that the third term in $3 \cdot 2 \cdot 2$ is 2 since the number of ways to get a spade on the third card provided the first two cards were spade and heart, respectively, is 2—remember, we have drawing without replacement here. Substitution from Eqs. (1) and (3) into Eq. (2) results in

$$P(S_1 \text{ and } H_2 \text{ and } S_3) = \frac{3 \cdot 2 \cdot 2}{5 \cdot 4 \cdot 3} = \frac{12}{60} = .200 \tag{4}$$

Thus, the answer to part (b) is 12/60 or 1/5 or .200.

Note that $P(S_1) = 3/5$, $P(H_2 \mid S_1) = 2/4$, and $P(S_3 \mid S_1 \cap H_2) = 2/3$, so that Eq. (4) can also be written as

$$P(S_1 \text{ and } H_2 \text{ and } S_3) = P(S_1) \cdot P(H_2 \mid S_1) \cdot P(S_3 \mid S_1 \cap H_2) \tag{5}$$

### A Review of the Concepts of Mutually Exclusive and Independent Events

**PROBLEM 3-15**  Give simplified expressions for $P(A \cup B)$, $P(B \mid A)$, and $P(A \mid B)$ if **(a)** $A$ and $B$ are mutually exclusive and **(b)** $A$ and $B$ are independent.

*Solution*

**(a)** If $A$ and $B$ are mutually exclusive, then $A \cap B = \varnothing$, and hence $P(A \cap B) = 0$. Thus, from Theorem 2.1 [or from Axiom 1.3'], we obtain

$$P(A \cup B) = P(A) + P(B) \tag{1}$$

Now, from Definition 3.1, we have $P(B \mid A) = P(A \cap B)/P(A)$ and $P(A \mid B) = P(A \cap B)/P(B)$—remember that $A \cap B = B \cap A$. Substitution of $P(A \cap B) = 0$ into these equations results in

$$P(B \mid A) = \frac{0}{P(A)} = 0, \qquad \text{assuming } P(A) > 0 \tag{2}$$

and

$$P(A \mid B) = \frac{0}{P(B)} = 0, \qquad \text{assuming } P(B) > 0 \tag{3}$$

**(b)** If $A$ and $B$ are independent, then from Definition 3.2, we have $P(A \cap B) = P(A) \cdot P(B)$. Thus, from Theorem 2.1, it follows that

$$P(A \cup B) = P(A) + P(B) - P(A) \cdot P(B) \tag{4}$$

Substituting $P(A \cap B) = P(A) \cdot P(B)$ into $P(B \mid A) = P(A \cap B)/P(A)$ and $P(A \mid B) = P(A \cap B)/P(B)$ leads to

$$P(B \mid A) = \frac{P(A \cap B)}{P(A)} = \frac{\cancel{P(A)} \cdot P(B)}{\cancel{P(A)}} = P(B), \qquad \text{assuming } P(A) > 0 \tag{5}$$

and

$$P(A \mid B) = \frac{P(A \cap B)}{P(B)} = \frac{P(A) \cdot \cancel{P(B)}}{\cancel{P(B)}} = P(A), \qquad \text{assuming } P(B) > 0 \tag{6}$$

## Supplementary Problems

**PROBLEM 3-16**  A nonfair coin [for example, one that is unsymmetrically weighted] is tossed twice. For the most subdivided sample space $S = \{TT, TH, HT, HH\}$, the probabilities for the simple events are $P(\{TT\}) = .16$; $P(\{TH\}) = P(\{HT\}) = .24$; $P(\{HH\}) = .36$. Let $A$ be the event "exactly one head," and let $B$ be the event "head on the first toss." **(a)** Represent $A$ and $B$ as subsets, and find $P(A)$ and $P(B)$. **(b)** Determine $P(B \mid A)$ and $P(A \mid B)$. **(c)** Determine if $A$ and $B$ are independent.

*Answer* (a) $A = \{TH, HT\}$; $P(A) = .48$; $B = \{HT, HH\}$; $P(B) = .60$ (b) First, $A \cap B = \{HT\}$, and thus $P(A \cap B) = .24$. Then, $P(B \mid A) = P(A \cap B)/P(A) = .50$, and $P(A \mid B) = P(A \cap B)/P(B) = .40$ (c) Since $P(A \cap B) \neq P(A) \cdot P(B)$, events $A$ and $B$ are not independent

**PROBLEM 3-17** Refer to the situation of Problem 3-16. Let $B$ be the event "head on the first toss," as before, and let $C$ be the event "tail on the second toss." (a) Represent $C$ as a subset, and find $P(C)$. (b) Determine if events $B$ and $C$ are independent.

*Answer* (a) $C = \{TT, HT\}$, and thus $P(C) = .40$ (b) Since $B \cap C = \{HT\}$, we see that $P(B \cap C) = .24$. Since $P(B \cap C) = P(B) \cdot P(C)$, events $B$ and $C$ are independent.

**PROBLEM 3-18** Given 100 chips in a container. Of these, 24 are stamped with all of the letters $A$, $B$, and $C$. Also, 24 are stamped with only the letters $A$ and $B$; 6 are stamped with only the letters $A$ and $C$; 14 are stamped with only the letters $B$ and $C$. Also, 6 are stamped only with the letter $A$; 18 are stamped only with the letter $B$; 6 are stamped only with the letter $C$. Finally, 2 chips have no letter at all stamped on them.

The experiment is to draw a single chip from the container. Now, let $A$ denote the event "a chip is stamped with the letter $A$," and similarly for the events $B$ and $C$. Show that $P(A \cap B \cap C) = P(A) \cdot P(B) \cdot P(C)$, but that events $A$, $B$, and $C$ are not independent. [Refer to Definition 3.3 for the definition of independence among three events.]

*Answer and Hints* Fill in probability values on a Venn diagram similar to that of Figure 3-6, which consists of a partition into eight disjoint subsets. On the Venn diagram, we have that $P(A \cap B \cap C) = .24$, $P(A \cap B \cap C') = .24$, $P(A \cap B' \cap C) = .06$, and so on. We see that $P(A) \cdot P(B) \cdot P(C) = (.6)(.8)(.5) = .24$, and that this is equal to $P(A \cap B \cap C)$. However, we see that $P(B \cap C) = .38$, but $P(B) \cdot P(C) = .40$. That is, $P(B \cap C) \neq P(B) \cdot P(C)$, which means that events $A$, $B$, and $C$ are not independent—see condition (ic) of Definition 3.3.

**PROBLEM 3-19** Adults in a community are classified as to whether they smoke [$A$] or don't smoke [$A'$], and whether they graduated from high school [$B$] or didn't graduate from high school [$B'$]. See Table 3-5. (a) Calculate $P(A \mid B)$ and $P(A \mid B')$. (b) Calculate $P(B \mid A)$ and $P(B \mid A')$. (c) Determine whether events $A$ and $B$ are independent.

TABLE 3-5: Table for Problem 3-19

| | High school graduate [$B$] | Not high school graduate [$B'$] | Row totals |
|---|---|---|---|
| Smokes [$A$] | 40% | 20% | 60% |
| Doesn't smoke [$A'$] | 30% | 10% | 40% |
| Column totals | 70% | 30% | 100% |

*Answer* (a) 4/7, 2/3 (b) 2/3, 3/4 (c) No

**PROBLEM 3-20** A fair die is tossed three times in succession. (a) How many elements are there in the most subdivided sample space? (b) Let $A$ be the event of obtaining an ace [face with one dot] on exactly the first and third tosses. Calculate $P(A)$. (c) Let $B$ be the event of obtaining fewer than three dots on exactly the first and second tosses. Calculate $P(B)$.

*Answer* (a) Let $S$ stand for "sample space." The number of elements in $S$, symbolized by $N(S)$, is given by $N(S) = 6 \cdot 6 \cdot 6 = 6^3 = 216$ (b) Here, $N(A) = 1 \cdot 5 \cdot 1 = 5$, and hence $P(A) = 5/216 = .02315$ (c) Here, $N(B) = 2 \cdot 2 \cdot 4 = 16$, and hence $P(B) = 16/216 = .07407$

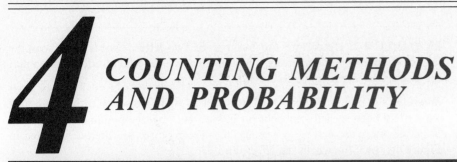

# 4 COUNTING METHODS AND PROBABILITY

## THIS CHAPTER IS ABOUT

☑ **More Counting Methods**

☑ **Permutations of *n* Objects, Where Some Are Alike**

☑ **The Binomial Theorem and Binomial Coefficients**

## 4-1. More Counting Methods

In this and the following chapters, we shall consider probability situations involving complex experiments. In order to be better able to describe such experiments, and to count the numbers of elements in the related sample spaces, and in the various events of those experiments, we shall need more sophisticated approaches to counting. We already introduced the topic of counting methods [or combinatorial methods, as they are often called] in Section 3-3. In that Section, we studied the following important counting methods topics: tree diagrams and the fundamental counting principle [Theorem 3.7].

### A. Permutations

Suppose we wish to determine how many distinct two-letter "words" can be formed using the letters *a*, *b*, *c*, and *d*, where no letter is used more than once in a word. Here, by "word," we mean an ordering of two letters in a line.

A tree diagram for this problem, shown in Figure 4-1, indicates that there are 12 words. [Refer to the final list on the right, in which the words are underlined.] The four initial [or first-level] branches indicate the four possible ways to determine the first letter of the word. [Here, we assume we are going in a left to right direction.] There are three second-level branches for each initial branch. These indicate the remaining three ways to determine the second letter of the word. The 12 possible words [*ab*, *ac*, etc.] are shown listed to the right of the 12 endpoints.

If one reads a path backwards from an endpoint to the starting point, one obtains the spelling of a word in reverse order.

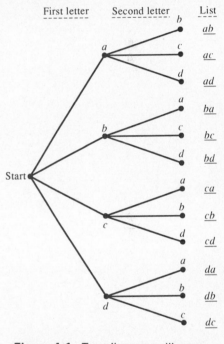

**Figure 4-1.** Tree diagram to illustrate permutations. [The 12 permutations are shown underlined.]

---

**EXAMPLE 4-1:** Use the fundamental counting principle [Theorem 3.7] to determine the number of two-letter words that can be formed from the letters *a*, *b*, *c*, and *d*. Assume that the repetition of a letter is not permitted.

*Solution:* Consider forming a word from left to right. The first position can be filled in any of four ways. Then the next [and last] position can be filled in three ways since repetition of a letter is not permitted. Thus, by the fundamental counting principle, there are $n_1 \cdot n_2 = 4 \cdot 3 = 12$ words.

Note that the 4 and the 3 correspond, respectively, to the number of initial branches, and the number of second-level branches associated with each initial branch in the tree diagram of Figure 4-1. Naturally, the number of endpoints in Figure 4-1, which is the same as the number of words, is equal to 4 times 3, that is, 12.

**EXAMPLE 4-2:** Determine the number of two-letter "words" that can be formed from four distinct letters, if repetition of a letter is allowed. [Assume that the four letters are *a*, *b*, *c*, and *d*.]

**Solution:** Forming a typical word from left to right, we see that there are four ways to fill both the left and right positions. Thus, by the fundamental counting principle, there are $4 \cdot 4 = 16$ words. The sixteen words are the twelve listed in Figure 4-1, plus the words *aa*, *bb*, *cc*, and *dd*.

The corresponding tree diagram [try picturing it in your mind] would be the same as that of Figure 4-1, except that there would be four second-level branches for each initial branch, and not three as in Figure 4-1. Such second-level branches would terminate at endpoints labeled *a*, *b*, *c*, and *d*.

Given a set of *n* objects [for example, letters, or digits, or other types of objects], an ordering or arrangement in a line of *r* of these objects is called a *permutation of the n objects taken r at a time*. Here, $r \leq n$, and repeated use of an object is not allowed. A symbol for the number of permutations of *n* distinct objects taken *r* at a time is $P(n, r)$.

In Example 4-1, and in the discussion preceding that Example, we had an illustration of the permutations of four distinct objects taken two at a time. There, the four objects were the letters *a*, *b*, *c*, *d*, and the twelve permutations were the twelve two-letter words listed next to the endpoints in Figure 4-1. Observe that a tree diagram approach and also the fundamental counting principle were used to find the number of permutations. We found that $P(4, 2) = n_1 \cdot n_2 = 4 \cdot 3 = 12$.

**EXAMPLE 4-3:** Given four distinct objects, determine the number of permutations taken (**a**) one at a time, (**b**) three at a time, and (**c**) four at a time, respectively. (*Hint:* Think of the objects as being the four letters *a*, *b*, *c*, and *d*, and the permutations as being "words" that can be formed from the letters, where repetition of letters is not allowed.)

**Solution:** For each of the following calculations, we shall employ the fundamental counting principle.

(**a**) $P(4, 1) = 4$. As words, these would be *a*, *b*, *c*, and *d*, namely the four original letters.
(**b**) $P(4, 3) = 4 \cdot 3 \cdot 2 = 24$. Illustrating, by using words formed from the four letters, several of the permutations are *abc*, *acb*, *bac*, *cad*, and *dcb*.
(**c**) $P(4, 4) = 4 \cdot 3 \cdot 2 \cdot 1 = 24$. Illustrating, by using words formed from the four letters, several of the permutations are *abcd*, *acbd*, *bacd*, *cadb*, and *dcba*.

**EXAMPLE 4-4:** Derive a general equation for $P(n, r)$, the number of permutations of *n* distinct objects taken *r* at a time. Here, $r \leq n$.

**Solution:** Let us think of the job of forming a permutation as being a combined operation consisting of *r* individual operations. The *r* individual operations consist of filling *r* positions [each position to be filled with a single object], as one works from left to right.

Suppose we label the positions—think of these as boxes—with the numbers 1, 2, 3, 4, ..., *r*, going from left to right. Refer to Figure 4-2.

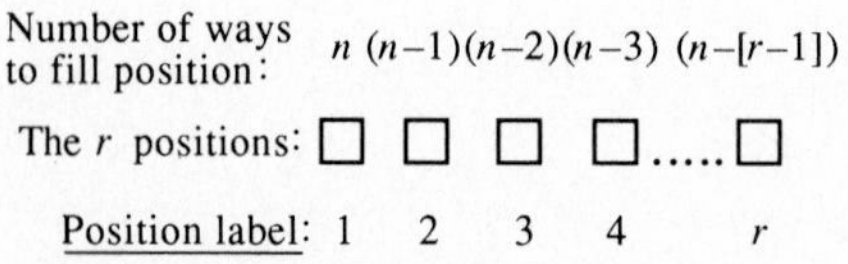

We can fill in position 1 in *n* possible ways, position 2 in $(n - 1)$ ways [recall that the object that filled position 1 can't be used again], position 3 in $(n - 2)$ ways, and position 4 in $(n - 3)$ ways. Continuing this pattern, we see that position *r* can be filled in $(n - [r - 1]) = (n - r + 1)$ ways. Notice that we subtract from

Number of ways to fill position: $\quad n \quad (n-1)(n-2)(n-3) \quad (n-[r-1])$

The *r* positions: ☐ ☐ ☐ ☐ ..... ☐

Position label: $\quad 1 \quad 2 \quad 3 \quad 4 \qquad r$

**Figure 4-2.** Deriving the equation for $P(n, r)$ in Example 4-4.

$n$ a number which is 1 less than the position label [here, this is $r$], just as we did for positions 1, 2, 3, and 4. Thus, by the fundamental counting principle, we have the following product expression for $P(n,r)$:

$$P(n,r) = n \cdot (n - 1) \cdot (n - 2) \cdot (n - 3) \cdot \ldots \cdot (n - r + 1)$$

**Theorem 4.1:** The number of permutations of $n$ distinct objects taken $r$ at a time, $P(n,r)$, is given by

$$P(n,r) = n \cdot (n - 1) \cdot (n - 2) \cdot (n - 3) \cdot \ldots \cdot (n - r + 1)$$

where there are $r$ terms in the product on the right. Also, $r \leq n$.

***note:*** Other symbols used instead of $P(n,r)$ are $P_r^n$ and $_nP_r$.

**EXAMPLE 4-5:** Determine the number of three-digit numbers that can be formed from the digits 1, 2, 3, 4, and 5 if no digit is repeated in any one number. Repeat for four- and five-digit numbers that can be formed from the same five digits.

***Solution:*** The number of three-digit numbers $= P(5,3) = 5 \cdot 4 \cdot 3 = 60$. The number of four-digit numbers $= P(5,4) = 5 \cdot 4 \cdot 3 \cdot 2 = 120$. The number of five-digit numbers $= P(5,5) = 5 \cdot 4 \cdot 3 \cdot 2 \cdot 1 = 120$.

***note:*** Observe that $P(5,4) = P(5,5)$. More generally, $P(n,n - 1) = P(n,n)$.

Another common equation for $P(n,r)$ uses factorial notation. The following definition deals with factorial notation.

**Definition 4.1:** Suppose that $k$ denotes a positive integer [whole number]. Then $k!$, which is read as "$k$ factorial," equals the product of the positive integers from 1 to $k$ inclusive. That is,

$$k! = k \cdot (k - 1) \cdot (k - 2) \cdot \ldots \cdot 3 \cdot 2 \cdot 1$$

For example, $1! = 1$, and $2! = 2 \cdot 1 = 2$, and $3! = 3 \cdot 2 \cdot 1 = 6$, and $4! = 4 \cdot 3 \cdot 2 \cdot 1 = 24$. Notice that $4!$ could be written as $4 \cdot 3!$. In general, we have the following relation if $k$ is a positive integer:

$$k! = k \cdot (k - 1)! \tag{I}$$

This result can be extended, and is useful when computing quotients involving factorials. For example, to compute $9!/6!$, we can first write $9! = 9 \cdot 8! = 9 \cdot 8 \cdot 7! = 9 \cdot 8 \cdot 7 \cdot 6!$, after repeated application of (I), and then calculate $9!/6!$ as follows:

$$\frac{9!}{6!} = \frac{9 \cdot 8 \cdot 7 \cdot 6!}{6!} = 9 \cdot 8 \cdot 7 = 504$$

**Theorem 4.2**

(a) An expression for $P(n,n)$ is $P(n,n) = n!$. Observe that $P(n,n)$ is read as the number of permutations of $n$ distinct objects taken $n$ at a time.
(b) Another expression for $P(n,r)$ is $P(n,r) = n!/(n - r)!$, where $r \leq n$.

**EXAMPLE 4-6:** Prove Theorem 4.2.

*Proof*

(a) In Theorem 4.1, if we let $r = n$, then there will be $n$ terms on the right. Moreover, the final term on the right becomes $(n - n + 1) = 1$, and we obtain

$$P(n, n) = n \cdot (n - 1) \cdot (n - 2) \cdot (n - 3) \cdot \ldots \cdot 2 \cdot 1$$

From Definition 4.1, we recognize the right side expression to be $n!$, and thus we obtain

$$P(n, n) = n!$$

(b) From Theorem 4.1, we have

$$P(n, r) = n \cdot (n - 1) \cdot (n - 2) \cdot (n - 3) \cdot \ldots \cdot (n - r + 1) \tag{1}$$

where there are $r$ terms on the right. Multiplying on the right side by 1 in the form $(n - r)!/(n - r)!$, we obtain

$$P(n, r) = n \cdot (n - 1) \cdot (n - 2) \cdot (n - 3) \cdot \ldots \cdot (n - r + 1) \cdot \frac{(n - r)!}{(n - r)!} \tag{2}$$

If we write out the $(n - r)!$ that occurs on top as

$$(n - r)! = (n - r) \cdot (n - r - 1) \cdot \ldots \cdot 3 \cdot 2 \cdot 1 \tag{3}$$

we see that Eq. (2) takes on the desired form, namely,

$$P(n, r) = \frac{n \cdot (n - 1) \cdot (n - 2) \cdot \ldots \cdot 3 \cdot 2 \cdot 1}{(n - r)!} = \frac{n!}{(n - r)!} \tag{4} \quad \Box$$

**EXAMPLE 4-7:** Compute $P(6, 2)$ and $P(6, 4)$ by using the $P(n, r)$ equation of (a) Theorem 4.1 and (b) Theorem 4.2.

*Solution*

(a) $P(6, 2) = 6 \cdot 5 = 30$, and $P(6, 4) = 6 \cdot 5 \cdot 4 \cdot 3 = 360$.

(b)
$$P(6, 2) = \frac{6!}{(6 - 2)!} = \frac{6!}{4!} = \frac{6 \cdot 5 \cdot 4!}{4!} = 6 \cdot 5 = 30$$

$$P(6, 4) = \frac{6!}{(6 - 4)!} = \frac{6!}{2!} = \frac{6 \cdot 5 \cdot 4 \cdot 3 \cdot 2!}{2!} = 360$$

From Theorem 4.2, part (a), we have

$$P(n, n) = n! \tag{1}$$

If we let $r = n$ in the $P(n, r)$ equation of Theorem 4.2, we obtain

$$P(n, n) = \frac{n!}{(n - n)!} = \frac{n!}{0!} \tag{2}$$

The symbol $0!$, which is as "zero factorial," has not been defined so far. [Recall that, in Definition 4.1, $k!$ was defined for $k$ equal to a positive integer.] If we are to have consistency between Eqs. (1) and (2) above, then the right sides must be equal. That is,

$$n! = \frac{n!}{0!} \quad \text{or} \quad 0! = \frac{n!}{n!} \quad \text{or} \quad 0! = 1 \tag{3}$$

Thus, we are led to the following definition for $0!$.

**Definition 4.1′:** Zero factorial, whose symbol is 0!, is defined as follows: 0! = 1.

The symbol 0! will arise in other equations pertaining to counting methods.

## B. Combinations

Remember that a permutation of *r* objects drawn from *n* distinct objects is an ordered arrangement of the *r* objects. [The symbol $P(n,r)$ denotes the number of such permutations. Each such permutation is also called a permutation of *n* distinct objects taken *r* at a time.] On the other hand, a combination of *r* objects drawn from *n* distinct objects is any selection [or choice] of *r* of the objects, where order does not matter. Such a combination is often referred to as a combination of *n* distinct objects taken *r* at a time. The main symbols we shall use for the number of combinations of *n* distinct objects taken *r* at a time are $C(n,r)$, and $\binom{n}{r}$. The latter symbol is also called the "binomial coefficient" symbol because it occurs in the binomial theorem [Theorem 4.7′, part (b)].

*note:* On occasion, we shall use the word "things" in place of "objects."

**EXAMPLE 4-8:** For the four letters *a*, *b*, *c*, and *d*, the 12 permutations of these distinct letters taken two at a time were listed in Figure 4-1. They were *ab*, *ac*, *ad*, *ba*, *bc*, *bd*, *ca*, *cb*, *cd*, *da*, *db*, and *dc*. Recall that order does matter for permutations. List the combinations of the four letters *a*, *b*, *c*, and *d*, taken two at a time.

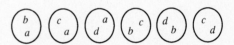

*Solution:* The six combinations of these four letters taken two at a time are listed in Figure 4-3. In particular, we see that $C(4,2) = 6$.

**Figure 4-3.** The combinations of the four letters *a*, *b*, *c*, and *d*, taken two at a time.

The random placement of the letters in the six combinations of Example 4-8 has been done deliberately in order to illustrate that order does not matter for combinations. Henceforth, when listing combinations, we shall usually list the objects [in Example 4-8, the "objects" are letters] side by side to save space. Thus, the first combination of Example 4-8 could be written as *ab* or *ba*.

**EXAMPLE 4-9:** Develop a general equation for $C(n,r)$ by determining how many permutations correspond to each combination.

*Solution:* First of all, let us consider $C(4,2)$. We know this is equal to 6, from Example 4-8. Now each of those combinations gives rise to $P(2,2) = 2! = 2$ permutations since each combination contains two distinct objects. For example, the first combination listed in Figure 4-3 gives rise to the permutations *ab* and *ba*. Thus, the total number of combinations $C(4,2)$ multiplied by 2! is equal to $P(4,2)$, the total number of permutations of four distinct objects taken two at a time. That is,

$$C(4,2) \cdot 2! = P(4,2) \qquad \text{or} \qquad C(4,2) = \frac{P(4,2)}{2!}$$

This checks out since $P(4,2) = 4 \cdot 3 = 12$, and thus the above equation for $C(4,2)$ yields $C(4,2) = 12/2 = 6$. The latter value is known to be correct from Example 4-8.

In the general case, matters are similar. In the case of $C(n,r)$, each combination contains $r$ distinct objects. Thus, each such combination gives rise to $P(r, r) = r!$ permutations. Thus, $C(n, r)$ times $r!$ is equal to the total number of permutations of $n$ distinct objects taken $r$ at a time. That is,

$$C(n, r) \cdot r! = P(n, r) \qquad \text{or} \qquad C(n, r) = \frac{P(n, r)}{r!} \qquad \textbf{(I)}$$

**Theorem 4.3:** The number of combinations of $n$ distinct objects taken $r$ at a time, denoted by $C(n, r)$, is given by the following equation:

$$C(n, r) = \frac{P(n, r)}{r!} \qquad \textbf{(a)}$$

Here, $n$ is a positive integer, as is $r$, and $r \leq n$. Using the equation for $P(n, r)$ from Theorem 4.1, we obtain

$$C(n, r) = \frac{n \cdot (n - 1) \cdot (n - 2) \cdot \ldots \cdot (n - r + 1)}{r!} \qquad \textbf{(b)}$$

where there are $r$ terms in the numerator. Using the equation for $P(n, r)$ from Theorem 4.2, part (b), we obtain

$$C(n, r) = \frac{n!/(n - r)!}{r!}$$

which reduces to

$$C(n, r) = \frac{n!}{r!(n - r)!} \qquad \textbf{(c)}$$

after multiplying the numerator and denominator by $(n - r)!$.

*notes*

(i) Other symbols used in place of $C(n, r)$ are $_nC_r$, $C_{n,r}$, $C_r^n$, and $\binom{n}{r}$. The last symbol is often referred to as the "binomial coefficient" symbol.
(ii) Equation (c) of Theorem 4.3 is one of the most frequently used equations of combinatorial analysis.
(iii) $C(n, r)$ or $\binom{n}{r}$ are often read as "$n$ over $r$" or "$n$ choose $r$."
(iv) Recall that we don't care about ordering when we form subsets from a set of $n$ distinct elements. Thus, we see that $C(n, r)$ also is equal to the number of subsets containing $r$ elements which can be formed from a set which contains $n$ elements $[r \leq n]$. Such subsets are often referred to as $r$-subsets.

For example, let us form all subsets with two elements from a set of four elements, say $S = \{a, b, c, d\}$. The list of these six subsets is as follows:

$$\{a, b\}, \quad \{a, c\}, \quad \{a, d\}, \quad \{b, c\}, \quad \{b, d\}, \quad \{c, d\}$$

Observe that these subsets are called the 2-subsets for the set $\{a, b, c, d\}$. Observe the correspondence between this list and the list of combinations in Figure 4-3. The number of subsets is clearly equal to $C(4, 2) = 6$.

Remember that, when we are dealing with sets, the set $\{a, b\}$ is considered to be equivalent to the set $\{b, a\}$. That is, order does not matter for the elements that belong to a set.

**EXAMPLE 4-10:** (a) Compute the number of, and list the combinations of the digits 1, 2, 3, and 4 taken three at a time. (b) Determine the number of permutations that corresponds to each combination. List the permutations that correspond to a particular combination.

*Solution*

(a) From Eq. (b) of Theorem 4.3,

$$C(4, 3) = \frac{4 \cdot 3 \cdot 2}{3!} = 4$$

The four combinations are 123, 124, 134, and 234. [Here, order does not matter. Thus, for example, we can just as well represent combination 134 as, say, 314 or 413, etc.]

(b) Since each combination consists of three distinct letters, that means that each one gives rise to $P(3, 3) = 3! = 6$ permutations; here, we are counting all possible orders in a line of three distinct objects. For example, the combination 123 gives rise to the six permutations 123, 132, 213, 231, 312, and 321. [In this latter list, order does matter, and, in fact, is crucial.]

---

The following property is not hard to illustrate or prove.

**Theorem 4.4:** $C(n, r) = C(n, n - r)$.

---

**EXAMPLE 4-11:** (a) Prove Theorem 4.4. (b) Illustrate for the case where $n = 5$ and $r = 3$.

*Solution*

(a) From Eq. (c) of Theorem 4.3,

$$C(n, r) = \frac{n!}{r!(n - r)!} \tag{1}$$

and

$$C(n, n - r) = \frac{n!}{(n - r)!(n - [n - r])!} = \frac{n!}{(n - r)!r!} \tag{2}$$

Since the ordering of terms in a product is immaterial, the denominators of Eqs. (1) and (2) are equal. Thus,

$$C(n, r) = C(n, n - r) \tag{3}$$

as was to be shown.

(b) For $n = 5$ and $r = 3$, we have $(n - r) = 2$. Also,

$$C(5, 3) = C(5, 2) = \frac{5!}{3!2!} = \frac{5 \cdot 4}{2 \cdot 1} = 10 \tag{4}$$

Let the five original distinct objects be the digits 1, 2, 3, 4, and 5. Now, for each combination of three digits that is selected, such as 145, there corresponds a combination of two digits that are not selected [in this case, 23]. As another example, corresponding to the combination 234, we have the unselected combination 15.

In the list below, we write, corresponding to each combination of three digits, an unselected combination of two digits—refer to the bracketed items.

123[45]    124[35]    125[34]    134[25]    135[24]

145[23]    234[15]    235[14]    245[13]    345[12]

Thus, it follows that the number of ways of choosing three objects out of five is the same as the number of ways of choosing two objects out of five objects.

In effect, each time we choose three objects [in other words, each time we form a combination containing three objects], we are automatically "choosing" the two objects that were left behind.

---

If, in the equation of Theorem 4.4, we replace $r$ by 0, we obtain

$$C(n,0) = C(n,n) \tag{1}$$

Now, from Eq. (a) or (b) of Theorem 4.3, we have $C(n,n) = 1$. This also follows from common sense since $C(n,n)$ indicates the number of combinations of $n$ objects taken $n$ at a time. Since there is only one way to select $n$ objects from a collection of $n$ objects, this number clearly is 1. The term $C(n,0)$ above appears to have no physical meaning. For convenience, we define

$$C(n,0) = 1 \tag{2}$$

For one thing, this leads to consistency in Eq. (1) above [since we know $C(n,n)$ equals 1]. For another, this leads to the proper result for $C(n,0)$ as calculated from Eq. (c) of Theorem 4.3 for the case where $r = 0$. That is, from Eq. (c), we have

$$C(n,0) = \frac{n!}{0!\,n!} = \frac{1}{0!} = 1 \tag{3}$$

because $0! = 1$. Also, we define

$$C(0,0) = 1 \tag{4}$$

Let us extend the definition of $C(n,r)$ to apply for all integer values of $n$ and $r$, even negative integers. This will simplify matters later when we write various formulas that involve $C(n,r)$.

**Definition 4.2:** If $n$ is negative, or $r$ is negative, or $(n - r)$ is negative [equivalently, if $r > n$], then

$$C(n,r) = 0$$

In all other cases, $C(n,r)$ is given by

$$C(n,r) = \frac{n!}{r!(n-r)!}$$

which is Eq. (c) of Theorem 4.3.

---

**EXAMPLE 4-12:** Given the set $V = \{1, 2, 3, 4, 5\}$, which has five elements. Calculate the number of—and list all subsets of—$V$ which have three elements. [*Hint:* Recall note (iv) following Theorem 4.3.]

***Solution:*** The number of subsets with three elements is $C(5,3)$, which equals 10. The subsets are as follows:

| | | | | |
|---|---|---|---|---|
| $\{1,2,3\}$, | $\{1,2,4\}$, | $\{1,2,5\}$, | $\{1,3,4\}$, | $\{1,3,5\}$, |
| $\{1,4,5\}$, | $\{2,3,4\}$, | $\{2,3,5\}$, | $\{2,4,5\}$, | $\{3,4,5\}$ |

---

*note:* Compare this list of ten subsets with the list of combinations of Example 4-11, part (b). There, the ten combinations were 123, 124, 125, ..., 245, and 345. The correspondence between those combinations and the above subsets is clear.

**EXAMPLE 4-13:** Given a miniature card deck consisting of three spades [ace, two, and three] and two hearts [ace and two]. (a) How many ways can three cards be dealt from this "mini-deck" of five cards? (b) List these three-card "hands." Note that a hand containing three cards is equivalent to a combination of five things taken three at a time. That is, order doesn't matter.

*Solution*

(a) The number of hands is equal to $C(5,3) = 5!/(3!2!) = 10$, since the "mini-deck" has $n = 5$ cards, and we are drawing $r = 3$ cards.

(b) Let us label the five cards as $AS$, $2S$, $3S$, $AH$, and $2H$, where the first three cards are the ace, two, and three of spades, and the last two cards are the ace and two of hearts. Then we have the numbered list below for the ten hands with three cards. Each hand is, in fact, a combination; our policy here is to separate the symbols for the individual cards by commas.

*List:*   **(1)** $AS, 2S, 3S$    **(2)** $AS, 2S, AH$    **(3)** $AS, 2S, 2H$    **(4)** $AS, 3S, AH$
       **(5)** $AS, 3S, 2H$    **(6)** $AS, AH, 2H$    **(7)** $2S, 3S, AH$    **(8)** $2S, 3S, 2H$
       **(9)** $2S, AH, 2H$    **(10)** $3S, AH, 2H$

**EXAMPLE 4-14:** (a) For the miniature deck of five cards of Example 4-13, develop a systematic way to determine the number of three-card hands with two spades and one heart. (b) List these hands.

*Solution*

(a) Two spades can be chosen from the three available spades in $C(3,2) = 3$ ways, namely $AS$, $2S$; $AS$, $3S$; and $2S$, $3S$. [Here, we have the three combinations of the three spades taken two at a time.]

     One heart can be chosen from the two available hearts in $C(2,1) = 2$ ways, namely $AH$ and $2H$.

     Thus, from the fundamental counting principle, the number of three-card hands with two spades and one heart is six, from

$$C(3,2) \cdot C(2,1) = 3 \cdot 2 = 6$$

(b) A list of these six hands is presented in the tree diagram of Figure 4-4. There, the three first-level branches indicate the three ways to select two spades, and the two second-level branches [for each first-level branch] indicate the two ways to select one heart.

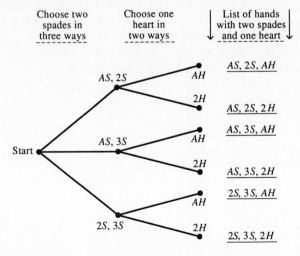

**Figure 4-4.** Tree for determining hands containing two spades and one heart. [Hands are underlined.]

The six hands are listed [and underlined] to the right of the six end-points. [Observe that a hand corresponding to a particular endpoint can be read in reverse order by going backwards from the endpoint to the starting point.]

The above six hands are labeled as hands (2), (3), (4), (5), (7), and (8) in the answer to Example 4-13.

---

**EXAMPLE 4-15:** Suppose three cards are drawn without replacement from a mini-deck of cards with three spades [ace, two, and three] and two hearts [ace and two]. What is the probability of obtaining two spades and one heart?

*Solution:* This problem ties in with Examples 4-13 and 4-14. In those, we had drawing without replacement even though we didn't explicitly say that in the wording of those Examples. That is, after drawing a hand of three cards, we ended up with two cards remaining in the original deck.

We assume that the $C(5, 3) = 10$ ways of drawing three cards from the five-card deck are all equally likely. [This, by the way, is what is meant by saying that the drawing of objects, without replacement, from a collection of objects is random.] Equivalently, the ten hands listed in Example 4-13 constitute an equally likely sample space for the experiment of drawing three cards, without replacement, from the deck.

Let us symbolize the event "two spades and one heart" by the letter $E$. Then, from Theorem 1.2, we have

$$P(E) = \frac{N(E)}{N(S)} \tag{1}$$

But, $N(S) = C(5, 3) = 10$ [the total number of hands listed in Example 4-13], and $N(E) = C(3, 2) \cdot C(2, 1) = 3 \cdot 2 = 6$, where the six favorable hands are listed in Figure 4-4 of Example 4-14. Thus,

$$P(E) = \frac{C(3, 2) \cdot C(2, 1)}{C(5, 3)} \quad \text{or} \quad P(E) = \frac{6}{10} \tag{2a, 2b}$$

---

*notes*

(a) Example 4-15 is an example of a hypergeometric probability problem. Typically, in a hypergeometric probability problem, we have drawing without replacement from a collection of objects, where the question deals with calculating the probability of obtaining definite numbers of particular types of items. In Example 4-15, the probability question dealt with obtaining exactly two spades and one heart in a draw of three cards without replacement from a deck of five cards.

The probability equation in Example 4-15 involving symbols for numbers of combinations, namely the equation

$$P(E) = \frac{C(3, 2) \cdot C(2, 1)}{C(5, 3)} \tag{I}$$

has a form similar to that which will occur in the solutions of other hypergeometric probability problems. The approach used in Examples 4-13, 4-14, and 4-15 will be referred to as a *batch approach* by us.

(b) A different approach to a similar problem was considered in part (d) of Example 3-17. There, we had a drawing without replacement of two cards from the same five-card deck considered in Examples 4-13, 4-14, and 4-15 above, and we asked for the probability of getting spade on the first card and heart on the second card. The approach there, in which we envisioned cards being drawn out one by one, will be called a *sequential approach* by us. Both batch and sequential approaches to the problem of calculating the probability of getting exactly one spade in a draw of two cards without replacement will be used in Problem 4-5 at the end of this chapter.

## 4-2. Permutations of *n* Objects, Where Some Are Alike

In dealing with permutations up to this point, we indicated that all the *n* objects were different [or distinct]. Sometimes, though, in practical situations such as probability applications, some of the *n* objects are like each other. For example, we might be concerned with the permutations of the four letters *a*, *a*, *b*, *b* instead of the four distinct letters *a*, *b*, *c*, *d*.

### A. The general equation for the number of permutations of *n* objects, where some are alike

Suppose, for purposes of generality, that the total of *n* objects contain *k* distinct kinds of objects, where there are $n_1$ of the first kind, $n_2$ of the second kind, and so on, and, finally, $n_k$ of the *k*th kind. Note that

$$n_1 + n_2 + n_3 + \ldots + n_k = n$$

We will always, unless otherwise noted, confine our attention to the total number of permutations of *n* such objects taken all [or *n*] at a time. Our main symbol for this will be $P(n; n_1, n_2, \ldots, n_k)$, although another frequently used symbol is

$$\binom{n}{n_1, n_2, \ldots, n_k}$$

The following theorem will be demonstrated for several special cases in this section.

**Theorem 4.5:** The equation for $P(n; n_1, n_2, \ldots, n_k)$ is

$$P(n; n_1, n_2, \ldots, n_k) = \frac{n!}{n_1! \cdot n_2! \cdot \ldots \cdot n_k!} \tag{1}$$

Here,

$$n_1 + n_2 + \ldots + n_k = n \tag{2}$$

---

**EXAMPLE 4-16:** Compute the total number of permutations of the four letters *a*, *a*, *b*, *b* taken all at a time. List the permutations.

*Solution:* We observe that $n = 4$, $n_1 = 2$ [since there are two *a*'s], and $n_2 = 2$ [since there are two *b*'s]. Thus, from Theorem 4.5, we have

$$\text{Number of permutations} = P(4; 2, 2) = \frac{4!}{2!2!} = \frac{4 \cdot 3}{2 \cdot 1} = 6$$

The six permutations are *aabb*, *abab*, *abba*, *bbaa*, *baba*, and *baab*. Note that if the four letters were all different from each other, then there would be $P(4, 4) = 4! = 24$ permutations.

---

**EXAMPLE 4-17:** Compute the total number of permutations of the five letters *S*, *S*, *S*, *F*, *F* taken all at a time. List the permutations.

*Solution:* Here, $n = 5$, $n_1 = 3$, and $n_2 = 2$ since there are three *S*'s and two *F*'s. Thus, from Theorem 4.5, we have

$$\text{Number of permutations} = P(5;3,2) = \frac{5!}{3!\,2!} = \frac{5 \cdot 4 \cdot 3!}{3! \cdot 2 \cdot 1} = 10$$

The list is as follows: *SSSFF, SSFSF, SFSSF, FSSSF, SSFFS, SFSFS, FSSFS, SFFSS, FSFSS, FFSSS.*

---

*note:* The letters *S* and *F* will subsequently often stand for "success" and "failure." In particular, this will be the case when we discuss the binomial distribution.

---

**EXAMPLE 4-18:** Use Example 4-17 as a basis for proving Theorem 4.5 for the special case where $n_1 = 3$ and $n_2 = 2$.

*Solution:* Here, we consider a total of five objects $[n = 5]$, where the objects are the five letters of Example 4-17. That is, there are three *S*'s and two *F*'s. Thus, we can say that $n_1 = 3$ and $n_2 = 2$.

Suppose we consider a particular permutation of the letters *S, S, S, F, F* taken all at a time, say the permutation *SSFSF*. If we replace the three identical *S*'s by three symbols different from each other and different from the *F*'s, such as $S_1$, $S_2$, and $S_3$, then the original particular permutation would give rise to 3! [or 6] new permutations. [Several of these would be $S_1 S_2 F S_3 F$; $S_1 S_3 F S_2 F$; $S_2 S_1 F S_3 F$.] Likewise, if the two *F*'s were replaced by two new symbols, different from each other [and from $S_1$, $S_2$, and $S_3$], such as $F_1$ and $F_2$, each of the six aforementioned new permutations would give rise to 2! [or 2] yet newer permutations. Now, by the fundamental counting principle, each original permutation gives rise to $3! \cdot 2! = 12$ permutations of five distinct symbols taken five [or all] at a time.

[In Figure 4-5, we see how the original permutation *SSFSF* gives rise to 12 new permutations. Among these are $S_1 S_2 F_1 S_3 F_2$; $S_1 S_2 F_2 S_3 F_1$; $S_1 S_3 F_1 S_2 F_2$; and $S_1 S_3 F_2 S_2 F_1$.]

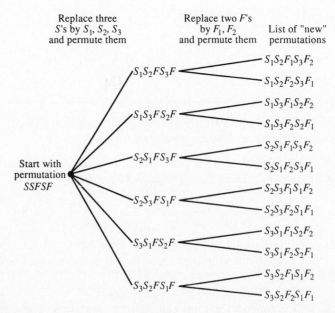

**Figure 4-5.** Counting tree for 12 permutations generated by *SSFSF*. [See Example 4-18.]

But, then the total number of new permutations would equal the number of permutations of five distinct symbols taken five at a time, that is, $P(5, 5)$. The value for the latter number is 5!. Thus, we have the equation

$$P(5; 3, 2) \cdot 3! \cdot 2! = 5! \tag{1}$$

Solving for $P(5; 3, 2)$, we obtain

$$P(5; 3, 2) = \frac{5!}{3!\,2!} \tag{2}$$

More generally, suppose there are $n$ objects of which $n_1$ are alike, another $n_2$ are alike, another $n_3$ are alike, ..., and finally the remaining $n_k$ are alike. [Naturally, $n = n_1 + n_2 + n_3 + \ldots + n_k$.] Then, we can find the number of permutations of these $n$ objects by using a similar argument to the one used in Example 4-18. Let $x$ tentatively denote the number of such permutations. Then, by the fundamental counting principle [or by an equivalent counting tree argument], we would have

$$x \cdot n_1! \cdot n_2! \cdot n_3! \cdot \ldots \cdot n_k! = n!$$

Observe that the $n!$ on the right side is the value of $P(n, n)$. Then, solving for $x$ and replacing it by the symbol $P(n; n_1, n_2, n_3, \ldots, n_k)$, we would have

$$P(n; n_1, n_2, \ldots, n_k) = \frac{n!}{n_1! \cdot n_2! \cdot \ldots \cdot n_k!}$$

This is the Theorem 4.5 equation.

**EXAMPLE 4-19:** How many permutations can be formed by using all the letters of the following words? List three of the permutations for each word: (**a**) kooks, (**b**) aardvark, (**c**) count.

*Solution*

(**a**) First of all, we see that

$$\text{Number} = P(5; 2, 2, 1) = \frac{5!}{2!\,2!\,1!} = \frac{5 \cdot 4 \cdot 3}{2 \cdot 1 \cdot 1} = 30$$

since there are a total of five letters, of which two are $k$, two are $o$, and one is an $s$. Three such permutations are *kooks, kkoos, sokok*.

(**b**) Here,

$$\text{Number} = P(8; 3, 2, 1, 1, 1) = \frac{8!}{3!\,2!\,1!\,1!\,1!} = 3360$$

since there are a total of eight letters, of which three are $a$, two are $r$, and one each are $d$, $v$, and $k$. Three such permutations are *aardvark, radavark, radarkav*.

(**c**) Here, all five letters are different. Using Eq. (a) of Theorem 4.2, we have

$$\text{Number} = P(5, 5) = 5! = 120$$

We could just as well use the equation of Theorem 4.5. Thus,

$$\text{Number} = P(5; 1, 1, 1, 1, 1) = \frac{5!}{1!\,1!\,1!\,1!\,1!} = 5! = 120$$

Three such permutations are *count, cnout, onuct*.

*note:* In part (c) of Example 4-19, we see that the equation from Theorem 4.5 reduces to the equation from Theorem 4.2, part (a) for the situation when all $n$ objects are distinct. That is,

$$P(n; 1, 1, \ldots, 1) = \frac{n!}{1!1!\ldots 1!} = n! = P(n, n)$$

## B. The observation that $P(n; r, n-r) = C(n, r)$

From Theorem 4.5, we have

$$P(n; n_1, n_2) = \frac{n!}{n_1! n_2!} \qquad (1)$$

if the total of $n$ objects consists of $n_1$ of one kind, and $n_2$ of another kind. Here, $n_1 + n_2 = n$, and thus $n_2 = n - n_1$.

If we replace $n_1$ by $r$, and $n_2$ by $(n - n_1)$, or $(n - r)$, we obtain

$$P(n; r, n-r) = \frac{n!}{r!(n-r)!} \qquad (2)$$

Now, from Eq. (c) of Theorem 4.3, we have

$$C(n, r) = \frac{n!}{r!(n-r)!} \qquad (3)$$

Thus, from Eqs. (2) and (3) above, we have the following theorem.

**Theorem 4.6:** The number of combinations of $n$ distinct objects taken $r$ at a time is equal to the number of permutations of $n$ objects taken all at a time, where the $n$ objects consist of $r$ objects of one kind, and $(n - r)$ objects of another kind. That is, $C(n, r) = P(n; r, n - r)$, where either quantity is equal to $n!/[r!(n-r)!]$.

---

**EXAMPLE 4-20:** For a physical demonstration as to why Theorem 4.6 is valid, consider the special case where $n = 5$ and $r = 3$, and hence, $(n - r) = 2$. Thus,

$$P(5; 3, 2) = \frac{5!}{3!2!} = 10$$

We have a listing of the ten permutations for this case in Example 4-17. There, $n_1 = r = 3$ and $n_2 = (n - r) = 2$ since there were a total of five letters of which three were $S$'s and two were $F$'s.

Now, suppose we allow the positions of the letters in these permutations to be labeled 1, 2, 3, 4, and 5, going from left to right. Show that there is a one-to-one correspondence between these ten permutations and the ten combinations of the five digits 1, 2, 3, 4, 5, taken three at a time.

*Solution:* First of all, note that $C(5, 3) = 5!/(3!2!) = 10$. Refer to Figure 4-6. The collection of position labels of the $S$'s in each permutation constitutes a combination of five distinct things taken three at a time. For example, in $SFSSF$, which is the third permutation in the list of Figure 4-6, the position labels for the $S$'s are 1, 3, and 4. Thus, the combination 134 corresponds to the single permutation $SFSSF$. Conversely, each combination of five things taken three at a time gives rise to a single permutation. For example, combination 245 gives rise to permutation $FSFSS$, in which the $S$'s are in positions 2, 4, and 5.

Thus, the above discussion and Figure 4-6 illustrate that there is a one-to-one correspondence between the permutations of the five letters $S$, $S$, $S$, $F$, $F$ taken all at a time, and the combinations of five distinct things taken three at a time. It follows that the numbers of such permutations and combinations are

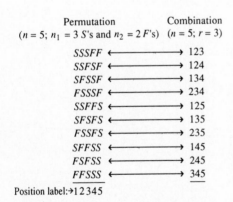

| Permutation ($n = 5$; $n_1 = 3$ $S$'s and $n_2 = 2$ $F$'s) | Combination ($n = 5$; $r = 3$) |
|---|---|
| $SSSFF$ ⟷ | 123 |
| $SSFSF$ ⟷ | 124 |
| $SFSSF$ ⟷ | 134 |
| $FSSSF$ ⟷ | 234 |
| $SSFFS$ ⟷ | 125 |
| $SFSFS$ ⟷ | 135 |
| $FSSFS$ ⟷ | 235 |
| $SFFSS$ ⟷ | 145 |
| $FSFSS$ ⟷ | 245 |
| $FFSSS$ ⟷ | 345 |

Position label: →1 2 3 4 5

**Figure 4-6.** Correspondence between permutations and combinations in Example 4-20.

identical. Thus, for the particular case discussed above, we have

$$P(5; 3, 2) = C(5, 3)$$

where each term is equal to 10. In general, we have

$$P(n; r, n - r) = C(n, r)$$

where in the first symbol $n_1 = r$ and $n_2 = n - r$.

## 4-3. The Binomial Theorem and Binomial Coefficients

### A. Introduction to the binomial theorem

A sum of two unlike symbols, such as $x + y$, is called a *binomial*. The binomial theorem, or binomial expansion, is an equation for the powers of a binomial. For example, consider the third power of $x + y$:

$$(x + y)^3 = (x + y)(x + y)(x + y)$$

The product of the three identical binomials on the right gives rise to eight terms, where each term is a product of three symbols selected, respectively, from the three binomials on the right. [From the fundamental counting principle, there will be $2 \cdot 2 \cdot 2 = 2^3 = 8$ terms.] For example, consider multiplication of the particular $x$'s and $y$ identified below by arrows:

$$\downarrow \qquad \downarrow \downarrow$$
$$(x + y)(x + y)(x + y)$$

That gives rise to the term $xyx$.

---

**EXAMPLE 4-21:** (a) Express the power $(x + y)^3$ as a sum of eight terms. (b) Collect like terms, and develop simplified expressions for $(x + y)^3$.

*Solution*

(a)
$$(x + y)^3 = (x + y)(x + y)(x + y) = xxx + xxy + xyx + xyy$$
$$+ yxx + yxy + yyx + yyy \qquad (1)$$

(b) After collecting like terms [i.e., terms that have equal value as products], Eq. (1) reduces to

$$(x + y)^3 = x^3 + 3x^2y + 3xy^2 + y^3 \qquad (2)$$

Observe that the three terms of Eq. (1) that gave rise to $3x^2y$ in Eq. (2) are $xxy$, $xyx$, and $yxx$. These are none other than the permutations of the symbols $x, x, y$ taken all at a time. Thus, it is no surprise that the coefficient of Eq. (2) is 3. Recall that

$$P(3; 2, 1) = \frac{3!}{2! \, 1!} = 3 \qquad (3)$$

Of course, from Theorem 4.6 in Section 4-2B, we see that we can write

$$P(3; 2, 1) = C(3, 2) \qquad (4)$$

We can also write

$$P(3; 2, 1) = C(3, 1) \qquad (5)$$

if we make use of $C(n, r) = C(n, n - r)$. (See Theorem 4.4.) Thus, we can rewrite Eq. (2) as

$$(x + y)^3 = x^3 + C(3, 1)x^2y + C(3, 2)xy^2 + y^3 \qquad (6)$$

*notes*

(a) The right side of Eq. (2) in Example 4-21 is called an *expansion* of the power $(x + y)^3$. Similarly, we shall be interested in expansions for other powers of binomials, such as the power $(x + y)^4$, $(x + y)^{10}$, etc.

(b) It is customary to employ combination symbols when expressing binomial expansions, as in Eq. (6) above. Recall that another symbol that is frequently used in place of $C(n, r)$ is $\binom{n}{r}$. The latter is known as the *binomial coefficient symbol*.

---

**EXAMPLE 4-22:** (a) Explain why the coefficient of $x^4 y^2$ in the expansion of the power $(x + y)^6$ is $C(6, 2)$. (b) Determine the coefficient of $x y^5$ in the expansion of $(x + y)^6$.

*Solution*

(a) The coefficient of $x^4 y^2$ is the number of ways of writing four $x$'s and two $y$'s in a row. Equivalently, it is the number of permutations of the six symbols $x, x, x, x, y$, and $y$ taken all at a time. [Three of these permutations are $xxxxyy$, $xxyxyx$, and $yxxyxx$.] Thus, the coefficient of $x^4 y^2$ is $P(6; 4, 2)$, which equals $C(6, 4)$, which, in turn, equals $C(6, 2)$. Each is equal to

$$\frac{6!}{4!\,2!} = \frac{6 \cdot 5}{2 \cdot 1} = 15$$

(b) Following the same reasoning as in part (a), the coefficient of $x y^5$ is $C(6, 5)$, which equals $6!/(5!\,1!) = 6$. Note that the $r$ in the $C(n, r)$ symbol is equal to the exponent of the power of $y$ [here, it is 5, and in part (a), it was 2].

---

## B. The binomial theorem stated and proved

**Theorem 4.7 (The Binomial Theorem):** The expression $(x + y)^n$, where $n$ is a positive integer, can be written as

$$(x + y)^n = x^n + C(n, 1)x^{n-1}y + C(n, 2)x^{n-2}y^2 + \ldots + C(n, r)x^{n-r}y^r + \ldots + y^n$$

*notes:* There are $(n + 1)$ terms on the right. A way to use the equation for a definite value of $n$ is to initially write the first $n$ terms, starting from the left; a typical term—that is, the $r$th term—has the form $C(n, r)x^{n-r}y^r$. Finally, the term $y^n$ is added on.

---

**EXAMPLE 4-23:** Expand $(x + y)^6$ by using the binomial theorem.

*Solution:* Since $n = 6$, we will have seven terms in the expansion. From Theorem 4.7, we have:

$$(x + y)^6 = x^6 + C(6, 1)x^5 y + C(6, 2)x^4 y^2 + C(6, 3)x^3 y^3$$
$$+ C(6, 4)x^2 y^4 + C(6, 5)x y^5 + y^6 \tag{1}$$

Calculating values for the $C(6, r)$ symbols, we obtain:

$$(x + y)^6 = x^6 + 6x^5 y + 15x^4 y^2 + 20x^3 y^3 + 15x^2 y^4 + 6x y^5 + y^6 \tag{2}$$

Observe that the coefficients of $x^5 y$ and $x y^5$ are the same, and likewise for the coefficients of $x^4 y^2$ and $x^2 y^4$. This is so because $C(n, r) = C(n, n - r)$. (See Theorem 4.4.)

**EXAMPLE 4-24:** Prove Theorem 4.7 [the binomial theorem].

***Proof:*** For the case where $n$ is any positive integer, the power $(x + y)^n$ is given by

$$(x + y)^n = (x + y)(x + y)(x + y)\cdot\ldots\cdot(x + y) \qquad \textbf{(1)}$$

where there are $n$ factors in the product on the right. In the expansion resulting from multiplying the factors, a product of the form $x^{(n-r)}y^r$ arises in the same number of ways as a group of $(n - r)$ $x$'s and a group of $r$ $y$'s can be written in a line. But, this number is equal to the number of permutations of $n$ things taken all at a time, where $(n - r)$ of them are $x$'s and $r$ of them are $y$'s. That is, the number is equal to $P(n; n - r, r)$, which in turn is equal to $C(n, r)$.

Also, a product of the form $x^n$ can arise in exactly one way, namely from using the $x$ in each of the $n$ factors on the right side of Eq. (1). Similarly, a product of the form $y^n$ can also arise in exactly one way. $\qquad \square$

It is appropriate to use the sigma notation ($\Sigma$) to express summation in the binomial theorem. The symbol $\Sigma$ is the capital Greek letter "sigma," which corresponds to the English "$S$," and stands for the word *sum*. For example,

$$\sum_{r=1}^{3} 4(r)^2$$

is read as the "the sum of $4(r)^2$ as $r$ runs from 1 to 3, inclusive." It is equal to

$$4(1)^2 + 4(2)^2 + 4(3)^2 = 4 + 16 + 36 = 56$$

Referring to the binomial theorem, and noting that $C(n, 0) = C(n, n) = 1$, we see that the $x^n$ and $y^n$ terms may be written as $C(n, 0)x^n$ and $C(n, n)y^n$, respectively. [See comments following Example 4-11.] If we note further that $y^0 = x^0 = 1$, we observe that every term in the binomial expansion can be put in the form $C(n, r)x^{n-r}y^r$. Thus, we have the following compact form for the binomial theorem.

**Theorem 4.7′ (The Binomial Theorem—Sigma Form):** The expansion $(x + y)^n$, where $n$ is a positive integer, can be written as

$$(x + y)^n = \sum_{r=0}^{n} C(n, r)x^{n-r}y^r \qquad \textbf{(a)}$$

Recall that another symbol for $C(n, r)$ is $\binom{n}{r}$, where the latter is often referred to as the binomial coefficient symbol. Thus, many authors rewrite the binomial theorem as

$$(x + y)^n = \sum_{r=0}^{n} \binom{n}{r}x^{n-r}y^r \qquad \textbf{(b)}$$

**EXAMPLE 4-25:** Use Theorem 4.7′ to expand $(x + y)^4$.

***Solution:*** From (b) of Theorem 4.7′, we have

$$(x + y)^4 = \sum_{r=0}^{4} \binom{4}{r}x^{4-r}y^r = \binom{4}{0}x^4y^0 + \binom{4}{1}x^3y^1 + \binom{4}{2}x^2y^2 + \binom{4}{3}x^1y^3 + \binom{4}{4}x^0y^4$$

$$= x^4 + 4x^3y + 6x^2y^2 + 4xy^3 + y^4$$

Here, observe that

$$\binom{4}{2} = \frac{4!}{2!2!} = \frac{4\cdot 3}{2\cdot 2} = 6 \qquad \text{and} \qquad \binom{4}{1} = \binom{4}{3} = 4$$

## C. Useful relations pertaining to combinations and the binomial theorem

**Theorem 4.8 (Pascal's Formula):** For any positive integer $n$ and for $r = 1, 2, \ldots, n - 1$,

$$C(n, r) = C(n - 1, r) + C(n - 1, r - 1)$$

*notes*

(a) Theorem 4.8 will be proved in Problem 4-9.
(b) Theorem 4.8 provides the basis for the construction of what is known as *Pascal's triangle*.

---

**EXAMPLE 4-26:** Verify Theorem 4.8 for the case where $n = 8$ and $r = 5$.

*Solution:* Observe that

$$C(8, 5) = \frac{8!}{5!3!} = \frac{8 \cdot 7 \cdot 6}{3 \cdot 2 \cdot 1} = 56 \tag{1}$$

$$C(7, 5) = \frac{7!}{5!2!} = \frac{7 \cdot 6}{2 \cdot 1} = 21 \tag{2}$$

$$C(7, 4) = \frac{7!}{4!3!} = \frac{7 \cdot 6 \cdot 5}{3 \cdot 2 \cdot 1} = 35 \tag{3}$$

We see that $C(8, 5) = C(7, 5) + C(7, 4)$ as predicted by Theorem 4.8 since both sides of this equation are equal to 56.

---

In applications of the next theorem, we will encounter terms of the form $C(n, r)$ with $n$ and $r$ positive, but with $r > n$. It follows from Definition 4.2 that

$$C(n, r) = 0 \quad \text{if } n > 0, r > 0, \quad \text{but } r > n$$

For example, $C(5, 7) = 0$.

**Theorem 4.9:** If $m$, $n$, and $k$ are positive integers, then

$$\sum_{r=0}^{k} C(m, r) \cdot C(n, k - r) = C(m + n, k)$$

A proof of Theorem 4.9 is given on page 16 of Freund and Walpole (1987).*

---

**EXAMPLE 4-27:** Verify Theorem 4.9 for the case where $m = 3$, $n = 4$, and $k = 5$.

*Solution:* Substituting these values into the equation of Theorem 4.9, we obtain:

$$C(3, 0) \cdot C(4, 5) + C(3, 1) \cdot C(4, 4) + C(3, 2) \cdot C(4, 3) + C(3, 3) \cdot C(4, 2)$$

$$+ C(3, 4) \cdot C(4, 1) + C(3, 5) \cdot C(4, 0) = C(7, 5) \tag{1}$$

Since $C(4, 5) = C(3, 4) = C(3, 5) = 0$, Eq. (1) reduces to

$$C(3, 1) \cdot C(4, 4) + C(3, 2) \cdot C(4, 3) + C(3, 3) \cdot C(4, 2) = C(7, 5) \tag{2}$$

---

* References are listed at the back of the book in the References section. Our method of referring to references is by listing the name(s) of the author(s), followed by the date of publication in parentheses.

Evaluating the various terms, we get

$$3 \cdot 1 + 3 \cdot 4 + 1 \cdot 6 = 21$$

which checks out.

---

**Theorem 4.10:** $C(n,0) + C(n,1) + C(n,2) + \ldots + C(n,n) = 2^n$. In sigma form, this becomes:

$$\sum_{r=0}^{n} C(n,r) = 2^n$$

---

**EXAMPLE 4-28:** Prove Theorem 4.10.

*Proof:* In Theorem 4.7′, let $x = y = 1$. The left side, which is $(x + y)^n$, becomes $(1 + 1)^n$ or $2^n$. On the right side, $x^{n-r}y^r$ becomes $1^{n-r}1^r$ or 1. Thus, we end up with

$$2^n = \sum_{r=0}^{n} C(n,r)$$

This is what we set out to show.  □

---

**Corollary to Theorem 4.10:** The number of subsets of a set with $n$ elements is $2^n$. The number of proper subsets is $2^n - 1$.

The Corollary is proved in Problem 4-10.

# SOLVED PROBLEMS

## More Counting Methods

**PROBLEM 4-1** (a) Determine how many distinct three-letter "words" can be formed by using the seven letters $d, e, f, g, h, i, j$, if no letter is used more than once in a word. By a three-letter "word" is meant any arrangement of three letters in a sequence, i.e., in a line. (b) Repeat part (a), but with the restriction that the first letter can't be a vowel.

*Solution*

(a) The answer is $P(7,3) = 7 \cdot 6 \cdot 5 = 210$ words.
(b) The two vowels in the seven letters are $e$ and $i$. Thus, the first letter in the word must be chosen from among five letters. Using the fundamental counting principle, and supposing that the letters are placed from left to right, we see that the answer is $5 \cdot 6 \cdot 5 = 150$ words.

**PROBLEM 4-2** (a) Compute the number of, and list the combinations of the five digits 3, 4, 5, 6, 7 taken three at a time. (b) Compute the number of permutations of the above five digits taken three at a time. (c) Determine how many permutations of part (b) correspond to each combination of part (a). (d) List the permutations that correspond to the particular combination 356.

*Solution*

(a) Number $= C(5,3) = 5!/(3!2!) = 10$. *Partial List:* 345, 357, 456, 567.
(b) Number $= P(5,3) = 5 \cdot 4 \cdot 3 = 60$.
(c) Since each combination contains three digits, then there are $P(3,3) = 3! = 6$ permutations corresponding to each combination.

(d) The six permutations that correspond to the combination 356 are 356, 365, 536, 563, 635, and 653.

**PROBLEM 4-3**   Given a collection of seven people, in which three are female and four are male. (a) How many committees with three people can be formed using people from the collection of seven people? (b) How many committees with one female and two males can be formed? (c) If a committee of three people is picked at random [say, by drawing lots], what is the probability that the committee contains one female and two males. (*Hint:* A committee is equivalent to a combination since order doesn't matter in a committee.)

*Solution*

(a) Number $= C(7, 3) = 7!/(3!4!) = (7 \cdot 6 \cdot 5)/(3 \cdot 2 \cdot 1) = 35$ committees.

(b) One female can be chosen from the three available females in $C(3, 1)$ ways, and two males can be chosen from the four available males in $C(4, 2)$ ways. Thus, using the fundamental counting principle, the number of ways to choose a committee consisting simultaneously of one female and two males is $n_1 \cdot n_2 = C(3, 1) \cdot C(4, 2) = 3 \cdot 6 = 18$ ways.

As an example, suppose the woman $W1$ has been chosen for a committee. Labeling the men as $M1$, $M2$, $M3$, and $M4$, we see that there are $C(4, 2) = 6$ ways that the two males can be chosen to be on the committee with $W1$. These six ways are: $M1$, $M2$; $M1$, $M3$; $M1$, $M4$; $M2$, $M3$; $M2$, $M4$; and $M3$, $M4$.

Likewise, six committees can be chosen containing $W2$ and two men, and also for $W3$ and two men. Adding, we obtain a total of $6 + 6 + 6 = 3 \cdot 6 = 18$ committees.

(c) Let the letter $E$ denote the event of forming a committee with one female and two males. Also, let $\mathscr{S}$ denote the sample space of all three-person committees [combinations] that can be formed from seven people. Thus, $N(\mathscr{S}) = C(7, 3) = 35$, and it is reasonable to assume that all 35 combinations are equally likely.

Now, the number of favorable combinations, $N(E)$, which is the number of those combinations containing one female and two males, is given by $N(E) = C(3, 1) \cdot C(4, 2) = 3 \cdot 6 = 18$. [Refer to part (b).] Thus, the desired probability, $P(E)$, is given by

$$P(E) = \frac{N(E)}{N(\mathscr{S})} = \frac{C(3, 1) \cdot C(4, 2)}{C(7, 3)} = \frac{3 \cdot 6}{35} = \frac{18}{35}$$

**PROBLEM 4-4**   Calculate $C(15, 12)$ from (a) Theorem 4.3, part (b) and (b) Theorem 4.3, part (c).

*Solution*

(a) From part (b) of Theorem 4.3, we have

$$C(15, 12) = \frac{15 \cdot 14 \cdot 13 \cdot \cancel{12} \cdot \cancel{11} \cdot \cancel{10} \cdot \cancel{9} \cdot \cancel{8} \cdot \cancel{7} \cdot \cancel{6} \cdot \cancel{5} \cdot \cancel{4}}{\cancel{12} \cdot \cancel{11} \cdot \cancel{10} \cdot \cancel{9} \cdot \cancel{8} \cdot \cancel{7} \cdot \cancel{6} \cdot \cancel{5} \cdot \cancel{4} \cdot 3 \cdot 2 \cdot 1} = 455$$

(b) From part (c) of Theorem 4.3, we have

$$C(15, 12) = \frac{15!}{12!3!} = \frac{15 \cdot 14 \cdot 13 \cdot \cancel{12}!}{\cancel{12}! \cdot 3 \cdot 2 \cdot 1} = 455$$

*notes*

(i) It appears that part (c) of Theorem 4.3 is more efficient to use. The best way to employ the expression of part (c) is to first write out the numerator factorial as a product of factors until the higher factorial of the denominator is encountered [above, that factorial is 12!]. Next, this higher factorial should be canceled, leaving a quotient involving the same number of factors in both numerator and denominator. This is $(15 \cdot 14 \cdot 13)/(3 \cdot 2 \cdot 1)$ in the above calculation.

(ii) We could have arrived at the same quotient in part (a) above if we had used Theorem 4.4 first. Thus, $C(15, 12) = C(15, 3)$, where the latter can be expressed as $(15 \cdot 14 \cdot 13)/(3 \cdot 2 \cdot 1)$, as we see by applying part (b) of Theorem 4.3.

**PROBLEM 4-5**   Refer to part (d) of Example 3-17 where two cards were drawn without replacement from a deck of cards containing three spades [Ace, Two, and Three] and two hearts

[Ace and Two]. (**a**) Compute the probability of getting one spade and one heart, with no regard to order, by using the method employed in Examples 4-13, 4-14, and 4-15. [The event here is also known as the event "exactly one spade" in a draw of two cards without replacement.] (**b**) Use a method based on the approach used in part (d) of Example 3-17.

*Solution*

(**a**) We visualize the two cards as being drawn out in a batch. That is, we do not consider one card as being drawn out after the other. Thus, we think of the two cards being drawn out simultaneously. Such a batch is equivalent to a combination [since in a combination order does not matter]. Thus, the number of ways two cards can be drawn [in a batch] from the deck of cards is

$$C(5, 2) = 10 \tag{1}$$

Here, the number of ways, or the number of hands, is the number of combinations of five things taken two at a time. It is reasonable to assume all 10 ways are equally likely. We can consider these ten combinations, among which are *AS-2S*, *AS-3S*, and *2S-2H*, to be the elements of a sample space $\mathscr{S}$ for the experiment. Thus, here, we have

$$N(\mathscr{S}) = 10 \tag{2}$$

for the number of elements in the sample space. Now, by the fundamental counting principle, the event $E$ of obtaining one spade and one heart can occur in $C(3, 1) \cdot C(2, 1) = 3 \cdot 2 = 6$ ways. Here, $C(3, 1)$ is the number of ways to draw one spade from the available three spades, and $C(2, 1)$ is the number of ways to draw one heart from the available two hearts. These six favorable combinations [or hands] are as follows:

$$AS\text{-}AH \qquad AS\text{-}2H \qquad 2S\text{-}AH \qquad 2S\text{-}2H \qquad 3S\text{-}AH \qquad 3S\text{-}2H$$

Thus, for $P(E)$, the probability of event $E$, we have:

$$P(E) = \frac{N(E)}{N(\mathscr{S})} = \frac{C(3, 1) \cdot C(2, 1)}{C(5, 2)} = \frac{3 \cdot 2}{10} = \frac{6}{10} \tag{3}$$

(**b**) Here, we envision the cards as being drawn out sequentially, that is, one after the other. Now, let us again indicate the event of getting one spade and one heart, with no regard to order, by the letter $E$. Now,

$$P(E) = P[(S_1 \text{ and } H_2) \text{ or } (H_1 \text{ and } S_2)] \tag{1}$$

Here, $(S_1 \text{ and } H_2)$ indicates the event "spade on the first card, followed by heart on the second card," and $(H_1 \text{ and } S_2)$ indicates "heart on the first card, followed by spade on the second card." Now, the events $(S_1 \text{ and } H_2)$ and $(H_1 \text{ and } S_2)$ are mutually exclusive—that is, these two events cannot happen simultaneously. [Note that in part (**d**) of Example 3-17, we used the intersection symbol $\cap$ in place of the word "and."] Thus, from Eq. (1), we obtain:

$$P(E) = P(S_1 \text{ and } H_2) + P(H_1 \text{ and } S_2) \tag{2}$$

Now, we have

$$P(S_1 \text{ and } H_2) = \frac{N(S_1 \text{ and } H_2)}{N(\hat{\mathscr{S}})} \qquad \text{and} \qquad P(H_1 \text{ and } S_2) = \frac{N(H_1 \text{ and } S_2)}{N(\hat{\mathscr{S}})} \tag{3a}, \text{(3b)}$$

after employing Theorem 1.2. Here, $N(\hat{\mathscr{S}})$ denotes the total number of ways to draw two cards, one after the other [in other words, sequentially] from the deck of cards. Also, $N(S_1 \text{ and } H_2)$ denotes the number of ways to draw a spade on the first card followed by a heart on the second card, and similarly for $N(H_1 \text{ and } S_2)$. Thus, from the fundamental counting principle, we have

$$N(\hat{\mathscr{S}}) = 5 \cdot 4 = 20, \qquad N(S_1 \text{ and } H_2) = 3 \cdot 2 = 6 \tag{4a}, \text{(4b)}$$

and

$$N(H_1 \text{ and } S_2) = 2 \cdot 3 = 6 \tag{4c}$$

Substituting from Eqs. (4a), (4b), and (4c) into Eqs. (3a) and (3b), and then from those into Eq. (2), results in

$$P(E) = \frac{3 \cdot 2}{5 \cdot 4} + \frac{2 \cdot 3}{5 \cdot 4} = \frac{6+6}{20} = \frac{12}{20} = \frac{6}{10} \qquad (5)$$

*note:* In effect, the sample space for part (a) consisted of 10 equally likely combinations [recall that $N(\mathscr{S}) = C(5,2) = 10$]. For part (b), we can think of there being a sample space of 20 equally likely orders. Here, these orders are the permutations of five distinct things taken two at a time. Observe that $P(5,2) = 5 \cdot 4 = 20$. [The "things" are the five cards in the miniature deck of cards.]

### Permutations of *n* Objects, Where Some Are Alike

**PROBLEM 4-6**   How may permutations can be formed by using all the letters of each of the following words? Give three examples of each. (a) mamma; (b) element; (c) borne.

*Solution*

(a)  We have five letters here, of which three are *m*'s and two are *a*'s. Thus,

$$\text{Number of Perms.} = P(5;3,2) = \frac{5!}{3!2!} = 10$$

*Three examples: mamma, mmmaa, amamm.*

(b)  We have seven letters here, of which three are *e*'s, and there is one each of *l*, *m*, *n*, and *t*. Thus,

$$\text{Number of Perms.} = P(7;3,1,1,1,1) = \frac{7!}{3!1!1!1!1!} = 840$$

*Three examples: element, eeemntl, menetel.*

(c)  We have five letters here, which are all different from one another. That is, the five letters are distinct. Thus, the number of permutations is equal to $P(5;1,1,1,1,1)$ or $P(5,5)$. From either symbol, we obtain $5! = 120$ as the number of permutations. *Three examples: borne, breon, roebn.*

**PROBLEM 4-7**   A die is tossed seven times. (a) Determine the total number of orders possible. By order is meant a succession of results on the seven tosses, such as 3, 1, 6, 5, 1, 2, 3. (b) Determine the number of orders in which exactly 3 aces, 2 threes, and 2 fives occur. [Recall that "ace" or "one" indicates the face with one dot.] (c) If a die is tossed seven times, compute the probability of getting exactly 3 aces, 2 threes, and 2 fives.

*Solution*

(a)  Each toss can occur in any of six ways. Thus, by the fundamental counting principle, there are $6 \cdot 6 \cdot \ldots \cdot 6 = 6^7 = 279{,}936$ orders possible.

(b)  The number of orders is the number of permutations of seven things [here, the "things" are digits] taken all at a time, where 3 are ones, 2 are threes, and 2 are fives. Thus,

$$\text{Number of Orders} = P(7;3,2,2) = \frac{7!}{3!2!2!} = 210$$

Four such orders are 1, 1, 1, 3, 3, 5, 5; 1, 3, 1, 5, 3, 1, 5; 5, 5, 1, 1, 3, 1, 3; and 1, 3, 5, 1, 3, 5, 1.

(c)  It is reasonable to assume that all $6^7$ orders of part (a) are equally likely. Taking the sample space $S$ to consist of all such orders, we have

$$N(S) = 6^7 = 279{,}936 \qquad (1)$$

Let the event of part (b) be denoted by $E$. From the part (b) computation, we have

$$N(E) = 210 \qquad (2)$$

Thus, using Theorem 1.2, the equally likely simple events theorem, we have

$$P(E) = N(E) = \frac{210}{279{,}936} = .00075017$$

## The Binomial Theorem and Binomial Coefficients

**PROBLEM 4-8**   Use the binomial theorem [Theorem 4.7] to write out $(x + y)^5$ in expanded form.

*Solution*

$$(x + y)^5 = x^5 + C(5,1)x^4y + C(5,2)x^3y^2 + C(5,3)x^2y^3 + C(5,4)xy^4 + y^5 \quad \textbf{(1)}$$

Since $C(5,1) = C(5,4) = 5$, and $C(5,2) = C(5,3) = 10$, Eq. (1) becomes

$$(x + y)^5 = x^5 + 5x^4y + 10x^3y^2 + 10x^2y^3 + 5xy^4 + y^5 \quad \textbf{(2)}$$

**PROBLEM 4-9**   Prove Theorem 4.8 [Pascal's formula], which states that

$$C(n,r) = C(n - 1, r) + C(n - 1, r - 1)$$

for $r = 1, 2, \ldots, n - 1$.

*Proof*  Recall that for a set of $n$ elements, the number of subsets with $r$ elements [these are called the so-called $r$-subsets] is $C(n,r)$.

Let us now focus on a set with $n$ elements, and the $r$-subsets of such a set. Let us single out a particular element of the original set, and name it $b$. Now, the $r$-subsets can be divided into two types of subsets, namely

(i) those that contain element $b$, and
(ii) those that don't contain element $b$.

The number of those $r$-subsets that contain element $b$ is $C(n - 1, r - 1)$, since along with $b$ in any $r$-subset there are $r - 1$ other elements selected from $n - 1$ elements.

The number of those $r$-subsets that do not contain element $b$ is $C(n - 1, r)$ since these $r$-subsets are selected from among the other $n - 1$ elements [other than $b$, that is].

Since the total number of $r$-subsets is $C(n,r)$, we have

$$C(n,r) = C(n - 1, r) + C(n - 1, r - 1)$$

*notes*

(a) The above proof appears in Niven (1965).
(b) Let us illustrate the above proof with respect to the 3-subsets of the set of five digits given by $\{1, 2, 3, 4, 5\}$. There are 10 such subsets since $C(5,3) = 10$. [Observe that these subsets are essentially equivalent to the 10 combinations of the five digits 1, 2, 3, 4, 5 taken three at a time, which are listed in Figure 4-6 of Example 4-20.]

If we focus on the element 1, we see that there are six 3-subsets that contain 1, namely, $\{1, 2, 3\}$, $\{1, 2, 4\}$, $\{1, 2, 5\}$, $\{1, 3, 4\}$, $\{1, 3, 5\}$, and $\{1, 4, 5\}$. Observe that the number of such subsets, 6, is identical to $C(4, 2)$. In fact, if we delete the 1 in the above six subsets we will have the six 2-subsets of the set $\{2, 3, 4, 5\}$.

Also, there are four 3-subsets that don't contain 1, and these are $\{2, 3, 4\}$, $\{2, 3, 5\}$, $\{2, 4, 5\}$, and $\{3, 4, 5\}$. Observe that the number of such subsets, 4, is identical to $C(4, 3)$. These subsets, of course, are the 3-subsets of the set $\{2, 3, 4, 5\}$.   □

**PROBLEM 4-10**   Prove the Corollary to Theorem 4.10. The corollary states that the number of subsets of a set with $n$ elements is $2^n$, and the number of proper subsets is $2^n - 1$.

*Proof*  From Theorem 4.10, we have

$$C(n, 0) + C(n, 1) + C(n, 2) + \ldots + C(n, n) = 2^n \quad \textbf{(1)}$$

Now, the number of subsets with $r$ elements of a set with $n$ elements [these are known as the $r$-subsets] is $C(n, r)$.

Now, consider all the subsets of a set with $n$ elements. Among these is the single subset with zero elements [the empty set]—thus, the number, here, can be expressed as $C(n, 0)$, since $C(n, 0) = 1$. Also, among the subsets of a set with $n$ elements are the 1-subsets of which there are $C(n, 1)$, the 2-subsets of which there are $C(n, 2)$, and so on, up to and including the single subset with $n$ elements. This last subset is the original set, itself. The number, here, can be expressed as $C(n, n)$, since $C(n, n) = 1$. Thus, we see that

$$\text{Total No. of Subsets} = C(n, 0) + C(n, 1) + C(n, 2) + \ldots + C(n, n) \qquad \textbf{(2)}$$

From Eqs. (1) and (2), we have that

$$\text{Total Number of Subsets} = 2^n \qquad \textbf{(3)}$$

We know that one of the subsets of a set with $n$ elements is the set itself. All the other subsets are called proper subsets. Thus, from Eq. (3) above, we have that

$$\text{Total Number of Proper Subsets} = 2^n - 1 \qquad \textbf{(4)} \quad \square$$

**PROBLEM 4-11**  Determine the number of, and list all the subsets of the set $\{a, b, c\}$.

*Solution*  The number of subsets is $2^3 = 8$ since the set has three elements. The subsets are $\varnothing$, $\{a\}$, $\{b\}$, $\{c\}$, $\{a, b\}$, $\{a, c\}$, $\{b, c\}$, and the original set itself, namely $\{a, b, c\}$. Note that there are three 1-subsets and three 2-subsets. Also, the first seven subsets are the proper subsets of $\{a, b, c\}$.

**PROBLEM 4-12**  Use the binomial theorem to expand $(2a - 4b)^3$.

*Solution*  In Theorem 4.7, the binomial theorem, let $x = 2a$ and $y = -4b$. Thus, the power we wish to expand can be written as

$$[(2a) + (-4b)]^3 \qquad \textbf{(1)}$$

Now, applying the binomial theorem to the expression in (1), we have

$$(2a - 4b)^3 = [(2a) + (-4b)]^3 = (2a)^3 + C(3, 1)(2a)^2(-4b) + C(3, 2)(2a)(-4b)^2 + (-4b)^3$$

$$= 8a^3 + 3(4a^2)(-4b) + 3(2a)(16b^2) + (-64b^3) \qquad \textbf{(2)}$$

Simplifying the last expression in Eq. (2), we see that the expanded form of $(2a - 4b)^3$ is

$$8a^3 - 48a^2b + 96ab^2 - 64b^3$$

**PROBLEM 4-13**  Verify Theorem 4.9 for the case where $m = 4$, $n = 2$, and $k = 5$.

*Solution*  For $m = 4$, $n = 2$, and $k = 5$, the theorem becomes:

$$C(4, 0) \cdot C(2, 5) + C(4, 1) \cdot C(2, 4) + C(4, 2) \cdot C(2, 3)$$
$$+ C(4, 3) \cdot C(2, 2) + C(4, 4) \cdot C(2, 1) + C(4, 5) \cdot C(2, 0) = C(6, 5) \qquad \textbf{(1)}$$

Since $C(2, 5) = C(2, 4) = C(2, 3) = C(4, 5) = 0$, Eq. (1) reduces to

$$C(4, 3) \cdot C(2, 2) + C(4, 4) \cdot C(2, 1) = C(6, 5) \qquad \textbf{(2)}$$

Evaluating the remaining $C(n, r)$ terms, we obtain

$$4 \cdot 1 + 1 \cdot 2 = 6 \qquad \textbf{(3)}$$

which checks out.

# Supplementary Problems

**PROBLEM 4-14**  Given the five digits 1, 3, 4, 6, and 7. In the following questions, it should be understood that repetition of a digit is not allowed. (a) How many three-digit numbers can be formed from the five digits? (b) How many three-digit numbers which are less than 600 can be formed from the five digits? (c) How many three-digit numbers which are even numbers can be formed from the five digits? (*Hint:* Consider the rightmost digit first.)

*Answer*  (a) $P(5, 3) = 60$   (b) $3 \cdot 4 \cdot 3 = 36$   (c) $4 \cdot 3 \cdot 2 = 24$

**PROBLEM 4-15** Show that $C(7,3) = C(7,4)$ by considering the set with the seven elements $a$, $b$, $c$, $d$, $e$, $f$, and $g$, and then pairing off each 3-subset with a 4-subset. (*Hint:* For example, the 3-subset $A = \{a, d, f\}$ is paired off with the 4-subset $A' = \{b, c, e, g\}$. Similarly, for the other pairings. Such a pairing procedure shows that the number of 3-subsets, which is $C(7,3)$, is equal to the number of 4-subsets, which is $C(7,4)$.)

**PROBLEM 4-16** A class has 30 students. In how many ways can a delegation of five be chosen to represent the class?

*Answer* $C(30,5) = 142{,}506$

**PROBLEM 4-17** A biologist is studying families with six children, and is classifying each child by sex, using the letter $M$ for a male, and $F$ for a female. Each grouping of six children is then coded according to an ordering by age. For example, $MFFMFF$ indicates the oldest is male, the next oldest is female, etc., and the youngest is female. (**a**) How many different codings are possible for groups of six children? (**b**) How many different codings are possible for groups of six children with four males?

*Answer* (**a**) $2^6 = 64$    (**b**) $P(6; 4, 2) = 15$

**PROBLEM 4-18** Prove that

$$C(n,0) - C(n,1) + C(n,2) - C(n,3) + \ldots + (-1)^n C(n,n) = 0.$$

(*Hint:* In the binomial theorem [Theorem 4.7], let $x = 1$ and $y = -1$. Note that $(-1)^n = 1$ if $n$ is even, and $(-1)^n = -1$ if $n$ is odd.)

**PROBLEM 4-19** A person is dealt five cards from an ordinary card deck. [The latter contains 13 spades, 13 hearts, 13 diamonds, and 13 clubs. Each of the four groupings is called a suit. The cards in a suit run from ace [one] to ten, and then Jack, Queen, and King.] (**a**) Determine the number of five-card hands. (**b**) What is the probability of being dealt a hand that contains exactly three spades and two hearts? (**c**) What is the probability of being dealt a hand that contains exactly three spades? (*Note:* Think of the ordering of cards as being immaterial here. That is, envision a hand of cards as being a combination.)

*Answer* (**a**) $C(52,5) = 2{,}598{,}960$    (**b**) $\dfrac{C(13,3) \cdot C(13,2)}{C(52,5)} = \dfrac{22{,}308}{2{,}598{,}960} = .008583$

(**c**) $\dfrac{C(13,3) \cdot C(39,2)}{C(52,5)} = \dfrac{211{,}926}{2{,}598{,}960} = .08154$

**PROBLEM 4-20** Suppose that five fair coins are tossed. (**a**) Determine the total number of outcomes possible. For example, $HTTHT$ symbolizes the outcome in which head occurs on coin 1, tail occurs on coin 2, tail occurs on coin 3, head occurs on coin 4, and, finally, tail on coin 5. (**b**) What is the probability of getting exactly three heads? (**c**) What is the probability of getting at least three heads?

*Answer* (**a**) $2^5 = 32$    (**b**) $\dfrac{P(5; 3, 2)}{2^5} = \dfrac{10}{32}$

(**c**) $[P(5; 3, 2) + P(5; 4, 1) + P(5; 5, 0)]/2^5 = (10 + 5 + 1)/32 = 1/2$

**PROBLEM 4-21** Use the binomial theorem to expand $(a - b)^7$. (*Hint:* Let $x = a$ and $y = -b$.)

*Answer* $a^7 - 7a^6 b + 21a^5 b^2 - 35a^4 b^3 + 35a^3 b^4 - 21a^2 b^5 + 7ab^6 - b^7$

# 5 INDEPENDENT TRIALS AND INTRODUCTION TO RANDOM VARIABLES

## 5-1. Independent Trials

Situations involving the performance of several experiments in succession occur frequently in the real world. It is often reasonable to suppose that the performance of any of the component experiments is not affected by the performances of those experiments that preceded it. We refer to such component experiments as being *independent*. To avoid confusing references to original experiments, and to a composite experiment [which consists of the original experiments], it is the usual practice of many practitioners to refer to the original experiments as **trials**. Then, the composite experiment is described as being an experiment [or new experiment] composed of several trials.

*note:* Section 5-1 contains some fairly technical material [dealing with Cartesian product sets, for example], which may safely be deemphasized by some readers. However, such individuals are urged to read the very significant Theorems 5.2 and 5.3, and also Examples 5-11, 5-12, and 5-13 of Section 5-1. In particular, Theorems 5.2 and 5.3 are important in the development of concepts relating to the binomial distribution [which is covered in detail in Section 5-3].

Thus, readers not overly concerned with a careful theoretical development should feel free to ignore most of the material in Section 5-1 and to skip ahead to Section 5-2.

---

**EXAMPLE 5-1:** Give some examples of a composite experiment consisting of independent trials.

*Solution*

(a) Consider the composite experiment of tossing a coin and then a die. There is no reason to expect that the tossing of a coin will have any effect on the probabilities associated with the tossing of a die.
(b) Consider the experiment of tossing a die two times in succession [or, more generally, $n$ times in succession, where $n$ is any positive integer]. Equivalently, we could have the tossing of two dice, or, more generally, of $n$ dice.
(c) Consider the experiment of drawing five cards from a standard deck of 52 cards, with replacement and shuffling after each draw.

---

## A. Identical experiment performed twice in succession

To keep matters simple at first, let us assume that the same experiment $\mathscr{E}$ is performed twice in succession, and that the two individual experiments are independent. [Put differently, we can say that we have a composite experiment consisting of two identical and independent trials.] An example of this is given in Example 5-1, part b.

One of our goals is to precisely define what is meant by "independent trials."

The development in this and other parts of Section 5-1 is similar to that found on pages 113 to 121 of Goldberg (1960).

Suppose that the sample space for the repeated trial is

$$S = \{e_1, e_2, \ldots, e_n\} \tag{1}$$

We assume that probabilities have been assigned to the simple events. That is, to each $\{e_i\}$, there is assigned a nonnegative number $P(\{e_i\})$ such that

$$P(\{e_1\}) + P(\{e_2\}) + \ldots + P(\{e_n\}) = 1 \tag{2a}$$

or

$$\sum_{i=1}^{n} P(\{e_i\}) = 1 \tag{2b}$$

*notes*

(a) Often we shall replace the above phrasing and refer to "probabilities corresponding to [or associated with] the elements $e_i$."

(b) Recall that a simple event is a subset to which belongs a single element; thus, for example, element $e_i$ belongs to simple event $\{e_i\}$.

The new experiment [which consists of the two successive, identical trials] is defined, as are all experiments, by a sample space. An appropriate choice for the elements of this new sample space is all ordered pairs of the form $(e_i, e_j)$. An ordered pair $(e_i, e_j)$ denotes the occurrence of the outcome symbolized by $e_i$ on the first trial, followed by the outcome symbolized by $e_j$ on the second trial.

*note:* Some authors regard elements and the outcomes they symbolize as synonymous. Thus, for example, they refer to "outcome $e_i$ of sample space $S$." At times, we also will slip into such casual language.

Suppose that the sample space used for each identical trial is $S$. Then, an appropriate sample space for the experiment consisting of two successive, identical trials is what is known as the **Cartesian product** set $S \times S$. We shall define and illustrate the concept Cartesian product set below.

## B. Cartesian product set

**Definition 5.1 (Cartesian Product Set):** Given sets $A$ and $B$. Then, the set of all ordered pairs $(a, b)$ such that element $a$ belongs to set $A$ and element $b$ belongs to set $B$ is called the Cartesian product [set] of $A$ and $B$, and it is denoted by $A \times B$. In symbols,

$$A \times B = \{(a, b) \mid a \text{ is in } A \text{ and } b \text{ is in } B.\}$$

Here the vertical line ($\mid$) is read as "such that" or "so that."

---

**EXAMPLE 5-2:** Suppose that $A = \{e, f\}$ and $B = \{1, 2, 3\}$. Express the following product sets: $A \times B$, $B \times A$, $A \times A$, and $B \times B$.

*Solution*

$$A \times B = \{(e, 1), (e, 2), (e, 3), (f, 1), (f, 2), (f, 3)\}$$

$$B \times A = \{(1, e), (1, f), (2, e), (2, f), (3, e), (3, f)\}$$

$$A \times A = \{(e, e), (e, f), (f, e), (f, f)\}$$

$$B \times B = \{(1, 1), (1, 2), (1, 3), (2, 1), (2, 2), (2, 3), (3, 1), (3, 2), (3, 3)\}$$

*notes*

(a) If $A$ has $m$ elements and $B$ has $n$ elements, then by the fundamental counting principle [Theorem 3.7] it follows that $A \times B$ will have $m \cdot n$ elements. Illustrations of this are provided by Example 5-2. For example, $A \times B$ has $2 \cdot 3 = 6$ elements since $A$ has 2 elements and $B$ has 3 elements.

(b) The concept of Cartesian product set can be extended to more than two sets. Thus, for example, the Cartesian product set $A \times B \times C$ is defined as the set of all ordered triples $(a, b, c)$, where $a$ belongs to set $A$, $b$ belongs to set $B$, and $c$ belongs to set $C$.

Another name for an ordered pair is ordered 2-tuple, and for an ordered triple it is ordered 3-tuple. An ordered $r$-tuple $(a_1, a_2, \ldots, a_r)$ indicates an ordering of the $r$ elements $a_1, a_2, \ldots,$ and $a_r$ with $a_1$ first, $a_2$ second, $\ldots,$ and $a_r$ as the $r$th [last] element. The Cartesian product [set] $A_1 \times A_2 \times \ldots \times A_r$ consists of all ordered $r$-tuples $(a_1, a_2, \ldots, a_r)$, where $a_1$ belongs to $A_1$, $a_2$ belongs to $A_2, \ldots,$ and $a_r$ belongs to $A_r$.

**EXAMPLE 5-3:** Say $E = \{e_1, e_2\}$, $F = \{f_1, f_2, f_3\}$, and $G = \{g_1, g_2\}$. Express the Cartesian product set $E \times F \times G$.

*Solution:* The Cartesian product set $E \times F \times G$ has $2 \cdot 3 \cdot 2 = 12$ elements since $E$ and $G$ have 2 elements, and $F$ has 3 elements. We have, after using a tree diagram for example, the following:

$$E \times F \times G = \{(e_1, f_1, g_1), (e_1, f_1, g_2), (e_1, f_2, g_1), (e_1, f_2, g_2), (e_1, f_3, g_1),$$

$$(e_1, f_3, g_2), (e_2, f_1, g_1), (e_2, f_1, g_2), (e_2, f_2, g_1), (e_2, f_2, g_2),$$

$$(e_2, f_3, g_1), (e_2, f_3, g_2)\}$$

### C. Assigning probabilities with respect to $S \times S$

Let us return now to our discussion of the Cartesian product set $S \times S$, which we shall regard as a suitable sample space for an experiment consisting of two identical trials performed in succession. [The sample space for each trial is, of course, $S$.] Let us suppose that $S$ has $n$ elements. Then there are $n^2$ ordered pairs in $S \times S$; these ordered pairs are the elements of $S \times S$. Each ordered pair has the form $(e_i, e_j)$ for $i, j = 1, 2, \ldots, n$.

In order to be able to answer probability questions pertaining to the experiment with sample space $S \times S$, we must assign probabilities to each simple event $\{(e_i, e_j)\}$ of $S \times S$, for each $i$ and $j$. For one thing, the sum of these $n^2$ probabilities must equal 1. Now, let us consider a reasonable choice for each probability $P(\{(e_i, e_j)\})$ if the two trials are independent.

**EXAMPLE 5-4:** When we say that we have independent trials we are saying that the performance of one trial is not affected by what happens on the other trial. Suppose we repeat the tossing of a pair of fair dice for 3600 times. Let us identify the dice as die one and die two. [We are, in essence, artificially introducing a succession here, by deciding to identify one die as die one, and the other as die two.] We would expect, based on intuition, that die one would show ace [the face with one dot] about $3600 \cdot P(\{Ace\}) = 3600 \cdot (1/6) = 600$ times.

Note that here we are appealing to the relative frequency interpretation of probability. For a review, refer to Section 1-2. (a) Approximately how many times out of the 3600 times should one expect to obtain an ace on die one and a three on die two? (b) Assign a probability to the simple event $\{(Ace, Three)\}$, where the ordered pair indicates ace on die one and three on die two.

*Solution*

(a) For the approximately $3600 \cdot (1/6) = 600$ times that die one will show ace, we would expect that die two would show three about $600 \cdot (1/6) = 100$ times.

(b) Since for the total of 3600 times that both dice are tossed, the simple event $\{(Ace, Three)\}$ occurs about 100 times, a reasonable assignment of probability to the event is given as follows:

$$P(\{(Ace, Three)\}) = \frac{100}{3600} = \frac{1}{36}$$

Note that this is equal to the product $P(\{Ace\}) \cdot P(\{Three\}) = (\frac{1}{6}) \cdot (\frac{1}{6})$.

---

**Definition 5.2:** Given an experiment consisting of two identical trials, where the sample space $S$ for each trial is $S = \{e_1, e_2, \ldots, e_n\}$. If the two trials are independent, then for any ordered pair $(e_i, e_j)$,

$$P(\{(e_i, e_j)\}) = P(\{e_i\}) \cdot P(\{e_j\}) \qquad \text{for } i \text{ and } j = 1, 2, \ldots, n$$

In Definition 5.2, the probability of the simple event $\{(e_i, e_j)\}$ of $S \times S$ is expressed as the product of the probabilities of the simple events $\{e_i\}$ and $\{e_j\}$ of $S$.

It can be shown that when $P(\{(e_i, e_j)\})$ is given by Definition 5.2, then the sum of the probabilities of all $n^2$ simple events is equal to 1. This is shown on pages 114–115 of Goldberg (1960). Also, the term $P(\{(e_i, e_j)\})$, as given in Definition 5.2, is nonnegative since it is the product of the two nonnegative numbers $P(\{e_i\})$ and $P(\{e_j\})$. Thus, Definition 5.2 provides an acceptable assignment of probabilities to the simple events $\{(e_i, e_j)\}$ of $S \times S$.

Thus, an experiment consisting of two independent trials, each with sample space $S$, has a sample space $S \times S$ [the Cartesian product set of $S$ with itself], and has probabilities assigned to simple events $\{(e_i, e_j)\}$ in accordance with Definition 5.2. [Recall that there are $n^2$ elements and simple events of $S \times S$. These are denoted by $(e_i, e_j)$ and $\{(e_i, e_j)\}$, respectively.]

---

**EXAMPLE 5-5:** A fair coin is tossed twice in succession. Each toss is a trial with [typical] sample space $S = \{H, T\}$, where $P(\{H\}) = P(\{T\}) = 1/2$. (a) Display a sample space for the experiment consisting of the two trials. (b) Determine probabilities for the simple events if the two tosses [i.e., trials] are assumed to be independent.

*Solution: Preliminary Note:* To keep matters simple, let us write ordered pairs here as $HH$, $HT$, etc., instead of $(H, H)$, $(H, T)$, etc.

(a) A convenient sample space for the experiment is $S \times S$, where $S \times S = \{TT, TH, HT, HH\}$.

(b) From Definition 5.2, we have $P(\{HT\}) = P(\{H\}) \cdot P(\{T\}) = (\frac{1}{2})(\frac{1}{2}) = \frac{1}{4}$, and similarly for the three other simple events $\{TT\}$, $\{TH\}$, and $\{HH\}$.

The following example will serve as a model for illustrating several key ideas.

---

**EXAMPLE 5-6:** A box contains many balls. Each ball is labeled with a single number, either 1, 2, 3, 4, or 5. The balls labeled with 1, 2, or 3 are painted red, and those labeled with 4 or 5 are painted green. The proportions of balls with numbers 1, 2, 3, 4, or 5 are .10, .20, .25, .20 and .25, respectively. For the experiment of drawing a ball from the container [here, the container is a box], let us use the sample space

$$S = \{1, 2, 3, 4, 5\} \tag{1}$$

Thus, we have the following probability assignments for the simple events:

$$P(\{1\}) = .10 \qquad P(\{2\}) = .20 \qquad P(\{3\}) = .25$$
$$P(\{4\}) = .20 \qquad P(\{5\}) = .25 \tag{2}$$

Suppose we draw two balls, with replacement. [That is, after the first ball is drawn, it is returned to the box before the second ball is drawn.] **(a)** Indicate the Cartesian product set $S \times S$ for the new experiment composed of the two trials [i.e., of the two draws]. **(b)** Determine probabilities of the simple events of $S \times S$. Observe that it is reasonable to assume that the two trials are independent.

*Solution*

**(a)** The set $S \times S$ has $5 \cdot 5 = 25$ elements and simple events. Using the abbreviated symbol 11 in place of $(1, 1)$, and likewise for the other elements of $S \times S$, we have

$$S \times S = \{11, 12, 13, 14, 15, 21, 22, 23, 24, 25, 31, 32, 33, 34, 35,$$
$$41, 42, 43, 44, 45, 51, 52, 53, 54, 55\}$$

**(b)** Denoting a typical simple event of $S \times S$ as $\{ij\}$ for $i = 1, 2, \ldots, 5$ and $j = 1, 2, \ldots, 5$, it follows from Definition 5.2 that

$$P(\{ij\}) = P(\{i\}) \cdot P(\{j\}) \qquad \text{for } i, j = 1, 2, \ldots, 5 \tag{1}$$

Thus, for example,

$$P(\{11\}) = P(\{1\}) \cdot P(\{1\}) = (.1)(.1) = .01;$$
$$P(\{12\}) = P(\{1\}) \cdot P(\{2\}) = (.1)(.2) = .02; \ldots;$$
$$P(\{23\}) = (.2)(.25) = .05; \ldots; \qquad P(\{54\}) = (.25)(.2) = .05$$

---

## D. Events determined by trials

To simplify our notation, we shall often write $P\{e_i\}$ instead of $P(\{e_i\})$. That is, we shall often omit the outer parentheses. Thus, in Example 5-6, we could have written $P\{2\}$ instead of $P(\{2\})$, say. Following a similar procedure, we shall often write $P\{(e_i, e_j)\}$ instead of $P(\{(e_i, e_j)\})$.

---

**EXAMPLE 5-7:** In the experiment of drawing two balls with replacement of Example 5-6, consider the events $E_1 = $ "red on first ball," and $E_2 = $ "green on second ball." Here, we shall call $E_1$ an event determined by the first trial, and $E_2$ an event determined by the second trial. Intuition tells us that $E_1$ and $E_2$ should be independent events. **(a)** Represent $E_1$, $E_2$, and $E_1 \cap E_2$ as subsets of $S \times S$. **(b)** Show that $E_1$ and $E_2$ are indeed independent events.

***Solution:*** Remember that the balls labeled 1, 2, or 3 are colored red, and those labeled 4 or 5 are colored green.

**(a)** Here, $E_1$ and $E_2$ are represented by the following subsets of $S \times S$:

$$E_1 = \{11, 12, 13, 14, 15, 21, 22, 23, 24, 25, 31, 32, 33, 34, 35\} \qquad \textbf{(1)}$$

$$E_2 = \{14, 24, 34, 44, 54, 15, 25, 35, 45, 55\} \qquad \textbf{(2)}$$

Observe that $E_1$ has $3 \cdot 5 = 15$ elements and $E_2$ has $5 \cdot 2 = 10$ elements. [This is consistent with what is predicted by Theorem 3.7, the Fundamental Counting Principle theorem. Observe that for $E_1$—"red on first ball"—we allow for any of balls 1, 2, 3, 4, or 5 as being the second ball, and, similarly, for $E_2$, we allow for any of those five balls as being the first ball.] From (1) and (2) above, or from common sense, we have the following for $E_1 \cap E_2$, which, in words, is "red on first ball, and green on second ball":

$$E_1 \cap E_2 = \{14, 15, 24, 25, 34, 35\} \qquad \textbf{(3)}$$

**(b)** To show that $E_1$ and $E_2$ are independent, we have to show that $P(E_1 \cap E_2) = P(E_1) \cdot P(E_2)$. First, observe that from (1) we have

$$E_1 = \{11\} \cup \{12\} \cup \{13\} \cup \ldots \cup \{34\} \cup \{35\} \qquad \textbf{(4)}$$

Here, we have partitioned $E_1$ into a union of 15 simple events. Then, since these simple events are disjoint, we have

$$P(E_1) = P(\{11\}) + P(\{12\}) + P(\{13\}) + \ldots + P(\{34\}) + P(\{35\}) \qquad \textbf{(5)}$$

Now, since the two trials are independent [see Example 5-6], we have $P(\{11\}) = P(\{1\}) \cdot P(\{1\})$, $P(\{12\}) = P(\{1\}) \cdot P(\{2\})$, etc., and thus, applying such results to Eq. (5), we obtain:

$$P(E_1) = P(\{1\})P(\{1\}) + P(\{1\})P(\{2\}) + P(\{1\})P(\{3\}) + \ldots$$
$$+ P(\{3\})P(\{4\}) + P(\{3\})P(\{5\}) \qquad \textbf{(6)}$$

Factoring out $P(\{1\})$, $P(\{2\})$, and $P(\{3\})$, respectively, from the first group of five, the second group of five, and the third group of five terms in Eq. (6), we obtain

$$P(E_1) = P(\{1\}) \cdot [P(\{1\}) + P(\{2\}) + \ldots + P(\{5\})]$$
$$+ P(\{2\}) \cdot [P(\{1\}) + P(\{2\}) + \ldots + P(\{5\})]$$
$$+ P(\{3\}) \cdot [P(\{1\}) + P(\{2\}) + \ldots + P(\{5\})] \qquad \textbf{(7)}$$

But,

$$P(\{1\}) + P(\{2\}) + P(\{3\}) + P(\{4\}) + P(\{5\}) = P(S) = 1 \qquad \textbf{(8)}$$

[In a comparable step for a general development, one would have $P(\{e_1\}) + P(\{e_2\}) + \ldots + P(\{e_n\}) = P(S) = 1$.] Thus, applying Eq. (8) to Eq. (7), we obtain

$$P(E_1) = P(\{1\}) + P(\{2\}) + P(\{3\}) = .10 + .20 + .25 = .55 \qquad \textbf{(9)}$$

Likewise, working from (2), and then using the fact that we have independent trials, we obtain

$$P(E_2) = P(\{1\})P(\{4\}) + P(\{2\})P(\{4\}) + \ldots + P(\{5\})P(\{4\})$$
$$+ P(\{1\})P(\{5\}) + P(\{2\})P(\{5\}) + \ldots + P(\{5\})P(\{5\})$$
$$= [P(\{1\}) + P(\{2\}) + \ldots + P(\{5\})]P(\{4\})$$
$$+ [P(\{1\}) + P(\{2\}) + \ldots + P(\{5\})]P(\{5\}) \qquad \textbf{(10)}$$

Now, after employing (8), we obtain

$$P(E_2) = P(\{4\}) + P(\{5\}) = .20 + .25 = .45 \qquad \textbf{(11)}$$

Likewise, working from (3), and using the fact that we have independent trials, we obtain

$$P(E_1 \cap E_2) = P(\{1\})P(\{4\}) + P(\{1\})P(\{5\}) + P(\{2\})P(\{4\})$$
$$+ P(\{2\})P(\{5\}) + P(\{3\})P(\{4\}) + P(\{3\})P(\{5\}) \qquad \textbf{(12)}$$

Factoring out $P(\{1\})$ from the first group of two terms, $P(\{2\})$ from the second group of two terms, and $P(\{3\})$ from the third group of two terms, we obtain

$$P(E_1 \cap E_2) = P(\{1\})[P(\{4\}) + P(\{5\})] + P(\{2\})[P(\{4\}) + P(\{5\})]$$
$$+ P(\{3\})[P(\{4\}) + P(\{5\})] \qquad \textbf{(13)}$$

Using Eq. (11) and then factoring out $P(E_2)$ from the resulting three terms, and finally using Eq. (9), we get

$$P(E_1 \cap E_2) = P(\{1\})P(E_2) + P(\{2\})P(E_2) + P(\{3\})P(E_2)$$
$$= [P(\{1\}) + P(\{2\}) + P(\{3\})]P(E_2) = P(E_1) \cdot P(E_2) \qquad \textbf{(14)}$$

That is, we have shown that $P(E_1 \cap E_2) = P(E_1) \cdot P(E_2)$. Thus, here, the numerical value of $P(E_1 \cap E_2)$ is given by

$$P(E_1 \cap E_2) = (.55)(.45) = .2475 \qquad \textbf{(15)}$$

---

### notes

(a) We see from Eq. (9) that, in calculating $P(E_1)$, we only have to pay attention to what is happening on the first draw [trial]. Likewise, from Eq. (11), we see that to calculate $P(E_2)$ we only have to pay attention to what is happening on the second draw [trial]. In either case, we see that we need work only with the sample space $S$, and that we can, in effect, disregard $S \times S$. These facts hinge on the assumption of independent trials.

(b) The results demonstrated in Example 5-7 are plausible because of the nature of the overall experiment, and the events $E_1$ and $E_2$. In the drawing with replacement of two balls from the container, the first draw *determines* whether or not $E_1$ occurs, and the second draw *determines* whether or not $E_2$ occurs. Also, because the two trials [draws] are independent, we would expect $E_1$ and $E_2$ to be independent events, and thus we would expect $P(E_1$ and $E_2)$ [which is the same as $P(E_1 \cap E_2)$] to equal $P(E_1)$ times $P(E_2)$.

For a general experiment consisting of two identical and independent trials, each with sample space $S$, a plausible sample space is $S \times S$. Each event $E$ of the experiment is a subset of $S \times S$, and is thus a set of ordered pairs. If we say that an event $E$ is determined by the first trial, we mean that the first member of each ordered pair in the subset representing $E$ is restricted by the requirement that $E$ occurs, while the second member is not restricted. [That is, the second member of each pair can be any element of $S$, as was the case in event $E_1$ of Example 5-7 above.] Likewise, if we say that an event $E$ is determined by the second trial, we mean that the first member of the ordered pair in the subset representing $E$ is not restricted, while the second member is restricted by the requirement that $E$ occurs. [For an example of an event determined by the second trial, refer to event $E_2$ in Example 5-7 above.]

In Definition 5.3, we shall state general versions of the concepts discussed in note (b) above.

**Definition 5.3:** Consider an experiment consisting of two independent trials, with each trial defined by sample space $S$. Thus, an acceptable sample space for the two-trial experiment is the Cartesian product set $S \times S$. An event $E_1$ of the experiment [thus, $E_1$ is a subset of $S \times S$] is said to be determined by the first trial if there is some subset $D_1$ of $S$ such that

$$E_1 = D_1 \times S$$

An event $E_2$ is said to be determined by the second trial if there is some subset $D_2$ of $S$ such that

$$E_2 = S \times D_2$$

---

**EXAMPLE 5-8:** Indicate $D_1$ for the event $E_1$ ["red on first ball"] and $D_2$ for the event $E_2$ ["green on second ball"] of Example 5-7.

*Solution:* It is easy to see that $E_1 = D_1 \times S$, where $D_1 = \{1, 2, 3\}$. Here, $E_1$ has $3 \cdot 5 = 15$ elements, where the first member of each ordered pair is either 1, or 2, or 3. [Note that in unrelaxed ordered pair notation, we would have $E_1 = \{(1, 1), (1, 2), \ldots, (1, 5), (2, 1), (2, 2), \ldots, (2, 5), (3, 1), (3, 2), \ldots, (3, 5)\}$.]

Also, $E_2 = S \times D_2$, where $D_2 = \{4, 5\}$. We see that $E_2$ has $5 \cdot 2 = 10$ elements, where the second member of each ordered pair is either 4 or 5. [Note that in unrelaxed ordered pair notation, we would have $E_2 = \{(1, 4), (2, 4), (3, 4), (4, 4), (5, 4), (1, 5), (2, 5), (3, 5), (4, 5), (5, 5)\}$.]

---

## E. Events $E_1$ and $E_2$ are independent if $E_1$ and $E_2$ are events determined by the first and second trials, respectively

The major theorem of Section 5-1 is Theorem 5.2. This theorem, which follows shortly, depends on Theorem 5.1, which is presented immediately below.

**Theorem 5.1:** Given an experiment consisting of two independent trials, where each trial has sample space $S$. [The Cartesian product set $S \times S$ is thus a plausible sample space for the two-trial experiment.]

Suppose that $D_1$ and $D_2$ are subsets of $S$. [That is, $D_1 \subset S$ and $D_2 \subset S$.] Then,

$$P(D_1 \times D_2) = P(D_1) \cdot P(D_2)$$

Instead of proving Theorem 5.1 [for a proof, see pages 118 and 119 of Goldberg (1960)], we will demonstrate its validity for a special case in Example 5-9.

---

**EXAMPLE 5-9:** Refer to the sample spaces $S$ and $S \times S$ of Examples 5-6, 5-7, and 5-8. [Recall that $S = \{1, 2, 3, 4, 5\}$.] Suppose that $D_1$ is the event "red on a ball" and $D_2$ is the event "green on a ball," where both $D_1$ and $D_2$ are subsets of sample space $S$ [not of $S \times S$]. **(a)** Express the Cartesian product set $D_1 \times D_2$. **(b)** Show that $P(D_1 \times D_2) = P(D_1) \cdot P(D_2)$.

*Solution*

**(a)** $D_1 = \{1, 2, 3\}$, and $D_2 = \{4, 5\}$ since the balls labeled 1, 2, and 3 are red and those labeled 4 and 5 are green. [We dealt with $D_1$ and $D_2$ in Example 5-8.] Thus, from Theorem 1.1, we have

$$P(D_1) = P(\{1\}) + P(\{2\}) + P(\{3\}) \tag{1}$$

and

$$P(D_2) = P(\{4\}) + P(\{5\}) \qquad (2)$$

Now,

$$D_1 \times D_2 = \{(1,4),(1,5),(2,4),(2,5),(3,4),(3,5)\} \qquad (3)$$

[Here, we use the unrelaxed ordered pair notation for the elements of $D_1 \times D_2$.]

Anticipating the next step, we partition $D_1 \times D_2$ into a union of six disjoint, simple events:

$$D_1 \times D_2 = \{(1,4)\} \cup \{(1,5)\} \cup \{(2,4)\} \cup \{(2,5)\} \cup \{(3,4)\} \cup \{(3,5)\} \quad (4)$$

**(b)** Thus, from Theorem 1.1, we have $P(D_1 \times D_2)$ expressed as the following sum:

$$P(D_1 \times D_2) = P(\{(1,4)\}) + P(\{(1,5)\}) + P(\{(2,4)\})$$
$$+ P(\{(2,5)\}) + P(\{(3,4)\}) + P(\{(3,5)\}) \qquad (5)$$

Since the trials are independent, we can, by Definition 5.2, express each $P(\{(i,j)\})$ term as $P(\{i\}) \cdot P(\{j\})$. Thus, Eq. (5) becomes

$$P(D_1 \times D_2) = P(\{1\})P(\{4\}) + P(\{1\})P(\{5\}) + P(\{2\})P(\{4\})$$
$$+ P(\{2\})P(\{5\}) + P(\{3\})P(\{4\}) + P(\{3\})P(\{5\}) \qquad (6)$$

Now, factoring and making use of the expressions for $P(D_1)$ and $P(D_2)$ from Eqs. (1) and (2), we have

$$P(D_1 \times D_2) = P(\{1\})[P(\{4\}) + P(\{5\})] + P(\{2\})[P(\{4\}) + P(\{5\})]$$
$$+ P(\{3\})[P(\{4\}) + P(\{5\})] \qquad (7)$$

Thus,

$$P(D_1 \times D_2) = P(\{1\})P(D_2) + P(\{2\})P(D_2) + P(\{3\})P(D_2)$$
$$= [P(\{1\}) + P(\{2\}) + P(\{3\})]P(D_2) = P(D_1)P(D_2) \qquad (8)$$

---

We are now ready for the statement of Theorem 5.2.

**Theorem 5.2:** Given an experiment consisting of two independent trials, where each trial has sample space $S$. [Thus, $S \times S$ is a valid sample space for the two-trial experiment.] Let $E_1$ and $E_2$ be two events of $S \times S$ such that $E_1$ is determined by the first trial, and $E_2$ is determined by the second trial. Then, $E_1$ and $E_2$ are independent events. Symbolically, this means that

$$P(E_1 \cap E_2) = P(E_1) \cdot P(E_2)$$

*notes*

**(a)** The proof of Theorem 5.2 depends upon establishing that

$$E_1 \cap E_2 = D_1 \times D_2 \qquad (1)$$

where $E_1 = D_1 \times S$ and $E_2 = S \times D_2$. Observe that the result indicated by Eq. (1) was established for a special case in Examples 5-7 and 5-9. That is, we see from Eq. (3) of Example 5-7 and Eq. (3) of Example 5-9 that

$$E_1 \cap E_2 = D_1 \times D_2 = \{(1,4),(1,5),(2,4),(2,5),(3,4),(3,5)\}$$

**(b)** A general proof that $E_1 \cap E_2 = D_1 \times D_2$ appears on pages 119 and 298 of Goldberg (1960).

---

**EXAMPLE 5-10:** Prove Theorem 5.2.

***Proof:*** In view of Definition 5.3, there exist sets $D_1$ and $D_2$, where each is a subset of $S$, such that

$$E_1 = D_1 \times S \quad \text{and} \quad E_2 = S \times D_2 \tag{1a), (1b}$$

Our goal is to prove that $P(E_1 \cap E_2) = P(E_1) \cdot P(E_2)$. From notes (a) and (b) preceding this example, we see that

$$E_1 \cap E_2 = D_1 \times D_2 \tag{2}$$

and thus,

$$P(E_1 \cap E_2) = P(D_1 \times D_2) \tag{3}$$

From Theorem 5.1, $P(D_1 \times D_2) = P(D_1) \cdot P(D_2)$, and thus

$$P(E_1 \cap E_2) = P(D_1) \cdot P(D_2) \tag{4}$$

Now, from (1a), (1b), Theorem 5.1, and Axiom 1.2 [which says that $P(S) = 1$], we have

$$P(E_1) = P(D_1 \times S) = P(D_1) \cdot P(S) = P(D_1) \tag{5}$$

and

$$P(E_2) = P(S \times D_2) = P(S) \cdot P(D_2) = P(D_2) \tag{6}$$

Thus, substituting from (5) and (6) into (4), we get

$$P(E_1 \cap E_2) = P(E_1) \cdot P(E_2) \tag{7} \quad \square$$

---

The results developed in this section on independent trials are very significant. Refer to the experiment of drawing two balls with replacement that was considered in Examples 5-6 through 5-9. We first interpreted "red on the first ball" as the event $\{1, 2, 3\} \times S$, which is a subset of $S \times S$ [see Example 5-7]. But the phrase "red on the first ball" also describes the event $\{1, 2, 3\}$, which is a subset of $S$. In general, the events $E_1 = D_1 \times S$ and $D_1$, though events of different sample spaces, are both determined by the first trial of the experiment. Thus, it is not surprising that $P(E_1) = P(D_1)$, a result proven as Eq. (5) in Example 5-10. This says, in effect, that in computing $P(E_1)$, "$E_1$ may be regarded merely as an event of $S$ alone."

Likewise, $E_2 = S \times D_2$ and $D_2$ are events of different sample spaces, but are determined by the second trial of the experiment. Thus, it is plausible to expect that $P(E_2) = P(D_2)$, a result proven as Eq. (6) in Example 5-10. This says, in effect, that in computing $P(E_2)$, "$E_2$ may be regarded merely as an event of $S$ alone."

To justify the final sentences of each of the preceding paragraphs, observe that we have proved that $P(E_1 \cap E_2)$, read as "the probability of $E_1$ and $E_2$," is equal to a product of probabilities of events [subsets] of sample space $S$. We proved this in Example 5-10 in Eq. (4), which we repeat here for clarity:

$$P(E_1 \cap E_2) = P(D_1) \cdot P(D_2) \tag{4}$$

Thus, in computing $P(E_1 \cap E_2)$, we really don't have to bother with calculations with respect to sample space $S \times S$.

---

**EXAMPLE 5-11:** Suppose we toss a fair die twice. Compute the probability of getting less than three on the first toss and an odd number on the second toss.

*Solution:* We assume that the tosses [trials] are independent. Thus, we can use Theorem 5.2 which states that $P(E_1 \cap E_2) = P(E_1) \cdot P(E_2)$, where, for this example, $E_1$ is "less than three on first toss" and $E_2$ is "odd number on second toss." Moreover, we can treat the events $E_1$ and $E_2$ as events of the equally likely sample space $S = \{1, 2, 3, 4, 5, 6\}$, which applies for a single toss of a die. Thus, we have

$$P(E_1) = P(\{1, 2\}) = 2/6 \qquad \qquad \textbf{(1)}$$

$$P(E_2) = P(\{1, 3, 5\}) = 3/6 \qquad \qquad \textbf{(2)}$$

$$P(E_1 \text{ and } E_2) = P(E_1) \cdot P(E_2) = \tfrac{2}{6} \cdot \tfrac{3}{6} = \tfrac{6}{36} = \tfrac{1}{6} \qquad \qquad \textbf{(3)}$$

---

### notes

**(a)** Previously, we used a different approach to this type of problem. Previously, we assumed, in effect, that all 36 elements of the overall sample space $S \times S$ were equally likely. [In relaxed notation, we have $S \times S = \{11, 12, \ldots, 16, \ldots, 61, 62, \ldots, 66\}$. Also, when we encountered this sample space before, we did not refer to it as $S \times S$.] Then, we expressed $E_1$ and $E_2$ as subsets of $S \times S$.

For example, $E_1 = \{11, 12, 13, 14, 15, 16, 21, 22, 23, 24, 25, 26\}$, and $E_2 = \{11, 21, \ldots, 61, 13, 23, \ldots, 63, 15, 25, \ldots, 65\}$. Then, $E_1 \cap E_2 = \{11, 13, 15, 21, 23, 25\}$, and hence

$$P(E_1 \cap E_2) = \frac{N(E_1 \cap E_2)}{36} = \frac{6}{36} = \frac{1}{6}$$

**(b)** Thus, we see that assuming that the trials are independent is, in effect, equivalent to assuming that all 36 elements of the overall sample space $S \times S$ are equally likely.

## F. Several independent trials

The definitions, theorems, and techniques of the prior development in Section 5-1 can be generalized to any finite number of repetitions of the same trial, or, yet more generally, to any number of trials performed sequentially, even when the individual trials are unlike one another. The key definition and theorem follow. Definition 5.4 is a generalization of Definition 5.2.

**Definition 5.4:** Given that $N$ is a positive integer. Let $S_j$ for $j = 1, 2, \ldots, N$ be the sample space for the $j$th trial of the experiment consisting of a sequence of $N$ trials. [$S_1$ is for trial 1, $S_2$ is for trial 2, etc.] Here,

$$S_j = \{e_1^j, e_2^j, \ldots, e_{n_j}^j\}$$

The sample space for the [overall] experiment is the Cartesian product set $S_1 \times S_2 \times \ldots \times S_N$, for which the elements are the $n_1 \cdot n_2 \cdot \ldots \cdot n_N$ ordered $N$-tuples $(e^1, e^2, \ldots, e^N)$, where $e^1$ is in $S_1$, $e^2$ is in $S_2$, $\ldots$, and $e^N$ is in $S_N$. Also, we suppose there is an acceptable assignment of probabilities to the simple events of each sample space $S_j$. The definition of independence for the $N$ trials means that the probability of every simple event for the sample space $S_1 \times S_2 \times \ldots \times S_N$ is given by the product rule:

$$P(\{(e^1, e^2, \ldots, e^N)\}) = P(\{e^1\}) \cdot P(\{e^2\}) \cdot \ldots \cdot P(\{e^N\})$$

Theorem 5.3 is a generalization of Theorem 5.2.

**Theorem 5.3:** Given the experiment consisting of $N$ independent successive trials corresponding, respectively, to the $N$ sample spaces $S_1, S_2, \ldots, S_N$. Suppose that event $E_j$ is an event determined by the $j$th trial, for $j = 1, 2, \ldots, N$. [For $E_1$, this means that there is a subset $D_1$ of $S_1$ such that $E_1 = D_1 \times S_2 \times S_3 \times \ldots \times S_N$. For $E_2$, this means that there is a subset $D_2$ of $S_2$ such that $E_2 = S_1 \times D_2 \times S_3 \times \ldots \times S_N$. And similarly for $E_3, E_4, \ldots, E_N$.]

Then the events are independent. In particular, this means that

$$P(E_1 \cap E_2 \cap \ldots \cap E_N) = P(E_1) \cdot P(E_2) \cdot \ldots \cdot P(E_N)$$

It is useful at this point to review the definition of independence for more than two events of a sample space. [See Definitions 3.3 and 3.4.]

Though the statements of Definition 5.4 and Theorem 5.3 are relatively cumbersome, the applications of them are simple.

*note:* Each term of the form $P(E_j)$ can be computed as if $E_j$ were a subset of $S_j$ alone. That is, we have $P(E_j) = P(D_j)$, where $D_j$ is a subset of $S_j$. For similar ideas for the $N = 2$ case, refer to Eqs. (5) and (6) of Example 5-10. [The latter contains the proof of Theorem 5.2.]

---

**EXAMPLE 5-12:** A fair die is tossed, a fair coin is tossed, and a card is drawn from a standard 52-card deck. What is the probability of getting less than 3 on the die, tail on the coin, and heart on the card?

*Solution:* We assume that the three trials are independent. [This is reasonable since we wouldn't expect the performance of one trial to affect the others.] We see that "less than three" is determined by the first trial, "tail" is determined by the second trial, and "heart" is determined by the third trial. From Theorem 5.3, it follows that "less than three," "tail," and "heart" are independent events. Thus, we have

$P(\text{"less than three" and "tail" and "heart"})$

$= P(\text{"less than three"}) \cdot P(\text{"tail"}) \cdot P(\text{"heart"}) = \frac{2}{6} \cdot \frac{1}{2} \cdot \frac{13}{52} = \frac{1}{24}$

Observe that, in computing $P(\text{"heart"})$, we consider "heart" to be an event of the sample space $S_3$, where the latter has 52 equally likely elements, one for each card of the deck. Similarly for the other two events. This is in accord with the note that precedes this example.

---

**EXAMPLE 5-13:** The situation in Problem 3-13 dealt with drawing three cards, with replacement, from a five-card deck containing three spades [ace, two, and three] and two hearts [ace and two]. Let us suppose that we have the same situation here. That is, suppose that three cards are drawn, with replacement, from this five-card deck. **(a)** Compute the probability of obtaining spade on card 1, heart on card 2, and spade on card 3. [This question is identical to that of part (b) of Problem 3-13.] **(b)** Compute the probability of obtaining exactly two spades when three cards are drawn with replacement from the five-card deck described above.

*Solution:* *Preliminaries:* Let $S_i$ stand for spade on draw $i$, and $H_j$ stand for heart on draw $j$. Thus, for part (a), the event whose probability is sought can be denoted as $S_1 H_2 S_3$.

It's reasonable to assume that the three draws [trials] are independent since we have drawing with replacement.

**(a)** From Theorem 5.3, we conclude that the events denoted by $S_1$, $H_2$, and $S_3$ are independent. Thus,

$$P(S_1 H_2 S_3) = P(S_1) \cdot P(H_2) \cdot P(S_3) \qquad (1)$$

Clearly, $P(S_1) = P(S_3) = 3/5$ and $P(H_2) = 2/5$ since three of the cards are spades and two are hearts. Thus,

$$P(S_1 H_2 S_3) = \frac{3}{5} \cdot \frac{2}{5} \cdot \frac{3}{5} = \frac{18}{125} = .144 \qquad (2)$$

In Problem 3-13, part (b), we used a different approach for calculating the probability $P(S_1 H_2 S_3)$.

**(b)** Here, when we say "exactly two spades, when drawing three cards," we imply that there is no regard to order. Now, exactly two spades in three cards can be obtained either in the order $S_1 H_2 S_3$ or in the order $S_1 S_2 H_3$ or in the order

$H_1 S_2 S_3$. Moreover, these orders are mutually exclusive. Thus, from Axiom 1.3′ as applied to the overall experiment of drawing three cards with replacement, we have

$$P(\text{``Exactly two spades''}) = P(S_1 H_2 S_3 \text{ or } S_1 S_2 H_3 \text{ or } H_1 S_2 S_3)$$

$$= P(S_1 H_2 S_3) + P(S_1 S_2 H_3) + P(H_1 S_2 S_3) \quad \textbf{(1)}$$

Applying Theorem 5.3 to each term on the right yields:

$$P(\text{``Exactly two spades''}) = P(S_1)P(H_2)P(S_3) + P(S_1)P(S_2)P(H_3)$$

$$+ P(H_1)P(S_2)P(S_3) \quad \textbf{(2)}$$

Applying reasoning similar to that used in part (a) reveals that each term on the right of Eq. (2) is equal to $(\frac{3}{5})^2(\frac{2}{5})$. Thus,

$$P(\text{``Exactly two spades''}) = 3(\tfrac{3}{5})^2(\tfrac{2}{5}) = \tfrac{54}{125} = .432 \quad \textbf{(3)}$$

*note:* If we drop the subscripts in the events $S_1 H_2 S_3$, $S_1 S_2 H_3$, and $H_1 S_2 S_3$, we observe that we have a list of the permutations of three objects taken all at a time, where two of the objects are of one kind and the other object is of another kind. [Here, the three "objects" are $S$, $S$, and $H$.]

If we refer to Theorem 4.5, we can confirm that the number of such permutations is indeed 3.

$$\text{Number} = P(3; 2, 1) = \frac{3!}{2!\,1!} = 3$$

## 5-2. Discrete Random Variables

### A. Preliminary concepts

Sometimes it is very useful to be able to refer to the outcomes of an experiment in terms of numbers. We can accomplish this by using a random variable, which is a function that assigns numerical values to the elements of a sample space. We shall use capital letters, such as $X$, $Y$, $Z$, $V$, and $W$, to denote random variables. The particular random variable to use in a given situation is often determined by the nature of the discussion.

**EXAMPLE 5-14:** Consider the "toss a fair coin twice" experiment. The experiment dealing with tossing a coin twice was first considered as Example 1-1. The most subdivided sample space for this experiment is $S = \{TT, TH, HT, HH\}$, where each element [outcome] has a probability associated with it. Let $X$ be a random variable which denotes the exact [or total] number of heads obtained in the two tosses. Determine the assignment of values to $X$ for the elements of $S$.

*Solution:* The assignment of values of $X$ is indicated by the arrows shown in Figure 5-1. Thus, for example, $X$ equals 1 for both $TH$ and $HT$ since for each of these elements, one head occurs. We see that the possible values of the random variable $X$ are 0, 1, and 2.

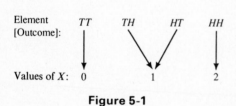

Figure 5-1

A generalization of the discussion of Example 5-14 is given in Definition 5.5.

**Definition 5.5:** Let $\mathscr{E}$ be an experiment and let $S$ be a sample space $\{e_1, e_2, e_3, \dots\}$ associated with the experiment. A function $X$ assigning a real number $X(e_i)$ to every element $e_i$ in $S$ is called a *random variable*.

*notes*

(a) Illustrating this terminology with respect to Example 5-14, we have $X(TT) = 0$, $X(TH) = X(HT) = 1$, and $X(HH) = 2$.

(b) The set of all possible values of $X$ [that is, those values that can occur when the experiment is performed] is called the range space of $X$, and it is denoted by $R_x$. Thus, in Example 5-14, we have $R_x = \{0, 1, 2\}$. Observe that, in general, we may consider $R_x$ to be another sample space for the experiment $\mathscr{E}$. In fact, if we refer back to Example 1-1, we see that $R_x$ is identical to the sample space $\hat{S}$ of Example 1-1.

**Definition 5.6:** For a random variable $X$, if the number of possible values of $X$ [these possible values constitute $R_x$, the range space] is finite or countably infinite, then $X$ is called a *discrete random variable*. Thus, the possible values of $X$ may be listed as $x_1, x_2, \ldots, x_m, \ldots$. In the finite case, the list terminates, and in the countably infinite case, the list goes on indefinitely.

To illustrate Definition 5.6 with respect to Example 5-14, we have, for instance, the terminating list of possible $X$ values: $x_1 = 0$, $x_2 = 1$, and $x_3 = 2$.

The values of the above random variable occur with certain probabilities. For example, $X = 0$ occurs with probability 1/4 since $X = 0$ corresponds to the element $TT$, which has the probability 1/4 associated with it. Also, $X = 1$ occurs with probability 1/2 since $X = 1$ corresponds to both $TH$ and $HT$, each of which has an associated probability of 1/4. [The probabilities associated with the elements of the "toss a coin twice experiment" were first listed in the note following Example 1-12.] Our symbol for the probability function of a random variable $X$ is $f(x)$ or $P(X = x)$. Here, lowercase $x$ refers to a typical value of the random variable $X$. The general definition for $f(x)$ is as follows:

**Definition 5.7:** Given a discrete random variable $X$. The probability function or probability distribution $f$ is defined as follows: $f(x)$ is equal to the probability of the event $E$ of $S$, where $E$ consists of those elements of $S$ which correspond to the value $X = x$. [Often, $P(X = x)$ is used instead of $f(x)$.]

For example, if $E = \{e_1, e_2, \ldots, e_k\}$, where $e_1, e_2, \ldots, e_k$ correspond to $X = x$ [that is, $X(e_i) = x$ for $i = 1, 2, \ldots, k$], then

$$f(x) = P(\{e_1, e_2, \ldots, e_k\}) = P(\{e_1\}) + P(\{e_2\}) + \ldots + P(\{e_k\})$$

*notes*

(a) We see that we can simply write $f(x) =$ Sum of the probabilities associated with those elements of $S$ which correspond to the value $X = x$.

(b) In going from $P(\{e_1, e_2, \ldots, e_k\})$ to the sum $P(\{e_1\}) + P(\{e_2\}) + \ldots + P(\{e_k\})$, we made use of Theorem 1.1.

(c) Remember that, for simplicity, we often replace $P(\{\ \ \})$ by $P\{\ \ \}$. That is, we drop the outer (smooth) parentheses. Thus, on the extreme right side of the $f(x)$ expression that appears in Definition 5.7, we could write $P\{e_1\} + P\{e_2\} + \ldots + P\{e_k\}$.

---

**EXAMPLE 5-15:** Determine all possible $f(x)$ values for Example 5-14.

*Solution*

$$f(0) = P(X = 0) = P\{TT\} = 1/4$$

$$f(1) = P(X = 1) = P\{TH, HT\} = P\{TH\} + P\{HT\} = 2/4 = 1/2$$

$$f(2) = P(X = 2) = P\{HH\} = 1/4$$

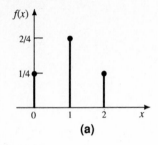

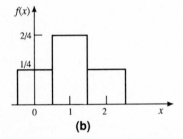

**Figure 5-2.** Bar chart and probability histogram for "toss a coin twice" experiment. (**a**) Bar chart. (**b**) Probability histogram.

The probability function is often displayed by means of a table consisting of values for $X$ and associated values for $f(x)$. The probability function table for our model "toss a fair coin twice" problem is as follows:

| $x$ | $f(x)$ |
|---|---|
| 0 | 1/4 |
| 1 | 2/4 |
| 2 | 1/4 |
| Sum = | 4/4 = 1 |

We can display probability functions graphically by means of a *bar chart* (also called a *line chart*) and a *probability histogram*. The bar chart and probability histogram for our model problem are given in Figure 5-2.

In the bar chart, each bar is located at a value of the random variable. The height of each bar [the so-called "bars" are shown as vertical line segments with large dots at the tops] is equal to the corresponding probability. In the probability histogram, each individual rectangular area is centered horizontally at a value of the random variable. Here, the area of each rectangle is set equal to the corresponding probability. In cases such as Figure 5-2b, where we take the width of each rectangle to equal one, the height of each rectangle is equal to the corresponding probability.

*notes*

(**a**) The above "toss a fair coin twice" example with random variable equal to the exact number of heads is a special case of the binomial probability function or binomial distribution. The binomial distribution will be explored in detail in Section 5-3.

(**b**) Other expressions that appear in the literature for probability function are probability distribution and probability density function. Sometimes, these expressions are preceded by the word discrete if the random variable is discrete.

(**c**) For any discrete random variable $X$, the sum of all $f(x)$ values is equal to 1.

**Theorem 5.4:** Let $f(x)$ be the probability function of a discrete random variable $X$. Then,

(**a**) $0 \leq f(x) \leq 1$ for each possible value of $x$, and
(**b**) $\sum_x f(x) = 1$.

Here, the summation is with respect to all possible values of the random variable $X$.

## B.  Tossing a pair of dice

Another example dealing with a random variable and the associated probability function follows.

---

**EXAMPLE 5-16:** Refer to the "toss a pair of dice" situation discussed previously [see, for instance, Examples 1-2, 1-13, and 1-16]. The most subdivided sample space $S$, which has 36 elements, is displayed below in abbreviated form:

$$S = \{11, 12, \ldots, 16, 21, 22, \ldots, 26, \ldots, 61, 62, \ldots, 66\}$$

We will suppose that the two dice are fair. Thus, the 36 elements above are equally likely, and the probability associated with any one of them is 1/36. Let $X$ be the random variable which assigns to each element the *sum* of the two numbers associated with the element. **(a)** Determine all possible values for $X$, and display the elements associated with each value of $X$. **(b)** Determine the probability function $f(x)$, and display its values in a table.

***Solution:*** **(a)** and **(b)** Corresponding to $X = 2$ is the single element 11 since $1 + 1 = 2$. Thus, $f(2) = 1/36$. Corresponding to $X = 3$ are the elements 12 and 21, and thus $f(3) = P\{12\} + P\{21\} = 2/36$.

Continuing in this fashion, we determine the following complete three-column table for $f(x)$.

| Elements | $x$ | $f(x)$ |
|---|---|---|
| 11 | 2 | 1/36 |
| 12, 21 | 3 | 2/36 |
| 13, 22, 31 | 4 | $1/36 + 1/36 + 1/36 = 3/36$ |
| 14, 23, 32, 41 | 5 | 4/36 |
| 15, 24, 33, 42, 51 | 6 | 5/36 |
| 16, 25, 34, 43, 52, 61 | 7 | 6/36 |
| 26, 35, 44, 53, 62 | 8 | 5/36 |
| 36, 45, 54, 63 | 9 | 4/36 |
| 46, 55, 64 | 10 | 3/36 |
| 56, 65 | 11 | 2/36 |
| 66 | 12 | 1/36 |
| | | Sum $= 36/36 = 1$ |

Observe that tossing a pair of dice is the basis for the popular gambling game "craps." The $f(x)$ probabilities listed above have a tremendous influence on what happens in this game. Note that the probability of "rolling a seven" [this means $X = 7$] is $6/36 = 1/6$.

***note:*** We can interpret the range space $R_x = \{2, 3, \ldots, 12\}$ as constituting another sample space for the "toss a pair of dice" experiment.

## 5-3. The Binomial Probability Distribution

### A. Deriving the binomial probability distribution equation

The next two examples illustrate the approach used in deriving the equation for the binomial probability function.

**EXAMPLE 5-17:** Given an ordinary 52-card deck. Suppose that five cards are drawn, with replacement after each draw. [Recall that this is taken to mean that each card is drawn from a well-shuffled deck of 52 cards.] Thus, the five draws constitute independent trials. **(a)** What is the probability of obtaining three spades and two nonspades in the order *SSSFF*? Here, *S* denotes spade or, more generally, success, and *F* denotes nonspade or, more generally, failure to get a success. **(b)** Repeat part (a) for the order *SSFSF*.

***Solution***

**(a)** Here, $P(\text{spade}) = P(S) = 13/52 = 1/4$, and $P(\text{nonspade}) = P(F) = 39/52 = 3/4$, for the draw of any card. Now, the order *SSSFF* can also be expressed as $S \cap S \cap S \cap F \cap F$. Then, using Theorem 5.3 in Section 5-1F, we obtain

$$P(SSSFF) = P(S)P(S)P(S)P(F)P(F) = (1/4)(1/4)(1/4)(3/4)(3/4)$$
$$= (1/4)^3(3/4)^2 \qquad \textbf{(1)}$$

**(b)** In the same way,

$$P(SSFSF) = (1/4)(1/4)(3/4)(1/4)(3/4) = (1/4)^3(3/4)^2 \tag{2}$$

**EXAMPLE 5-18:** Say we have a draw of five cards with replacement as in Example 5-17. Suppose that now we wish to calculate the probability of obtaining exactly three spades [or obtaining a total of three spades] in a draw of five cards, with no regard to order. Let us denote this probability as $P(X = 3)$. This is read as the "probability that $X$ equals 3."

***Solution:*** Here, we are asking for the probability of getting exactly three spades, which means either in order $A$ or $B$ or $C$, or some other one of the orders (labeled with letters) that are listed below.

A: *SSSFF*    B: *SSFSF*    C: *SSFFS*    D: *FSSSF*    E: *SFSSF*

F: *SFSFS*    G: *FSSFS*    H: *SFFSS*    I: *FSFSS*    J: *FFSSS*

Thus,

$$P(X = 3) = P(A \text{ or } B \text{ or } C \text{ or } \ldots) \tag{1}$$

where the "or" is equivalent to $\cup$. Now, the events $A$, $B$, $C$, and so on are mutually exclusive with respect to the experiment of drawing five cards from the 52-card deck. Thus, the addition rule applies, and we obtain the following from Eq. (1):

$$P(X = 3) = P(A) + P(B) + P(C) + \ldots \tag{2}$$

Now, above we have listed all possible favorable orders for getting three spades and two nonspades. [Refer to the list right in front of Eq. (1).] But, we could also calculate the number of such orders, because they are the *permutations* of the five letters $S$, $S$, $S$, $F$, $F$ taken all at a time. (See Theorem 4.5 for a review.) Letting $K$ denote the total number of favorable orders, we have

$$K = P(5; 3, 2) = \frac{5!}{3!2!} = \frac{5 \cdot 4 \cdot 3!}{3! \cdot 2 \cdot 1} = 10 \tag{3}$$

[Recall from Theorem 4.5 that $P(5; 3, 2)$ denotes the total number of permutations of five objects taken all at a time, where there are three of one kind—here, three are $S$'s—and two are of a second kind—here, two are $F$'s. Also, recall from Theorem 4.6 that $P(5; 3, 2) = C(5, 3)$, where the latter symbol denotes the number of combinations of 5 distinct objects taken 3 at a time.]

Each of the orders $A$, $B$, $C$, and so on has the same probability. In fact, orders $A$ and $B$ were considered in Example 5-17. Thus,

$$P(A) = P(B) = P(C) = \ldots = (1/4)^3(3/4)^2 \tag{4}$$

Thus, in Eq. (2), we have the adding together of $K = P(5; 3, 2) = 10$ terms, each of which is equal to $(1/4)^3(3/4)^2$. But, this equals $K$ times $(1/4)^3(3/4)^2$. Thus,

$$P(X = 3) = \frac{5!}{3!2!} \cdot \left(\frac{1}{4}\right)^3\left(\frac{3}{4}\right)^2 = 10 \cdot \left(\frac{1}{64}\right)\left(\frac{9}{16}\right) = \frac{90}{1024} = .0879 \tag{5}$$

Thus, the probability of getting exactly three spades when drawing five cards with replacement is .0879.

It is useful to rewrite the first part of Eq. (5) as below in Eq. (6), because then a general way of handling similar problems is suggested. The two terms on the right are labeled $K$ and $P(A)$.

$$P(X = 3) = \overbrace{\frac{5!}{3!2!}}^{K} \cdot \overbrace{\left(\frac{1}{4}\right)^3\left(\frac{3}{4}\right)^2}^{P(A)} \tag{6}$$

The $K$ term denotes the number of orders in which exactly three successes (here, spades) and two failures (here, nonspades) occur in five trials, while the $P(A)$ term denotes the probability of getting exactly three successes and two failures in a particular order, namely the order $SSSFF$. [Some authors use the word "specific" instead of "particular."] The terms $1/4$ and $3/4$ denote the probabilities of success and failure, respectively, in a single trial.

---

In Example 5-18, we had a specific example of a binomial probability situation in the calculation of $P(X = 3)$—the probability of getting exactly three spades when drawing five cards with replacement from a standard 52-card deck. The characteristics of the general binomial situation are as follows:

**Definition 5.8 (Characteristics of the Binomial Probability Situation):**

(a) Given an overall experiment consisting of $n$ independent and identical trials.
(b) For any trial, we are concerned with some event occurring [success], or not occurring [failure]. In general, $p$ is the probability of success in a single trial, and $q = 1 - p$ is the probability of failure in a single trial. Note that $p$ does not vary from trial to trial.
(c) The binomial random variable $X$ denotes the exact or total number of successes in the $n$ trials. The $(n + 1)$ possible values for $X$ are $0, 1, 2, \ldots,$ $(n - 1)$, $n$. The symbol $b(x; n, p)$ [we also will use $f(x)$ and $P(X = x)$ from time to time] denotes the probability of getting exactly $x$ successes in $n$ trials, with no regard to order. Here, $b(x; n, p)$ is known as the binomial probability function or binomial distribution.

The word "binomial" stems partly from the fact that *bi* is a prefix which stands for the word "two"; in each trial, we are concerned with two possibilities, namely the success or failure of some event.

The independent and identical trials, such that for each trial we focus on either success or failure, are often referred to as *Bernoulli trials* [after James Bernoulli, 1654–1705].

The general equation for $f(x)$ for the binomial probability situation is as simple to derive as was the equation for $P(X = 3)$ in Example 5-18. Refer, in particular, to Eq. (6). In the example, the specific values for the general symbols are $n = 5$ since there are five draws with replacement, and $p = 1/4$ since the probability of spade (success) in a single draw (trial) is $1/4$. Clearly, $q = 1 - p = 3/4$, the probability of nonspade (failure) in a single draw. In the above, $X = x$ was $X = 3$ since we were interested in three spades (successes) in five trials, with no regard to order.

For the general binomial situation, all we have to do is first develop equations for $P(A)$ and $K$, and then multiply them. The order $A$ is a particular order in which there are $x$ successes in $n$ trials. For example, suppose $A$ is the order in which first there are $x$ successes and then $(n - x)$ failures.

$$\text{Order } A: \quad \underbrace{SSS\ldots SS}_{x \text{ successes}} \ \underbrace{FF\ldots F}_{\substack{(n - x) \\ \text{failures}}}$$

Since the trials are independent, it follows from Theorem 5.3 that

$$P(A) = \underbrace{p \cdot p \cdot p \cdot \ldots \cdot p \cdot}_{x \text{ terms}} \underbrace{q \cdot q \cdot \ldots \cdot q}_{(n - x) \text{ terms}} = p^x q^{(n-x)} \qquad \textbf{(1)}$$

The term $K$ stands for the total number of favorable orders; this is equal to the total number of permutations of $n$ items taken all at a time, where $x$ of them are $S$'s and $(n - x)$ of them are $F$'s. The symbol for this is $P(n; x, n - x)$. Thus, we have

$$K = P(n; x, n - x) = \frac{n!}{x!(n - x)!} \tag{2}$$

When we recall that $P(n; x, n - x) = C(n, x)$—see Theorem 4.5—and that another symbol for the latter is $\binom{n}{x}$, we may also write

$$K = C(n, x) = \binom{n}{x} \tag{2'}$$

Now, the probability of getting exactly $x$ successes in $n$ trials is equal to the product of $K$ times $P(A)$. Thus, expressing $P(A)$ from Eq. (1) and $K$ from Eq. (2) or (2'), we obtain the equation for $b(x; n, p)$, the binomial probability function, which is the symbol for the probability of getting $x$ successes in $n$ trials. The $p$ is the symbol for the probability of a success in a single trial.

**Theorem 5.5 (Binomial Probability Function):**

$$b(x; n, p) = \frac{n!}{x!(n - x)!} p^x q^{(n-x)} \tag{1a}$$

or

$$b(x; n, p) = \binom{n}{x} p^x q^{(n-x)} \tag{1b}$$

Each of the above applies for $x = 0, 1, 2, \ldots, (n - 1), n$, and for $0 < p < 1$.

*notes*

(a) The term $b(x; n, p)$ above is often replaced by $f(x)$ or $P(X = x)$. The binomial probability function is also called the *binomial probability distribution*, or, briefly, the *binomial distribution*. [The latter expression is probably the one that is most often used in practice.]

(b) A random variable whose probability function is given by Theorem 5.5 is called a *binomial random variable*.

(c) Sometimes, the phrase "exactly $x$ successes, with no regard to order" is replaced by "exactly $x$ successes, in some order." Also, the brief phrase "exactly $x$ successes" is sometimes used. The words "a total of $x$ successes" are often used in place of "exactly $x$ successes."

The Bernoulli probability distribution is a binomial distribution for which $n = 1$. Thus, here we are concerned with the single performance of an experiment in which we focus on two possible outcomes, namely, success and failure. The associated probabilities are, of course, $p$ and $(1 - p)$, as in the usual binomial situation.

**Theorem 5.5, Corollary (A):**  A random variable $X$ has a *Bernoulli distribution* (or *Bernoulli probability function*) and is referred to as a Bernoulli random variable, if its probability distribution is given by

$$f(x) = \begin{cases} p & \text{if } x = 1 \\ q = 1 - p & \text{if } x = 0 \end{cases}$$

This information can also be expressed by the following single equation:

$$f(x) = p^x(1 - p)^{1-x} \qquad \text{for } x = 0 \text{ and } 1$$

Observe that $f(1) = p$ is simply the probability of obtaining a success in a single trial, and $f(0) = 1 - p$ is the probability of *not* obtaining a success in a single trial.

The deck with five cards of Example 5-19 [below] was seen before in Examples 3-16, 3-17, 4-13, 4-14, 4-15, and 5-13, and in Problems 3-13, 3-14, and 4-4.

---

**EXAMPLE 5-19:** Given a five-card deck with 3 spades (ace, two, and three—denoted as $AS$, $2S$, and $3S$) and two hearts (ace and two—denoted as $AH$ and $2H$). Suppose that three cards are drawn, with replacement after each draw. Let $X$ stand for the exact number of spades obtained. Thus, $X$ is a binomial random variable for which $n = 3$ and $p = 3/5$, and $f(x)$ is the binomial probability of obtaining exactly $x$ spades. Determine $f(x)$ values for $X = 0, 1, 2$, and 3 and verify that the sum is equal to 1.

*Solution:* From Theorem 5.5, we have

$$P(X = x) = f(x) = \frac{3!}{x!(3 - x)!}(3/5)^x(2/5)^{3-x} \qquad \text{for } x = 0, 1, 2, 3 \qquad \textbf{(1)}$$

Thus, we have $f(0) = (2/5)^3 = 8/125$, $f(1) = 3(3/5)(2/5)^2 = 36/125$, $f(2) = 3(3/5)^2(2/5) = 54/125$, and $f(3) = (3/5)^3 = 27/125$. We then arrive at the following table for this binomial probability function:

| $x$ | $f(x)$ |
|---|---|
| 0 | $8/125 = .064$ |
| 1 | $36/125 = .288$ |
| 2 | $54/125 = .432$ |
| 3 | $27/125 = .216$ |

$$\text{Sum} = 125/125 = 1$$

It is important that we not think mechanically, and merely substitute numbers into equations. Thus, for example, it should be clear that $P(X = 3) = P(SSS) = (3/5)^3$, and $P(X = 2) = P(SSF) + P(SFS) + P(FSS)$—here, $F$ stands for heart. The latter reduces to $3(3/5)^2(2/5)$ since each of the three orders cited has a probability equal to $(3/5)^2(2/5)$. The calculation for $P(X = 2)$ was also done in detail in Example 5-13.

---

**EXAMPLE 5-20:** Use the binomial theorem [Theorem 4.7′] to show that the sum of binomial probabilities from $x = 0$ to $x = n$, inclusive, is equal to 1.

*Solution:* From Theorem 4.7′, we have

$$(x + y)^n = \sum_{r=0}^{n} \frac{n!}{r!(n - r)!} x^{n-r} y^r \qquad \textbf{(1)}$$

Now, replace $x$ and $y$ by $q$ and $p$, respectively, and then replace the index symbol $r$ by $x$ to get

$$(q + p)^n = \sum_{x=0}^{n} \frac{n!}{x!(n - x)!} p^x q^{n-x} \qquad \textbf{(2)}$$

Now, $(q + p)^n = 1^n = 1$, and the term after the $\Sigma$ on the right is $b(x; n, p)$, as we know from Theorem 5.5. Thus, Eq. (2) reduces to

$$\sum_{x=0}^{n} b(x; n, p) = 1 \qquad \textbf{(3)}$$

---

## B. Binomial probability examples

Since the binomial probability function has so many practical applications, extensive tables have been developed for it. For example, Table I in Appendix

A has binomial probabilities tabulated for $p = .05, .10, .15, \ldots, .40, .45,$ and $.50,$ and for $n = 1, 2, 3, \ldots, 19,$ and 20.

---

**EXAMPLE 5-21:** A manufacturer of bolts knows from experience that 10 percent of the bolts are defective. If he sells boxes containing eight bolts each, find the probability that **(a)** a box will contain exactly two defective bolts, **(b)** a box will contain at most two defective bolts, and **(c)** a box will contain at least three defective bolts.

*Solution:* The usual assumption in this type of situation is that the binomial probability model applies. Here, we have $n = 8$ and $p = .1$. Note that we translate from "10 percent of the bolts are defective" to "the probability that a single bolt is defective is .1." Notice that "defective" is taken as "success" here.

**(a)** Applying the binomial probability equation, we have

$$P(X = 2) = \frac{8!}{2!6!}(.1)^2(.9)^6 = .1488$$

since we want the probability of exactly two successes (defectives) out of eight bolts. Here, it is easier to use the Binomial Probability Table (Table I). Refer to the subtable below, which shows the part of Table I containing the relevant $n$, $x$, and $p$ values.

**TABLE 5-1: Subtable Formed from Main Binomial Probability Table**

| $n$ | $x$ | .05 | $p$ ↓ .10 | .15 |
|---|---|---|---|---|
| 8 | 0 | | .4305 | |
| | 1 | | .3826 | |
| → | 2 | .0515 | .1488 | .2376 |
| | 3 | | .0331 | |

First, locate $n = 8$ in the left margin, and then the $x = 2$ row for $n = 8$. [Refer to the horizontal arrow in Table 5-1.] Next, locate $p = .10$ in the top margin [refer to the vertical arrow]. Then, the value for $P(X = 2) = b(2; 8, .1)$ is found where the $x = 2$ row and $p = .10$ column meet. [Refer to the rectangle in Table 5-1.]

**(b)** At most two successes mean $X \leq 2$, or, equivalently, that $X = 0$ or 1 or 2. Now, these $X$ values indicate mutually exclusive events. Applying the addition rule, and data from Table I, we get

$$P(\text{at most 2 defectives}) = P(X \leq 2) = P(X = 0) + P(X = 1) + P(X = 2)$$

$$= .4305 + .3826 + .1488 = .9619$$

**(c)** *Method 1:*

$$P(\text{at least 3 defs.}) = P(X \geq 3) = P(X = 3) + P(X = 4) + \ldots + P(X = 8)$$

$$= .0331 + .0046 + \ldots + .0000 = .0381$$

*Method 2:* Recall the general result $P(\text{not } A) = 1 - P(A)$. Now, "at least 3 defectives" means "not at most 2 defectives." Thus,

$$P(\text{at least 3 defs.}) = P(\text{not at most 2 defectives}) = 1 - P(\text{at most 2 defectives})$$

Thus, using the value from part (b), we obtain

$$P(\text{at least 3 defs.}) = 1 - .9619 = .0381$$

---

Recall that $b(x; n, p)$ is a major symbol for the binomial probability of getting exactly $x$ successes in $n$ trials, where the probability of success in a single trial is $p$.

Shortly, we will show how we can use Table I in Appendix A together with Theorem 5.6 [below] to calculate binomial probabilities for $p > .5$. See, for instance, Example 5-23.

**Theorem 5.6:** The following is true:

$$b(x; n, p) = b(n - x; n, 1 - p)$$

---

**EXAMPLE 5-22:** Prove Theorem 5.6.

*Proof:* From Theorem 5.5, we have

$$b(x; n, p) = \frac{n!}{x!(n - x)!} p^x (1 - p)^{n - x} \tag{1}$$

To obtain the expression for $b(n - x; n, 1 - p)$, we merely replace $x$ and $p$ on both sides of Eq. (1) by $(n - x)$ and $(1 - p)$, respectively. Thus, we obtain

$$b(n - x; n, 1 - p) = \frac{n!}{(n - x)![n - (n - x)]!} (1 - p)^{n - x} [1 - (1 - p)]^{n - (n - x)} \tag{2}$$

But, $n - (n - x) = x$ and $1 - (1 - p) = p$, and so Eq. (2) becomes

$$b(n - x; n, 1 - p) = \frac{n!}{(n - x)!x!} (1 - p)^{n - x} p^x \tag{3}$$

The right side of Eq. (1) equals the right side of Eq. (3), and so the theorem is proved. □

---

**EXAMPLE 5-23:** Ninety percent of the disk drives manufactured by the COMDISK Company are known to function properly. What is the probability, for a collection of 18 disk drives, that (**a**) exactly 15 function properly, (**b**) at least 15 function properly, and (**c**) at most 14 function properly?

*Solution: Preliminaries:* The binomial distribution model is assumed to apply here, with $n = 18$ and $p = .90$, where $p$ is interpreted as the probability that a single disk drive will function properly.

(**a**) *Method 1:* Here, we have

$P(\text{exactly 15 out of 18 disk drives function properly}) = b(15; 18, .90)$

Thus, we wish to calculate $b(15; 18, .90)$. Since Table I does not have a $p = .90$ entry in the top margin, we make use of Theorem 5.6. Thus, we have $b(15; 18, .90) = b(3; 18, .10)$. Looking up the latter probability in Table I, we see that our answer is .1680.

*Method 2:* By common sense,

$P(\text{exactly 15 out of 18 disk drives function properly})$

$= P(\text{exactly 3 out of 18 disk drives do not function properly}) \quad (1)$

Now, the probability $\hat{p}$ that a single disk drive does not function properly is equal to $1 - p = 1 - .90 = .10$. Thus, we wish to calculate the probability of $\hat{x} = 3$ binomial "successes" in 18 trials, where "success" means that a disk drive doesn't function properly. This probability is given as .1680 in Table I of Appendix A. [Here, the probability of "success" in a single trial is $\hat{p} = .10$.]

**(b)** Making use of Theorem 5.6 and Table I of Appendix A, we have

$$P(\text{at least 15 out of 18 function properly})$$

$$= \sum_{x=15}^{18} b(x; 18, .90)$$

$$= \sum_{x=15}^{18} b(18 - x; 18, .10) = b(3; 18, .10) + \ldots + b(0; 18, .10)$$

$$= .1501 + .3002 + .2835 + .1680 = .9018$$

**(c)**    $$P(\text{at most 14 out of 18 function properly})$$

$$= 1 - P(\text{not at most 14 out of 18 function properly})$$

$$= 1 - P(\text{at least 15 out of 18 function properly})$$

Using the result from part (b), we get $1 - .9018 = .0982$ as the answer.

## 5-4. Drawing Without Replacement; The Hypergeometric Probability Distribution

### A. Preliminaries

In Example 3-17 and Problem 3-14, we considered drawing cards without replacement from a five-card deck containing three spades [ace, two, and three] and two hearts [ace and two]. These cards were symbolized as $AS$, $2S$, $3S$, $AH$, $2H$. We shall use situations involving the drawing without replacement of cards from this deck as models for our general approach.

In Example 3-17 and Problem 3-14, we pictured the cards as being drawn out one by one, in sequence. [This is also referred to as drawing the cards out *sequentially*.]

**EXAMPLE 5-24:** In Problem 3-14, we calculated the probability of obtaining spade, heart, and spade, in that order, when drawing three cards without replacement from the model five-card deck described above. From Theorem 3.7 [Fundamental Counting Principle], we obtained

$$N(\mathscr{S}) = 5 \cdot 4 \cdot 3 \qquad \text{and} \qquad N(S_1 \cap H_2 \cap S_3) = 3 \cdot 2 \cdot 2$$

and thus,

$$P(S_1 \cap H_2 \cap S_3) = \frac{N(S_1 \cap H_2 \cap S_3)}{N(\mathscr{S})} = \frac{3 \cdot 2 \cdot 2}{5 \cdot 4 \cdot 3} \qquad (1)$$

Recall that we read the symbol $\cap$ as "and"; in fact, on occasion, we shall use the word "and" in place of $\cap$. Now, it is also true [see Theorem 3.2] that

$$P(S_1 \cap H_2 \cap S_3) = P(S_1) \cdot P(H_2 \mid S_1) \cdot P(S_3 \mid S_1 \cap H_2) \qquad (2)$$

Show that $P(S_1) = \frac{3}{5}$, $P(H_2 \mid S_1) = \frac{2}{4}$, and $P(S_3 \mid S_1 \cap H_2) = \frac{2}{3}$. Thus, we wish to verify [see Figure 5-3] that the quotient of the first pair of numbers lined up vertically in Eq. (1) above is $P(S_1)$, the quotient of the second pair is $P(H_2 \mid S_1)$, and the quotient of the third pair of numbers lined up vertically is $P(S_3 \mid S_1 \cap H_2)$.

*Solution:* First, from applying Theorem 3.7 to both numerator and denominator, we have

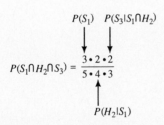

**Figure 5-3.** Interpreting vertical quotients in Example 5-24.

$$P(S_1) = \frac{N(S_1)}{N(\mathscr{S})} = \frac{3 \cdot 4 \cdot 3}{5 \cdot 4 \cdot 3} \qquad (3)$$

and

$$P(S_1 \cap H_2) = \frac{N(S_1 \cap H_2)}{N(\mathscr{S})} = \frac{3 \cdot 2 \cdot 3}{5 \cdot 4 \cdot 3} \qquad (4)$$

Here, for example, for $N(S_1 \cap H_2)$, we have the number of ways to draw spade on card one [namely, 3] times the number of ways to then draw heart on card two [namely, 2] times the number of ways to then draw any card on card three [namely, 3], if we have the drawing of three cards in sequence, without replacement. Next, from Definition 3.1 or Theorem 3.1, we have

$$P(H_2 \mid S_1) = \frac{P(S_1 \cap H_2)}{P(S_1)} \qquad (5)$$

Then, after substituting Eqs. (3) and (4) into (5), and then simplifying, we obtain

$$P(H_2 \mid S_1) = \frac{2}{4} \qquad (6)$$

Now,

$$P(S_3 \mid S_1 \cap H_2) = \frac{P(S_1 \cap H_2 \cap S_3)}{P(S_1 \cap H_2)} \qquad (7)$$

since $(S_1 \cap H_2) \cap S_3 = S_1 \cap H_2 \cap S_3$. Substituting from Eqs. (1) and (4) into (7), and simplifying, we get

$$P(S_3 \mid S_1 \cap H_2) = \frac{2}{3} \qquad (8)$$

Now, the use of an equation like Eq. (2) above, namely

$$P(S_1 \cap H_2 \cap S_3) = P(S_1) \cdot P(H_2 \mid S_1) \cdot P(S_3 \mid S_1 \cap H_2) \qquad (2)$$

with the probabilities on the right computed from Eqs. (3), (6), and (8) above, namely from

$$P(S_1) = \frac{3}{5}, \qquad P(H_2 \mid S_1) = \frac{2}{4}, \qquad \text{and} \qquad P(S_3 \mid S_1 \cap H_2) = \frac{2}{3} \qquad (3), (6), (8)$$

is quite common in the literature for sequential approaches to drawing without replacement situations. [Observe that we first simplified the expression given for $P(S_1)$ to get 3/5 in the original Eq. (3).]

Note the intuitive plausibility of Eqs. (3), (6), and (8). Thus, we reason that $P(S_1)$, the probability that the first card is a spade, is equal to 3/5 since three out of five original cards are spades. It is plausible to think of $P(H_2 \mid S_1)$ as being equal to 2/4 since if the first card were a spade, then two out of the remaining four cards would be hearts. It is plausible to think of $P(S_3 \mid S_1 \cap H_2)$ as being equal to 2/3 since if the first two cards were spade and heart, then two out of the remaining three cards would be spades.

---

Let's now work through the calculations for a problem similar to Example 5-18, where we had a drawing of five cards, with replacement, from a standard 52-card deck. In Example 5-25 [below], however, the drawing is without replacement.

**EXAMPLE 5-25:** Five cards are drawn without replacement from a standard 52-card deck. Compute the probability of getting exactly three spades, with no regard to order. Observe that after the five cards are drawn, 47 cards remain in the deck of cards.

*Solution:* Let's use a sequential approach, as in Example 5-18. Thus, we suppose that the cards are drawn out, one by one, in sequence, from the deck of cards. [Another approach to the current example, in which we think of the five cards as being drawn out simultaneously—this will be referred to by us as a *batch approach*—will be explored in Examples 5-26 and 5-27.] Let us denote the probability we wish to calculate as $P(X = 3)$. Here, the "$X = 3$" symbolism refers to the event of obtaining exactly three spades in a draw of five cards.

Now, getting exactly three spades means getting three spades either in order $A$ or $B$ or $C$, or some other one of the orders that were listed in the solution of Example 5-18. Orders $A$, $B$, and $C$ are as follows: *SSSFF*, *SSFSF*, *SSFFS*. Here, $S$ denotes spade or success, and $F$ denotes nonspade or failure to get spade. Thus,

$$P(X = 3) = P(A \text{ or } B \text{ or } C \text{ or} \ldots) \tag{1}$$

where the "or" is equivalent to $\cup$. Now, the events $A$, $B$, $C$, and so on, are mutually exclusive with respect to the experiment of drawing five cards from the 52-card deck. Thus, as in the reasoning of Example 5-18, the addition rule applies, and Eq. (1) becomes

$$P(X = 3) = P(A) + P(B) + P(C) + \ldots \tag{2}$$

Now, let's work out the calculations for orders $A$ and $B$. First, we see that we can write

$$P(A) = P(SSSFF) = P(S_1 \cap S_2 \cap S_3 \cap F_4 \cap F_5) \tag{3a}$$

Applying the multiplication rule [Theorem 3.3] to the right side of Eq. (3a), we get

$$P(A) = P(S_1)P(S_2 \mid S_1)P(S_3 \mid S_1 \cap S_2)P(F_4 \mid S_1 \cap S_2 \cap S_3)$$
$$\times P(F_5 \mid S_1 \cap S_2 \cap S_3 \cap F_4) \tag{3b}$$

This becomes

$$P(A) = \frac{13}{52} \cdot \frac{12}{51} \cdot \frac{11}{50} \cdot \frac{39}{49} \cdot \frac{38}{48} \tag{3c}$$

if we use an approach similar to that discussed in the text preceding this example. [Of course, we could proceed as in Problem 3-14. Thus, from Eq. (3a), we would have

$$P(S_1 \cap S_2 \cap S_3 \cap F_4 \cap F_5) = \frac{N(S_1 \cap S_2 \cap S_3 \cap F_4 \cap F_5)}{N(\mathscr{S})} = \frac{13 \cdot 12 \cdot 11 \cdot 39 \cdot 38}{52 \cdot 51 \cdot 50 \cdot 49 \cdot 48}$$

Here, we have applied Theorem 3.7 to both the numerator and denominator of the second expression above.] Proceeding in similar fashion for $P(B)$, we find that

$$P(B) = P(SSFSF) = \frac{13}{52} \cdot \frac{12}{51} \cdot \frac{39}{50} \cdot \frac{11}{49} \cdot \frac{38}{48} \tag{3d}$$

Observe that $P(A) = P(B)$. In fact, each of the favorable orders $A$, $B$, $C$, and so on, has the same probability. That is

$$P(A) = P(B) = P(C) = \frac{13}{52} \cdot \frac{12}{51} \cdot \frac{11}{50} \cdot \frac{39}{49} \cdot \frac{38}{48} \tag{4}$$

Now the number of terms on the right side of Eq. (2) above is the same as the number of favorable orders. This quantity, call it $K$ as in Example 5-18, is the number of permutations of the five letters $S$, $S$, $S$, $F$, $F$, taken all at a time.

Thus, again as in Example 5-18, we have

$$K = P(5; 3, 2) = \frac{5!}{3!2!} = 10 \tag{5}$$

Now, all the terms on the right of Eq. (2) are equal to $P(A)$, and there are $K$ of them. Thus,

$$P(X = 3) = K \cdot P(A) = \underbrace{\frac{5!}{3!2!}}_{K} \cdot \underbrace{\frac{13}{52} \cdot \frac{12}{51} \cdot \frac{11}{50} \cdot \frac{39}{49} \cdot \frac{38}{48}}_{P(A)} = .0815 \tag{6}$$

It is useful to compare the next-to-last expression in Eq. (6) here with the right side of Eq. (6) of Example 5-18. There, we had drawing with replacement, which led to independent trials. The only difference in our calculations is in the different values calculated for $P(A)$.

---

**EXAMPLE 5-26:** Suppose that we have the same situation as in Example 5-25, but now let us visualize that the five cards are drawn out simultaneously [that is, in a *batch*] from the 52-card deck. Recalculate $P(X = 3)$, the probability of getting exactly three spades in a draw of five cards [with no regard to order]. Use an approach similar to that employed in Example 4-15.

*Solution:* If we draw a batch of five cards from the 52-card deck, then the total number of possible hands containing five cards is equal to the number of combinations of 52 things taken 5 at a time. That is, using the binomial coefficient symbol for the number of combinations, we have

$$N(S) = \binom{52}{5} \tag{1}$$

These hands [or combinations] are equally likely. Here, the symbol $S$ refers to a sample space of all possible equally likely hands. [In the approach we are using here, as in Example 4-15, we don't care about order within a hand. That is, a hand containing five cards chosen from 52 cards is equivalent to a combination of 52 things taken 5 at a time. Also, we would expect that all $\binom{52}{5}$ combinations (hands) are *equally likely*. In essence, we are considering a sample space which consists of $\binom{52}{5}$ equally likely elements, where each element is a combination of five items, chosen from 52 items.]

Now, as in Example 4-15, the probability $P(X = 3)$ is given by

$$P(X = 3) = \frac{N(X = 3)}{N(S)} \tag{2}$$

Here we are making use of Theorem 1.2. Also, $N(S)$ is given by Eq. (1). Now, $N(X = 3)$, the number of favorable hands [that is, the number of hands, or combinations, that contain exactly three spades, which is to say three spades and two nonspades] is given by the product

$$N(X = 3) = \binom{13}{3}\binom{39}{2} \tag{3}$$

Here, $\binom{13}{3}$ is the number of ways to choose 3 spades from the 13 spades available in the deck, and $\binom{39}{2}$ is the number of ways to choose 2 nonspades from the 39 nonspades available in the deck. In Eq. (3), as in Example 4-15, we have made use of Theorem 3.7 [the fundamental counting principle]. Substituting from Eqs. (1) and (3) into Eq. (2), we get

$$P(X = 3) = \frac{\binom{13}{3}\binom{39}{2}}{\binom{52}{5}} \tag{4}$$

Now, for the values of various terms of Eq. (4), we have

$$\binom{13}{3} = \frac{13!}{3! \cdot 10!} = \frac{13 \cdot 12 \cdot 11}{3 \cdot 2 \cdot 1} = 286 \qquad \binom{39}{2} = \frac{39 \cdot 38}{2 \cdot 1} = 741$$

$$\binom{52}{5} = \frac{52 \cdot 51 \cdot 50 \cdot 49 \cdot 48}{5 \cdot 4 \cdot 3 \cdot 2 \cdot 1} = 2{,}598{,}960 \tag{5a), (5b), (5c)}$$

Now, substituting from Eqs. (5a), (5b), and (5c) into Eq. (4) yields

$$P(X = 3) = \frac{286 \cdot 741}{2{,}598{,}960} = .0815 \tag{6}$$

Observe that we got the same final result in Example 5-25 by using a different approach.

---

Next, we shall consider the hypergeometric probability situation, in which we have a generalization of the drawing without replacement probability situation, such as was encountered in Examples 5-25 and 5-26.

**Definition 5.9 (Characteristics of the Hypergeometric Probability Situation):**

(a) Suppose that $n$ objects are drawn without replacement from a set containing $M$ objects, which is divided up into $a$ objects having a certain property, and $(M - a)$ objects not having that property. Sometimes, the $a$ objects are called "successes," and the $(M - a)$ other objects are called "failures."

(b) The hypergeometric random variable $X$ denotes the exact number of successes that occur in the $n$ objects drawn. The $(n + 1)$ possible values for $X$ are $0, 1, 2, \ldots, n$. The symbol $h(x; n, a, M)$ [we shall also use $f(x)$ and $P(X = x)$ from time to time] denotes the probability of getting exactly $x$ successes [with no regard to order] when drawing $n$ objects.

The proper equation for $h(x; n, a, M)$ can easily be derived from a generalization of the batch approach of Example 5-26. First, observe that the total number of different [and equally likely] ways of drawing $n$ objects from the available $M$ objects is:

$$\binom{M}{n} \tag{1}$$

Since the original set contains $a$ successes, there are $\binom{a}{x}$ possible ways to choose $x$ successes from the available $a$ successes. Since the set contains $(M - a)$ failures, there are $\binom{M - a}{n - x}$ possible ways to choose $(n - x)$ failures from the available $(M - a)$ failures. From the fundamental counting principle [Theorem 3.7], there are thus

$$\binom{a}{x} \cdot \binom{M - a}{n - x} \tag{2}$$

different ways to simultaneously choose $x$ successes and $(n - x)$ failures when drawing $n$ objects without replacement from a set of $M$ objects. Thus, the probability of getting exactly $x$ successes in a draw of $n$ objects is obtained by dividing the expression in (2) by the expression in (1). Doing this we obtain the equation for $h(x; n, a, M)$, the hypergeometric probability function, which is the symbol for the probability of exactly $x$ successes in a draw of $n$ objects, where the drawing of objects is without replacement.

**Theorem 5.7 (Hypergeometric Probability Function):**

$$h(x; n, a, M) = \frac{\binom{a}{x} \cdot \binom{M-a}{n-x}}{\binom{M}{n}}$$

The above equation applies for $x = 0, 1, 2, \ldots, (n-1), n$. For realistic situations, we have $n \le M$, $x \le a$, and $n - x \le M - a$. [In unrealistic situations, such as with $n - x > M - a$, we would expect that $h(x; n, a, M)$ would equal zero. This would indeed be the case from the above equation since Definition 4.2 indicates that $\binom{M-a}{n-x} = 0$ if $n - x > M - a$. Definition 4.2 is also relevant in other unrealistic situations.]

*notes*

(a) The term $h(x; n, a, M)$ is often replaced by $f(x)$ or $P(X = x)$. Also, we often will use the alternate form for the number of combinations symbol; that is, for example, we could use $C(a, x)$ instead of $\binom{a}{x}$. The hypergeometric probability function is also called the *hypergeometric probability distribution*, or, briefly, the *hypergeometric distribution*.

(b) A random variable whose probability function is given by Theorem 5.7 is called a *hypergeometric random variable*.

---

**EXAMPLE 5-27:** Given the model five-card deck with 3 spades and 2 hearts. [See, for instance, Examples 5-19 and 5-24.] Suppose that three cards are drawn without replacement. Let $X$ stand for the exact number of spades obtained. Thus, $X$ is a hypergeometric random variable for which $M = 5$, $a = 3$, and $n = 3$, and $f(x)$ [or $h(x; 3, 3, 5)$] is the hypergeometric probability of obtaining $x$ spades. [Refer to Theorem 5.7. Here, spade is identified with success, and nonspade or heart with failure.] Determine $f(x)$ values for $x = 0, 1, 2$, and 3, and verify that the sum is equal to 1.

*Solution:* From Theorem 5.7, we have

$$f(x) = \frac{\binom{3}{x} \cdot \binom{2}{3-x}}{\binom{5}{3}} = \frac{\binom{3}{x} \cdot \binom{2}{3-x}}{10} \qquad \text{for } x = 0, 1, 2, 3 \qquad \textbf{(1)}$$

We already did the calculation for $f(2)$ in Example 4-15, and found that $f(2) = 6/10$. It is useful to recall that $\binom{k}{0} = 1$ and $\binom{k}{k} = 1$. [See Section 4-1B.] Thus, from Eq. (1), or from common sense, we have

$$f(1) = \frac{\binom{3}{1} \cdot \binom{2}{2}}{10} = \frac{3 \cdot 1}{10} = \frac{3}{10} \qquad \textbf{(2)}$$

$$f(3) = \frac{\binom{3}{3} \cdot \binom{2}{0}}{10} = \frac{1 \cdot 1}{10} = \frac{1}{10} \qquad \textbf{(3)}$$

Now, $f(0)$ has to be zero because it is impossible to draw exactly three hearts [this is the same as zero spades] from a deck of cards that contains a total of two hearts. Also, $f(0)$ is correctly calculated from Eq. (1) to be zero since $\binom{2}{3} = 0$. [See Definition 4.2 in Section 4-1B.]

The table for this hypergeometric probability distribution follows:

| $x$ | $f(x)$ |
|---|---|
| 0 | 0 |
| 1 | 3/10 |
| 2 | 6/10 |
| 3 | 1/10 |

Sum = 10/10 = 1

## B. Hypergeometric probability examples

We say we have *sampling without replacement* if we do drawing without replacement of subsets of size $n$ from a set [or population] of size $M$. The selected subset of size $n$ is often referred to a sample of size $n$. The probability associated with such a sample is computed from the hypergeometric probability equation [Theorem 5.7] if we are concerned with the sample containing $x$ successes and $(n - x)$ failures. Table II of Appendix A is useful for calculating hypergeometric probabilities. So also are the *Tables of Hypergeometric Probability Distribution* by G. J. Lieberman and D. B. Owen [Stanford University Press, Stanford, California, 1961].

---

**EXAMPLE 5-28:** A small town has 20 voters of whom 12 are registered as Democrats, with the rest having other political affiliations. If a sample of five voters is drawn, find the probability (a) that exactly two are Democrats, (b) that at most two are Democrats, and (c) that at least three are Democrats.

*Solution:* *Preliminaries:* Here, we have a classical hypergeometric situation, for which $M = 20$, $n = 5$, and $a = 12$. [Democrats are denoted as successes here.] Note that in this and similar problems, the word "drawn" or an equivalent word [such as "chosen" or "selected"] is to be interpreted as "drawn without replacement."

**(a)** Applying the hypergeometric distribution equation [Theorem 5.7], we have

$$P(X = 2) = h(2; 5, 12, 20) = \frac{\binom{12}{2} \cdot \binom{8}{3}}{\binom{20}{5}} = \frac{66 \cdot 56}{15,504} = .238 \qquad \textbf{(1)}$$

**(b)** At most two Democrats means $X \leq 2$, or equivalently, that $X = 0$ or 1 or 2. Thus, using the addition rule for mutually exclusive events [Axiom 1.3′], we have

$$P(\text{at most 2 are Dems.}) = P(X \leq 2) = P(X = 0) + P(X = 1) + P(X = 2) \qquad \textbf{(2)}$$

Now,

$$P(X = 0) = \frac{\binom{8}{5}}{\binom{20}{5}} = \frac{56}{15,504} \qquad (3)$$

and

$$P(X = 1) = \frac{\binom{12}{1} \cdot \binom{8}{4}}{\binom{20}{5}} = \frac{12 \cdot 70}{15,504} = \frac{840}{15,504} \qquad (4)$$

Substituting from (1), (3), and (4) into (2) yields

$$P(\text{at most 2 are Dems.}) = \frac{56 + 840 + 66 \cdot 56}{15,504} = \frac{4592}{15,504} = .2962 \qquad (5)$$

(c) Here, we have that

$$P(\text{at least 3 are Dems.}) = P(X \geq 3) = 1 - P(X \leq 2) \qquad (6)$$

But the term $P(X \leq 2)$ is the same as $P(\text{at most 2 are Dems.})$. Thus, substituting from Eq. (5) into Eq. (6), we obtain

$$P(\text{at least 3 are Dems.}) = 1 - \frac{4592}{15,504} = .7038 \qquad (7)$$

---

In a drawing without replacement situation where $n$, the sample size, is small with respect to $M$, the population size, it is possible to approximate a hypergeometric probability calculation by a corresponding binomial probability calculation. [This is particularly useful for cases where $M$ is known to be very large relative to $n$, even though the exact value of $M$ is unknown.] The basis for this is the idea that drawing a few objects without replacement from a large population is very similar, from a probability standpoint, to the drawing of the objects with replacement.

The literature indicates that the approximation of the hypergeometric by the binomial distribution will be good if $n/M \leq .05$. In the approximating binomial distribution, $p$ is estimated by $a/M$, the proportion of successes in the original population.

---

**EXAMPLE 5-29:** A collection of 500 items from a manufacturing process is known to contain 20% defectives. If a sample of five is selected from this collection, (a) what is the probability that the sample contains two defectives? (b) Determine an estimate [approximation] for the probability of part (a) by using a binomial approximation.

*Solution: Preliminaries:* We are given that $M = 500$ and that $a/M = .20$. Thus, $a = (.20)(500) = 100$. Here, we are considering that the defectives are "successes."

(a) We have drawing without replacement here, with $M = 500$, $a = 100$, $n = 5$, and $x = 2$. Thus, from Theorem 5.7, we have

$$P(X = 2) = h(2; 5, 100, 500) = \frac{\binom{100}{2}\binom{400}{3}}{\binom{500}{5}} = \frac{(4950)(10,586,800)}{(2.5524468 \cdot 10^{11})} = .20531$$

$$(1)$$

Note the use of scientific notation in the denominator of the fourth expression. [Let us consider some examples of scientific notation. Observe that $123.456 \cdot 10^2$ is equivalent to 12,345.6—here, the $10^2$ is an indication that we should shift the decimal point two places to the right. Also, $123.456 \cdot 10^{-2}$ is equivalent to 1.23456—here, the $10^{-2}$ is an indication that we should shift the decimal point two places to the left.]

(b) Here, we can envision the drawing as being of the with replacement variety. Then, the probability $P(X = 2)$ will be a binomial probability. For $p$, the probability of defective ["success"] in a single trial, we use $p = a/M = .20$. Then, from Theorem 5.5, we have

$$P(X = 2) \approx b(2; 5, .2) = \frac{5!}{2!3!}(.2)^2(.8)^3 = .20480 \tag{2}$$

---

*notes*

(a) In studies dealing with the estimation of some quantity by an approximate value for that quantity, the relative error is defined by

$$\text{Relative error} = \left| \frac{A - E}{E} \right| \tag{1}$$

where $E$ denotes the exact value of a quantity, and $A$ denotes an approximate value for the quantity. In practice, the relative error is often given as a percentage, and for this one uses

$$\text{Percentage relative error} = 100 \cdot (\text{Relative error}) \tag{2}$$

Here, the percentage relative error is expressed as a percent. Thus, in Example 5-29, we have

$$\text{Percentage relative error} = 100 \cdot \left( \frac{.00051}{.20531} \right) = .248\% \tag{3}$$

(b) Observe that $n/M = 5/500 = .01$ in Example 5-29. Since $.01 < .05$, we expect the binomial probability from part (b) of Example 5-29 to be close in value to the hypergeometric probability from part (a). Indeed, this is the case as the low percentage relative error of note (a) indicates.

(c) An easy way of seeing why the binomial calculation will yield a good approximation in Example 5-29 is by using a sequential approach in part (a), similar to what was done in Example 5-25. Thus, for our hypergeometric probability $P_H(X = 2)$, we have

$$P_H(X = 2) = K \cdot P(A)$$

where $A$ is a particular order with exactly two successes in a sample of five items, say the order $SSFFF$. Also, $K$ denotes the number of permutations of five letters taken all at a time, where two of the letters are $S$'s and three are $F$'s. Thus, for $P_H(X = 2)$, we have

$$P_H(X = 2) = \overbrace{\frac{5!}{2!3!}}^{K} \cdot \overbrace{\frac{100}{500} \cdot \frac{99}{499} \cdot \frac{400}{498} \cdot \frac{399}{497} \cdot \frac{398}{496}}^{P(A)} \tag{i}$$

where the symbols $K$ and $P(A)$ are clearly indicated above Eq. (i). The value for $P_H(X = 2)$ from Eq. (i) is .20531 as in part (a) of Example 5-29. [Observe that the $P(A)$ term in Eq. (i) clearly indicates the drawing without replacement aspect of the situation.]

Let us use $P_B(X = 2)$ to indicate the corresponding binomial probability, which applies if we have drawing with replacement. From Eq. (2) of Example 5-29, we had

$$P_B(X = 2) = \frac{5!}{2!3!} \cdot (.2)^2 \cdot (.8)^3 \qquad \textbf{(ii)}$$

The closeness of $P_H(X = 2)$ and $P_B(X = 2)$ follows from the fact that in Eq. (i) each of the factors 100/500 and 99/499 is close to .2 [actually, 100/500 is equal to .2], and each of the last three factors in Eq. (i) is close to .8.

---

**EXAMPLE 5-30:** Suppose it is known that 10% of the people in a large city [population of more than 200,000] have high blood pressure. If a sample of twenty people is chosen, find the probability **(a)** that exactly three people in the sample have high blood pressure, and **(b)** that at most three people have high blood pressure.

*Solution: Preliminaries:* Here, we have an example where $n = 20$, and $M$, the population size, is large though unknown. Certainly, we can assume that $n$ is very small relative to $M$, and thus we may use the binomial distribution to approximate the hypergeometric distribution.

The proper $p$ value to use is $p = .10$ since the data indicate that there is a 10% or .10 probability that a person picked at random has high blood pressure. Here, we equate success with having high blood pressure.

**(a)** From Table I of Appendix A, we find that

$$P(X = 3) \approx b(3; 20, .10) = .1901$$

**(b)** Again using Table I, we have

$$P(X \le 3) \approx \sum_{x=0}^{3} b(x; 20, .10) = b(0; 20, .10) + \ldots + b(3; 20, .10)$$

$$= .1216 + .2702 + .2852 + .1901 = .8671$$

This high probability indicates that it is highly likely that at most three [i.e., three or fewer] people in the sample of twenty people have high blood pressure.

---

# SOLVED PROBLEMS

### Independent Trials

**PROBLEM 5-1** Suppose we have the experiment consisting of three independent tosses of a fair die. Clearly, each toss has a most subdivided sample space given by $S = \{1, 2, 3, 4, 5, 6\}$. Also, one has $P\{i\} = 1/6$ for each simple event, where $i = 1, 2, \ldots, 6$. **(a)** List the elements of the sample space $S \times S \times S$ of the experiment, and determine $P\{(i, j, k)\}$ for each simple event of $S \times S \times S$. **(b)** Let the event "odd number on the first toss" be denoted by $E_1$, let the event "more than four on the second toss" be denoted by $E_2$, and let the event "one on the third toss" be denoted by $E_3$. Calculate the probability $P(E_1 \text{ and } E_2 \text{ and } E_3)$, where the "and" can be interpreted as $\cap$.

*Solution*

**(a)** The sample space $S \times S \times S$ of the experiment has $6 \cdot 6 \cdot 6 = 216$ elements.

$$S \times S \times S = \{(1, 1, 1), (1, 1, 2), \ldots, (1, 1, 6), (1, 2, 1), (1, 2, 2), \ldots, (1, 2, 6), \ldots, (6, 6, 1),$$

$$(6, 6, 2), \ldots, (6, 6, 6)\} \qquad \textbf{(1a)}$$

Or, using symbols, we have

$$S \times S \times S = \{(i,j,k) \mid i,j,k = 1,2,3,4,5,6\} \tag{1b}$$

For each simple event of $S \times S \times S$, we have

$$P\{(i,j,k)\} = P\{i\} \cdot P\{j\} \cdot P\{k\} = \left(\frac{1}{6}\right)^3 = \frac{1}{216}$$

(b) We have $P(E_1) = \frac{3}{6}$, $P(E_2) = \frac{2}{6}$, and $P(E_3) = \frac{1}{6}$. Also, since the trials [tosses] are independent, we have

$$P(E_1 \text{ and } E_2 \text{ and } E_3) = P(E_1)P(E_2)P(E_3) = (\tfrac{3}{6})(\tfrac{2}{6})(\tfrac{1}{6}) = \tfrac{6}{216} = \tfrac{1}{36}$$

**PROBLEM 5-2** Suppose that two cards are drawn with replacement from the five-card deck which contains the ace, two, and three of spades, and the ace and two of hearts, and then a fair die is tossed. (a) Describe the most subdivided sample space for this three-trial experiment. How many elements are in the sample space? (b) Compute the probability of the event "spade followed by heart followed by even number on the die." (c) Compute the probability of getting exactly one spade on the two cards, followed by an even number on the die.

*Solution*

(a) $S = S_1 \times S_2 \times S_3$, where $S_1 = S_2 = \{AS, 2S, 3S, AH, 2H\}$, and $S_3 = \{1,2,3,4,5,6\}$. Here, $S$ has $5 \cdot 5 \cdot 6 = 150$ elements.

   The probability corresponding to any element of $S$ is $\frac{1}{5} \cdot \frac{1}{5} \cdot \frac{1}{6} = \frac{1}{150}$ since we assume the three trials are independent. Typical elements of $S$ are the ordered triples $(AS, 3S, 2)$ and $(AH, 2S, 4)$.

(b) Let $Sp$, $Ht$, and $Ev$ symbolize "spade," "heart," and "even number," respectively. Since we assume the three trials are independent, we can employ Theorem 5.3. Then from that theorem, we have

$$P(Sp_1 \cap Ht_2 \cap Ev_3) = P(Sp_1) \cdot P(Ht_2) \cdot P(Ev_3) = \tfrac{3}{5} \cdot \tfrac{2}{5} \cdot \tfrac{3}{6} = \tfrac{3}{25} = .12$$

   Here, $Sp_1$ means "spade on the first card," $Ht_2$ means "heart on the second card," and $Ev_3$ means "even number on the die."

(c) Here,

   $P(\text{exactly one spade from the two cards drawn, and even number on the die})$

$$= P[(Sp_1 \cap Ht_2 \cap Ev_3) \text{ or } (Ht_1 \cap Sp_2 \cap Ev_3)] \tag{1}$$

if we use the notation of part (b). The events on the right side of Eq. (1) are mutually exclusive. Thus, employing Axiom 1.3' and then Theorem 5.3 twice, we get

   $P(\text{exactly one spade from the two cards drawn, and even number on the die})$

$$= P(Sp_1 \cap Ht_2 \cap Ev_3) + P(Ht_1 \cap Sp_2 \cap Ev_3) = 2 \cdot \tfrac{3}{5} \cdot \tfrac{2}{5} \cdot \tfrac{3}{6} = .24 \tag{2}$$

### Discrete Random Variables

**PROBLEM 5-3** A box contains 4 red and 2 black balls. Two balls are drawn without replacement. Let the random variable $X$ equal the number of red balls obtained. (a) Find the table for the probability function $f(x)$. (b) Sketch the bar chart [line chart] of $f(x)$ versus $x$. (c) Sketch the histogram for $f(x)$.

*Solution* Preliminaries: Let us proceed as in Example 4-13, 4-14, and 4-15. We assume the $\binom{6}{2} = 15$ ways [combinations] of drawing two balls from the six balls are equally likely. Suppose we employ a sample space consisting of the 15 equally likely combinations.

(a), (b), and (c): The possible values of $X$ are 0, 1, and 2. We have

$$P(X = 0) = P(\text{two black balls}) = \frac{\binom{2}{2}}{15} = \frac{1}{15}$$

$$P(X = 1) = P(\text{one red and one black ball}) = \frac{\binom{4}{1}\binom{2}{1}}{15} = \frac{4 \cdot 2}{15} = \frac{8}{15}$$

$$P(X = 2) = P(\text{two red balls}) = \frac{\binom{4}{2}}{15} = \frac{6}{15}$$

The probability table, the bar chart, and the histogram for $f(x) = P(X = x)$ are shown in Figure 5-4. For the histogram, each rectangular area equals the corresponding probability since the width of each rectangle equals one.

**note:** The random variable here is a *hypergeometric random variable*. [See also Section 5-4; using the symbols of Section 5-4, we have $n = 2$, $a = 4$, and $M = 6$.]

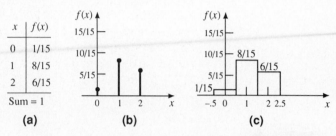

| $x$ | $f(x)$ |
|-----|--------|
| 0 | 1/15 |
| 1 | 8/15 |
| 2 | 6/15 |
| Sum = 1 | |

(a)    (b)    (c)

**Figure 5-4.** Table of probability function, bar chart, and histogram for Problem 5-3. (**a**) Table for $f(x)$. (**b**) Bar chart. (**c**) Histogram [total area = 1].

**PROBLEM 5-4** A box contains a nickel, two dimes, and a quarter. Suppose two coins are selected from the box. If random variable $X$ equals the sum of the values of the two coins, determine the probability function of $X$.

*Solution* In a situation like this, it is valid to assume that the two coins are drawn without replacement, if no other information is provided. Thus, we may take the sample space to consist of the $\binom{4}{2} = 6$ equally likely combinations that can result when two coins are chosen from four coins. Let us symbolize the four coins in the box as 5, 25, $10_a$, and $10_b$, where the subscripts $a$ and $b$ distinguish between the two dimes. The list of the 6 equally likely combinations, and the value of $X$ for each are now presented.

| Element [Combination] | Value of $X$ |
|-----------------------|--------------|
| $5-10_a$ | 15 |
| $5-10_b$ | 15 |
| $5-25$ | 30 |
| $10_a-10_b$ | 20 |
| $10_a-25$ | 35 |
| $10_b-25$ | 35 |

For example, for the element $5-10_a$, $X = 5 + 10 = 15$. Now, let us calculate $f(x)$ values for the possible values of $X$, namely for $X = 15, 20, 30$, and 35. For $f(15)$, we have

$$f(15) = P(X = 15) = P(\{5 - 10_a\}) + P(\{5 - 10_b\}) = \tfrac{1}{6} + \tfrac{1}{6} = \tfrac{2}{6}$$

and similarly for the other $f(x)$ values. Thus, the table for the probability function $f(x)$ is as follows:

| $x$ | $f(x)$ |
|-----|--------|
| 15 | 2/6 |
| 20 | 1/6 |
| 30 | 1/6 |
| 35 | 2/6 |

Sum = 6/6 = 1

**PROBLEM 5-5** Suppose the annual incomes in thousands of dollars for the Edwards, Francis, Gray, Harris, and Icklar families are 30, 35, 30, 25, and 35, respectively. If two families are chosen from the five, and if $X$ denotes the average [or mean] annual income in thousands of dollars, determine the probability function of $X$.

**Solution** We assume that the "choosing" of the two families is equivalent to drawing without replacement. Thus, we can take a sample space to consist of $\binom{5}{2} = 10$ equally likely elements [or combinations]. Now, the average income $x$ of two incomes $I_i$ and $I_j$ is given by $x = (I_i + I_j)/2$.

We let the equal incomes for the Edwards and Gray families be denoted by $30_a$ and $30_b$, respectively, and similarly for the equal incomes of the Francis and Icklar families. Thus, we obtain the following table:

| Element [Combination] | Value of $x$ |
|---|---|
| $30_a$–$35_a$ | $(30 + 35)/2 = 32.5$ |
| $30_a$–$30_b$ | $(30 + 30)/2 = 30$ |
| $30_a$–$25$ | $(30 + 25)/2 = 27.5$ |
| $30_a$–$35_b$ | $(30 + 35)/2 = 32.5$ |
| $35_a$–$30_b$ | $32.5$ |
| $35_a$–$25$ | $30$ |
| $35_a$–$35_b$ | $35$ |
| $30_b$–$25$ | $27.5$ |
| $30_b$–$35_b$ | $32.5$ |
| $25$–$35_b$ | $30$ |

Thus, we obtain the following table for $f(x)$, the probability function of random variable $X$.

| $x$ | $f(x)$ |
|---|---|
| 27.5 | 2/10 |
| 30 | 3/10 |
| 32.5 | 4/10 |
| 35 | 1/10 |

Sum $= 10/10 = 1$

**PROBLEM 5-6** Given a coin for which the probability of head on any toss is 3/5. The coin is tossed three times. Determine the probability function **(a)** for $X$, where $X$ stands for the number of heads, and **(b)** for $Y$, where $Y$ stands for the absolute value of the number of heads minus the number of tails.

**Solution** *Preliminaries:* The most subdivided sample space, whose elements are indicated in the table below, contains $2^3 = 8$ elements. The three tosses constitute independent and identical trials, and we can employ Theorem 5.3 to compute probabilities of simple events. For example,

$$P(\{hht\}) = (\tfrac{3}{5})^2(\tfrac{2}{5}) = \tfrac{18}{125}$$

**(a)** and **(b)**: We have the following preliminary table which indicates elements, corresponding probabilities, and values of $X$ and $Y$.

| Element | $P(\{\text{Element}\})$ | $x$ | $y$ |
|---|---|---|---|
| $ttt$ | 8/125 | 0 | 3 |
| $tth$ | 12/125 | 1 | 1 |
| $tht$ | 12/125 | 1 | 1 |
| $thh$ | 18/125 | 2 | 1 |
| $htt$ | 12/125 | 1 | 1 |
| $hth$ | 18/125 | 2 | 1 |
| $hht$ | 18/125 | 2 | 1 |
| $hhh$ | 27/125 | 3 | 3 |

For probabilities associated with the values of $X$, we have

$$P(X = 0) = P(\{ttt\}) = \tfrac{8}{125} \qquad P(X = 1) = P(\{tth, tht, htt\}) = 3 \cdot (\tfrac{12}{125}) = \tfrac{36}{125}$$

$$P(X = 2) = P(\{thh, hth, hht\}) = 3 \cdot (\tfrac{18}{125}) = \tfrac{54}{125} \qquad P(X = 3) = P(\{hhh\}) = \tfrac{27}{125}$$

For probabilities associated with the values of $Y$, we have

$$P(Y = 1) = P(\{tth, tht, \ldots, hht\}) = \frac{3 \cdot 12 + 3 \cdot 18}{125} = \frac{90}{125}$$

and

$$P(Y = 3) = P(\{ttt, hhh\}) = \frac{8 + 27}{125} = \frac{35}{125}$$

Thus, tables for the probability functions [or probability distributions] of $X$ and $Y$ are given below:

| $x$ | $f(x)$ | $y$ | $f(y)$ |
|-----|--------|-----|--------|
| 0 | 8/125 | 1 | 90/125 |
| 1 | 36/125 | 3 | 35/125 |
| 2 | 54/125 | | |
| 3 | 27/125 | Sum = 125/125 = 1 | |
| Sum = 125/125 = 1 | | | |

Observe that random variable $X$ is a binomial random variable for which $p = 3/5$ and $n = 3$. (See Section 5-3.)

## The Binomial Probability Distribution

**PROBLEM 5-7** Five fair dice are tossed. What is the probability of obtaining exactly three aces?

*Solution* Let us distinguish among the dice by denoting them as $die_1$, $die_2$, $die_3$, $die_4$, and $die_5$. Then, we may regard the situation as being a binomial situation with five independent trials [i.e., $n = 5$], for which the probability of success [or ace] on a single trial is $p = 1/6$. Thus,

$$P(X = 3) = b(3; 5, 1/6) = \frac{5!}{3!2!} \left(\frac{1}{6}\right)^3 \left(\frac{5}{6}\right)^2 = \frac{250}{7776} = .03215$$

**PROBLEM 5-8** A carton of 20 printer ribbons contains 7 that are defective. If a sample of eight ribbons is drawn with replacement, compute the probability that (a) exactly three will be defective, (b) at most three will be defective, and (c) at least four will be defective.

*Solution Preliminaries:* Since the drawing is with replacement, the sample of eight is equivalent to 8 independent and identical trials. Thus, the relevant probabilities are binomial probabilities with $n = 8$ and $p = 7/20 = .35$, where $p$ is the probability of a defective on any ribbon. We can look up the key probabilities in Table I of Appendix A. Let us let $X$ stand for the number of defective ribbons in a sample of size 8. Clearly, $X$ is a binomial random variable.

(a) Here, we wish to compute $P(X = 3)$, where $P(X = 3) = b(3; 8, .35) = .2786$.
(b) Here, we wish to compute $P(X \leq 3)$. We have

$$P(X \leq 3) = b(0; 8, .35) + \ldots + b(3; 8, .35) = .0319 + .1373 + .2587 + .2786 = .7065$$

(c) Here, we wish to compute $P(X \geq 4)$. We have

$$P(X \geq 4) = 1 - P(\text{not } X \geq 4) = 1 - P(X \leq 3)$$

Then, using the result from part (b) in the preceding equation, we obtain:

$$P(X \geq 4) = 1 - .7065 = .2935$$

**PROBLEM 5-9**  A box contains 4 red and 2 black balls, as in Problem 5-3. Three balls are drawn with replacement. Let the random variable $X$ equal the number of red balls obtained. **(a)** Explain why $X$ is a binomial random variable, and determine $n$ and $p$. **(b)** Determine the table for the probability function of $X$.

*Solution*

**(a)** Since the drawing is with replacement, we have an overall experiment consisting of three identical and independent trials. Each trial consists of drawing a ball from a box containing four red and two black balls. Also, the random variable $X$ denotes the number of successes [here, red balls] obtained in three trials. Thus, we have a binomial experiment with $n = 3$ and $p = 4/6 = 2/3$, the probability of obtaining a red ball on a single trial.

**(b)** From Theorem 5.5, the binomial probability function is given by

$$f(x) = b\left(x; 3, \frac{2}{3}\right) = \frac{3!}{x!(3-x)!}\left(\frac{2}{3}\right)^x\left(\frac{1}{3}\right)^{3-x} \quad \text{for } x = 0, 1, 2, 3$$

For example, we have

$$f(1) = 3\left(\frac{2}{3}\right)\left(\frac{1}{3}\right)^2 = \frac{6}{27} \quad \text{and} \quad f(2) = 3\left(\frac{2}{3}\right)^2\left(\frac{1}{3}\right) = \frac{12}{27}$$

Thus, we obtain the following table for this binomial probability function [or binomial distribution]:

| $x$ | $f(x)$ |
|---|---|
| 0 | 1/27 |
| 1 | 6/27 |
| 2 | 12/27 |
| 3 | 8/27 |

Sum $= 27/27 = 1$

**PROBLEM 5-10**  Given a coin for which the probability of head on a single toss is $3/5$. Determine the minimum number of times one should toss the coin so that the probability of getting heads one or more times is at least .999.

*Solution*  Here,

$$p = P(\text{head on single toss}) = \tfrac{3}{5} \tag{1}$$

Let $X$ be the number of heads obtained in $n$ tosses [or trials]. Then, $X$ is a binomial random variable. Here, we want to find the minimum $n$ such that

$$P(X \geq 1) \geq .999 \tag{2}$$

Now,

$$P(X \geq 1) = 1 - P(\text{not } X \geq 1) = 1 - P(X = 0 \text{ heads}) \tag{3}$$

Also,

$$P(X = 0 \text{ heads in } n \text{ trials}) = P(n \text{ tails in } n \text{ trials}) = (1 - p)^n = (\tfrac{2}{5})^n \tag{4}$$

Now, substituting from Eq. (4) into Eq. (3), we obtain

$$P(X \geq 1) = 1 - (\tfrac{2}{5})^n \tag{5}$$

Now, if we substitute from Eq. (5) into (2), we see that our problem then involves the determination of the minimum $n$ such that

$$1 - (\tfrac{2}{5})^n \geq .999 \quad \text{or} \quad .001 \geq (.4)^n \tag{6a}, (6b)$$

Taking logarithms to the base 10 [other bases, such as base $e = 2.71828\ldots$, are also useful here] on both sides of inequality (6b), we get

$$-3 \geq n(-.39794) \tag{7}$$

[Here, we used the algebraic fact that $\log a^n = n \cdot \log a$.] Now, if we divide both sides by $-.39794$ [and reverse the sense of the inequality], we obtain

$$n \geq \frac{3}{.39794} = 7.539 \qquad (8)$$

Thus, from Inequality (8), we have $n \geq 8$ since $n$ must be an integer. Thus, the answer to Problem 5-10 is that the minimum number of times to toss the coin is 8 times.

## Drawing Without Replacement; The Hypergeometric Probability Distribution

**PROBLEM 5-11** Refer to Example 5-27, where three cards were drawn without replacement from our model five-card deck containing three spades and two hearts. The hypergeometric random variable $X$ represented the number of spades in the resulting hand of three cards. Use a sequential approach to compute $P(X = 2) = h(2; 3, 3, 5)$, the probability of exactly two spades in a three-card hand.

*Solution* In the sequential approach, we envision the cards as being drawn out, one by one, in sequence. Then,

$$P(X = 2) = P(A \text{ or } B \text{ or } C) \qquad (1)$$

where the three mutually exclusive, favorable orders are $A = SSF$, $B = SFS$, and $C = FSS$.

Here, $S$ means "success" or spade, and $F$ means "failure" or heart. The orders $A$, $B$, and $C$ are the three permutations of three objects [here, letters] taken all at a time, where two are of one kind [here, $S$] and one is of another kind [here, $F$]. That is, $P(3; 2, 1) = 3$. [On other occasions, we could use the symbol $K$ for $P(3; 2, 1)$; the symbol $K$ has stood for the number of favorable orders in similar situations.]

From Eq. (1), we have

$$P(X = 2) = P(A) + P(B) + P(C) \qquad (2)$$

since $A$, $B$, and $C$ are mutually exclusive. Now,

$$P(A) = P(SSF) = P(S_1 \cap S_2 \cap F_3) = P(S_1) \cdot P(S_2 \mid S_1) \cdot P(F_3 \mid S_1 \cap S_2) \qquad (3)$$

where on the right we have made use of the multiplication rule as given in Theorem 3.2. Then,

$$P(A) = \left(\tfrac{3}{5}\right)\left(\tfrac{2}{4}\right)\left(\tfrac{2}{3}\right) = \tfrac{12}{60} \qquad (4)$$

It's not hard to see that $P(B)$ and $P(C)$ are each equal to $P(A)$. Thus, Eq. (2) becomes

$$P(X = 2) = 3 \cdot P(A) = 3 \cdot \left(\tfrac{12}{60}\right) = \tfrac{36}{60} = .60 \qquad (5)$$

This is the same result obtained in Example 5-27 from the hypergeometric probability equation [Theorem 5.7].

**PROBLEM 5-12** A school board has fifteen members, of whom ten are women and five are men. A committee of five people is to be selected by chance [by a lottery method, for example]. Find the probability that (a) exactly three members of the committee will be women, (b) at least three members of the committee will be women, and (c) at most two members of the committee will be women.

*Solution Preliminaries:* This is a hypergeometric probability problem for which $M = 15$, $a = 10$ [the number of women on the school board], and $n = 5$.

(a) $\quad P(\text{exactly 3 out of 5 will be women}) = h(3; 5, 10, 15) = \dfrac{\dbinom{10}{3}\dbinom{5}{2}}{\dbinom{15}{5}} = \dfrac{120 \cdot 10}{3003} = .3996$

Here, one could make use of Table II of Appendix A to look up values for numbers of combinations.

**(b)**   $P(\text{at least 3 out of 5 will be women}) = h(3; 5, 10, 15) + h(4; 5, 10, 15) + h(5; 5, 10, 15)$

$$= \frac{(1200 + 1050 + 252)}{3003} = \frac{2502}{3003} = .8332$$

**(c)**   $P(\text{at most 2 out of 5 will be women}) = 1 - P(\text{at least 3 out of 5 will be women})$

$$= 1 - .8332 = .1668$$

Alternatively,

$$P(\text{at most 2 out of 5 will be women}) = h(0; 5, 10, 15) + h(1; 5, 10, 15) + h(2; 5, 10, 15)$$

$$= \frac{(1 + 50 + 450)}{3003} = \frac{501}{3003} = .1668$$

**PROBLEM 5-13**   Use Theorem 4.9 to show that the sum of probability values for a hypergeometric distribution [i.e., of values of $h(x; n, a, M)$] from $x = 0$ to $x = n$, inclusive, is equal to 1.

*Solution*   The expression for the sum of hypergeometric probabilities is

$$\sum_{x=0}^{n} h(x; n, a, M) = \sum_{x=0}^{n} \frac{\binom{a}{x}\binom{M-a}{n-x}}{\binom{M}{n}} = \frac{1}{\binom{M}{n}} \cdot \sum_{x=0}^{n} \binom{a}{x}\binom{M-a}{n-x} \qquad \textbf{(1)}$$

Here, we have made use of Theorem 5.7. Now, we make use of Theorem 4.9 with $m$, $n$, $r$, and $k$ replaced, respectively, by $a$, $M - a$, $x$, and $n$. $\left[ \text{Recall that } C(m, r) \text{ is an alternate form for } \binom{m}{r}, \text{ etc.} \right]$ Thus, from Theorem 4.9, we have

$$\sum_{x=0}^{n} \binom{a}{x}\binom{M-a}{n-x} = \binom{M}{n} \qquad \textbf{(2)}$$

Now, if we substitute from Eq. (2) into the right side of Eq. (1), we obtain

$$\sum_{x=0}^{n} h(x; n, a, M) = \frac{\binom{M}{n}}{\binom{M}{n}} = 1 \qquad \textbf{(3)}$$

**PROBLEM 5-14**   A quality control technician inspects a sample of three transistor radios from each incoming lot of 60 radios. He will accept the lot if all three radios are working properly. Otherwise, he will order the entire lot to be inspected, with the cost of inspection to be charged to the vendor. What is the probability the lot will be accepted without further inspection **(a)** if the lot contains six radios that don't work properly and **(b)** if the lot contains 20 radios that don't work properly?

*Solution*   *Preliminaries:* Observe that in this hypergeometric situation we have $n = 3$ and $M = 60$ since each lot contains 60 radios, and a sample of size 3 is selected from the lot for inspection purposes. Also, we will let $a$ stand for the number of radios that do work properly.

**(a)**   Here, $a = 60 - 6 = 54$ since we are given that six radios don't work properly. Now,

$$P(\text{accept without further inspection}) = P(3 \text{ out of 3 work properly})$$

$$= h(3; 3, 54, 60) = \frac{\binom{54}{3}}{\binom{60}{3}} = \frac{54 \cdot 53 \cdot 52}{60 \cdot 59 \cdot 58} = .7248$$

**(b)** Here, $a = 40$. Thus,

$$P(\text{accept without further inspection}) = P(3 \text{ out of } 3 \text{ work properly})$$

$$= h(3; 3, 40, 60) = \frac{\binom{40}{3}}{\binom{60}{3}} = \frac{40 \cdot 39 \cdot 38}{60 \cdot 59 \cdot 58} = .2887$$

*note:* Since $n/M = 3/60 = .05$, binomial approximations should be fairly accurate. For example, for part (a), estimating $p$ with $a/M = 54/60 = .9$, we have

$$P(\text{accept without further inspection}) \approx b(3; 3, .9) = (.9)^3 = .7290$$

and this is quite close to the value previously obtained.

**PROBLEM 5-15** Refer to Examples 5-25 and 5-26. In both, we calculated the probability of obtaining exactly three spades when drawing five cards without replacement from a standard 52-card deck. In Example 5-25, we used a sequential approach [we envisioned the cards being drawn out one by one, in sequence], while in Example 5-26, we used a batch approach [we envisioned the five cards as being drawn out all at once—that is, simultaneously, as in a batch]. We obtained the same value for the hypergeometric probability $P(x = 3) = h(3; 5, 13, 52)$, namely .0815.

Show, by algebraic manipulation, and by recognition of the correspondence between $P(n, r)$ and $\binom{n}{r}$, the symbols for the numbers of permutations and combinations of $n$ objects taken $r$ at a time, that Eq. (4) of Example 5-26 can be derived from Eq. (6) of Example 5-25.

*Solution* If, in Eq. (6) of Example 5-25, we group 3! with $13 \cdot 12 \cdot 11$, 2! with $39 \cdot 38$, and 5! with $52 \cdot 51 \cdot 50 \cdot 49 \cdot 48$, we get

$$P(X = 3) = \frac{\left(\dfrac{13 \cdot 12 \cdot 11}{3!}\right)\left(\dfrac{39 \cdot 38}{2!}\right)}{\left(\dfrac{52 \cdot 51 \cdot 50 \cdot 49 \cdot 48}{5!}\right)} \qquad \textbf{(a)}$$

Now,

$$\binom{n}{r} = \frac{P(n, r)}{r!} \qquad \textbf{(b)}$$

Recall that

$$P(n, r) = n \cdot (n - 1) \cdot (n - 2) \cdot \ldots \cdot (n - r + 1) \qquad \textbf{(c)}$$

where there are $r$ factors on the right side of Eq. (c). Using (c), we can rewrite Eq. (a) as

$$P(X = 3) = \frac{\left(\dfrac{P(13, 3)}{3!}\right)\left(\dfrac{P(39, 2)}{2!}\right)}{\left(\dfrac{P(52, 5)}{5!}\right)} \qquad \textbf{(d)}$$

Next, applying Eq. (b) to the three quotients of Eq. (d), we obtain

$$P(X = 3) = \frac{\binom{13}{3}\binom{39}{2}}{\binom{52}{5}} \qquad \textbf{(e)}$$

We see that Eq. (3) preceding is identical to Eq. (4) of Example 5-26.

## Supplementary Problems

**PROBLEM 5-16**   Given an experiment consisting of the following two independent trials: a toss of a fair die followed by a draw of a card from a five-card deck which contains the ace, two, and three of spades, and the ace and two of hearts. **(a)** List the elements of the sample space $S_1 \times S_2$, where $S_1$ and $S_2$ are the most subdivided sample spaces for the two trials which comprise the experiment. **(b)** Let $E_1$ be "ace on the die," and $E_2$ be "ace on the card." Determine $P(E_1 \text{ and } E_2)$. **(c)** Let $F_1$ be "even number on the die," and $F_2$ be "spade on the card." Determine $P(F_1 \text{ and } F_2)$.

*Answer* **(a)** $S_1 \times S_2$ has $6 \cdot 5 = 30$ elements. $S_1 \times S_2 = \{(1, AS), \dots, (1, 2H), \dots, (6, AS), \dots, (6, 2H)\}$
**(b)** $\frac{1}{6} \cdot \frac{2}{5} = \frac{1}{15}$    **(c)** $\frac{3}{6} \cdot \frac{3}{5} = \frac{3}{10}$

**PROBLEM 5-17**   Given the same experiment as in Problem 5-6. That is, a coin for which $P(\text{head}) = 3/5$ is tossed three times. Let the random variable $W$ denote the product of the number of heads times the number of tails. Determine the probability function for $W$.

*Answer* $P(W = 0) = 35/125$ and $P(W = 2) = 90/125$

**PROBLEM 5-18**   A box contains 12 red balls and four green balls. Three balls are drawn with replacement. Let the random variable $X$ denote the number of red balls obtained. **(a)** Determine the table for the probability function of random variable $X$. **(b)** Sketch the bar [line] chart and histogram for the probability function of $X$.

*Answer (Partial)* **(a)** Here, $f(x)$ denotes $P(X = x)$. $f(0) = 1/64$, $f(1) = 9/64$, $f(2) = 27/64$, and $f(3) = 27/64$. Here, $X$ is a binomial random variable with $p = 3/4$ and $n = 3$.

**PROBLEM 5-19**   Repeat Problem 5-18, but this time suppose that the three balls are drawn without replacement.

*Answer (Partial)* **(a)** Let $f(x)$ denote the probability function. We find that $f(0) = 4/560$, $f(1) = 72/560$, $f(2) = 264/560$, $f(3) = 220/560$. Here, $X$ is a hypergeometric random variable with $n = 3$, $a = 12$, and $M = 16$.

**PROBLEM 5-20**   A box contains a nickel, two dimes, and a quarter. Suppose two coins are drawn, with replacement, from the box. If random variable $Y$ equals the sum of values of the two coins, determine the probability function of $Y$.

*Answer* $P(Y = 10) = 1/16$, $P(Y = 15) = 4/16$, $P(Y = 20) = 4/16$, $P(Y = 30) = 2/16$, $P(Y = 35) = 4/16$, and $P(Y = 50) = 1/16$

**PROBLEM 5-21**   In a certain country, it has been found over many years that 55% of the babies born there are males. For a family in that country with five children, what is the probability that **(a)** the two youngest children will be female and the three oldest children male, **(b)** exactly three children will be male, and **(c)** at least three children will be male?

*Answer* **(a)** .033691    **(b)** .33691    **(c)** .5931

**PROBLEM 5-22**   A shipment of 100 burglar alarms contains 10 that are defective. If four of these are shipped to a particular customer, what is the probability that **(a)** the customer will get exactly two defective burglar alarms and **(b)** the customer will get at least two defective burglar alarms?

*Answer* **(a)** .045961    **(b)** .048769

**PROBLEM 5-23**   Rework Problem 5-22 by using a binomial approximation.

*Answer* **(a)** .0486    **(b)** .0523

# 6 DISCRETE PROBABILITY FUNCTIONS CONTINUED

## 6-1. Cumulative Distribution Function for a Discrete Random Variable

It is often of interest to know the probability that the value of a random variable $X$ is less than or equal to some fixed number $x$. This quantity is called the *cumulative distribution function* [or the *distribution function* or the *cumulative distribution*] for the random variable $X$.

**Definition 6.1 (Cumulative Distribution Function):** Given a random variable $X$ with probability function $f(x)$. The function of $x$ given by

$$F(x) = P(X \leq x) \tag{a}$$

is called the cumulative distribution function for the random variable $X$.

The definition as given by Eq. (a) is general and applies also to random variables that are continuous [see Section 8-1], or that are a mixture of discrete and continuous. For discrete random variables, since $P(X \leq x)$ equals the sum of discrete probabilities for all numbers less than or equal to $x$, that is since $P(X \leq x) = \sum_{t \leq x} f(t)$, we have

$$F(x) = \sum_{t \leq x} f(t) \tag{b}$$

as the equation for $F(x)$ for a discrete random variable $X$ with probability function $f(x)$.

*notes*

(a) The reader should not be confused by the "$t$" that appears in Eq. (b). This $t$ is merely a symbol for a variable whose values are less than or equal to the fixed value $x$. In Example 6-1 below, we shall use $x_0$ to indicate a fixed value of the random variable $X$, and we shall replace Eq. (b) by the following equivalent form, in which, additionally, $t$ is replaced by $x$:

$$F(x_0) = \sum_{x \leq x_0} f(x) \tag{b'}$$

(b) Often, we shall abbreviate cumulative distribution function as cdf. Also, we shall occasionally abbreviate probability function as pf, and random variable as rv.

(c) The cumulative distribution function evaluated at $x_0$, namely $F(x_0)$, should be interpreted as a measure of the accumulation of probability up to and including $x_0$.

**EXAMPLE 6-1:** Refer to Examples 5-14 and 5-15, which dealt with the binomial random variable $X$ for $n = 2$ and $p = 1/2$. In those examples, $X$ represented the number of heads obtained in two tosses of a fair coin. The table for the probability function of $X$ [known as $f(x)$ or $b(x; 2, 1/2)$] is as follows:

| $x$ | 0 | 1 | 2 |
|---|---|---|---|
| $f(x)$ | 1/4 | 2/4 | 1/4 |

Here, we laid out this table horizontally, while in Example 5-15 it was laid out vertically. The corresponding bar [line] chart is reproduced below in Figure 6-1a. Determine the cumulative distribution function $F(x)$ in terms of $x$ for all $x$. Also, draw the graph for $F(x)$ in terms of $x$.

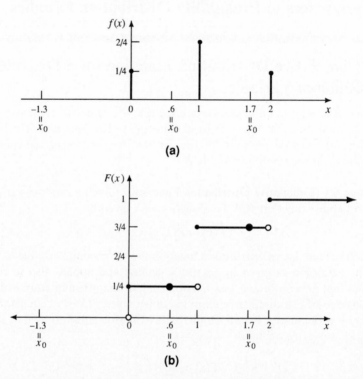

**(a)**

**(b)**

**Figure 6-1.** Bar chart and graph for cumulative distribution function for Example 6-1. (a) Bar chart. (b) Cumulative distribution function graph.

*Solution:* Let us tentatively indicate a fixed value of random variable $X$ by the symbol $x_0$. Thus, we have

$$F(x_0) = P(X \leq x_0) \tag{1}$$

where the right side is read as "the probability that random variable $X$ is less than or equal to $x_0$." As an example, let us choose $x_0$ equal to $-1.3$. Refer to Figure 6-1a, where the number $-1.3$ is marked on the $x$ scale. We see that

$$F(-1.3) = P(X \leq -1.3) = 0 \tag{2}$$

That is, no probability has been accumulated up to and including $-1.3$. Now, an equation like Eq. (2) holds for all $x_0$ values which are less than 0. For other examples, consider $x_0 = -10$, $x_0 = -6.25$, $x_0 = -.03$, etc. Summarizing, we can write

$$F(x_0) = 0 \qquad \text{for all } x_0 < 0 \tag{3}$$

Now, let $x_0 = 0$. Then, we have

$$F(0) = P(X \leq 0) = f(0) = 1/4 \qquad \textbf{(4)}$$

That is, $F(0) = 1/4$ since for $x_0$ equal to zero the probability accumulated up to and including zero is the probability at zero, and that probability is 1/4. Now, let us set $x_0$ equal to .6. Refer again to Figure 6-1a, where $x_0 = .6$ is marked off on the $x$-axis. We have

$$F(.6) = P(X \leq .6) = f(0) = 1/4 \qquad \textbf{(5)}$$

since the only value of $X$ less than or equal to .6 where a positive probability occurs is $X = 0$. Equations like Eqs. (4) and (5) hold for all $x_0$'s such that $0 \leq x_0 < 1$. For example, consider $x_0 = .25$, $x_0 = .436$, and $x_0 = .9994$. Summarizing, we can write

$$F(x_0) = 1/4 \qquad \text{for} \qquad 0 \leq x_0 < 1 \qquad \textbf{(6)}$$

If we set $x_0$ equal to 1, we have

$$F(1) = P(X \leq 1) = f(0) + f(1) = 1/4 + 2/4 = 3/4 \qquad \textbf{(7)}$$

since those values of $X$ less than or equal to 1 which have positive probabilities are $X = 0$ and $X = 1$. Suppose we let $x_0$ equal 1.7. Refer to Figure 6-1a, where the number 1.7 is marked on the $x$ scale. We see that

$$F(1.7) = P(X \leq 1.7) = f(0) + f(1) = 3/4 \qquad \textbf{(8)}$$

Again, we observe that those values of $X$ less than or equal to 1.7 which have positive probabilities are $X = 0$ and $X = 1$. Now, equations like Eqs. (7) and (8) hold for all $x_0$'s such that $1 \leq x_0 < 2$. For example, consider $x_0 = 1.36$, $x_0 = 1.525$, and $x_0 = 1.99996$.

Summarizing, we can write

$$F(x_0) = 3/4 \qquad \text{for} \qquad 1 \leq x_0 < 2 \qquad \textbf{(9)}$$

If we set $x_0$ equal to 2, we have

$$F(2) = P(X \leq 2) = f(0) + f(1) + f(2) = 1/4 + 2/4 + 1/4 = 1 \qquad \textbf{(10)}$$

since those values of $X$ less than or equal to 2 which have positive probabilities are $X = 0$, $X = 1$, and $X = 2$. Suppose we let $x_0$ equal 3.4. We see that

$$F(3.4) = P(X \leq 3.4) = f(0) + f(1) + f(2) = 1 \qquad \textbf{(11)}$$

Again, we observe that those values of $X$ less than or equal to 3.4 which have positive probabilities are $X = 0$, $X = 1$, and $X = 2$. Now, equations like Eqs. (10) and (11) hold for all $x_0$'s such that $2 \leq x_0$. For example, consider $x_0 = 2.13$, $x_0 = 6.84$, and $x_0 = 1356.2$.

Summarizing, we can write

$$F(x_0) = 1 \qquad \text{for} \qquad 2 \leq x_0 \qquad \textbf{(12)}$$

Now, if we take the key equations (3), (6), (9), and (12) and replace $x_0$ by $x$ throughout, we get the following table for the cumulative distribution function, $F(x)$:

$$F(x) = \begin{cases} 0 & \text{for} \quad x < 0 \\ 1/4 & \text{for } 0 \leq x < 1 \\ 3/4 & \text{for } 1 \leq x < 2 \\ 1 & \text{for } 2 \leq x \end{cases} \qquad \textbf{(13)}$$

The graph of the cdf is shown in Figure 6-1b. Note that $F(x)$ is discontinuous at $x = 0$, 1, and 2. It provides an example of a *step function*. The holes [o] at $x = 0$, 1, and 2 indicate points that are not on the graph, while the "filled-in" dots [●] at $x = 0$, 1, and 2 indicate points that are on the graph of $F(x)$. For example, the point $(1, 1/4)$ is not on the graph of $F(x)$ [there is a hole at that point], while the point (1, 3/4) is on the graph of $F(x)$. [There is a filled-in dot at that point.]

---

*note:* For purposes of uniformity, we can introduce the $\infty$ symbol, and write the first and last items of (13) as

$$F(x) = 0 \qquad \text{for } -\infty < x < 0$$

and

$$F(x) = 1 \qquad \text{for } 2 \le x < \infty$$

Thus, each of the four intervals in (13) would then have $x$ bounded both below and above.

---

**EXAMPLE 6-2:** Determine the cumulative distribution function $F(x)$ for the random variable of Problem 5-4. Sketch the bar chart for $f(x)$, and the graph of the cdf.

*Solution:* The table for the probability function of the random variable $X$ is as follows:

| $x$ | 15 | 20 | 30 | 35 |
|---|---|---|---|---|
| $f(x)$ | 2/6 | 1/6 | 1/6 | 2/6 |

The bar chart of $f(x)$ is shown in Figure 6-2a. Let us now work with the general equation for $F(x_0)$ given by $F(x_0) = P(X \le x_0)$, as in Example 6-1. We

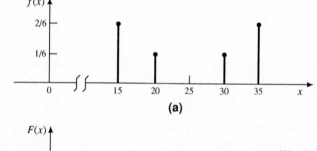

**(a)**

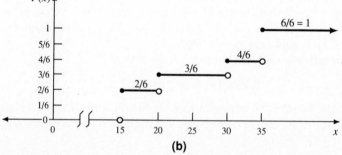

**(b)**

**Figure 6-2.** Bar chart and graph of cumulative distribution function for Example 6-2. **(a)** Bar chart for $f(x)$. **(b)** Graph of cumulative distribution function $F(x)$.

can then build up to our final results by first considering specific values of $x_0$, again as in Example 6-1. Refer to Table 6-1 below.

**TABLE 6-1: Derivation of Results for $F(x_0)$ in Terms of $x_0$ for Example 6-2**

| Sample calculations | Summary result for $F(x_0)$ |
|---|---|
| $F(6.2) = P(X \le 6.2) = 0$ | $F(x_0) = 0$ for $x_0 < 15$ |
| $F(15) = P(X \le 15) = f(15) = 2/6$<br>$F(18.3) = P(X \le 18.3) = f(15) = 2/6$ | $F(x_0) = 2/6$ for $15 \le x_0 < 20$ |
| $F(20) = P(X \le 20) = f(15) + f(20)$<br>$\quad = 2/6 + 1/6 = 3/6$<br>$F(26.8) = P(X \le 26.8) = f(15) + f(20)$<br>$\quad = 2/6 + 1/6 = 3/6$ | $F(x_0) = 3/6$ for $20 \le x_0 < 30$ |
| $F(30) = P(X \le 30) = f(15) + f(20) + f(30)$<br>$\quad = 2/6 + 1/6 + 1/6 = 4/6$<br>$F(34.99) = P(X \le 34.99) = f(15) + f(20) + f(30)$<br>$\quad = 2/6 + 1/6 + 1/6 = 4/6$ | $F(x_0) = 4/6$ for $30 \le x_0 < 35$ |
| $F(35) = P(X \le 35) = f(15) + f(20) + f(30) + f(35)$<br>$\quad = 2/6 + 1/6 + 1/6 + 2/6 = 1$<br>$F(98) = P(X \le 98) = f(15) + f(20) + f(30) + f(35)$<br>$\quad = 2/6 + 1/6 + 1/6 + 2/6 = 1$ | $F(x_0) = 1$ for $35 \le x_0$ |

The summary results for $F(x_0)$ in terms of $x_0$ are presented in the right column of Table 6-1. For the summary results for $F(x)$ in terms of $x$, one merely has to replace $x_0$ by $x$ throughout. The graph of $F(x)$ in terms of $x$ is shown in Figure 6-2b. As with the cdf of Example 6-1, the current $F(x)$ function is discontinuous, and is known as a step function.

Theorem 6.1 below expresses some of the main properties of a cumulative distribution function of a discrete random variable. It can be derived fairly easily from the axioms and early theorems of probability.

**Theorem 6.1:** Values of the cumulative distribution function $F(x)$ satisfy the following conditions:

(a) $0 \le F(x) \le 1$ for any $x$.
(b) If $z$ is less than the lowest value of $x$ for which $f(x)$ is positive, then $F(z) = 0$. If $w$ is greater than the highest value of $x$ for which $f(x)$ is positive, then $F(w) = 1$.
(c) $F(-\infty) = 0$ and $F(\infty) = 1$.
(d) If $s < t$, then $F(s) \le F(t)$ for any real numbers $s$ and $t$.

Suppose the range space $R_x$ of a random variable $X$ consists of the finite collection of values $x_1, x_2, \ldots, x_n$. [See Definition 5.6 for the definition of *range space*. Recall that if $x_i$ is in the range space of a random variable $X$, then it is most often the case that $f(x_i)$ is a positive number. We will suppose that is the situation that applies here.] Then, $F$ is a step function with $n$ steps, with steps occurring at each of $x_1, x_2, \ldots, x_n$. The graph of $F$ is comprised of $(n + 1)$ horizontal line segments.

*note:* For instance, in Example 6-1, the range space consisted of the $X$ values 0, 1, and 2. In Figure 6-1, we observe the steps that occur at those three values of the graph of $F(x)$.

**Theorem 6.2:** Suppose that the only values of random variable $X$ where $f(x)$ is positive are the values $x_1, x_2, \ldots, x_n$ where $x_1 < x_2 < \ldots < x_n$.

(a) Then,

$$F(x) = \begin{cases} 0 & \text{for } -\infty < x < x_1, \\ f(x_1) & \text{for } x_1 \leq x < x_2, \\ f(x_1) + f(x_2) & \text{for } x_2 \leq x < x_3, \\ \vdots \\ f(x_1) + f(x_2) + \ldots + f(x_n) = 1 & \text{for } x_n \leq x < \infty \end{cases}$$

(b) From (a), it follows that $f(x_1) = F(x_1)$, $f(x_2) = F(x_2) - F(x_1)$, and so on. That is,

$$f(x_i) = F(x_i) - F(x_{i-1}) \qquad \text{for } i = 2, 3, \ldots, n$$

---

**EXAMPLE 6-3:** Verify Theorem 6.2 with respect to Example 6-2.

*Solution:* First, we observe that the $x$ values where the probability function is positive are $x_1 = 15$, $x_2 = 20$, $x_3 = 30$, and $x_4 = 35$. [These $x_i$ values comprise the $x$ values that are in the range space $R_x$.] We see that the table for $F(x)$ in terms of $x$ as given in Table 6-1 verifies part (a) of Theorem 6.2 [replace $x_0$ by $x$].

Verification of part (b) is provided by the observations that $f(15) = F(15) = 2/6$, $f(20) = F(20) - F(15) = 3/6 - 2/6 = 1/6$, $f(30) = F(30) - F(20) = 4/6 - 3/6 = 1/6$, and $f(35) = F(35) - F(30) = 6/6 - 4/6 = 2/6$.

---

## 6-2. More Discrete Probability Functions

In this section, we shall focus on several frequently occurring discrete random variables and the probability functions [probability distributions] that correspond to them.

### A. Poisson probability function

For a calculation of a binomial probability for which $n$ is large and $p$ is small, the use of Theorem 5.5 can prove to be fairly awkward. For example, let us say we wish to calculate the probability that exactly six people out of 50,000 people who fly from New York to Chicago will be involved in an airplane accident. Suppose the probability is .0001 that a single individual flying the New York to Chicago route will be involved in an airplane accident.

Using Theorem 5.5, we would attempt to calculate the following:

$$b(6; 50{,}000, .0001) = \frac{50{,}000!}{6! \cdot (49{,}994!)} (.0001)^6 (.9999)^{49{,}994} \tag{1}$$

The calculation here is quite messy. Fortunately, we can [and will] approximate the above binomial probability by employing the easier-to-use Poisson probability function.

In particular, the Poisson probability function arises as a limiting form of the binomial probability function when $n \to \infty$ and $p \to 0$, with the product $np$ held constant. In the approximation of the binomial probability function by the Poisson probability function, the product $np$ is set equal in value to the quantity $\lambda$, which occurs in the expression for the Poisson probability function. [More will be said about $\lambda$ shortly.]

**Definition 6.2 (Poisson Probability Function):** Below, $\lambda$ denotes a positive constant. The probability function given by

$$p(x; \lambda) = \frac{\lambda^x e^{-\lambda}}{x!} \qquad \text{for } x = 0, 1, 2, \ldots$$

is called the *Poisson probability function* [or *Poisson probability distribution* or merely the *Poisson distribution*] with "parameter" $\lambda$. The term $p(x; \lambda)$ above is often replaced by $f(x)$ or $P(X = x)$.

### notes

(a) A random variable with Poisson probability function is called a Poisson random variable. This is consistent with the terminology used for binomial and hypergeometric random variables, and, in fact, for any random variable whose associated probability function has been named.

(b) Here, the Poisson random variable takes on a countably infinite number of values, and this is indicated in Definition 6.2 above by writing $x = 0$, $1, 2, \ldots$.

(c) Some practitioners allow for $\lambda = 0$. In this limiting case, the Poisson probability function reduces to the following: $p(x; \lambda) = 1$ for $x = 0$ and $p(x; \lambda) = 0$ for $x = 1, 2, 3, \ldots$.

(d) We'll have more to say about the term "parameter" in Section 6-3.

In the literature [see, for example, Strait (1989), Hogg and Tanis (1983), and Freund and Walpole (1987)], it is indicated that the Poisson distribution will provide a good approximation to the binomial distribution when $n \geq 20$ and $p \leq .05$, and a very good approximation when $n \geq 100$ and $np \leq 10$. In the approximation by the Poisson distribution, $\lambda$ is set equal to the product $np$.

---

**EXAMPLE 6-4:** Refer to Eq. (1) preceding Definition 6.2. Compare the values of $b(6; 50,000, .0001)$ and the corresponding value of the Poisson probability function with the same value of $X$, namely $X = 6$, and with $\lambda = np = (50,000)(.0001) = 5$.

*Solution:* Let's use some of the symbolism of Chapter 5. Thus, we can write:

$$b(6; 50,000, .0001) = K \cdot P(A) \tag{1}$$

where

$$K = \frac{50,000!}{6!\,49,994!} = \frac{50,000 \cdot 49,999 \cdot \ldots \cdot 49,995}{6!} \tag{2}$$

and

$$P(A) = (.0001)^6 (.9999)^{49,994} \tag{3}$$

We find that

$$b(x; n, p) = b(6; 50,000, .0001) = .146230 \tag{4}$$

Now, for $x = 6$ and $\lambda = 5$, we see that

$$p(x, \lambda) = p(6; 5) = \frac{5^6 e^{-5}}{6!} = .146223 \tag{5}$$

Thus, in this situation, we see that the Poisson probability function provides an approximation of the binomial probability function which is accurate to four significant digits. It should be clear that the calculation involved in Eq. (5) is much easier and quicker than the calculation involved in Eq. (4). The very good accuracy of the approximation is expected since $n \geq 100$ and $np = \lambda \leq 10$.

---

We can look up Poisson probability values in Appendix A, Table IV.

An illustration of the limiting behavior of binomial probabilities with $np$ held constant at 10, and $x$ held constant at 11 is provided in Problem 6-2.

In that problem, $n$ is allowed to increase without bound with $p$ simultaneously approaching zero, but with the product $np$ held constant at 10.

**Theorem 6.3 (Poisson Probability Function as a Limit of the Binomial Probability Function):** The limit of the binomial probability function $b(x; n, p)$ as $n \to \infty$ and $p \to 0$, with $np$ held constant and $x$ held constant, is the Poisson probability function $p(x; \lambda)$ with the same value of $x$, and with $\lambda = np$.

The proof of Theorem 6.3 that is done in Example 6-5 depends on employing the limit concept of calculus.

---

**EXAMPLE 6-5:** Prove Theorem 6.3.

*Proof:* In the proof that follows, we shall make use of the following two limit results [from calculus], which are indicated as Eqs. (a) and (b).

$$\lim_{z \to 0} (1 - z)^{-1/z} = e \qquad \qquad \textbf{(a)}$$

$$\lim_{z \to 0} (1 - z)^x = 1, \qquad \text{where } x \text{ is a constant} \qquad \textbf{(b)}$$

[Recall that $e$ denotes a universal constant whose value is 2.71828 ....]

From the equation for $b(x; n, p)$ as given in Theorem 5.5, we obtain the following:

$$b(x; n, p) = \frac{n(n - 1)(n - 2) \cdot \ldots \cdot (n - x + 1)}{x!} (p)^x (1 - p)^{n - x} \qquad \textbf{(1)}$$

[Note that in the fractional expression of Eq. (1), in which $x!$ occurs in the denominator, there are $x$ factors in the numerator.] Now, if we factor $n$ out of each of the $x$ factors in the numerator of the fractional expression of Eq. (1), we obtain

$$n(n - 1)(n - 2) \cdot \ldots \cdot (n - x + 1) = n^x \cdot 1 \cdot \left(1 - \frac{1}{n}\right)\left(1 - \frac{2}{n}\right) \cdot \ldots \cdot \left(1 - \frac{x - 1}{n}\right) \qquad \textbf{(2)}$$

Now, forming the product of $n^x$ with $p^x$, we get $(np)^x$. Also, let us write $(1 - p)^{n-x}$ as $(1 - p)^n / (1 - p)^x$. With these modifications, Eq. (1) becomes

$$b(x; n, p) = \frac{1\left(1 - \dfrac{1}{n}\right)\left(1 - \dfrac{2}{n}\right) \cdot \ldots \cdot \left(1 - \dfrac{x - 1}{n}\right)}{(1 - p)^x} \frac{(np)^x}{x!} (1 - p)^n \qquad \textbf{(3)}$$

Now, let us take the limit in Eq. (3) as $n \to \infty$ and $p \to 0$, but with the product $np$ held constant, and with $x$ held constant. First, let

$$\lambda = np \qquad \qquad \textbf{(4)}$$

Also, let us rewrite the last term of Eq. (3) as follows:

$$(1 - p)^n = [(1 - p)^{-1/p}]^{-np} = [(1 - p)^{-1/p}]^{-\lambda} \qquad \textbf{(5)}$$

[Here, we made use of the algebraic result that $x^{ab} = (x^a)^b$, for the case where $x = (1 - p)$, $a = -1/p$, $b = -np$, and $ab = n$.] Now, if we take the limit in Eq. (3) in the way indicated above [that is, let $n \to \infty$ and $p \to 0$, but with the product $np$ held constant, and with $x$ held constant], we obtain

$$\lim b(x; n, p) = \frac{\lim\left[1\left(1 - \dfrac{1}{n}\right)\left(1 - \dfrac{2}{n}\right) \cdot \ldots \cdot \left(1 - \dfrac{x - 1}{n}\right)\right]}{\lim (1 - p)^x} \frac{\lambda^x}{x!} \lim (1 - p)^n \qquad \textbf{(6)}$$

Now, employing Eqs. (a) and (b) and the final form for $(1 - p)^n$ as given in Eq. (5), we get

$$\lim_{n \to \infty} \left[ 1 \left( 1 - \frac{1}{n} \right) \left( 1 - \frac{2}{n} \right) \cdots \left( 1 - \frac{x-1}{n} \right) \right] = 1 \tag{7}$$

$$\lim_{p \to 0} (1 - p)^x = 1 \tag{8}$$

and

$$\lim_{p \to 0} (1 - p)^n = \lim_{p \to 0} [(1 - p)^{-1/p}]^{-\lambda} = e^{-\lambda} \tag{9}$$

Observe that Eqs. (8) and (9) follow from applying Eqs. (b) and (a) above with $z = p$. Now, applying the limit results from Eqs. (7), (8), and (9) to Eq. (6) results in

$$\lim b(x; n, p) = \frac{\lambda^x e^{-\lambda}}{x!} \tag{10}$$

Hence, the limiting probability function is the Poisson probability function $p(x, \lambda)$. ☐

**EXAMPLE 6-6:** In the production of fuses in a factory, it is known that 5% of them will be defective. If a box of 20 fuses is selected, what is the probability that **(a)** exactly two of the fuses are defective, **(b)** at most two of the fuses are defective, and **(c)** at least three of the fuses are defective? Use a Poisson approximation to the binomial distribution here.

*Solution: Preliminaries:* Recall that we actually have a hypergeometric situation here with $n = 20$, and $a/M = .05$. However, it is plausible to assume that the population size $M$ [which is unknown] is very large with respect to $n$; thus, as a first step, we can suppose that our probability function is approximately a binomial probability function with $n = 20$ and $p = .05$. [See Section 5-4B for a review of these ideas.]

Furthermore, with these values of $n$ and $p$, the binomial probability function is well approximated by a Poisson probability function with $\lambda = np = (20)(.05) = 1.0$.

**(a)** Here, we wish to calculate $P(X = 2) \approx p(2; 1)$, where random variable $X$ denotes the number of defectives. We see that

$$p(2; 1) = \frac{1^2 \cdot e^{-1}}{2!} = .1839 \tag{1}$$

For practice, let us look up this value in Table IV of Appendix A. Refer to the subtable below [which is a portion of Table IV], which indicates that we should find the location in the body of the table which is at the intersection of the column labeled $\lambda = 1.0$ and the row labeled $x = 2$.

| $x$ | $\lambda$ 0.9 | ↓ 1.0 |
|---|---|---|
| 0 | .4066 | .3679 |
| 1 | .3659 | .3679 |
| → 2 | .1647 | .1839 |
| 3 | .0494 | .0613 |

**(b)** $P$(at most 2 defectives) $\approx p(0; 1) + p(1; 1) + p(2; 1) = .3679 + .3679 + .1839 = .9197$ after using Table IV.

**(c)** $P$(at least 3 defectives) $= 1 - P$(at most 2 defectives) $\approx 1 - .9197 = .0803$, after using the result from part (b).

*note:* We could easily solve the above example by using the binomial probability function with $n = 20$ and $p = .05$. For example, from Table I of Appendix A, we see that $P(X = 2) = b(2; 20, .05) = .1887$, and $P(X \leq 2) = \Sigma_{x=0}^{2} b(x; 20, .05) = .3585 + .3774 + .1887 = .9246$. These are preferred answers to parts (a) and (b) of Example 6-6.

---

**EXAMPLE 6-7:** Verify that $\Sigma_{x=0}^{\infty} p(x; \lambda) = 1$. That is, verify that the sum of all Poisson probabilities is equal to 1. Note that we have the sum of an infinite series here, and that means we will have to make use of calculus.

*Solution:* From calculus, it is known that the infinite series for $e^z$ in powers of $z$ is given by

$$e^z = 1 + z + \frac{z^2}{2!} + \ldots + \frac{z^x}{x!} + \ldots \tag{1}$$

and this is valid for all values of $z$.

Now, for a Poisson distribution, $p(x; \lambda) = (e^{-\lambda} \lambda^x)/(x!)$, and thus,

$$\sum_{x=0}^{\infty} p(x; \lambda) = e^{-\lambda} + e^{-\lambda} \lambda + \frac{e^{-\lambda} \lambda^2}{2!} + \ldots + \frac{e^{-\lambda} \lambda^x}{x!} + \ldots \tag{2}$$

Factoring out $e^{-\lambda}$ from the right side of Eq. (2), we obtain

$$\sum_{x=0}^{\infty} p(x; \lambda) = e^{-\lambda} \left[ 1 + \lambda + \frac{\lambda^2}{2!} + \ldots + \frac{\lambda^x}{x!} + \ldots \right] \tag{3}$$

Comparing the expression in the square brackets of Eq. (3) with the right side of Eq. (1), we see that the expression in the square brackets of Eq. (3) is equal to $e^{\lambda}$ [replace $z$ in Eq. (1) by $\lambda$]. Thus, we can rewrite Eq. (3) as follows:

$$\sum_{x=0}^{\infty} p(x; \lambda) = e^{-\lambda} \cdot e^{\lambda} = e^{(-\lambda + \lambda)} = e^0 = 1 \tag{4}$$

We see that Eq. (4) is that which we wished to show.

---

We shall deal again with the Poisson probability function when we cover the topic of Poisson processes.

## B. The discrete uniform probability function

Probably the simplest type of discrete probability function is the *discrete uniform probability function*.

**Definition 6.3 (Discrete Uniform Probability Function):** Here, the discrete random variable can take on any of the distinct values $x_1, x_2, \ldots, x_{n-1}, x_n$. If the probability $P(X = x) = f(x)$ is given by

$$f(x) = \frac{1}{n} \qquad \text{for each of } x = x_1, x_2, \ldots, x_{n-1}, x_n$$

then $f(x)$ is a discrete uniform probability function [or discrete uniform probability distribution]. The associated random variable is called a discrete uniform random variable.

An example of a discrete random variable is the random variable that represents the number of dots that occurs when a fair die is tossed. Here, $f(x) = \frac{1}{6}$ for $X = 1, 2, 3, 4, 5,$ or 6.

**EXAMPLE 6-8:** A box contains 50 chips where ten are marked with the number 2, ten with the number 4, ten with the number 5, ten with the number 6, and ten with the number 9. An experiment consists of drawing a single chip from the box. Let random variable $X$ indicate the number on the chip that is drawn. Determine and identify $f(x)$, the probability function for $X$.

***Solution:*** $f(x) = \frac{1}{5}$ for $X = 2, 4, 5, 6, 9$. Thus, $f(x)$ is a discrete uniform probability function.

## C. The geometric probability function

Recall that the repeated independent and identical trials in a binomial experiment are often referred to as *Bernoulli trials*.

**Definition 6.4 (Geometric Probability Function):** Given a sequence of Bernoulli trials for which the probability of success on a single Bernoulli trial is $p$ [which is constant]. Let the random variable $X$ denote the number of the trial on which the first success occurs. This $X$ is called a *geometric random variable*, and it takes on the values 1, 2, 3, 4, .... The probability function for this $X$ is denoted by $g(x; p)$, and it is called the *geometric probability function* or the *geometric probability distribution* or the *geometric distribution*.

**Theorem 6.4:** The geometric probability function $g(x; p)$ is given by

$$g(x; p) = (1 - p)^{x-1} p, \qquad \text{where } X = 1, 2, 3, 4, \ldots$$

[Sometimes, we shall use the symbol $q$ for $(1 - p)$, as was the case with the binomial probability function.] Observe that $X$ takes on a countably infinite number of values. Also, $0 < p < 1$. [Note that some practitioners allow for $p = 1$. In this special case, one would have $g(x; 1) = 1$ for $X = 1$, and $g(x; 1) = 0$ for $X = 2, 3, 4, \ldots$.]

**EXAMPLE 6-9:** Prove Theorem 6.4.

***Proof:*** Suppose that $X$ denotes the number of the trial on which the first success occurs. If the first success occurs on the first trial, then we have $X = 1$ and

$$g(1; p) = P(\text{success on first trial}) = p \tag{1}$$

Now, let's focus on deriving the equation for $g(x; p)$ for $X = 2, 3, 4$, and so on. Now, $g(x; p)$ is the probability that there is a sequence of $x$ trials consisting of $(x - 1)$ failures in a row followed by the first success on the last trial. That is,

$$P(X = x) = g(x; p) = P(\overbrace{FFF\ldots F}^{(x-1)\text{ failures}}S)$$

$$= P(F_1)P(F_2) \cdot \ldots \cdot P(F_{x-1})P(S_x) = \underbrace{(1 - p)(1 - p) \cdot \ldots \cdot (1 - p)}_{(x-1)\text{ factors}} p \tag{2}$$

Thus, we have

$$g(x; p) = (1 - p)^{x-1} p \qquad \text{for } X = 2, 3, 4, \ldots \tag{3}$$

If we let $X = 1$ in Eq. (3), we obtain Eq. (1). $\qquad\square$

A verification that $\sum_{x=1}^{\infty} g(x; p) = 1$ is presented in Problem 6-4.

**EXAMPLE 6-10:** Cards are drawn with replacement from a standard deck of cards containing 52 cards. What is the probability that the first spade occurs on the seventh card?

*Solution:*  *Method 1:* We wish to compute

$$P(F_1 F_2 F_3 F_4 F_5 F_6 S_7) = P(F_1)P(F_2)\cdot \ldots \cdot P(F_6)P(S_7) \tag{1}$$

Here, $F_i$ indicates a nonspade on the $i$th trial and $S_7$ indicates spade on the seventh trial. Now, $P(F_i) = P(\text{nonspade on } i\text{th card}) = \frac{3}{4}$ and $P(S_7) = P(\text{spade on 7th card}) = \frac{1}{4}$. Thus, Eq. (1) leads to the following result:

$$(\tfrac{3}{4})^6(\tfrac{1}{4}) = .044495 \tag{2}$$

*Method 2:* The probability we seek is the geometric probability function evaluated for $p = 1/4$ and $X = 7$. This quantity is $g(x; p) = g(7; \frac{1}{4})$. By Theorem 6.4, this equals

$$(\tfrac{3}{4})^6(\tfrac{1}{4})$$

as in Method 1.

---

In calculus, one learns that a geometric series is an infinite series symbolized by

$$\sum_{x=1}^{\infty} ar^{(x-1)} = a + ar + ar^2 + \ldots + ar^x + \ldots \tag{I}$$

**Theorem 6.5 (Sum of a Geometric Series):** A geometric series converges provided that $-1 < r < 1$. In such case, its sum is given by

$$\frac{a}{1 - r} \tag{a}$$

[For $r \geq 1$ or $r \leq -1$, the geometric series diverges; that is, it doesn't converge and hence no finite value can be assigned to its sum.] Briefly, we write

$$\sum_{x=1}^{\infty} ar^{(x-1)} = \frac{a}{1 - r} \qquad \text{when } -1 < r < 1 \tag{b}$$

*notes*

(a) Observe that $\sum_{x=1}^{\infty} ar^{(x-1)}$ may also be written as $\sum_{x=0}^{\infty} ar^x$.
(b) The statement $-1 < r < 1$ is often written as $|r| < 1$.

---

**EXAMPLE 6-11:** Suppose that $X$ is a geometric random variable. Suppose in the following that $k$ denotes a nonnegative integer. (a) Prove that $P(X > k) = (1 - p)^k$ or $q^k$. (b) Prove that $P(X \leq k) = 1 - q^k$.

*Proof*

(a) First of all, we know that

$$P(X > k) = P(X \geq k + 1) = g(k + 1; p) + g(k + 2; p) + g(k + 3; p) + \ldots \tag{1}$$

From Theorem 6.4, we have $g(k + 1; p) = (1 - p)^k p$, $g(k + 2; p) = (1 - p)^{(k+1)}p$, and so on. Thus, substituting equations like these into Eq. (1), we obtain

$$P(X > k) = (1 - p)^k p + (1 - p)^{(k+1)}p + (1 - p)^{(k+2)}p + \ldots \tag{2}$$

Factoring out $(1 - p)^k p$ from all terms of the right side of Eq. (2), we get

$$P(X > k) = (1 - p)^k p[1 + (1 - p) + (1 - p)^2 + \ldots] \tag{3}$$

or

$$P(X > k) = (1 - p)^k p \sum_{x=1}^{\infty} (1 - p)^{x-1} \tag{4}$$

Let us apply the equation for the sum of a geometric series [see, for example, Eq. (b) of Theorem 6.5] to the sigma expression of Eq. (4). Here, $a = 1$ and $r = 1 - p$. [Note that it is valid to do this since $0 < p < 1$ implies $0 < r < 1$, where $r = 1 - p$.] Thus, we obtain

$$P(X > k) = (1 - p)^k p \frac{1}{1 - (1 - p)} = (1 - p)^k = q^k \tag{5}$$

**(b)** Here, we have

$$P(X \leq k) = 1 - P(\text{not } X \leq k) = 1 - P(X > k) \tag{6}$$

If we substitute the result from Eq. (5) into Eq. (6), we obtain

$$P(X \leq k) = 1 - q^k \tag{7} \quad \square$$

---

**EXAMPLE 6-12:** Given the situation of Example 6-10. What is the probability that the first spade occurs on or before the seventh card drawn?

*Solution:* The desired probability is $P(X \leq 7)$, where $X$ is a geometric random variable [which stands for the number of the trial, or card, on which the first spade occurs]. Thus, from part (b) of Example 6-11, and the fact that $q = 1 - p = 1 - \frac{1}{4} = \frac{3}{4}$, we obtain

$$P(X \leq 7) = 1 - q^k = 1 - (\tfrac{3}{4})^7 = 1 - .13348 = .86652$$

---

## D. The negative binomial probability function

Concerning repeated Bernoulli trials, sometimes we are interested in the number of the trial on which the $k$th success occurs. [Here, $k = 1$ or $2$ or $3$, etc. If $k$ were to equal $1$, we would have the geometric distribution of Section 6-2C.] For example, we may want to know the probability that the eighth toss of a fair die will yield the third ace, or the probability that the twentieth card drawn with replacement from a standard deck of cards will result in the sixth spade drawn.

**Definition 6.5 (Negative Binomial Probability Function):** Given a sequence of Bernoulli trials for which the probability of success on a single trial is $p$ [which is constant]. Let the random variable $X$ denote the number of the trial on which the $k$th success occurs. [The integer $k$ has the property that $k \geq 1$.] This random variable $X$ is called a *negative binomial random variable* and it takes on the values $k$, $k + 1$, $k + 2$, $\ldots$. Its probability function, denoted by $b^*(x; k, p)$, is called a *negative binomial probability function*, or a *negative binomial probability distribution*, or, merely, a *negative binomial distribution*.

**Theorem 6.6:** The negative binomial probability function $b^*(x; k, p)$ is given by

$$b^*(x; k, p) = \binom{x - 1}{k - 1} p^k (1 - p)^{x-k}$$

for $x = k$, $k + 1$, $k + 2$, $\ldots$. Observe that $x$ takes on a countably infinite number of values. Also, $0 < p < 1$. [The special case $b^*(x; k, p)$ for $p = 1$ is given by $b^*(x; k, 1) = 1$ for $x = k$ and $b^*(x; k, 1) = 0$ for $x = k + 1$, $k + 2$, $k + 3$, and so on.]

*notes*

(a) Sometimes, $q$ is used instead of $1 - p$.

(b) Recall that the geometric distribution is the negative binomial distribution for $k = 1$. Thus, we have $b^*(x; 1, p) = g(x; p)$.

(c) Negative binomial distributions are also known as binomial waiting-time distributions or as Pascal distributions.

---

**EXAMPLE 6-13:** Prove Theorem 6.6.

*Proof:* If the $k$th success occurs on the $x$th trial [where $x$ is equal to either $k$ or $k + 1$ or $k + 2$, etc.], that means that there must be $(k - 1)$ successes on the first $(x - 1)$ trials, followed by a success on the $k$th trial. The probability of the former is the binomial probability of $(k - 1)$ successes in $(x - 1)$ trials, and it is given by

$$\binom{x - 1}{k - 1} p^{k-1}(1 - p)^{x-k} \tag{1}$$

as we see from Theorem 5.5 [replace $n$ by $(x - 1)$ and $x$ by $(k - 1)$]. The probability of success on the $k$th trial is merely $p$. Now, because the $x$ trials are independent, it follows that the event of obtaining $(k - 1)$ successes on the first $(x - 1)$ trials is independent of the event of getting a success on the $x$th trial. Thus, by Definition 3.2, the probability $b^*(x; k, p)$ is the product of $p$ times the expression given in Eq. (1). That is, we have

$$b^*(x; k, p) = \binom{x - 1}{k - 1} p^k(1 - p)^{x-k} \qquad \text{for } x = k, k + 1, k + 2, \ldots \tag{2} \quad \square$$

---

A proof that

$$\sum_{x=k}^{\infty} b^*(x; k, p) = 1$$

based on the properties of the binomial infinite series [another calculus topic], is presented on page 222 of Larsen and Marx (1986).

---

**EXAMPLE 6-14:** What is the probability that the twentieth card drawn with replacement from a standard deck of cards will be the sixth spade drawn?

*Solution:* First, note that the probability of spade on a single card is .25. That is, $p = .25$.

*Method 1:* If the sixth spade occurs on the twentieth card drawn, then there must be exactly 5 spades on the first 19 cards drawn, followed by a spade on the twentieth card drawn. The probability of the former event is the binomial probability

$$b(5; 19, .25) = \frac{19!}{5!\,14!}(.25)^5(.75)^{14} = .20233 \tag{1}$$

This value is available from Table I of Appendix A as .2023. The probability of spade on the twentieth card drawn is .25. Thus, the probability we seek is the product of the probabilities of these two independent events, namely

$$\frac{19!}{5!\,14!}(.25)^5(.75)^{14} \cdot (.25) = .05058 \tag{2}$$

*Method 2:* The answer we seek is just the negative binomial probability for $X = 20$ trials, where $k = 6$ and $p = .25$. Thus, from Theorem 6.6, we have

$$b^*(20; 6, .25) = \binom{19}{5}(.25)^6(.75)^{14}$$

and this is equal to .05058.

---

The following theorem expresses the negative binomial probability function in terms of the binomial probability function. Theorem 6.7 enables one to use the table of binomial probabilities [Table I of Appendix A] in the calculation of negative binomial probabilities.

**Theorem 6.7:** $b^*(x; k, p) = (k/x) \cdot b(k; x, p)$, where $x = k, k + 1, k + 2, \ldots$. On the left is the negative probability function for success number $k$ occurring on trial number $x$. On the right, we have the binomial probability function for exactly $k$ successes in $x$ trials. [It is important to keep the symbols straight here. The general binomial probability function symbol is $b(x; n, p)$, which represents the binomial probability of exactly $x$ successes in $n$ trials.]

---

**EXAMPLE 6-15:** Redo Example 6-14 by making use of Theorem 6.7 and Table I of Appendix A.

*Solution:* Here, we wish to calculate $b^*(20; 6, .25)$. Let us substitute $x = 20$, $k = 6$, and $p = .25$ into the equation of Theorem 6.7 to get

$$b^*(20; 6, .25) = \tfrac{6}{20} b(6; 20, .25) \tag{1}$$

In Table I, we find that $b(6; 20, .25) = .1686$. Thus, we have

$$b^*(20; 6, .25) = \tfrac{6}{20}(.1686) = .05058 \tag{2}$$

as before.

---

**EXAMPLE 6-16:** Prove Theorem 6.7.

*Proof:* In the usual expression for the binomial probability function, namely in

$$b(x; n, p) = \frac{n!}{x!(n - x)!} p^x (1 - p)^{(n-x)} \tag{1}$$

let us replace $x$ by $k$ and $n$ by $x$. Thus, we get

$$b(k; x, p) = \frac{x!}{k!(x - k)!} p^k (1 - p)^{(x-k)} \tag{2}$$

Now,

$$\frac{k}{x} \cdot \frac{x!}{k!(x - k)!} = \frac{(x - 1)!}{(k - 1)!(x - k)!} \tag{3}$$

since $a!/a = (a - 1)!$ if $a$ is a positive integer. Now, from Theorem 6.6, we have

$$b^*(x; k, p) = \frac{(x - 1)!}{(k - 1)!(x - k)!} p^k (1 - p)^{x-k} \tag{4}$$

Thus, if we multiply both sides of Eq. (2) by $k/x$, and take note of Eqs. (3) and (4), we see that

$$b^*(x; k, p) = \frac{k}{x} b(k; x, p) \tag{5}$$

Here, $x = k, k + 1, k + 2, \ldots$. $\square$

## 6-3.  Parameters of Probability Distribution Families

Each probability distribution that we covered in this chapter and in Chapter 5 actually consisted of a family of individual probability distributions. For example, for the binomial distribution as given by Theorem 5.5, namely for

$$b(x; n, p) = \frac{n!}{x!(n-x)!} p^x (1-p)^{(n-x)}$$

there is an individual probability distribution for each pair of values of $n$ and $p$. The quantities $n$ and $p$ are known as parameters for the binomial distribution family. For example, we have a particular binomial distribution in Example 6-1 [there, $n = 2$ and $p = .5$], and we have another binomial distribution in Example 5-21 [there, $n = 8$ and $p = .10$]. Each of these individual binomial distributions may be considered as members of the binomial distribution family.

Speaking in general, we say that the parameters for a probability distribution are quantities which are constants for a particular member of the probability distribution family, but which take on different sets of values for different members of the probability distribution family. Thus, for the binomial distribution family cited above, the parameters are $n$ and $p$, where $n$ denotes a positive integer [and $n \geq 2$ for meaningful situations], and $p$ can vary continuously on the interval given by $0 < p < 1$.

---

**EXAMPLE 6-17:** Specify the members of the binomial distribution family for which the parameters take on the following values: (a) $p = 1/2$, $n = 3$; (b) $p = .6$, $n = 20$.

*Solution*

(a)
$$b\left(x; 3, \frac{1}{2}\right) = \frac{3!}{x!(3-x)!} \left(\frac{1}{2}\right)^3 \qquad \text{for } x = 0, 1, 2, 3$$

(b)
$$b(x; 20, .6) = \frac{20!}{x!(20-x)!} (.6)^x (.4)^{20-x} \qquad \text{for } x = 0, 1, 2, \ldots, 19, 20$$

---

**EXAMPLE 6-18:** Indicate sets of parameters for the following probability distribution families: (a) hypergeometric, (b) Poisson, (c) geometric, (d) negative binomial. Refer to the equations given thus far in this book for these probability distribution families.

*Solution*

(a)  For the hypergeometric distribution family [Theorem 5.7], we have

$$h(x; n, a, M) = \frac{\binom{a}{x}\binom{M-a}{n-x}}{\binom{M}{n}} \qquad \text{for } x = 0, 1, 2, \ldots, (n-1), n$$

The three parameters here are $n$, $a$, and $M$.

(b)  For the Poisson distribution family [Definition 6.2], we have

$$p(x; \lambda) = \frac{\lambda^x e^{-\lambda}}{x!} \qquad \text{for } x = 0, 1, 2, 3, \ldots$$

Here, the single parameter is $\lambda$.

(c)  For the geometric distribution family [Theorem 6.4], we have

$$g(x; p) = (1-p)^{x-1} p \qquad \text{for } x = 1, 2, 3, \ldots$$

Here, the single parameter is $p$.

**(d)** For the negative binomial distribution family [Theorem 6.6], we have

$$b^*(x; k, p) = \binom{x-1}{k-1} p^k (1-p)^{x-k} \qquad \text{for } x = k, k+1, k+2, \ldots$$

Here, the two parameters are $p$ and $k$.

# SOLVED PROBLEMS

**Cumulative Distribution Function for a Discrete Random Variable**

**PROBLEM 6-1** Determine the cumulative distribution function $F(x)$ for the binomial random variable for which $n = 3$ and $p = 3/5$. Sketch the bar chart for $f(x) = b(x; 3, 3/5)$, and the graph for the cumulative distribution function.

*Solution* The table for this binomial probability function was presented in Problem 5-6, part (a), and it is reproduced below:

| $x$ | 0 | 1 | 2 | 3 |
|---|---|---|---|---|
| $f(x)$ | $8/125 = .064$ | $36/125 = .288$ | $54/125 = .432$ | $27/125 = .216$ |

The bar chart of $f(x)$ is shown in Figure 6-3a. To find $F(x_0)$, we merely have to compute the total probability accumulated up to and including the fixed value $x_0$. In symbols, we work with

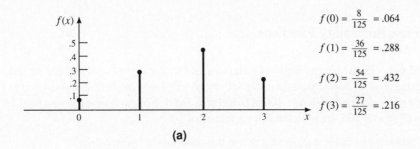

$$f(0) = \frac{8}{125} = .064$$

$$f(1) = \frac{36}{125} = .288$$

$$f(2) = \frac{54}{125} = .432$$

$$f(3) = \frac{27}{125} = .216$$

**(a)**

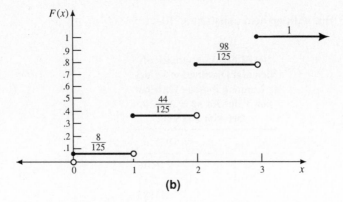

**(b)**

**Figure 6-3.** Bar chart and cumulative distribution function graph for Problem 6-1. **(a)** Bar chart for $f(x) = b(x; 3, .6)$. **(b)** Graph of $F(x)$ in terms of $x$.

$F(x_0) = P(X \leq x_0)$. Some sample calculations and the final results for $F(x_0)$ are presented in Table 6-2.

**TABLE 6-2: Derivation of Results for $F(x_0)$ in Terms of $x_0$ for Problem 6-1**

| Sample calculations | Summary result for $F(x_0)$ |
|---|---|
| $F(-3.2) = P(X \leq -3.2) = 0$ | $F(x_0) = 0$ for $x_0 < 0$ |
| $F(0) = P(X \leq 0) = f(0) = 8/125$ <br> $F(.42) = P(X \leq .42) = f(0) = 8/125$ | $F(x_0) = 8/125$ for $0 \leq x_0 < 1$ |
| $F(1) = P(X \leq 1) = f(0) + f(1)$ <br> $\quad = 8/125 + 36/125 = 44/125$ <br> $F(1.73) = P(X \leq 1.73) = f(0) + f(1)$ <br> $\quad = 8/125 + 36/125 = 44/125$ | $F(x_0) = 44/125$ for $1 \leq x_0 < 2$ |
| $F(2) = P(X \leq 2) = f(0) + f(1) + f(2)$ <br> $\quad = 8/125 + 36/125 + 54/125 = 98/125$ <br> $F(2.6) = P(X \leq 2.6) = f(0) + f(1) + f(2)$ <br> $\quad = 8/125 + 36/125 + 54/125 = 98/125$ | $F(x_0) = 98/125$ for $2 \leq x_0 < 3$ |
| $F(3) = P(X \leq 3) = f(0) + f(1) + f(2) + f(3)$ <br> $\quad = 8/125 + 36/125 + 54/125 + 27/125 = 1$ <br> $F(49.2) = P(X \leq 49.2)$ <br> $\quad = f(0) + f(1) + f(2) + f(3)$ <br> $\quad = 1$ | $F(x_0) = 1$ for $3 \leq x_0$ |

The summary results for $F(x_0)$ in terms of $x_0$ are given on the right of Table 6-2. To obtain $F(x)$ in terms of $x$, one merely replaces $x_0$ by $x$. The graph of $F(x)$ versus $x$ is presented in Figure 6-3b.

## More Discrete Probability Functions

**PROBLEM 6-2**  Illustrate the limiting behavior of binomial probabilities for $np$ held constant at 10 by letting $n \to \infty$ and simultaneously letting $p \to 0$. Let $x$ be held constant at 11. Start with $n = 20$ and $p = .5$, and end with $n = 100,000$ and $p = .0001$. For each new calculation of a binomial probability after $n = 100$, let $n$ increase by a factor of 10.

*Solution*  Following is Table 6-3, in which the tabulated data indicate that

$$\lim b(11; n, p) = .11374$$

where $n \to \infty$ and $p \to 0$, but with $np$ held constant at 10.

**TABLE 6-3: Approach of Binomial Distribution Values to Limiting Poisson Distribution Value for $np = \lambda = 10$, and with $x = 11$**

| $n$ | $p$ | $b(11; n, p)$ |
|---|---|---|
| 20 | .5 | .16018 |
| 50 | .2 | .12711 |
| 100 | .1 | .11988 |
| 1,000 | .01 | .11431 |
| 10,000 | .001 | .11379 |
| 100,000 | .0001 | .11374 |

Observe that the Poisson probability $p(x;\lambda)$ for $x = 11$ and $\lambda = 10$ is equal to .11374, as expected.

***note:*** Recall that $p(11;10) = (10^{11}e^{-10})/(11!)$. A value for $p(11;10)$, which is accurate to four significant digits, namely .1137, is available from Table IV of Appendix A.

**PROBLEM 6-3** Over a long period of time, 2.3% of the calls received by a switchboard are wrong numbers. Use the Poisson approximation to the binomial probability function to estimate the probability that among 200 calls received by the switchboard, **(a)** exactly four are wrong numbers, **(b)** at most two are wrong numbers, and **(c)** at least four are wrong numbers.

***Solution*** *Preliminaries:* In the approximating Poisson probability function, we use $\lambda = np = 200 \cdot (.023) = 4.60$. We can look up the relevant Poisson probabilities in Table IV of Appendix A.

**(a)** $P(X = 4) = p(x;\lambda) = p(4;4.60) = .1875$.
**(b)** $P(X \le 2) = p(0;4.60) + p(1;4.60) + p(2;4.60) = .1626$.
**(c)** $P(X \ge 4) = 1 - P(X \le 3) = 1 - [p(0;4.60) + p(1;4.60) + p(2;4.60) + p(3;4.60)] = 1 - .3257 = .6743$.

**PROBLEM 6-4** Verify that $\Sigma_{x=1}^{\infty} g(x;p) = 1$. Recall that $g(x;p)$ denotes the geometric probability function for which the probability of success in a single Bernoulli trial is $p$.

***Proof*** Since $g(x;p) = (1-p)^{x-1}p$, we have

$$\sum_{x=1}^{\infty} g(x;p) = \sum_{x=1}^{\infty} (1-p)^{x-1}p \tag{1}$$

From Theorem 6.5, for the sum of a geometric series, we have

$$\sum_{x=1}^{\infty} ar^{x-1} = \frac{a}{1-r} \qquad \text{for } -1 < r < 1 \tag{2}$$

The series of Eq. (1) matches that of Eq. (2) if we let $a = p$, and $r = 1 - p$. Thus, making use of the right side of Eq. (2), we see that the sum in Eq. (1) is given by

$$\sum_{x=1}^{\infty} g(x;p) = \frac{p}{1-(1-p)} = \frac{p}{p} = 1 \tag{3} \quad \square$$

**PROBLEM 6-5** It is known that within a certain species of laboratory mice, 25% have black spots on their fur. Suppose that mice are examined randomly. **(a)** What is the probability that the sixth mouse examined is the first one that has black spots on its fur? **(b)** What is the probability that the first spotted mouse occurs on or before the fourth mouse examined? **(c)** What is the probability that the tenth mouse is the fourth one with black spots on its fur?

***Solution*** *Preliminaries:* We assume that the examining of the mice is equivalent to a sequence of Bernoulli trials, for which the probability of success [i.e., of black spots] is given by $p = .25$.

**(a)** Here, we have to calculate a geometric probability. That is, the answer is given by

$$P(X = 6) = g(6;.25) = (1-p)^5 p = (.75)^5(.25) = .05933$$

**(b)** Here, we make use of a result applicable to a geometric probability distribution that was demonstrated in part (b) of Example 6-11. That result is

$$P(X \le k) = 1 - (1-p)^k \tag{1}$$

In the current situation, we wish to calculate $P(X \le 4)$ for a geometric distribution in which $p = .25$. Thus, our answer is given by

$$P(X \le 4) = 1 - (.75)^4 = 1 - .31641 = .68359 \tag{2}$$

**(c)** Here, we have to calculate a value of a negative binomial probability function for which $p = .25$, $k = 4$, and $x = 10$. Thus, after using Theorem 6.6, our answer is given by

$$b^*(10;4,.25) = \binom{9}{3}(.25)^4(.75)^6 = .058399 \tag{3}$$

**PROBLEM 6-6** (a) A discrete random variable $X$ is said to have a "memoryless" property if

$$P(X = x + s \mid X > s) = P(X = x) \tag{I}$$

The expression on the left side is the *conditional probability* that random variable $X$ is equal to $x + s$, given that $X$ is greater than $s$. Show that a geometric random variable has a memoryless property, where $x$ and $s$ are *integers* and are such that $x \geq 1$ and $s \geq 0$. (b) Suppose that a fair die is repeatedly tossed. What is the probability that the first ace occurs on the eighth trial, given that the first ace occurs on a trial higher than the third trial?

*Solution*

(a) Suppose that $X$ is a geometric random variable. Then the right side of Eq. (I) can be expressed as follows:

$$P(X = x) = (1 - p)^{x-1}p \qquad \text{for } x = 1, 2, 3, \ldots \tag{1}$$

Using Definition 3.1, we see that the conditional probability term on the left can be expressed as

$$P(X = x + s \mid X > s) = \frac{P(X = x + s \text{ and } X > s)}{P(X > s)} \tag{2}$$

From part (a) of Example 6-11, the denominator of Eq. (2) can be expressed as

$$P(X > s) = (1 - p)^s \tag{3}$$

Consider the event $(X = x + s \text{ and } X > s)$ of the numerator of Eq. (2). At the outset, recall that $x$ and $s$ are integers such that $x \geq 1$ and $s \geq 0$. The collection of values of random variable $X$ such that $X > s$ consists of $1 + s, 2 + s, 3 + s, \ldots, x - 1 + s, \underline{x + s}, x + 1 + s, \ldots$ . Thus, the collection of values of random variable $X$ such that $X = x + s$ and $X > s$ consists only of the single value $X = x + s$. Thus, we have

$$P(X = x + s \text{ and } x > s) = P(X = x + s) = (1 - p)^{x+s-1}p \tag{4}$$

The last term of Eq. (4) holds since $X$ is a geometric random variable. Substituting from Eqs. (3) and (4) into Eq. (2), and canceling out $(1 - p)^s$, yields

$$P(X = x + s \mid X > s) = (1 - p)^{x-1}p \tag{5}$$

Thus, since the right sides of Eqs. (1) and (5) are identical, we see that

$$P(X = x + s \mid X > s) = P(X = x) \tag{6}$$

which is what we intended to show [see Eq. (I) in the statement of the problem].

(b) Let $X$ denote the trial on which the first ace occurs. Thus, $X$ is a geometric random variable for which $p = 1/6$. In this problem, we are seeking the value of the conditional probability given by the following expression:

$$P(X = 8 \mid X > 3) \tag{7}$$

Now, the result proved in part (a), which is given by Eq. (6) above, applies here with $s = 3$ and $x + s = 8$. Thus, it follows that $x = 5$. Now, for $s = 3$ and $x = 5$, Eq. (6) becomes

$$P(X = 8 \mid X > 3) = P(X = 5) \tag{8}$$

For $P(X = 5)$, we have the geometric probability $g(5; 1/6)$. Thus, we have

$$P(X = 8 \mid X > 3) = g(5; 1/6) = \left(\tfrac{5}{6}\right)^4\left(\tfrac{1}{6}\right) = .080376 \tag{9}$$

**PROBLEM 6-7** A fair die has the faces with 1, 2, and 3 dots painted red, the faces with 4 and 5 dots painted green, and the face with 6 dots painted white. The die is tossed repeatedly until either a red or green face occurs [if white occurs, the die is tossed again]. What is the probability that a red face occurs before a green face occurs?

*Solution* For a single toss of the die, we have

$$P(R) = 3/6, \qquad P(G) = 2/6, \qquad \text{and } P(W) = 1/6 \qquad \textbf{(1a), (1b), (1c)}$$

where $R$, $G$, and $W$ indicate faces that are "red," "green," and "white," respectively.

Let $R_i$ indicate "red occurs on toss $i$," and, similarly, let $W_j$ indicate "white occurs on toss $j$." Thus, we have

$$P(\text{red occurs first}) = P(R_1 \text{ or } W_1 R_2 \text{ or } W_1 W_2 R_3 \text{ or } W_1 W_2 W_3 R_4 \text{ or } \ldots) \qquad (2)$$

where, for example, the symbol $W_1 W_2 R_3$ means that "white" occurs on the first two tosses followed by "red" on the third toss. The events cited on the right side of Eq. (2) are mutually exclusive, and thus

$$P(\text{red occurs first}) = P(R_1) + P(W_1 R_2) + P(W_1 W_2 R_3) + \ldots \qquad (3)$$

Since the individual tosses constitute independent trials, the events in the individual terms are independent. [See Theorem 5.3. For example, in the individual term $W_1 W_2 R_3$, the events $W_1$, $W_2$, and $R_3$ are independent.] Thus, Eq. (3) becomes

$$P(\text{red occurs first}) = P(R_1) + P(W_1)P(R_2) + P(W_1)P(W_2)P(R_3) + \ldots \qquad (4)$$

Now, using Eqs. (1a) and (1c) in Eq. (4), and noting that $P(R_i) = P(R)$, etc., we obtain

$$P(\text{red occurs first}) = \tfrac{3}{6} + (\tfrac{1}{6})\tfrac{3}{6} + (\tfrac{1}{6})^2\tfrac{3}{6} + \ldots \qquad (5)$$

That is,

$$P(\text{red occurs first}) = \sum_{x=1}^{\infty} \tfrac{3}{6}(\tfrac{1}{6})^{x-1} \qquad (6)$$

Thus, after using Theorem 6.5 to determine the sum of this geometric series [here, $a = \tfrac{3}{6}$, and $r = \tfrac{1}{6}$], we obtain

$$P(\text{red occurs first}) = \frac{\tfrac{3}{6}}{1 - \tfrac{1}{6}} = \frac{\tfrac{3}{6}}{\tfrac{5}{6}} = \tfrac{3}{5} \qquad (7)$$

Because of the nature of the termination rule [which is to stop whenever red or green occurs], we have

$$P(\text{green occurs first}) = 1 - P(\text{red occurs first}) = \tfrac{2}{5} \qquad (8)$$

## Parameters of Probability Distribution Families

**PROBLEM 6-8** Specify the members of the hypergeometric distribution family for which the parameters take on the following values: (a) $n = 20$, $a = 30$, $M = 100$; (b) $n = 5$, $a = 3$, $M = 25$.

*Solution*

(a) Here

$$h(x; n, a, M) = h(x; 20, 30, 100) = \frac{\binom{30}{x}\binom{70}{20 - x}}{\binom{100}{20}} \qquad \text{for } x = 0, 1, 2, \ldots, 19, 20$$

(b) Here,

$$h(x; n, a, M) = h(x; 5, 3, 25) = \frac{\binom{3}{x}\binom{22}{5 - x}}{\binom{25}{5}} \qquad \text{for } x = 0, 1, 2, 3, 4, 5$$

*note:* Probabilities for $x = 4$ and $5$ are equal to zero.

## Supplementary Problems

**PROBLEM 6-9**  Determine the cumulative distribution function $F(x)$ for the hypergeometric random variable for which $n = 3$, $a = 4$, and $M = 10$. Sketch the bar chart for $f(x) = h(x; 3, 4, 10)$ and the graph for the cumulative distribution function, $F(x)$, in terms of $x$.

*Answer (Partial)*  $F(x) = 0$ for $x < 0$, $F(x) = 20/120$ for $0 \le x < 1$, $F(x) = 80/120$ for $1 \le x < 2$
$F(x) = 116/120$ for $2 \le x < 3$, and $F(x) = 1$ for $3 \le x$

**PROBLEM 6-10**  In the Johnstown lottery, it costs one dollar to place a bet to win a prize of \$500. For a single bet, the probability of winning the \$500 prize is .00075. Let random variable $X$ denote the number of \$500 prizes won by a gambler after placing $n$ bets of one dollar each. [Thus, $X$ is a binomial random variable.] (a) If a gambler places $n = 2{,}400$ bets of one dollar each, state the condition on $X$ such that the gambler will be behind [i.e., losing money overall]. (b) Refer to part (a). Use the Poisson distribution to approximate the probability that the gambler will be behind if he places 2,400 bets of one dollar each.

*Answer*  (a) $P$(Gambler is behind.) $= P(X \le 4)$      (b) .9636

**PROBLEM 6-11**  The probability is .025 that a defective item is produced by a machine. Suppose that an inspector randomly checks items produced by the machine. (a) What is the probability that the first defective item is the tenth item checked? (b) What is the probability that the fourth defective item occurs on the twentieth item checked?

*Answer*  (a) .01991      (b) .0002524

**PROBLEM 6-12**  Given the situation of Problem 6-11. Compute the probability that more than 50 items must be checked until the first defective one is located. [*Hint:* See Example 6-11.]

*Answer*  .2820

**PROBLEM 6-13**  The random variable for the negative binomial distribution is taken by some authors to be the number of failures that precedes the $k$th success. Let this random variable be denoted by $Y$. (a) Determine $Y$ in terms of $X$, where $X$ is the random variable used in prior discussions of the negative binomial distribution in this book—see Definition 6.5 and Theorem 6.6, for example. Find the range space of $Y$. Here, a value $Y = y$ is in the range space if the probability corresponding to $y$ is positive. (b) Determine the equation for the probability function of $Y$. Use $\hat{b}(y; k, p)$ as its symbol.

*Answer*  (a) Here, $Y = X - k$, where $X$ denotes the number of the trial on which the $k$th success occurs. Now, the range space for $X$ is $k$, $k + 1$, $k + 2$, $k + 3$, .... Thus, the range space for $Y$ is 0, 1, 2, 3, ....
(b) First, observe that $b^*(x; k, p)$ is equal to $b^*(y + k; k, p)$, since $x = y + k$. Then, $\hat{b}(y; k, p) = b^*(y + k; k, p)$, and thus,

$$\hat{b}(y; k, p) = \binom{y + k - 1}{k - 1} p^k (1 - p)^y \qquad \text{for } y = 0, 1, 2, 3, \ldots$$

**PROBLEM 6-14**  The game of craps involves the repeated tossing of a pair of dice. In the game of craps, Nancy throws a five on her first toss of a pair of dice. ["Five" means that the sum of the number of dots on both dice is equal to five.] The rules call for her to keep on tossing the dice until she gets a five again, in which case she wins, or until she gets a seven, in which case she loses. What is the probability that she will win? [*Hint:* See Problem 6-7.]

*Answer*  2/5

# 7 DISCRETE PROBABILITY PROBLEMS AND BAYES' FORMULA

## THIS CHAPTER IS ABOUT

☑ **Elementary Probability Applications**
☑ **Total Probability Theorem and Bayes' Formula**

## 7-1. Elementary Probability Applications

### A. The two-stage approach

We shall often make use of a *two-stage approach* for determining the probability of an event, $E$.

**Stage One** [Or/Addition]  First, we express the event $E$ as an "or" statement [equivalently, a "union" statement] which is in terms of mutually exclusive events $A_1, A_2, \ldots, A_k$. That is,

$$E = A_1 \text{ or } A_2 \text{ or } \ldots \text{ or } A_k \tag{1a}$$

Then, we take the probability of both sides of Eq. (1a), where on the right side we apply the addition rule. [See Axiom 1.3′.] Thus, we obtain Eq. (1b).

$$P(E) = P(A_1) + P(A_2) + \ldots + P(A_k) \tag{1b}$$

**Stage Two** [And/Multiplication] Next, we express each $A_i$ as an "and" statement [equivalently, an "intersection" statement] in terms of sequentially occurring events such as $B_1, B_2, \ldots, B_r$. That is,

$$A_i = B_1 \text{ and } B_2 \text{ and } B_3 \text{ and } \ldots \text{ and } B_r \tag{2a}$$

Then, we take the probability of both sides of Eq. (2a), where on the right side we apply the general multiplication rule [see Theorem 3.3]. Thus, we obtain Eq. (2b).

$$P(A_i) = P(B_1) \cdot P(B_2 \mid B_1) \cdot P(B_3 \mid B_1 \text{ and } B_2) \cdot \ldots \cdot \tag{2b}$$

Now, Eq. (2b) reduces to the following equation if the events $B_1, B_2, B_3, \ldots, B_r$ are independent. [See Definition 3.3.]:

$$P(A_i) = P(B_1) \cdot P(B_2) \cdot P(B_3) \cdot \ldots \cdot P(B_r) \tag{2c}$$

[This will be the case, for example, if $B_1, B_2, B_3, \ldots, B_r$ are events which are respectively determined by $r$ independent trials $1, 2, \ldots, r$. Refer to Definitions 5.3 and 5.4 and Theorems 5.2 and 5.3 for a review of the concepts of independent trials and of events determined by independent trials. A key point is that such events are then independent events, in accordance with Definitions 3.2, 3.3, and 3.4.]

The second stage thus consists of expressing each $P(A_i)$ probability by an equation of the type (2b) or (2c). Finally, we solve for $P(E)$ by substituting the expressions thus obtained for $P(A_1), P(A_2), \ldots, P(A_k)$ back into Eq. (1b).

**EXAMPLE 7-1:** Tom wants to go on a date Saturday with either Alice or Barbara. He estimates [subjectively] that if he asks for a date, the probability that Alice will say yes is .4, while the probability Barbara will say yes is .54. Tom decides to flip a coin—he'll call Alice if it comes up a head and call Barbara if it comes up a tail. **(a)** What is the probability Tom will get a date, assuming he makes only one phone call? **(b)** What is the probability Tom will get a date, assuming he decides to make a call to the other girl if the first girl says no?

*Solution: Preliminaries:* Let $A$ stand for the event "Alice says yes," and let $B$ stand for "Barbara says yes," so that $P(A) = .40$ and $P(B) = .54$.

**(a)** The event $E$ of getting a date after making one phone call is equivalent to getting a head $[H]$ on the coin, and having Alice say yes, or getting a tail $[T]$ on the coin, and having Barbara say yes. Thus, in symbols, we have

$$P(E) = P[(H \text{ and } A) \text{ or } (T \text{ and } B)] \tag{1}$$

and hence

$$P(E) = P(H \text{ and } A) + P(T \text{ and } B) \tag{2}$$

since the events separated by the "or" in Eq. (1) are mutually exclusive. Next, using the multiplication rule on $P(H \text{ and } A)$ and $P(T \text{ and } B)$, we get

$$P(E) = P(H)P(A \mid H) + P(T)P(B \mid T) = P(H)P(A) + P(T)P(B)$$
$$= (.5)(.4) + (.5)(.54) = .47 \tag{3}$$

Thus, the probability is .47 or 47% that Tom will get a date after only one phone call.

**(b)** Letting $F$ be the symbol for the event "Tom will get a date," we have

$$P(F) = P[(\text{Tom will get "yes" on first call}) \text{ or}$$

$$(H \text{ and Alice says "no" and Barbara says "yes"}) \text{ or}$$

$$(T \text{ and Barbara says "no" and Alice says "yes"})] \tag{1}$$

Here, the three events in parentheses on the right side of Eq. (1) are mutually exclusive. Thus, we add the probabilities of the events separated by "or." Also, the first right side event listed in parentheses is the event $E$ from part (a). Thus, we have

$$P(F) = P(E) + P(H \text{ and } A' \text{ and } B) + P(T \text{ and } B' \text{ and } A) \tag{2}$$

and, after using the multiplication rule for the events within parentheses [in Eq. (2)] which are separated by "and," we get

$$P(F) = P(E) + P(H)P(A')P(B) + P(T)P(B')P(A)$$
$$= .47 + (.5)(.6)(.54) + (.5)(.46)(.4) = .47 + .162 + .092 = .724 \tag{3}$$

So we see that, by making a second phone call [provided the first girl says "no"], Tom will increase his chances of getting a date by about 25%.

*note:* In part (a) of Example 7-1, we regarded the events $H$ and $A$ to be independent events [and, likewise, for $T$ and $B$], and we wrote $P(H \text{ and } A) = P(H) \cdot P(A)$. This is so because the processes of tossing a coin and calling a particular girl constitute independent actions. We don't have a sequence of independent trials here since the result on the coin determines what the next process will be. That is, Tom will call Alice if $H$ occurs, and he will call Barbara if $T$ occurs. [On the other hand, with independent trials, the second process doesn't depend on what *happens* on the first process. As an example of independent trials, consider the tossing of a coin followed by the tossing of a die. Refer to Section 5-1 for a review of the concept of "independent trials."] But, it should be clear that the result of making a phone call to a particular girl should truly be independent of the result of a coin toss. After all, the girl is not going to base her answer on what happens on the coin; there would be no reason to expect that she is even aware that a coin was tossed.

A subtle topic similar to that discussed above is treated more elaborately on pages 78–80 of Goldberg (1960). The techniques of parts (a) and (b) of Example 7-1 will be used extensively throughout this book.

---

**EXAMPLE 7-2:** Two defective bolts have been mixed in with four good bolts. Suppose that bolts are tested, one by one, until both defective bolts are found. What is the probability that the testing of exactly four bolts is required?

***Solution:*** Let $D_i$ be the event of finding a defective bolt on the $i$th bolt tested, and let $G_i$ be the event of finding a good bolt on the $i$th one tested, where $i = 1, 2, 3, 4$. An example of a particular favorable order is $D_1 G_2 G_3 D_4$, which indicates that defective bolts were found on the first and fourth, while good bolts were found on the second and third bolts tested. It should be clear that the testing of exactly four bolts was required for this particular order. We see that

$$P(\text{test exactly 4 bolts}) = P(D_1 G_2 G_3 D_4 \text{ or } G_1 D_2 G_3 D_4 \text{ or } G_1 G_2 D_3 D_4 \text{ or } G_1 G_2 G_3 G_4) \tag{1}$$

Note that each of the first three orders has a defective occurring on the fourth bolt. [Clearly, if the second defective bolt were found on the second or third bolt tested, then testing of a fourth bolt wouldn't be required.] The reason for the order $G_1 G_2 G_3 G_4$ is that if the first four bolts tested are good, then the remaining two must be defective by default. After all, we are given that exactly two of the six bolts are defective.

The four events connected by "or" on the right side of Eq. (1) are mutually exclusive. Thus, letting $E$ stand for "test exactly 4 bolts," and using the addition rule, we obtain

$$P(E) = P(D_1 G_2 G_3 D_4) + P(G_1 D_2 G_3 D_4) + P(G_1 G_2 D_3 D_4) + P(G_1 G_2 G_3 G_4) \tag{2}$$

Now, we can compute each of the four terms on the right side of Eq. (2) without difficulty since we have a *drawing without replacement situation*. For a review of the basic ideas, see Example 5-24. For example, for $P(D_1 G_2 G_3 D_4)$, we have

$$P(D_1 G_2 G_3 D_4) = P(D_1)P(G_2 \mid D_1)P(G_3 \mid D_1 \text{ and } G_2)P(D_4 \mid D_1 \text{ and } G_2 \text{ and } G_3)$$

$$= (\tfrac{2}{6})(\tfrac{4}{5})(\tfrac{3}{4})(\tfrac{1}{3}) = \tfrac{1}{15} \tag{3}$$

Employing a similar process for the other probability terms on the right side of Eq. (2), we get

$$P(E) = (\tfrac{2}{6})(\tfrac{4}{5})(\tfrac{3}{4})(\tfrac{1}{3}) + (\tfrac{4}{6})(\tfrac{2}{5})(\tfrac{3}{4})(\tfrac{1}{3}) + (\tfrac{4}{6})(\tfrac{3}{5})(\tfrac{2}{4})(\tfrac{1}{3}) + (\tfrac{4}{6})(\tfrac{3}{5})(\tfrac{2}{4})(\tfrac{1}{3})$$

$$= 4(\tfrac{1}{15}) = \tfrac{4}{15} = .266\overline{6} \tag{4}$$

[Note that each of the probabilities on the right side of Eq. (2) is equal to $\tfrac{1}{15}$.] Thus, the probability that the testing of exactly four bolts is required in order to find the two defective bolts is $\tfrac{4}{15}$.

---

**EXAMPLE 7-3:** Box I contains two red and three white balls, and box II contains four red and two white balls. The boxes look identical to an observer who is viewing the boxes from the outside; that is, an external observer can't tell which box is which. A box is chosen at random [say, with a coin toss, with a head leading an observer to choose the closer box, and a tail leading to the choice of the other box], and then a ball is drawn from that box. What is the probability that a red ball is obtained?

***Solution:*** The event "red ball is obtained" can occur in either of two mutually exclusive ways:

(**A**)  Select box I and draw a red ball, or
(**B**)  Select box II and draw a red ball.

Thus,

$$P(\text{red}) = P[(\text{I and red}) \text{ or } (\text{II and red})] = P(\text{I and red}) + P(\text{II and red}) \quad \textbf{(1)}$$

Then, using the multiplication rule for both the probabilities on the right, we obtain

$$P(\text{red}) = P(\text{I})P(\text{red} \mid \text{I}) + P(\text{II})P(\text{red} \mid \text{II}) \quad \textbf{(2)}$$

Thus, we obtain

$$P(\text{red}) = (\tfrac{1}{2})(\tfrac{2}{5}) + (\tfrac{1}{2})(\tfrac{4}{6}) = \tfrac{8}{15} \quad \textbf{(3)}$$

*notes*

(**a**)  This example is similar to Example 7-1. The processes of randomly choosing a box and then drawing a ball from the chosen box are *independent* actions. Thus, for example, if we rewrite "I and red" from Eq. (1) as "I and red$_\text{I}$," and rewrite (red | I) from Eq. (2) as red$_\text{I}$, we would have

$$P(\text{I and red}_\text{I}) = P(\text{I})P(\text{red}_\text{I}) \quad \textbf{(4)}$$

Here, the events symbolized by I and red$_\text{I}$ [with the latter symbol meaning red on the ball chosen from box I] are *independent events*. Then, we have

$$P(\text{I and red}_\text{I}) = (\tfrac{1}{2})(\tfrac{2}{5}) \quad \textbf{(5)}$$

The same reasoning applies to the probability $P(\text{II and red})$. Such ideas are discussed in detail on pages 78–80 of Goldberg (1960).

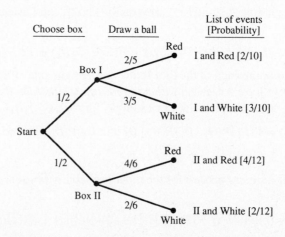

**Figure 7-1.** Probability tree diagram for Example 7-3.

(**b**)  The tree diagram of Figure 7-1 indicates paths for generating the various possible events and their corresponding probabilities. For example, the probabilities of the event "I and red" is obtained by multiplying the probability numbers in the tree along the path corresponding to the event. Thus, $P(\text{I and red}) = (\tfrac{1}{2})(\tfrac{2}{5})$. The four overall events and their probabilities [in parentheses] are listed to the right of the four paths through the tree. Next, since the event red can be expressed as [(I and red) or (II and red)], where the events (I and red) and (II and red) are mutually exclusive, we have

$$P(\text{red}) = P(\text{I and red}) + P(\text{II and red}) = \tfrac{2}{10} + \tfrac{4}{12} = \tfrac{8}{15}$$

## B.  The birthday problem

We shall now consider the birthday problem, which is a classical problem in probability studies. Examples 7-4 and 7-5, which follow, will serve as lead-in examples for the birthday problem.

**EXAMPLE 7-4:** There are four balls in a box, labeled 1, 2, 3, and 4. Suppose that three balls are drawn, with replacement. (**a**) Find the probability of obtaining a different number on each ball? (**b**) What is the probability that at least two balls have the same number?

*Solution: Preliminaries:* Since each draw can result in any of 4 numbered balls, there are, by the fundamental counting principle [Theorem 3.7], $4 \cdot 4 \cdot 4 = 4^3 = 64$ possible equally likely orders. Several of these orders are 132, 244, 312, 323, 431, and 222. In the first, third, and fifth of these orders, a different number occurs on each ball. Let sample space $S$ consist of the $4^3 = 64$ equally likely orders. Thus, $N(S)$, which stands for the number of elements in $S$, is given by

$$N(S) = 64 \tag{1}$$

(**a**) Let $A$ symbolize the event of getting a different number on each ball. The number on the first ball can be any of 4 numbers, that on the second ball can be any of 3 numbers, and that on the third ball can be any of 2 numbers. That is, $N(A)$ is equal to the number of permutations of 4 distinct objects taken 3 at a time [Theorem 4.1]. That is,

$$N(A) = P(4, 3) = 4 \cdot 3 \cdot 2 = 24 \tag{2}$$

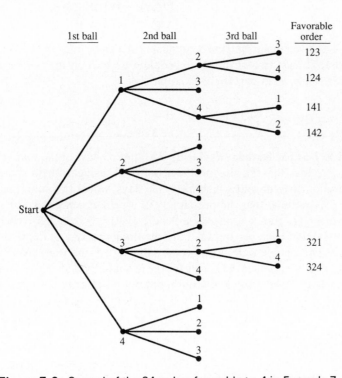

**Figure 7-2.** Several of the 24 orders favorable to $A$ in Example 7-4.

Refer to Figure 7-2 for a partial tree diagram, which indicates how to display the $4 \cdot 3 \cdot 2 = 24$ orders favorable to $A$. Thus,

$$P(A) = \frac{N(A)}{N(S)} = \frac{P(4, 3)}{4^3} = \frac{4 \cdot 3 \cdot 2}{4 \cdot 4 \cdot 4} = \frac{24}{64} = .375 \tag{3}$$

**(b)** The event such that at least two balls in an order of three balls have the same number is $A'$, the complement of $A$. [Some of the orders in $A'$ are 244, 323, 222, 112, and 212.] Thus,

$$P(A') = 1 - P(A) = 1 - \frac{P(4, 3)}{4^3} = 1 - \frac{3}{8} = \frac{5}{8} = .625 \tag{4}$$

---

**EXAMPLE 7-5:** Let the quarters of the year the labeled 1, 2, 3, and 4, where quarter 1 refers to the months January, February, and March, quarter 2 to April, May, and June, etc. Assume that the probability of a person being born in any quarter of the year is 1/4. For a group of three people chosen randomly from the overall population, what is the probability **(a)** that each person was born in a different quarter of the year, and **(b)** that at least two of the three people were born in the same quarter of the year?

*Solution: Preliminaries:* This example is equivalent to Example 7-4. There are $N(S) = 4^3 = 64$ possible orders, all equally likely, where an order pertains to the labeling of quarters [in which a person is born] for the first, second, and third persons chosen randomly from the overall population. Here, for example, the order 132 indicates that the first person chosen was born in quarter 1, the second in quarter 3, and the third in quarter 2. The order 414 indicates that the first and third persons chosen were born in quarter 4, while the second was born in quarter 1.

**(a)** Let $A$ symbolize the event that each of three people was born in a different quarter. Thus, as in part (a) of Example 7-4, the number of elements [orders] in $A$ is given by $N(A) = 4 \cdot 3 \cdot 2 = 24$, and, hence,

$$P(A) = \frac{N(A)}{N(S)} = \frac{P(4, 3)}{4^3} = \frac{4 \cdot 3 \cdot 2}{4 \cdot 4 \cdot 4} = \frac{3}{8} \tag{1}$$

**(b)** With $A$ as given in part (a), the complement $A'$ can be interpreted as the event such that at least two of the three people were born in the same quarter of the year. Thus, as in part (b) of Example 7-4,

$$P(A') = 1 - P(A) = 1 - \frac{P(4, 3)}{4^3} = \frac{5}{8} \tag{2}$$

---

**EXAMPLE 7-6 (The Birthday Problem):** Let the 365 days of the year be labeled 1, 2, 3, 4, ..., 364, 365. [In this problem, we shall disregard leap years, which occur every fourth year and which have 366 days, where the additional day is February 29.] Assume that the probability of a person being born on any day of the year is $\frac{1}{365}$. For a group of $r$ people [with $2 \le r \le 365$], what is the probability **(a)** that each person was born on a different day of the year, and **(b)** that at least two people were born on the same day of the year? (*Note:* In part (b), the question is equivalent to asking for the probability that at least two people have the same birthday [day and month, but not necessarily the same year].)

*Solution: Preliminaries:* The current solution will parallel that of Example 7-5, except that here we shall deal with 365 equally likely days instead of 4 equally likely quarters, and the number of people in the group [$r$] is unspecified in the current problem.

**(a)** Let $A$ symbolize the event that each of the $r$ people was born on a different day of the year. Thus, as in part (a) of Examples 7-4 and 7-5, the number of elements [orders] in $A$ is given by

$$N(A) = P(365, r) = 365 \cdot 364 \cdot 363 \cdot \ldots \cdot [365 - (r - 1)] \qquad (1)$$

where there are $r$ factors in the product on the right of Eq. (1). Here, the symbol $P(365, r)$ is that for the number of permutations of 365 distinct objects [here, days] taken $r$ at a time.

The total number of orders possible for a group of $r$ people [equivalently, the number of equally likely elements in the sample space $S$ of all possible orders of birthdays for a group of $r$ people] is given by

$$N(S) = 365^r \qquad (2)$$

since each person can have a birthday on any of 365 days. [Recall that a similar calculation for $N(S)$ was done in Examples 7-4 and 7-5; in those examples, of course, $N(S)$ was of the form $4^r$.] Thus, we have

$$P(A) = \frac{N(A)}{N(S)} = \frac{P(365, r)}{365^r} \qquad (3)$$

**(b)** With $A$ as given in part (a), the complement event $A'$ can be interpreted as the event such that at least two of the $r$ people were born on the same day of the year; that is, that "at least two of the $r$ people have the same birthday." Thus,

$$P(A') = 1 - P(A) = 1 - \frac{P(365, r)}{365^r} \qquad (4)$$

Values of $P(A')$ as a function of $r$ are given in Table 7-1.

**TABLE 7-1: Probabilities for at Least Two People to Have the Same Birthday [$P(A')$ in Terms of $r$]**

| $r$ | $P(A')$ | $r$ | $P(A')$ |
|-----|---------|-----|---------|
| 2 | .00274 | 24 | .538 |
| 5 | .0271 | 25 | .569 |
| 10 | .117 | 30 | .706 |
| 15 | .253 | 40 | .891 |
| 20 | .411 | 50 | .970 |
| 21 | .444 | 60 | .9941 |
| 22 | .476 | 80 | .99914 |
| 23 | .507 | 100 | .9999997 |

Thus, we see that for a group of 23 or more people [that is, for $r \geq 23$], the probability that at least two people will have the same birthday is larger than .5. For 60 or more people in a group [$r \geq 60$] it is virtually certain that two or more people will have the same birthday.

## C. The general addition rule

The addition rules pertaining to $P(A \cup B)$ and $P(A \cup B \cup C)$ were given in Theorems 2.1 and 2.2. The generalization for $P(A_1 \cup A_2 \cup A_3 \cup \ldots \cup A_n)$, where $n$ is any positive integer, is stated as follows:

**Theorem 7.1 (General Addition Rule):**

$$P(A_1 \cup A_2 \cup A_3 \cup \ldots \cup A_n) = P(A_1) + P(A_2) + \ldots + P(A_n)$$

$$- [P(A_1 \cap A_2) + P(A_1 \cap A_3)$$

$$+ \ldots + P(A_1 \cap A_n) + P(A_2 \cap A_3)$$
$$+ \ldots + P(A_{n-1} \cap A_n)]$$
$$+ [P(A_1 \cap A_2 \cap A_3) + P(A_1 \cap A_2 \cap A_4) + \ldots]$$
$$- \ldots + (-1)^{n-1} P(A_1 \cap A_2 \cap A_3 \cap \ldots \cap A_n)$$

The subscript pairs in the $P(A_i \cap A_j)$ terms constitute all possible combinations of $n$ things taken 2 at a time [also, note that $i \neq j$]. Thus, there are $\binom{n}{2}$ such terms. The subscript trios in the $P(A_i \cap A_j \cap A_k)$ terms constitute all possible combinations of $n$ things taken 3 at a time. [Also, note that $i$, $j$, and $k$ are all different numbers.] Thus, there are $\binom{n}{3}$ such terms, and so on.

---

**EXAMPLE 7-7:** Write out Theorem 7.1 for the case where $n = 4$.

*Solution*

$$P(A_1 \cup A_2 \cup A_3 \cup A_4) = P(A_1) + P(A_2) + P(A_3) + P(A_4)$$
$$- [P(A_1 \cap A_2) + P(A_1 \cap A_3) + P(A_1 \cap A_4)$$
$$+ P(A_2 \cap A_3) + P(A_2 \cap A_4) + P(A_3 \cap A_4)]$$
$$+ [P(A_1 \cap A_2 \cap A_3) + P(A_1 \cap A_2 \cap A_4)$$
$$+ P(A_1 \cap A_3 \cap A_4) + P(A_2 \cap A_3 \cap A_4)]$$
$$- P(A_1 \cap A_2 \cap A_3 \cap A_4)$$

Note that there are $\binom{4}{2} = 6$ terms of the $P(A_i \cap A_j)$ type, and $\binom{4}{3} = 4$ terms of the $P(A_i \cap A_j \cap A_k)$ type. The subscript pairs in the former terms, namely 12, 13, 14, 23, 24, and 34, constitute the six combinations of the four integers 1, 2, 3, and 4 taken two at a time. Likewise, the subscript trios in the latter terms, namely 123, 124, 134, and 234, constitute the four combinations of the four integers 1, 2, 3, and 4 taken three at a time.

---

A clear discussion which leads to a result similar to Theorem 7.1 appears in Chapter Five of Niven (1965).

---

**EXAMPLE 7-8:** Given four chips numbered 1, 2, 3, and 4, which are in a container, and four envelopes labeled, respectively, with the numbers 1, 2, 3, and 4. The chips are drawn out one at a time from the container and placed in the envelopes, such that the first chip drawn out goes to envelope 1, the second chip drawn out goes to envelope 2, and so on. What is the probability that at least one of the chips will be placed into an envelope labeled with the same number as the chip?

*Solution:* This problem is easy if we do a list of all $4! = 24$ permutations of the four integers 1, 2, 3, and 4, taken all at a time. [Remember that $P(4, 4) = 4!$; see Theorem 4.2.] However, let us do the problem in a general way which will allow us to deduce a correct approach for the case of $n$ chips and $n$ envelopes.

Let $A_i$ stand for the event that the chip numbered $i$ is put into the envelope labeled with $i$. Thus, our goal is to calculate the probability

$$P(A_1 \cup A_2 \cup A_3 \cup A_4) \tag{1}$$

[The event $A_1 \cup A_2 \cup A_3 \cup A_4$ here is read as "at least one of $A_1$, $A_2$, $A_3$, $A_4$ occurs." In the current situation, this translates to "at least one of the chips will be placed in an envelope labeled with the same number as the chip."] The general expression for this probability appears in the solution of Example 7-7. Now, $P(A_1)$ denotes the probability that the chip numbered with the integer 1 will be placed in the envelope labeled with 1. [This probability is not conditional on the placements of the other chips.] Thus,

$$P(A_1) = \tfrac{1}{4} \tag{2a}$$

Also, for a similar reason,

$$P(A_2) = P(A_3) = P(A_4) = \tfrac{1}{4} \tag{2b), (2c), (2d}$$

Now, $P(A_1 \cap A_2)$ denotes the probability that the chip numbered with 1 will be placed in the envelope labeled 1, and that the chip numbered with 2 will be placed in the envelope labeled 2. Thus,

$$P(A_1 \cap A_2) = P(A_1)P(A_2 \mid A_1) = (\tfrac{1}{4})(\tfrac{1}{3}) \tag{3}$$

[Here, $P(A_2 \mid A_1)$ is read as "the probability that the chip numbered 2 is placed into the envelope labeled with 2 given that the chip numbered with 1 is placed into the envelope labeled with 1."] It should be clear that

$$P(A_i \cap A_j) = (\tfrac{1}{4})(\tfrac{1}{3}) \tag{4}$$

for each of the six $P(A_i \cap A_j)$ terms. Similarly, each of the four $P(A_i \cap A_j \cap A_k)$ is given by

$$P(A_i \cap A_j \cap A_k) = (\tfrac{1}{4})(\tfrac{1}{3})(\tfrac{1}{2}) \tag{5}$$

and

$$P(A_1 \cap A_2 \cap A_3 \cap A_4) = (\tfrac{1}{4})(\tfrac{1}{3})(\tfrac{1}{2})(\tfrac{1}{1}) \tag{6}$$

Thus, from the equation of Example 7-7 [which is a special case of Theorem 7.1 for $n = 4$], we have

$$P(A_1 \cup A_2 \cup A_3 \cup A_4) = 4 \cdot (\tfrac{1}{4}) - \binom{4}{2} \cdot (\tfrac{1}{4})(\tfrac{1}{3}) + \binom{4}{3} \cdot (\tfrac{1}{4})(\tfrac{1}{3})(\tfrac{1}{2}) - (\tfrac{1}{4})(\tfrac{1}{3})(\tfrac{1}{2})(\tfrac{1}{1}) \tag{7}$$

This leads us to Eq. (8) below:

$$P(A_1 \cup A_2 \cup A_3 \cup A_4) = 1 - \frac{4 \cdot 3}{2!} \cdot \left(\frac{1}{4}\right)\left(\frac{1}{3}\right) + \frac{4 \cdot 3 \cdot 2}{3!} \cdot \left(\frac{1}{4}\right)\left(\frac{1}{3}\right)\left(\frac{1}{2}\right) - \frac{1}{4!}$$

$$= 1 - \frac{1}{2!} + \frac{1}{3!} - \frac{1}{4!} \tag{8}$$

Thus,

$$P(A_1 \cup A_2 \cup A_3 \cup A_4) = 1 - \tfrac{1}{2} + \tfrac{1}{6} - \tfrac{1}{24} = .625 \tag{9}$$

Thus, the probability is .625 that at least one of the chips will be placed into an envelope labeled with the same number as the chip.

---

**EXAMPLE 7-9:** Consider the extension of Example 7-8 to the case of $n$ chips in a container numbered 1, 2, 3, …, $n$, which are drawn out randomly, one at a time, and placed into $n$ envelopes [one chip per envelope], which have been labeled, respectively, 1, 2, 3, …, $n$. Deduce the correct expression for $P(A_1 \cup A_2 \cup A_3 \cup \ldots \cup A_n)$. [*Hint:* Refer to the development of the expression for $P(A_1 \cup A_2 \cup A_3 \cup A_4)$ in Eq. (8) of Example 7-8.]

*Solution:* If we followed the same type of process as in Example 7-8, with particular attention to the manipulations in Eq. (8), we would obtain the following equation, in which the expression on the right is a generalization of the final expression of Eq. (8):

$$P(A_1 \cup A_2 \cup A_3 \cup \ldots \cup A_n) = 1 - \frac{1}{2!} + \frac{1}{3!} - \frac{1}{4!} + \frac{1}{5!} - \ldots + (-1)^{n-1}\frac{1}{n!} \quad \textbf{(1)}$$

For example, for the case of six chips and six envelopes [i.e., $n = 6$], we would have

$$P(A_1 \cup A_2 \cup A_3 \cup \ldots \cup A_6) = 1 - \frac{1}{2!} + \frac{1}{3!} - \frac{1}{4!} + \frac{1}{5!} - \frac{1}{6!} = .631944 \quad \textbf{(2)}$$

Consider the $n!$ permutations of the integers $1, 2, 3, \ldots, n$ taken all at a time. Let us suppose that a permutation is picked at random. In effect, in Examples 7-8 and 7-9, we were asking for the probability that in such a permutation at least one integer occurs in its natural place. For example, consider the permutation of the four integers 1, 2, 3, 4 given by 3214. In this permutation, both 2 and 4 are in their natural places since 2 is in the second place and 4 is in the fourth place. Neither 1 nor 3 is in its natural place.

Thus, in Examples 7-8 and 7-9, we developed equations for the probability that a randomly chosen permutation of the integers $1, 2, 3, \ldots, n$ has at least one integer in its natural place. [In Example 7-8, we dealt with the special case where $n = 4$.] In Table 7-2, we have a tabulation of this probability [which, symbolically, is given as $P(A_1 \cup A_2 \cup A_3 \cup \ldots \cup A_n)$ as in Example 7-9] in terms of $n$ for $n$ between 2 and 10, inclusive. Note that the probability is well approximated by an apparently limiting value, for fairly small $n$. The first four significant digits are the same [namely, 6321] for $n \geq 7$. In fact, it can be shown, using calculus, that the limit of $P(A_1 \cup A_2 \cup A_3 \cup \ldots \cup A_n)$ as $n$ increases without bound is $1 - e^{-1} = 1 - 1/e$, which equals .63212056, to eight significant digits. Table 7-2 follows:

**TABLE 7-2:**
**$P(A_1 \cup A_2 \cup \ldots \cup A_n)$ in**
**Terms of $n$ [Example 7-9]**

| $n$ | $P(A_1 \cup A_2 \cup \ldots \cup A_n)$ |
|---|---|
| 2 | .500000 |
| 3 | .666667 |
| 4 | .625000 |
| 5 | .633333 |
| 6 | .631944 |
| 7 | .632143 |
| 8 | .632118 |
| 9 | .632121 |
| 10 | .632121 |

$P(A_1 \cup A_2 \cup \ldots \cup A_n)$ is the probability that a randomly chosen permutation of $1, 2, 3, \ldots, n$ has at least one integer in its natural place.

*note:* The demonstration that $1 - e^{-1}$ is the limit of the probability expression of Eq. (1) of Example 7-9 is as follows. An infinite series for $e^x$ is the following:

$$e^x = 1 + x + x^2/(2!) + x^3/(3!) + \ldots \quad \textbf{(1)}$$

This series representation is valid for any value of $x$. Thus, for $x = -1$, we have

$$e^{-1} = 1 - 1 + 1/(2!) - 1/(3!) + 1/(4!) - \ldots \quad \textbf{(2)}$$

Thus, we see that

$$1 - e^{-1} = 1 - 1/(2!) + 1/(3!) - 1/(4!) + \ldots \quad \textbf{(3)}$$

Now, if we compare Eq. (3) with Eq. (1) of Example 7-9, we see that

$$\lim_{n \to \infty} P(A_1 \cup A_2 \cup A_3 \cup \ldots \cup A_n) = 1 - e^{-1} = 1 - \frac{1}{e} \qquad \textbf{(4)}$$

A permutation of the integers $1, 2, 3, \ldots, n$ in which none of the $n$ integers is in its natural place is called a derangement of the integers. Thus, for example, the permutation 3412 is a derangement of the integers 1, 2, 3, 4. A question dealing with derangements occurs in Problem 7-3. The topics of derangements and the inclusion-exclusion principle [which is closely related to Theorem 7.1, the general addition rule] are covered in Niven (1965).

# 7-2. Total Probability Theorem and Bayes' Formula

Suppose we have a sample space $S$, which is partitioned into a collection of events. [The latter events are, of course, mutually exclusive.] Our goal will be to develop an equation for $P(E)$ for some event $E$ of $S$ by making use of information dealing with the partition.

## A. Theorem of total probability

Suppose that sample space $S$ is partitioned into $k$ events $B_1, B_2, B_3, \ldots, B_k$. [That is, $S = B_1 \cup B_2 \cup B_3 \cup \ldots \cup B_k$ and $B_i \cap B_j = \varnothing$ for $i \neq j$. For a review of the partition concept, see Definition 1.8.] We also suppose that the probabilities $P(B_1), P(B_2), P(B_3), \ldots, P(B_k)$ are given or can easily be determined.

For example, we demonstrate the key ideas for the case where we have a partition of $S$ into three events; that is, $k = 3$. Refer to Figure 7-3, where the three mutually exclusive events are $B_1$, $B_2$, and $B_3$.

Thus, in Figure 7-3a, we have

$$S = B_1 \cup B_2 \cup B_3, \qquad \text{where } B_1 \cap B_2 = B_1 \cap B_3 = B_2 \cap B_3 = \varnothing \qquad \textbf{(1)}$$

Now, suppose that $E$ is any event [subset] of sample space $S$, as shown in Figure 7-3b. Thus, $E$ can be expressed as a union of three mutually exclusive events as follows:

$$E = (B_1 \cap E) \cup (B_2 \cap E) \cup (B_3 \cap E) \qquad \textbf{(2)}$$

***note:*** In specific cases, one or more of the $(B_i \cap E)$ terms may be the empty set. Now if, say, $(B_1 \cap E) = \varnothing$, that means that $E$ doesn't intersect the subset $B_1$.

From the addition rule for mutually exclusive events [see, for example, Axiom 1.3'], we have

$$P(E) = P(B_1 \cap E) + P(B_2 \cap E) + P(B_3 \cap E) \qquad \textbf{(3)}$$

Now, if we know [or can easily determine] the conditional probabilities $P(E \mid B_i)$ for $i = 1, 2,$ and 3, then we can use the multiplication rule [Theorem 3.1] to write

$$P(B_i \cap E) = P(B_i)P(E \mid B_i) \qquad \text{for } i = 1, 2, \text{ and } 3 \qquad \textbf{(4)}$$

Finally, if we substitute from Eq. (4) for $i = 1, 2,$ and 3 into Eq. (3), we obtain

$$P(E) = P(B_1)P(E \mid B_1) + P(B_2)P(E \mid B_2) + P(B_3)P(E \mid B_3) \qquad \textbf{(5)}$$

The generalization of Eq. (5) is known as the Theorem [or Law] of Total Probability.

**Theorem 7.2 (Theorem of Total Probability):** If the sample space $S$ is partitioned into $k$ events $B_1, B_2, \ldots, B_k$ [that is, $S = B_1 \cup B_2 \cup \ldots \cup B_k$, where $B_i \cap B_j = \varnothing$ for $i \neq j$], and $E$ is an event of the sample space $S$, then the probability of the event $E$ can be expressed as follows:

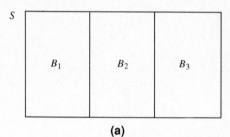

(a)

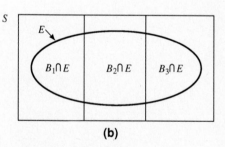

(b)

**Figure 7-3.** Diagrams for the Total Probability Theorem [for $k = 3$].

$$P(E) = P(B_1)P(E \mid B_1) + P(B_2)P(E \mid B_2) + \ldots + P(B_k)P(E \mid B_k)$$

Examples 7-10 and 7-11 illustrate Theorem 7.2 for cases in which the sample space is partitioned into two mutually exclusive events [i.e., $k = 2$].

> *note:* Actually, we have already applied the total probability theorem on several occasions in this book, without referring to it by name. For example, consider Examples 7-1 and 7-3.

---

**EXAMPLE 7-10:** Suppose we choose two cards without replacement from a standard deck of 52 cards. Find the probability that the second card is a spade; denote the probability of this event by $P(S_2)$. [*Note:* We solved a similar problem in part (c) of Example 3-17.]

*Solution:* Let's try to use common sense here, instead of just plugging numbers into an equation. Spade on card 2 can arise if card 1 is a spade [$S_1$] and card 2 is a spade or if card 1 is a nonspade [$S_1'$] and card 2 is a spade. Thus, in symbols, where $S_2$ indicates spade on card 2, we have

$$S_2 = (S_1 \text{ and } S_2) \text{ or } (S_1' \text{ and } S_2) = (S_1 \cap S_2) \cup (S_1' \cap S_2) \tag{1}$$

The two events on the right of Eq. (1) which are enclosed in parentheses are mutually exclusive, and so

$$P(S_2) = P(S_1 \cap S_2) + P(S_1' \cap S_2) \tag{2}$$

Then, applying the multiplication rule to the two terms on the right of Eq. (2), we obtain

$$P(S_2) = P(S_1)P(S_2 \mid S_1) + P(S_1')P(S_2 \mid S_1') \tag{3}$$

[Note that Eq. (3) is a special case of Theorem 7.2 for $k = 2$. Here, $E$ is $S_2$, $B_1$ is $S_1$, and $B_2$ is $S_1'$.]

Now, the probability of spade on card 1 [i.e., of $P(S_1)$] is $\dfrac{13}{52}$, and the probability of nonspade on card 1 [i.e., of $P(S_1')$] is $\dfrac{39}{52}$. The term $P(S_2 \mid S_1)$, which denotes the conditional probability of spade on card 2 given a spade on card 1, is equal to $\dfrac{(13-1)}{(52-1)} = \dfrac{12}{51}$. Similarly, $P(S_2 \mid S_1') = \dfrac{13}{(52-1)} = \dfrac{13}{51}$, since the condition here is for a nonspade on card 1. [For an important relevant discussion, refer to Example 5-24.] Substituting these probability values into the right side of Eq. (3) leads to

$$P(S_2) = (\tfrac{13}{52})(\tfrac{12}{51}) + (\tfrac{39}{52})(\tfrac{13}{51}) = (\tfrac{1}{4})(\tfrac{12}{51}) + (\tfrac{3}{4})(\tfrac{13}{51}) = \tfrac{51}{204} = \tfrac{1}{4} \tag{4}$$

Thus, we have obtained the surprising result that the unconditional probability of spade on card 2 is $\tfrac{1}{4}$, even though we have been drawing without replacement. In fact, it can be shown that if three, four, or more cards were drawn without replacement, then the probability of spade on the last card would still be $\tfrac{1}{4}$; it is just as though only one card were drawn.

---

**EXAMPLE 7-11:** Suppose a box contains three coins, of which two are fair coins and one is a coin with two heads. A coin is selected at random from the box and tossed once. What is the probability that the result of that toss is a head?

*Solution:* Again, let's use a commonsense approach in which we express the event of interest [namely, head] in terms of two or more mutually exclusive events. Head [$H$] can arise if the coin drawn is fair and it yields a head upon tossing or if the coin drawn is two-headed and it yields a head upon tossing. Thus, denoting the two-headed "bad" coin by $B$ and a fair coin by $F$, we have

$$H = (F \text{ and } H) \text{ or } (B \text{ and } H) = (F \cap H) \cup (B \cap H) \qquad \textbf{(1)}$$

The two events on the right separated by the $\cup$ are mutually exclusive. Thus, by the addition rule, we get

$$P(H) = P(F \cap H) + P(B \cap H) \qquad \textbf{(2)}$$

Now, using the multiplication rule for the two probabilities on the right, we get

$$P(H) = P(F)P(H \mid F) + P(B)P(H \mid B) \qquad \textbf{(3)}$$

Now, since there are two fair coins and one bad coin in the box, we know that $P(F) = \frac{2}{3}$ and $P(B) = \frac{1}{3}$. We also know that $P(H \mid F) = \frac{1}{2}$ and $P(H \mid B) = 1$. Thus, substituting these values into Eq. (3), we get

$$P(H) = (\tfrac{2}{3})(\tfrac{1}{2}) + (\tfrac{1}{3})(1) = \tfrac{2}{3} \qquad \textbf{(4)}$$

---

Example 7-12 can be worked out by using a commonsense approach or by making use of Theorem 7.2 for $k = 3$.

---

**EXAMPLE 7-12:** The Great Idea [GI] Company manufactures a certain type of light bulb at factories I, II, and III. GI has in stock a large batch of bulbs, 25% of which were made in factory I, 35% in factory II, and 40% in factory III. Also, the GI track record shows that 3% of all bulbs produced by factory I are defective, 5% of all bulbs produced by factory II are defective, and 4% of all bulbs produced by factory III are defective. If a single light bulb is picked at random from the batch, what is the probability that it is defective?

*Solution:* Let's solve the problem by using the equation of Theorem 7.2 for $k = 3$. Let $E$ denote the event that "the bulb picked is defective," and let $B_1$, $B_2$, and $B_3$ denote that the bulb came from factory I, or II, or III, respectively. We wish to calculate $P(E)$. Now, the given data can be expressed in terms of probabilities.

Since 25% of the bulbs come from factory I, we have $P(B_1) = .25$. Similarly, $P(B_2) = .35$ and $P(B_3) = .40$.

Since 3% of the bulbs from factory I are defective, that means that $P(E \mid B_1) = .03$. Likewise, $P(E \mid B_2) = .05$ and $P(E \mid B_3) = .04$. Substituting these six known probabilities into the right side of the equation of Theorem 7.2, we get

$$P(E) = P(B_1)P(E \mid B_1) + P(B_2)P(E \mid B_2) + P(B_3)P(E \mid B_3)$$

$$= (.25)(.03) + (.35)(.05) + (.40)(.04) = .041$$

Thus, the probability is .041 [or 4.1%] that a bulb picked at random is defective.

---

## B. Bayes' Formula

The terminology and assumptions used in the development of Bayes' Formula are the same as those introduced in the discussion pertaining to Theorem 7.2. Let $B_i$ denote any of the quantities $B_1$, $B_2$, ..., $B_k$. Suppose that the quantities $P(B_1)$, $P(B_2)$, ..., $P(B_k)$, and $P(E \mid B_1)$, $P(E \mid B_2)$, ..., $P(E \mid B_k)$ are all known or can be determined easily.

It is often important in practical situations to be able to calculate $P(B_i \mid E)$, the conditional probability of $B_i$ given $E$. The following equation, known as Bayes' Formula, expresses $P(B_i \mid E)$ in terms of $P(B_i)$, $P(E \mid B_i)$, and $P(E)$.

**Theorem 7.3 (Bayes' Formula):**

$$P(B_i \mid E) = \frac{P(B_i)P(E \mid B_i)}{P(E)}$$

An equation for the denominator term, $P(E)$, is given in Theorem 7.2.

**EXAMPLE 7-13:** Prove Bayes' Formula.

***Proof:*** From the definition for conditional probability [Definition 3.1], we have

$$P(B_i \mid E) = \frac{P(E \cap B_i)}{P(E)} \tag{1}$$

We can replace the numerator term by $P(B_i \cap E)$ since $E \cap B_i = B_i \cap E$. Then, we apply the multiplication rule [Theorem 3.1] to $P(B_i \cap E)$ as follows:

$$P(B_i \cap E) = P(B_i)P(E \mid B_i) \tag{2}$$

Replacing the numerator of Eq. (1) by the right side of Eq. (2) results in Bayes' Formula as given by Theorem 7.3 above.   □

Example 7-14 illustrates the usefulness of Bayes' Formula.

**EXAMPLE 7-14:** Refer to Example 7-12, which deals with light bulbs manufactured at three factories of the GI Company. Suppose we have picked out a defective light bulb from the batch of light bulbs discussed in Example 7-12. What is the probability that the light bulb came **(a)** from factory I, **(b)** from factory II, **(c)** from factory III?

***Solution:*** Let us continue to use the symbols introduced in Example 7-12.

**(a)** Here, we wish to calculate the conditional probability $P(B_1 \mid E)$. Observe that $P(B_1) = .25$, $P(E \mid B_1) = .03$, and $P(E) = .041$ from Example 7-12. Substitution of these values into Bayes' Formula yields

$$P(B_1 \mid E) = \frac{P(B_1)P(E \mid B_1)}{P(E)} = \frac{(.25)(.03)}{.041} = .183$$

Thus, the probability is .183 [or 18.3%] that the light bulb came from factory I given that it is defective.

**(b)** Proceeding as in part (a), we have

$$P(B_2 \mid E) = \frac{P(B_2)P(E \mid B_2)}{P(E)} = \frac{(.35)(.05)}{.041} = .427$$

**(c)** Again proceeding as in part (a), we have

$$P(B_3 \mid E) = \frac{P(B_3)P(E \mid B_3)}{P(E)} = \frac{(.40)(.04)}{.041} = .390$$

Observe that the sum of the probabilities from parts (a), (b), and (c) in Example 7-13 is 1. This is no accident, but is a confirmation of Theorem 7.4, which follows. The symbols that occur in Theorem 7.4 are those described and employed in Theorems 7.2 and 7.3.

**Theorem 7.4:** Suppose that sample space $S$ is partitioned into $k$ events $B_1$, $B_2, \ldots, B_k$, and that $E$ is any event of sample space $S$. Then

$$P(B_1 \mid E) + P(B_2 \mid E) + \ldots + P(B_k \mid E) = 1$$

***notes***

**(a)** A special case of Theorem 7.4 occurs in Theorem 3.4, which states that $P(B' \mid A) = 1 - P(B \mid A)$. In this equation, $B$, $B'$, and $A$ replace $B_1$, $B_2$, and $E$, respectively, of Theorem 7.4. Here, a partition of $S$ is formed by $B$ and $B'$, and $k = 2$.

(b) Another special case arises when $E$ is $S$, the sample space. In such a case, $P(B_i \mid E) = P(B_i \mid S) = P(B_i)$, and Theorem 7.4 reduces to

$$P(B_1) + P(B_2) + \dots + P(B_k) = 1 \qquad \textbf{(I)}$$

Equation (I) provides us with a consistency check since the collection of $B_i$'s constitutes a partition of $S$. [That is, we have $S = B_1 \cup B_2 \cup \dots \cup B_k$, where the $B_i$'s are mutually exclusive.]

---

**EXAMPLE 7-15:** Prove Theorem 7.4.

**Proof:** Here, we have a partition of $S$ into $k$ events $B_1$, $B_2$, ..., $B_k$. That is, $S = B_1 \cup B_2 \cup \dots \cup B_k$, where $B_i \cap B_j = \varnothing$ whenever $i \neq j$. Since the diagram for the $k = 3$ case is given in Figure 7-3, let's do the proof for that case. First of all, we have

$$P(B_i \mid E) = \frac{P(E \cap B_i)}{P(E)} = \frac{P(B_i \cap E)}{P(E)} \qquad \text{for } i = 1, 2, 3 \qquad \textbf{(1)}$$

Thus,

$$P(B_1 \mid E) + P(B_2 \mid E) + P(B_3 \mid E) = \frac{P(B_1 \cap E) + P(B_2 \cap E) + P(B_3 \cap E)}{P(E)} \qquad \textbf{(2)}$$

From Figure 7-3, we see that $E = (B_1 \cap E) \cup (B_2 \cap E) \cup (B_3 \cap E)$, where the three events on the right of the form $(B_i \cap E)$ are mutually exclusive [disjoint]. Thus, from Axiom 1.3', we have

$$P(E) = P(B_1 \cap E) + P(B_2 \cap E) + P(B_3 \cap E) \qquad \textbf{(3)}$$

Substituting $P(E)$ from Eq. (3) into the denominator of Eq. (2) yields

$$P(B_1 \mid E) + P(B_2 \mid E) + P(B_3 \mid E) = 1 \qquad \textbf{(4)} \quad \square$$

---

# SOLVED PROBLEMS

## Elementary Probability Applications

**PROBLEM 7-1** Carla and Dave each toss a coin twice. The one who tosses the greater number of heads wins a prize. Suppose that Dave has a fair coin $[P_D(H) = .5]$, while Carla has a coin for which the probability of head on a single toss is .4 $[P_C(H) = .4]$. What is the probability that (a) Carla will win the prize, (b) a tie results, (c) Dave will win the prize?

**Solution** *Preliminaries:* Let $C_1$ stand for Carla's getting exactly one head, and similarly for the symbols $C_0$ and $C_2$. Let $D_2$ stand for Dave's getting exactly two heads, and similarly for the symbols $D_0$ and $D_1$. Now, let us determine the individual probabilities for $C_i$ and $D_i$ for $i = 0, 1$, and 2. For example, $P(C_1) = P_C(HT) + P_C(TH) = 2(.4)(.6) = .48$, and $P(D_2) = P_D(HH) = (.5)^2 = .25$. All the $P(C_i)$ and $P(D_i)$ values are presented in Table 7-3, which follows. Observe that the $P(C_i)$ values are binomial probabilities for $n = 2$ and $p = .4$, while the $P(D_i)$ values are binomial probabilities for $n = 2$ and $p = .5$. [See Theorem 5.5.]

(a)
$$P(\text{Carla wins}) = P[(C_1 \text{ and } D_0) \text{ or } (C_2 \text{ and } D_0) \text{ or } (C_2 \text{ and } D_1)]$$

$$= P(C_1 \text{ and } D_0) + P(C_2 \text{ and } D_0) + P(C_2 \text{ and } D_1)$$

$$= P(C_1)P(D_0) + P(C_2)P(D_0) + P(C_2)P(D_1)$$

$$= (.48)(.25) + (.16)(.25) + (.16)(.50) = .24$$

**(b)**
$$P(\text{Tie}) = P[(C_0 \text{ and } D_0) \text{ or } (C_1 \text{ and } D_1) \text{ or } (C_2 \text{ and } D_2)]$$
$$= P(C_0 \text{ and } D_0) + P(C_1 \text{ and } D_1) + P(C_2 \text{ and } D_2)$$
$$= P(C_0)P(D_0) + P(C_1)P(D_1) + P(C_2)P(D_2)$$
$$= (.36)(.25) + (.48)(.50) + (.16)(.25) = .37$$

**(c)**
$$P(\text{Dave wins}) = P[(D_1 \text{ and } C_0) \text{ or } (D_2 \text{ and } C_0) \text{ or } (D_2 \text{ and } C_1)]$$
$$= P(D_1 \text{ and } C_0) + P(D_2 \text{ and } C_0) + P(D_2 \text{ and } C_1)$$
$$= P(D_1)P(C_0) + P(D_2)P(C_0) + P(D_2)P(C_1)$$
$$= (.50)(.36) + (.25)(.36) + (.25)(.48) = .39$$

An alternate approach for $P(\text{Dave wins})$ is the following:

$$P(\text{Dave wins}) = 1 - [P(\text{Carla wins}) + P(\text{Tie})] = 1 - [.24 + .37] = .39$$

**TABLE 7-3: Probabilities when Carla [C] and Dave [D] Each Toss a Coin Twice [Problem 7-1]**

| $i$ | $P(C_i)$ | $P(D_i)$ |
|---|---|---|
| 0 | $.6^2 = .36$ | $.5^2 = .25$ |
| 1 | $2(.4)(.6) = .48$ | $2(.5)^2 = .50$ |
| 2 | $.4^2 = .16$ | $.5^2 = .25$ |
| Sum | 1.00 | 1.00 |

**PROBLEM 7-2** For a machine consisting of four separate components, the probability of successful operation for each component is .95. What is the probability the machine will work **(a)** if the machine will work only if all the components operate successfully; **(b)** if the machine will work when any three of the components operate successfully?

Assume for both parts (a) and (b) that the components function independently of each other.

***Solution*** *Preliminaries:* Let $C_i$ for $i = 1, 2, 3, 4$ stand for "component $i$ operates successfully." Thus, $P(C_1) = P(C_2) = P(C_3) = P(C_4) = .95$.

**(a)**
$$P(\text{machine will work}) = P(C_1 \text{ and } C_2 \text{ and } C_3 \text{ and } C_4) = P(C_1 C_2 C_3 C_4)$$
$$= P(C_1)P(C_2)P(C_3)P(C_4) = (.95)^4 = .8145 \qquad \textbf{(1)}$$

**(b)** Let $C_i'$ indicate that component $i$ doesn't operate successfully, for $i = 1, 2, 3, 4$. Since the machine will work when any three of its components operate successfully, there are four new orders for which the machine will work. These are as follows: $C_1 C_2 C_3 C_4'$, $C_1 C_2 C_3' C_4$, $C_1 C_2' C_3 C_4$, $C_1' C_2 C_3 C_4$. Now, $P(C_i') = 1 - P(C_i) = .05$. The probability for $C_1 C_2 C_3 C_4'$ is calculated as follows:

$$P(C_1 C_2 C_3 C_4') = P(C_1)P(C_2)P(C_3)P(C_4') = (.95)^3(.05) \qquad \textbf{(2)}$$

The probability is also $(.95)^3(.05)$ for each of $C_1 C_2 C_3' C_4, C_1 C_2' C_3 C_4$, and $C_1' C_2 C_3 C_4$. Thus, we have

$$P(\text{machine will work}) = P(C_1 C_2 C_3 C_4) + 4 \cdot P(C_1 C_2 C_3 C_4') = (.95)^4 + 4(.95)^3(.05)$$
$$= .8145 + .17148 = .9860 \qquad \textbf{(3)}$$

**PROBLEM 7-3** **(a)** Given the integers 1, 2, 3, ..., $n$. Develop an equation for the probability that a randomly chosen permutation of these integers is a derangement. [Recall that, in such a permutation, no integer appears in its natural place. For a review of these and related ideas, refer to Examples 7-8, 7-9, and the discussion following Example 7-9.] **(b)** Suppose that seven chips, numbered 1, 2, 3, 4, 5, 6, 7 are randomly drawn out one at a time, from a container, and placed in envelopes labeled, respectively, with the numbers 1, 2, 3, 4, 5, 6, 7. [The first chip drawn out goes to envelope 1, the second chip to envelope 2, etc.] What is the probability that every envelope will contain a chip whose number is different from the number labeled on the envelope?

***Solution***

(a) In Example 7-9, we stated the equation for $P(A_1 \cup A_2 \cup \ldots \cup A_n)$, which is equal to the probability that a randomly chosen permutation of $1, 2, 3, \ldots, n$ has at least one integer in its natural place. Now, the complement of the event $A_1 \cup A_2 \cup \ldots \cup A_n$ is an event that consists of all the derangements of the integers $1, 2, 3, \ldots, n$. Thus,

$$P(\text{derangement of } 1, 2, \ldots, n) = 1 - P(A_1 \cup A_2 \cup A_3 \cup \ldots \cup A_n)$$

$$= 1 - \left[ 1 - \frac{1}{2!} + \frac{1}{3!} - \frac{1}{4!} + \frac{1}{5!} - \ldots + (-1)^{n-1} \frac{1}{n!} \right]$$

$$= \frac{1}{2!} - \frac{1}{3!} + \frac{1}{4!} - \frac{1}{5!} + \ldots + (-1)^n \frac{1}{n!} \qquad (1)$$

[Note that the $(-1)^n$ in the last term arises from $-(-1)^{n-1}$.] The limit of this probability as $n$ becomes infinite is $e^{-1}$, which equals .36787944 to eight significant digits. The probability expression of Eq. (1) agrees with the value of $e^{-1}$ to at least six significant digits for $n \geq 9$. In this regard, see Table 7-2, which follows Example 7-9. Table 7-2 presents values of $P(A_1 \cup A_2 \cup \ldots \cup A_n)$ as a function of $n$ for $n$ between 2 and 10, inclusive.

(b) The probability here is given by the Eq. (1) result for $n = 7$. Thus, the probability that every envelope will contain a chip with a number different than the number on the envelope is given by

$$\frac{1}{2!} - \frac{1}{3!} + \frac{1}{4!} - \frac{1}{5!} + \frac{1}{6!} - \frac{1}{7!} = .367857$$

## Total Probability Theorem and Bayes' Formula

**PROBLEM 7-4**  An appliance store receives shipments of a certain type of radio from three of its warehouses: 20% come from warehouse I, 50% from warehouse II, and 30% from warehouse III. It is known from past experience that 5% of the radios from warehouse I are defective, while the corresponding percentages of defectives are 4% and 7% for warehouses II and III, respectively. (a) If a single radio is picked at random from the inventory at the store, what is the probability it is defective? [Equivalently, what is the fraction of defective radios among all the radios shipped to the appliance store from the three warehouses?] (b) What is the probability that a radio came from warehouse I if [equivalently, given that] it turns out to be defective?

***Solution***  *Preliminaries:*  Let $D$ stand for a "defective radio," and let $R_1$ indicate that a radio comes from warehouse I, with similar interpretations for $R_2$ and $R_3$.

(a) From the theorem of total probability [Theorem 7.2] and the given data, we have

$$P(D) = P(R_1)P(D \mid R_1) + P(R_2)P(D \mid R_2) + P(R_3)P(D \mid R_3)$$

$$= (.2)(.05) + (.5)(.04) + (.3)(.07) = .051$$

Thus, the probability of picking a defective ratio is .051 or 5.1%. Equivalently, the fraction of defective radios among all the radios received by the appliance store is .051.

(b) Here, we want to calculate $P(R_1 \mid D)$. From Bayes' Formula [Theorem 7.3] and the given data, we have

$$P(R_1 \mid E) = \frac{P(R_1)P(E \mid R_1)}{P(E)} = \frac{(.2)(.05)}{.051} = .196$$

Here, the denominator value is from part (a). Thus, the probability is .196 that a radio came from warehouse I, given that it is defective.

**PROBLEM 7-5**  A new test for detecting cancer has been developed. Suppose that 90% of the cancer patients in a large hospital reacted positively to the test, while 15% of the remaining patients in the hospital reacted positively to the test. [Assume that none of these remaining patients has cancer. That is, assume that the only patients in the hospital who have cancer are the cancer patients. Also, observe that a positive reaction to the test means that the test indicates that a person has cancer. That is, it is possible for a person who does not have cancer to show a positive reaction to the test. This is known as a *false positive*.] Suppose that 4% of the patients in the hospital are cancer patients [equivalently, 4% of the patients in the hospital actually have cancer]. What is the probability that a hospital patient picked at random who reacts positively to the test is a cancer patient?

*Solution* Let $C$ and $C'$ denote, respectively, the events that a patient is a cancer patient, or is not a cancer patient [i.e., is a "remaining patient"]. Let Pos denote that a patient reacts positively to the test. We wish to calculate $P(C \mid \text{Pos})$. From Bayes' Formula [Theorem 7.3], we have

$$P(C \mid \text{Pos}) = \frac{P(C)P(\text{Pos} \mid C)}{P(\text{Pos})} \tag{1}$$

From Theorem 7.2, and the given data, we have

$$P(\text{Pos}) = P(C)P(\text{Pos} \mid C) + P(C')P(\text{Pos} \mid C') = (.04)(.90) + (.96)(.15) = .18 \tag{2}$$

Thus, substituting the given data and the value for $P(\text{Pos})$ from Eq. (2) into Eq. (1), we get

$$P(C \mid \text{Pos}) = \frac{(.04)(.90)}{(.18)} = .20 \tag{3}$$

which means that there is a 20% probability that a patient who reacts positively to the test is a cancer patient.

**PROBLEM 7-6** This problem deals with the subject of Bayesian decision theory, which is based, in part, on Bayes' Formula.

Having commissioned a survey of oil deposits on adjoining regions, a landowner estimates that the probability of an oil discovery on his land is .2. Let $S_1$ and $S_2$ indicate, respectively, the presence or the absence of oil on his land. Thus,

$$P(S_1) = .2 \quad \text{and} \quad P(S_2) = .8 \tag{1}, (2)$$

These are known as *a priori* or *prior* probabilities.

The landowner decides to have expensive sonar testing [involving sound waves] done on his land. Let $\theta_1$ designate the event that the sonar testing indicates the presence of oil, and $\theta_2$ designate the event that the sonar testing indicates that no oil is present. The engineer doing the sonar testing admits that 10% of the time the test will indicate no oil when, in fact, oil is present. That is,

$$P(\theta_2 \mid S_1) = .1, \quad \text{and hence} \quad P(\theta_1 \mid S_1) = .9 \tag{3}, (4)$$

Also, when oil is not present, the test indicates no oil 60% of the time. That is,

$$P(\theta_2 \mid S_2) = .6, \quad \text{and hence} \quad P(\theta_1 \mid S_2) = .4 \tag{5}, (6)$$

[The results derived in Eqs. (4) and (6) follow directly from Eqs. (3) and (5), respectively, as a consequence of applying Theorem 7.4 for the case where $k = 2$.] Find $P(S_1 \mid \theta_1)$ and $P(S_1 \mid \theta_2)$, which are the *a posteriori* or *posterior* probabilities that oil is present, given, respectively, positive and negative results of sonar testing.

*Solution* First, we find the [total] probability $P(\theta_1)$ that sonar testing will show the presence of oil. From the total probability theorem [Theorem 7.2], or from common sense,

$$P(\theta_1) = P(S_1)P(\theta_1 \mid S_1) + P(S_2)P(\theta_1 \mid S_2) = (.2)(.9) + (.8)(.4) = .50 \tag{7}$$

Thus, by Bayes' Formula [Theorem 7.3],

$$P(S_1 \mid \theta_1) = \frac{P(S_1)P(\theta_1 \mid S_1)}{P(\theta_1)} = \frac{(.2)(.9)}{(.5)} = .36 \tag{8}$$

and thus,

$$P(S_2 \mid \theta_1) = 1 - P(S_1 \mid \theta_1) = .64 \tag{9}$$

Now, it should be clear that $P(\theta_2) = 1 - P(\theta_1)$, and hence that $P(\theta_2) = .50$. As a check, after using Theorem 7.2, we have

$$P(\theta_2) = P(S_1)P(\theta_2 \mid S_1) + P(S_2)P(\theta_2 \mid S_2) = (.2)(.1) + (.8)(.6) = .50 \qquad \textbf{(10)}$$

Thus, by Bayes' Formula,

$$P(S_1 \mid \theta_2) = \frac{P(S_1)P(\theta_2 \mid S_1)}{P(\theta_2)} = \frac{(.2)(.1)}{(.5)} = .04 \qquad \textbf{(11)}$$

and thus,

$$P(S_2 \mid \theta_2) = 1 - P(S_1 \mid \theta_2) = .96 \qquad \textbf{(12)}$$

Note that the posterior probability of finding oil given a positive test result, namely, .36, is greater than the prior probability of finding oil, namely .2. The posterior probability of finding oil given a negative test result, namely .04, is lower than the prior probability of finding oil. Here, we have a realistic example of subjective probabilities being influenced by the appearance of further information.

# Supplementary Problems

**PROBLEM 7-7** At Milton College, it is known that 75% of the students who took three years of high school mathematics will pass the Elementary Statistics course, while 45% of those who didn't take three years of high school mathematics will pass the Elementary Statistics course. [These figures apply to the first attempt at taking the Elementary Statistics course. If a student has not passed the course that means that the student has either failed the course or has not completed the requirements for the course.] It is also known that 40% of the students at Milton College who attempt to take the Elementary Statistics course have taken three years of high school mathematics. A student is picked at random from among those students at Milton College who have already attempted to take the Elementary Statistics course. **(a)** What is the probability that the student has passed the Elementary Statistics course on the first attempt? **(b)** What is the probability that a student has taken three years of high school mathematics given that the student has passed the Elementary Statistics course on the first attempt?

*Answer* **(a)** .570  **(b)** $30/57 = .526$

**PROBLEM 7-8** Refer to Example 7-2. What is the probability that testing is required of **(a)** exactly two bolts, **(b)** exactly three bolts, and **(c)** exactly five bolts?

*Answer* **(a)** $\frac{1}{15}$  **(b)** $\frac{2}{15}$  **(c)** $\frac{8}{15}$

*note:* Suppose that random variable $X$ denotes the exact number of bolts that are required to be tested. Then, the probability function for $X$ is given as follows: $f(2) = \frac{1}{15}, f(3) = \frac{2}{15}, f(4) = \frac{4}{15}$ [from the solution to Example 7-2], and $f(5) = \frac{8}{15}$.

**PROBLEM 7-9** Of all the job applicants at the Apex Company who come for a job interview, 30% are given the interview right away [say, within 5 minutes of arriving]. The other 70% are asked to wait in a waiting room. About 40% of the time, people who are asked to wait will leave the waiting room before being given the job interview, and will not return. **(a)** What percent of the job applicants who come for a job interview will actually have the interview? **(b)** Given that an applicant is given an interview, what is the probability that the applicant had to wait in the waiting room?

*Answer* **(a)** 72%  **(b)** $42/72 = .58\overline{3}$

**PROBLEM 7-10**   Refer to Example 7-5. For a group of three people chosen randomly from the overall population, what is the probability that (**a**) exactly two were born in the same quarter of the year and (**b**) all three were born in the same quarter of the year?

*Answer* (**a**) 36/64 = 9/16 = .5625     (**b**) 4/64 = 1/16 = .0625

**PROBLEM 7-11**   Each of three boxes contains two balls. One box contains two green balls, another contains a green and a red ball, and the third contains two red balls. A box is chosen at random, and one ball is taken out. What is the probability that the second ball in the box is green, given that the ball that is taken out is green?

*Answer* 2/3

**PROBLEM 7-12**   Box I contains two green and three red balls, box II contains four green and two red balls, and box III contains three green and three red balls. Suppose one ball is drawn from each box. What is the probability that a green ball was drawn from box I, given that two green balls were drawn?

*Answer* 1/2

**PROBLEM 7-13**   The Slobovian army believes that their enemy, the Grimalkin army, will attack at locations 1, 2, and 3 with probabilities .4, .25, and .35, respectively. [Note that the probabilities add up to 1.0.] In symbols, $P(A_1) = .4$, $P(A_2) = .25$, and $P(A_3) = .35$.

A Slobovian reconnaissance team then observes the Grimalkin army setting up long range cannons. Let this action be designated as action $B$. The Slobovian intelligence experts believe that the probabilities of action $B$, given a plan to attack at locations 1, 2, and 3, respectively, are .2, .5, and .3. Use Bayes' Formula [Theorem 7.3] to determine posterior probabilities that the Grimalkin army will attack at locations 1, 2, and 3, respectively.

*Answer* $P(A_1 \mid B) = .258$, $P(A_2 \mid B) = .403$, $P(A_3 \mid B) = .339$

**PROBLEM 7-14**   Suppose Bertha holds two tickets from a lottery in which 1000 tickets were sold. The tickets have been numbered from 1 to 1000. Ten prizes are to be awarded. [That is, ten of the 1000 tickets will be designated as prize-winning tickets.] Let $X$ be the number of prizes that Bertha will win. Determine the probability function of $X$.

*Answer*

| $x$ | 0 | 1 | 2 |
|---|---|---|---|
| $f(x)$ | .98009 | .01982 | .00009 |

**PROBLEM 7-15**   Refer to Problem 7-1 again. In the experiment of that problem, Dave tosses a fair coin twice while Carla also tosses a coin twice. For Carla's coin, the probability of head on a single toss is .4. What is the probability that Dave will win the prize provided that the experiment is repeated whenever a tie occurs? [*Suggestion:* Make use of Theorem 6.5, which gives the equation for the sum of a geometric series.]

*Answer* 13/21 = .61905

# MIDTERM EXAM (Chapters 1-7)

1. Given a container with three red balls, labeled 1, 2, 3, and two green balls, labeled 4 and 5. Suppose two balls are drawn in succession, resulting in three balls remaining in the container [i.e., we have drawing without replacement here]. Describe a sample space if (a) one wishes to record the number on each ball, if (b) one wishes to record the color on each ball, or if (c) one wishes to record the total number of red balls obtained.

2. For the sample space $S_1$ of Problem 1 above [see solution to part (a)], indicate the following events as subsets of $S_1$: (a) Obtaining an odd number on both draws. [Call this event $A$.] (b) Obtaining a sum of numbers equal to six. [Call this event $B$.] (c) Obtaining exactly one red ball. [Call this event $C$.]

3. Determine the probabilities of the events $A$, $B$, and $C$ of Problem 2. Note that it is reasonable to assume that all 20 elements of sample space $S_1$ are equally likely.

4. For people who visit Eddie's Fast Food Restaurant, 40% order a meal containing a hamburger, 35% order a meal containing a milkshake, and 10% order a meal containing both a hamburger and a milkshake. What is the probability that a person visiting the restaurant will (a) have either a hamburger or a milkshake or both, (b) will have exactly one of a hamburger or a milkshake, and (c) will have neither a hamburger nor a milkshake? (d) What are fair odds in favor of the event of part (a)?

5. For customers who come to a restaurant, 45% are seated immediately. Of those who are asked to wait, 40% become impatient and leave the restaurant before being seated. (a) What percent of all customers who come to the restaurant are actually seated? (b) For those customers who are seated, what is the probability that they had previously been asked to wait?

6. A person is dealt five cards from an ordinary deck of 52 cards. What is the probability of being dealt a hand with (a) four spades and one heart, (b) exactly four spades, (c) exactly four cards of the same suit, and (d) at least four spades?

7. A pair of fair dice are tossed. Let $A$ be the event "even number on both tosses," and let $B$ be the event "sum of the values on the dice equals six." (a) Display $A$ as a subset, and compute $P(A)$. (b) Display $B$ as a subset, and compute $P(B)$. (c) Compute $P(B \mid A)$. (d) Compute $P(A \mid B)$. (e) Determine if $A$ and $B$ are independent events.

8. Angie and Bobbie play for the handball championship. The one who wins two games first is the champion. [This is referred to as a "best two out of three" series of games.] The probability that Angie will beat Bobbie in an individual game is 3/5. What is the probability that Angie will win the championship?

9. Given a container with six red and four white balls. Three balls are drawn without replacement. Let $X$ denote the number of red balls drawn. (a) Determine the table of values for $f(x)$, the probability function for $X$. (b) Determine values of the cumulative distribution function $F(x)$ for all $x$ values.

10. In an office, 25% of the 200 employees were smokers. A set of five employees is randomly selected. (a) What is the probability that exactly three of them are smokers? (b) Use the binomial distribution to approximate the answer to part (a).

11. A large shipment [several thousand] of radios contains 2.5% that are defective. Suppose that 100 of the radios are randomly selected. What is the probability that (a) exactly three will be defective, (b) at most three will be defective, and (c) at least four will be defective? (d) Rework parts (a), (b), and (c) by using a Poisson approximation to the binomial distribution.

**12.** In a container, 40% of the chips are red and the rest are of other assorted colors. Chips are drawn from the container, with replacement after each draw. **(a)** What is the probability that the first red chip occurs on the eighth chip drawn? **(b)** What is the probability that the first red chip occurs on or before the eighth chip drawn? **(c)** What is the probability that the third red chip occurs on the eighth chip drawn?

**13.** Carriers of a certain disease constitute 2.5% of a certain population. There exists a blood test for which the probability is .98 that a carrier of the disease will test positive, and for which the probability is .99 that a noncarrier will test negative. **(a)** For a person selected at random from the population, what is the probability that the person will test positive? **(b)** If a person tests positive, what is the probability that the person is a carrier of the disease?

**14.** Two defective batteries have been mixed in with four good ones. Suppose that batteries are tested one by one, until both defective batteries are found. **(a)** What is the probability that the testing of exactly five batteries is required? **(b)** Let random variable $X$ denote the exact number of batteries to test so as to find both defective batteries. Determine the pf [probability function] for $X$.

# *Solutions to Midterm Exam*

**1. (a)** Here, for example, element 25 indicates two on the first ball and five on the second ball. Thus, using the label $S_1$ for the sample space, we have

$$S_1 = \{12, 13, 14, 15, 21, 23, 24, 25, 31, 32, 34, 35, 41, 42, 43, 45, 51, 52, 53, 54\}$$

Note that $S_1$, which contains 20 elements, is the most subdivided sample space for the experiment of drawing two balls without replacement.
   **(b)** Here, $S_2 = \{RG, RR, GG, GR\}$, if the ordering is of interest, and $\hat{S}_2 = \{rg, rr, gg\}$, if ordering is not of interest. Here, the element $rg$ corresponds to $RG$ and $GR$ of $S_2$.
   **(c)** Here, $S_3 = \{0, 1, 2\}$, where the elements refer to the number of red balls obtained.

**2. (a)** $A = \{13, 15, 31, 35, 51, 53\}$      **(b)** $B = \{15, 24, 42, 51\}$
   **(c)** $C = \{14, 15, 24, 25, 34, 35, 41, 42, 43, 51, 52, 53\}$

**3.** From Theorem 1.2 [Section 1-5B], we know that if all the elements of a sample space $S$ are equally likely, then $P(E) = N(E)/N(S)$, for any event [subset] $E$ of $S$. Thus, since $N(S_1) = 20$,

$$P(A) = N(A)/N(S_1) = 6/20 = .30 \qquad P(B) = N(B)/N(S_1) = 4/20 = .20 \qquad \text{and}$$

$$P(C) = N(C)/N(S_1) = 12/20 = .60$$

**4.** *Preliminaries:* Refer to the Venn diagram of Figure M-1, where $H$ refers to "hamburgers," and $M$ refers to "milkshakes." For the disjoint regions, we have $R_1 = H \cap M$, $R_2 = H \cap M'$, etc. Thus, $P(R_1) = .10$, $P(R_2) = .40 - .10 = .30$, $P(R_3) = .35 - .10 = .25$, and $P(R_4) = 1 - (.10 + .30 + .25) = .35$.

**(a)** This event is $H \cup M$, and $P(H \cup M) = P(H) + P(M) - P(H \cap M) = P(R_1) + P(R_2) + P(R_3) = .65$.
**(b)** In terms of disjoint regions, this event is $R_2 \cup R_3$. Now, $P(R_2 \cup R_3) = P(R_2) + P(R_3) = .55$.
**(c)** In terms of disjoint regions, this event is $R_4$. Thus,

$$P(\text{"neither hamburger nor milkshake"}) = P(R_4) = .35$$

**(d)** Fair odds for the event $H \cup M$ are equal to the ratio of $P(H \cup M)$ to $P[(H \cup M)']$, that is, of .65 to .35, or of 65 to 35, or of 13 to 7.

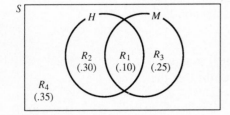

**Figure M-1.** Venn diagram for Problem 4.

**5.** *Preliminaries:* Refer to Figure M-2. From Figure M-2, we see that

$$B = C \cup D \qquad \text{where } C \cap D = \varnothing \text{ [i.e., } C \text{ and } D \text{ are disjoint]} \tag{1}$$

**(a)**
$$P(\text{``seated''}) = P(A \cup C) = P(A) + P(C) \tag{2}$$

Now, $C = B \cap C$ since $C \subset B$, and thus,

$$P(C) = P(B \cap C) = P(B)P(C \mid B) \tag{3}$$

But,

$$P(B) = 1 - P(A) = .55 \qquad \text{and} \tag{4}$$

$$P(C \mid B) = 1 - P(D \mid B) = 1 - .4 = .60 \tag{5}$$

Thus, substituting Eqs. (4) and (5) in Eq. (3),

$$P(C) = (.55)(.60) = .33 \tag{6}$$

Substituting the values for $P(C)$ and $P(A)$ into Eq. (2), we get

$$P(\text{``seated''}) = .45 + .33 = .78 \tag{7}$$

**(b)** First of all,

$$P(\text{wait} \mid \text{seated}) = P(\text{wait} \cap \text{seated})/P(\text{seated}) \tag{8}$$

From part (a),

$$P(\text{seated}) = P(A \cup C) = .78 \tag{9}$$

Also, from part (a),

$$P(\text{wait} \cap \text{seated}) = P(C) = .33 \tag{10}$$

Thus, substituting from Eqs. (9) and (10) into Eq. (8),

$$P(\text{wait} \mid \text{seated}) = .33/(.78) = .423$$

**6.** In all parts of this problem, we have drawing without replacement of five cards.

**(a)**
$$P(4 \text{ spades and } 1 \text{ heart}) = \binom{13}{4}\binom{13}{1}\Big/\binom{52}{5} = (715)(13)/(2{,}598{,}960) = .0035764$$

**(b)**
$$P(\text{exactly } 4 \text{ spades}) = P(4 \text{ spades and } 1 \text{ nonspade})$$

$$= \binom{13}{4}\binom{39}{1}\Big/\binom{52}{5} = .010729$$

**(c)** $\quad P(\text{exactly } 4 \text{ cards from same suit}) = 4 \cdot [\text{answer to part (b)}] = .042917$

**(d)** $\quad P(\text{at least } 4 \text{ spades}) = P(\text{exactly } 4 \text{ spades}) + P(\text{exactly } 5 \text{ spades})$

$$= \left[\binom{13}{4}\binom{39}{1} + \binom{13}{5}\right]\Big/\binom{52}{5} = 29{,}172\Big/\binom{52}{5} = .011224$$

**7.** The most subdivided sample space is $S = \{11, 12, 13, \ldots, 65, 66\}$, and $N(S) = (6)(6) = 36$. Also, all the 36 simple events of $S$ are equally likely.

**(a)** $A = \{22, 24, 26, 42, 44, 46, 62, 64, 66\}$, and $P(A) = N(A)/N(S) = 9/36 = .25$. Alternately, since the two tosses constitute independent experiments,

$$P(A) = P(\text{even}_1 \text{ and even}_2) = [P(\text{even})]^2 = \left(\frac{1}{2}\right)^2 = \frac{1}{4}$$

**(b)** Here, $B = \{15, 24, 33, 42, 51\}$, and $P(B) = N(B)/N(S) = 5/36$.

**(c)** Since $A \cap B = \{24, 42\}$, $P(A \cap B) = 2/36$, hence $P(B \mid A) = P(A \cap B)/P(A) = 2/9$.

**(d)** $P(A \mid B) = P(A \cap B)/P(B) = 2/5$.

**(e)** Since $P(A \cap B) = 2/36 = .05\overline{5}$, and $P(A)P(B) = (\frac{1}{4})(\frac{5}{36}) = .0347\overline{2}$, events $A$ and $B$ are not independent.

**8.** Let $A_i$ stand for the event that Angie wins game $i$, for $i = 1, 2, 3$, with a similar meaning for $B_i$, with respect to Bobbie. Thus,

| | Seated immediately ($A$) | Asked to wait ($B$) |
|---|---|---|
| $S$ | $A$ <br><br> [.45] | Waits and is seated later—$C$ [.33] <br>--- <br> Waits and later leaves—$D$ [.22] |

**Figure M-2.** Venn diagram for Problem 5 [$B = C \cup D$, where $C$ and $D$ are mutually exclusive].

$$P(\text{Angie wins champ.}) = P(A_1 A_2) + P(A_1 B_2 A_3) + P(B_1 A_2 A_3) \tag{1}$$

We assume the individual games are equivalent to independent trials. Thus,

$$P(A_1 B_2 A_3) = P(A_1)P(B_2)P(A_3) = (\tfrac{3}{5})^2(\tfrac{2}{5}) = \tfrac{18}{125} \tag{2}$$

and similarly for $P(A_1 A_2)$ and $P(B_1 A_2 A_3)$. Thus,

$$P(\text{Angie wins champ.}) = (\tfrac{3}{5})^2 + 2(\tfrac{3}{5})^2(\tfrac{2}{5}) = \tfrac{81}{125} \tag{3}$$

**9. (a)** Here, we have a hypergeometric pf given by

$$f(x) = h(x; n, a, M) = h(x; 3, 6, 10) = \frac{\dbinom{6}{x}\dbinom{4}{3-x}}{\dbinom{10}{3}}$$

$$= \frac{\dbinom{6}{x}\dbinom{4}{3-x}}{120} \quad \text{for } x = 0, 1, 2, 3 \tag{1}$$

Thus, we generate the following table of values of $f(x)$:

| $x$ | 0 | 1 | 2 | 3 | Sum |
|---|---|---|---|---|---|
| $f(x)$ | 4/120 | 36/120 | 60/120 | 20/120 | 1 |

**(b)** The table for the cdf, $F(x)$, is as follows:

| $x$ interval | $x < 0$ | $0 \le x < 1$ | $1 \le x < 2$ | $2 \le x < 3$ | $3 \le x$ |
|---|---|---|---|---|---|
| $F(x)$ | 0 | 4/120 | 40/120 | 100/120 | 1 |

**10. (a)** Here, we have a hypergeometric probability function with $M = 200$, $a = (.25)(200) = 50$, $n = 5$, and $X = 3$, where $X$ stands for the number of smokers in a set of 5 employees. Thus,

$$P(X = 3) = \binom{50}{3}\binom{150}{2} \Big/ \binom{200}{5} \tag{1}$$

Here, from a computational standpoint, it is more efficient to use a sequential approach. [See Section 5-4A, especially Example 5-25.] Thus,

$$P(X = 3) = KP(A) \tag{2}$$

where $A$ is the particular favorable order $SSSFF$. [$S$ indicates "success," which means a "smoker" here.] Then, $K$, the number of favorable orders, is given by $K = P(5; 3, 2) = 5!/(3!2!) = 10$, and thus,

$$P(X = 3) = 10 \cdot \frac{50}{200} \cdot \frac{49}{199} \cdot \frac{48}{198} \cdot \frac{150}{197} \cdot \frac{149}{196} = .08638 \tag{3}$$

**(b)** Here, the binomial approximation should be fairly accurate since $n/M = 5/200 = .025 < .05$. Thus,

$$P(X = 3) \approx \binom{5}{3}(.25)^3(.75)^2 = .08789 \tag{4}$$

Refer also to Table I of Appendix A.

**11.** *Preliminaries:* Here, we have a hypergeometric situation, which is well approximated by a binomial situation since $n/M$ is clearly small. For the relevant binomial distribution, $n = 100$ and $p = .025$. For parts (a), (b), and (c), we use computer/calculator software to obtain probability values.

**(a)**
$$P(X = 3) = \binom{100}{3}(.025)^3(.975)^{97} = .2168$$

**(b)**
$$F(3) = P(X \le 3) = \sum_{x=0}^{3} b(x; 100, .025) = .7590$$

**(c)**
$$P(X \ge 4) = 1 - P(X \le 3) = 1 - .7590 = .2410$$

**(d)** The Poisson approximation should be fairly accurate since $n = 100$ and $\lambda = np = 2.5 < 10$. From a calculator or from Table IV,

$$P(X = 3) = (\lambda^x e^{-\lambda})/(x!) = (2.5^3 e^{-2.5})/(6) = .2138$$

Also,

$$F(3) = P(X \le 3) = \sum_{x=0}^{3} p(x; 2.5) = .7576 \quad \text{and}$$

$$P(X \ge 4) = 1 - P(X \le 3) = 1 - .7576 = .2424$$

**12.** Parts (a) and (b) involve the geometric pf [Section 6-2C].

**(a)** $P(\text{first red chip on eighth chip drawn}) = g(8; .40) = (.60)^7(.40) = .011197$.

**(b)** In Example 6-11, we showed that $P(X \le k) = 1 - (1 - p)^k$ for a geometric pf. Thus, $P(\text{first red chip on or before eighth chip drawn}) = 1 - (.60)^8 = .98320$.

**(c)** Here, we have a situation involving the negative binomial probability function $b^*$ [Section 6-2D]. Thus,

$$P(\text{third red chip occurs on eighth chip}) = b^*(8; 3, .40)$$

$$= P(\text{two red chips on first seven chips,} \\ \text{and red chip on eighth chip})$$

$$= \binom{7}{2}(.4)^2(.6)^5(.4) = .10451$$

**13.** *Preliminaries:* We shall make use of the Total Probability Theorem and Bayes' Formula of Section 7-2.

**(a)** Let the symbols $C$, Pos, and Neg, respectively, denote that a person selected at random is a carrier, tests positive, or tests negative. Thus, from Theorem 7.2 in Section 7-2A,

$$P(\text{Pos}) = P(C)P(\text{Pos} \mid C) + P(C')P(\text{Pos} \mid C') \tag{1}$$

Here, $P(C) = .025$, $P(C') = 1 - P(C) = .975$, $P(\text{Pos} \mid C) = .98$, and $P(\text{Pos} \mid C') = 1 - P(\text{Neg} \mid C') = 1 - .99 = .01$. Thus, substituting these values into Eq. (1),

$$P(\text{Pos}) = (.025)(.98) + (.975)(.01) = .03425 \tag{2}$$

**(b)** From Theorem 7.3 [Bayes' Formula],

$$P(C \mid \text{Pos}) = P(C \cap \text{Pos})/P(\text{Pos}) = [P(C)P(\text{Pos} \mid C)]/P(\text{Pos}) \tag{3}$$

Substituting in the data from part (a),

$$P(C \mid \text{Pos}) = [(.025)(.98)]/(.03425) = .7153 \tag{4}$$

**14. (a)** Let $D$ and $G$ stand for a defective and good battery, respectively. There are four favorable orders in which one $D$ is found on exactly one of the first 4 batteries, and the other $D$ is found on the fifth battery. These orders are $GGGDD$, $GGDGD$, $GDGGD$, and $DGGGD$. The probability for each of these is 1/15. For example,

$$P(GGGDD) = (\tfrac{4}{6})(\tfrac{3}{5})(\tfrac{2}{4})(\tfrac{2}{3})(\tfrac{1}{2}) = \tfrac{1}{15} \tag{1}$$

There are four favorable orders in which one $D$ is found on exactly one of the first 4 batteries, and a $G$ is found on the fifth battery. For each such order, the second $D$ is clearly the battery that remains *untested*. The probability for each of these orders is also 1/15. For example,

$$P(GDGGG) = (\tfrac{4}{6})(\tfrac{3}{5})(\tfrac{3}{4})(\tfrac{2}{3})(\tfrac{1}{2}) = \tfrac{1}{15} \tag{2}$$

Thus, from Eqs. (1) and (2), we have

$$P(\text{test exactly 5 batteries to find both } D\text{'s}) = 8(\tfrac{1}{15}) = \tfrac{8}{15} \tag{3}$$

**(b)** From part (a), we have $f(5) = 8/15$. From Example 7-2 in Section 7-1A, we have $f(4) = 4/15$. For the other positive $f(x)$ values, we have $f(2) = P(DD) = 1/15$, and $f(3) = P(GDD) + P(DGD) = 2/15$. Thus, the table of values of $f(x)$ is as follows:

| $x$ | 2 | 3 | 4 | 5 | Sum |
|---|---|---|---|---|---|
| $f(x)$ | 1/15 | 2/15 | 4/15 | 8/15 | 1 |

# *CONTINUOUS RANDOM VARIABLES*

## THIS CHAPTER IS ABOUT

☑ **Continuous Random Variables**
☑ **Several Major Probability Density Functions**
☑ **The Gamma Function and the Gamma Probability Distribution**

## 8-1. Continuous Random Variables

### A. Motivational discussion

Continuous random variables are usually associated with situations that involve a measurement which might, in theory, take on any value on an interval of real numbers. For example, cereal boxes labeled as containing a weight of 16 ounces of product could, in actuality, contain any weight between, say, 15 ounces and 17 ounces. Batteries supposed to last for about two months when in continuous use could, in actuality, last for any length of time between zero months and, say, three months.

Suppose a 1000 yard stretch of highway is studied for the purpose of determining locations of auto accidents. The location for an accident, denoted by random variable $X$, could, in actuality, occur at any point on the 1000 yard interval.

In general, if $X$ denotes a continuous random variable, the set of possible values for $X$ is usually either an interval of real numbers [of the form $(a, b)$, say], a union of such intervals, the interval of all real numbers $(-\infty, \infty)$, or an interval of the form $(k, \infty)$, or $[k, \infty)$, where the real number $k$ is often 0.

Let us now consider the problem of associating probabilities with continuous random variables. Consider a purchaser of Geewhiz cereal, which comes in boxes labeled "Net Weight 16 Ounces." Clearly, the actual weight of cereal varies from box to box—thus, the actual amount in a given box is a continuous random variable. If one rounds off the weights of cereal to the nearest .4 ounce, one will be dealing with a discrete random variable which has a probability function. Such a probability function may be graphically represented by a histogram, in which probabilities are given by the areas of rectangles. See Figure 8-1a, for example. [For a review of histograms for discrete random variables, see Section 5-2.]

For example, we might have the following table of values for the probability function $p(x)$.

| $x$ | 15.2 | 15.6 | 16.0 | 16.4 | 16.8 |
|------|------|------|------|------|------|
| $p(x)$ | .138 | .234 | .330 | .209 | .089 |
| $f^*(x)$ | .345 | .585 | .825 | .5225 | .2225 |

$$f^*(x) = \frac{p(x)}{.4} = 2.5p(x)$$
$$w = .4$$

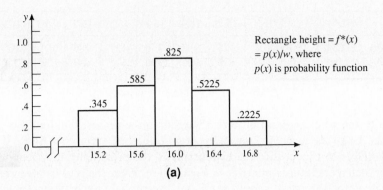

**(a)**

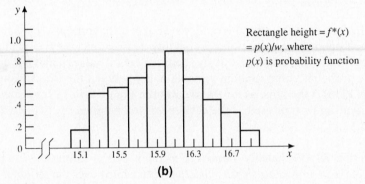

**(b)**

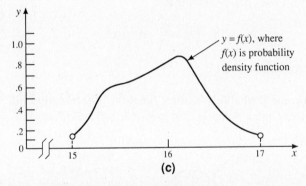

**(c)**

**Figure 8-1.** Motivation of probability density function and continuous random variable concepts. (**a**) Histogram if rounding is to nearest .4 ounce [$w = .4$]. (**b**) Histogram if rounding is to nearest .2 ounce [$w = .2$]. (**c**) Limiting case as $w \to 0$.

In Figure 8-1a, the width of each rectangle $w$ is given by $w = .4$. Thus, each rectangle has height equal to $p(x)/w = p(x)/(.4) = 2.5p(x)$. Also, the height of a rectangle is denoted by $f^*(x)$. Thus, the area of each rectangle will equal the corresponding probability [for example, the area of the first rectangle is .138, which is the probability function $p(x)$ evaluated at $x = 15.2$], and the total area of all rectangles will equal 1.

If one rounds off weights of cereal in the boxes to the nearest .2 ounce, one would be dealing with a different probability function, in which probabilities are represented in a histogram by areas of rectangles with base width $w$ equal to .2. Thus, we might have the table below of values for the probability function $p(x)$ if $w = .2$. The corresponding probability histogram is shown in Figure 8-1b.

| x | 15.1 | 15.3 | 15.5 | 15.7 | 15.9 | 16.1 | 16.3 | 16.5 | 16.7 | 16.9 |
|---|---|---|---|---|---|---|---|---|---|---|
| p(x) | .038 | .100 | .108 | .126 | .152 | .178 | .123 | .086 | .059 | .030 |
| f*(x) | .190 | .500 | .540 | .630 | .760 | .890 | .615 | .430 | .295 | .150 |

$$f*(x) = p(x)/w \quad \text{and} \quad w = .2$$

If one were to round off cereal weights to the nearest hundredth of an ounce, or to the nearest thousandth of an ounce, or even more finely, it is reasonable to expect that the histograms of the probability functions will approach a continuous curve as in Figure 8-1c. Then, the area between two x values for the curve should represent the probability between those x values.

In the case of a continuous random variable x, the concept of probability is based upon the existence of a probability density function $f(x)$ such that the area under the graph of $y = f(x)$ between two X values a and b is equal to the probability that random variable X lies between the two X values a and b. [This is the same as for the situation in Figure 8-1c.] From calculus, this area can be expressed as the *definite integral* between the two X values a and b.

**Definition 8.1 (Probability Density Function):** For a continuous random variable X, a *probability density function* $f(x)$ defined over the set of all real numbers [i.e., over $(-\infty, \infty)$] is such that

$$P(a \leq X \leq b) = \int_a^b f(x)\,dx$$

Here, a is less than or equal to b.

Other names for probability density function [abbreveiated pdf] are *probability density*, *density function*, and *probability distribution*.

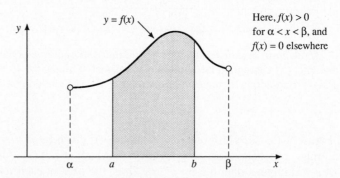

**Figure 8-2.** $P(a \leq X \leq b)$ equals $\int_a^b f(x)\,dx$, the area between $x = a$ and $x = b$ under $y = f(x)$.

In Figure 8-2, the area between the curve $y = f(x)$ and the x-axis, and between vertical lines at $x = a$ and $x = b$ is equal to $\int_a^b f(x)\,dx$.

*notes*

(a) It is assumed that $f(x)$ is integrable for the interval from a to b so that $\int_a^b f(x)\,dx$ will exist.

(b) Recall for a discrete random variable that $f(c)$ was equal to $P(X = c)$ and that either term denoted the probability that $x = c$. This is not the case for a continuous random variable x. The interpretation of $f(c)$ as being a sort of probability density [per unit change in x] will be explored in Example 8-3. The fact that $P(X = c) = 0$ for a continuous random variable $X$ is not too hard to demonstrate. Now, $P(X = c)$ can be interpreted as $P(c \leq x \leq c)$, and the latter is equal to $\int_c^c f(x)\,dx$, as we see from Definition 8.1. But, the latter is clearly 0, as we know from calculus—in particular, that definite integral represents the area of a vertical line segment.

(c) In Figure 8-2, the vertical broken lines at $x = \alpha$ and $x = \beta$ are guide lines.

In Definition 8-1, we can replace either or both of the $\leq$ signs by $<$.

**Theorem 8.1:** Given that $a \leq b$. We have that

$$P(a \leq X \leq b) = P(a \leq X < b) = P(a < X \leq b)$$

$$= P(a < X < b) = \int_a^b f(x)\,dx$$

The following properties hold for a probability density function $f(x)$.

**Theorem 8.2:** A function $f(x)$ is an acceptable probability density function of a continuous random variable $X$ if

(a) $f(x) \geq 0$ for $-\infty < x < \infty$

and

(b) $\int_{-\infty}^{\infty} f(x)\,dx = 1$

The latter expression, which can be written as $P(-\infty < X < \infty)$, corresponds to the expression in part (b) of Theorem 5.4.

*notes*

(i) Very often in realistic situations, one has $f(x) \geq 0$ for an interval $\alpha < x < \beta$, and $f(x) = 0$ elsewhere. [That is, $f(x) = 0$ for $x \leq \alpha$ and for $x \geq \beta$.] Moreover, it is most often the case that $f(x) > 0$—that is, $f(x)$ is strictly positive—for $\alpha < x < \beta$. In such situations, the interval $(\alpha, \beta)$ is called the range space; this will be denoted in this book as $R_x$. [The range space for discrete random variables was introduced in Section 5-2.] At times, we will refer to the range space as the *region of interest* for random variable $X$.

(ii) Sometimes, for the range space described in note (i), either one of the $<$ symbols will be replaced by $\leq$. In many cases, the range space is given by $\alpha < x < \infty$ or $\alpha \leq x < \infty$.

---

**EXAMPLE 8-1:** Given that

$$f(x) = \begin{cases} kx & \text{for } 1 < x < 3 \\ 0 & \text{elsewhere} \end{cases}$$

(a) Determine $k$ so that $f(x)$ is a probability density function. (b) Determine $P(2 < X < 3)$.

*Solution*

(a) From Theorem 8.2, we have $\int_{-\infty}^{\infty} f(x)\,dx = 1$, which becomes

$$\int_{-\infty}^{1} 0\,dx + k \int_{1}^{3} x\,dx + \int_{3}^{\infty} 0\,dx = 1 \qquad \textbf{(1)}$$

for the current $f(x)$. The first and third definite integrals equal zero, and we have

$$k\left(\frac{x^2}{2}\right)\Bigg]_1^3 = k\left[\frac{3^2}{2} - \frac{1^2}{2}\right] = 1 \qquad (2)$$

and thus, $k = 1/4$. The graph of $f(x)$ appears in Figure 8-3.

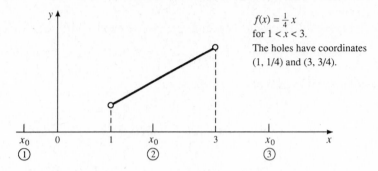

$f(x) = \frac{1}{4}x$
for $1 < x < 3$.
The holes have coordinates
$(1, 1/4)$ and $(3, 3/4)$.

**Figure 8-3.** Graph of $y = f(x)$ for Example 8-1. [The $x_0$ symbols refer to Example 8-2.]

**(b)** From Definition 8.1, we have

$$P(2 < X < 3) = \left(\frac{1}{4}\right)\int_2^3 x\,dx = \left(\frac{1}{4}\right)\frac{x^2}{2}\Bigg]_2^3 = \frac{5}{8} \qquad (3)$$

Though $f(x)$ is positive only for $1 < x < 3$ in Example 8-1, we consider the domain of $f(x)$ to include all real numbers. [Thus, $1 < x < 3$ is the range space of random variable $X$, but not the domain of $f(x)$.] This is fairly standard practice, and we shall usually follow it in this book.

## B.  Cumulative distribution function

**Definition 8.2:** As with a discrete random variable, we have $F(x) = P(X \le x)$. (See Definition 6.1.) Since $P(X \le x)$ is the same as $P(-\infty < X \le x)$, we have $F(x) = \int_{-\infty}^x f(t)\,dt$ for a continuous random variable. [It is just as correct to write $F(x_0) = \int_{-\infty}^{x_0} f(x)\,dx$.]

Here, $F(x)$ is called the *cumulative distribution function* of random variable $X$, and it is abbreviated as cdf. Some authors refer to $F(x)$ as the *distribution function* or the *cumulative distribution*.

**EXAMPLE 8-2:** For the probability density function of Example 8-1, **(a)** determine $F(x)$ in terms of $x$ for all $x$, and **(b)** sketch the graph of $F(x)$.

**Solution:** Let us work with $F(x_0) = \int_{-\infty}^{x_0} f(x)\,dx$, where $x_0$ denotes a fixed value of random variable $X$.

**(a)** Refer to Figure 8-3. The $x_0$ labeled ① indicates a typical random variable value which is less than or equal to 1. For all $x$ values to the left of such an $x_0$, we have $f(x) = 0$. Thus,

$$F(x_0) = \int_{-\infty}^{x_0} 0\,dx = 0 \qquad \text{for } x_0 \le 1 \qquad (1)$$

The $x_0$ labeled ② denotes a typical random variable value between 1 and 3. For $1 < x < 3$, $f(x) = x/4$. Thus,

$$F(x_0) = \int_{-\infty}^{x_0} f(x)\,dx = \int_{-\infty}^{1} 0\,dx + \left(\frac{1}{4}\right)\int_{1}^{x_0} x\,dx$$

$$= \left(\frac{1}{4}\right)\frac{x^2}{2}\bigg]_{1}^{x_0} \qquad \text{for } 1 < x_0 < 3 \qquad \textbf{(2)}$$

Here, note that $f(x) = 0$ for $x \le 1$ and $f(x) = x/4$ for $1 < x < x_0$. In Eq. (2), we are computing the probability [or area] accumulated from $x = -\infty$ until $x = x_0$, where $x_0$ is some value between 1 and 3. Thus,

$$F(x_0) = \left(\frac{1}{8}\right)[x_0^2 - 1] \qquad \text{for } 1 < x_0 < 3 \qquad \textbf{(3)}$$

Note that this equation yields $F(3) = 1$. The $x_0$ labeled ③ in Figure 8-3 denotes a typical random variable value which is at least 3. Now, for $x \ge 3$, we have $f(x) = 0$. Thus,

$$F(x_0) = \int_{-\infty}^{1} 0\,dx + \left(\frac{1}{4}\right)\int_{1}^{3} x\,dx + \int_{3}^{x_0} 0\,dx = \left(\frac{1}{4}\right)\int_{1}^{3} x\,dx$$

$$= 1 \text{ for } 3 \le x_0 \qquad \textbf{(4)}$$

To obtain $F(x)$ in terms of $x$, one merely replaces $x_0$ by $x$ in Eqs. (1), (3), and (4).

**(b)** Replacing $x_0$ by $x$, and plotting $F(x)$ versus $x$ yields Figure 8-4.

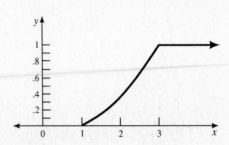

**Figure 8-4.** Graph of cumulative distribution function $F(x)$ for Example 8-2. $F(x) = 0$ for $x \le 1$; $F(x) = \frac{1}{8}[x^2 - 1]$ for $1 \le x \le 3$; $F(x) = 1$ for $x \ge 3$.

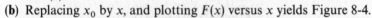

*notes*

**(i)** In Example 8-2, observe that $F(x)$ is a *continuous function* of $x$ [continuous, here, in a mathematical sense]. From a practical standpoint, this means that the $F(x)$ graph has no breaks in it. One consequence of the continuity of $F(x)$ is that $\lim_{x \to c^-} F(x) = \lim_{x \to c^+} F(x)$, which says that the left- and right-hand limits of $F(x)$ are equal at $x = c$. Let us illustrate this last limit statement for $c = 3$ in Example 8-2. Thus, we see that $\lim_{x \to 3^-} (\frac{1}{8})[x^2 - 1]$ and $\lim_{x \to 3^+} 1$ are both equal to 1.

**(ii)** The situation of Example 8-2 is that which applies in almost all cases involving continuous random variables. That is, $F(x)$ is a continuous function of $x$ [in a mathematical sense] for all $x$.

As with the cumulative distribution function for a discrete random variable, we have

$$F(-\infty) = 0; \qquad F(\infty) = 1 \qquad \textbf{(1a), (1b)}$$

and

$$F(x_1) \le F(x_2) \qquad \text{if } x_1 < x_2 \qquad \textbf{(2)}$$

It follows from Definition 8.2 that the following important properties hold for $F(x)$.

**Theorem 8.3:**

**(a)** Suppose that $a$ is less than or equal to $b$. Then,

$$P(a \le X \le b) = F(b) - F(a)$$

Also, since $P(a \le X \le b) = P(a < X < b)$, we have

$$P(a < X < b) = F(b) - F(a)$$

**(b)** $f(x) = dF(x)/dx$ for any $x$ value for which the derivative $dF(x)/dx$ exists.

**EXAMPLE 8-3:** (**a**) Prove part (a) of Theorem 8.3. (**b**) Prove part (b) of Theorem 8.3. (**c**) Demonstrate that

$$f(x_0) = \lim_{\Delta x \to 0} \frac{P(x_0 < X < x_0 + \Delta x)}{\Delta x}$$

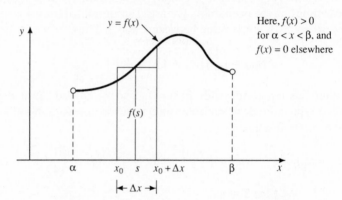

**Figure 8-5.** Demonstrating that $f(x_0) = \left. \dfrac{dF(x)}{dx} \right|_{x=x_0}$.

***Solution:*** We will postulate that $f(x) > 0$ for $\alpha < x < \beta$ and that $f(x) = 0$ elsewhere. Refer to Figure 8-5, in which the $f(x)$ of the graph conforms to these requirements.

(**a**) Clearly,

$$F(b) = \int_{-\infty}^{b} f(x)\,dx \quad \text{and } F(a) = \int_{-\infty}^{a} f(x)\,dx \qquad \textbf{(1), (2)}$$

The integral for $F(b)$ may be interpreted as the area under $y = f(x)$ for the interval from $-\infty$ to $b$. A similar interpretation applies to $F(a)$, only here the interval goes up to $x = a$. Now, from a calculus rule for definite integrals, we have

$$\int_{-\infty}^{b} f(x)\,dx - \int_{-\infty}^{a} f(x)\,dx = \int_{a}^{b} f(x)\,dx \qquad \textbf{(3)}$$

The integral on the right in Eq. (3) is equal to $P(a \le X \le b)$, as we see from Definition 8.1. Then, applying Eqs. (1) and (2) to the left side of Eq. (3), we obtain part (a) of Theorem 8.3.

(**b**) This result follows directly from the fundamental theorem of calculus. However, for motivational purposes, we shall use the mean value theorem of calculus in our proof. Thus, consider Figure 8-5, where $f(x)$ is shown to be continuous for $\alpha < x < \beta$. [In addition, $f(x) = 0$ for $x \le \alpha$ and $x \ge \beta$.] We note that Figure 8-5 has the same characteristics as Figure 8-2.

Let $x_0$ be a fixed value of $X$ which is strictly between $\alpha$ and $\beta$, as in Figure 8-5. Let $\Delta x$ denote a change in $x$ relative to $x_0$. From Theorem 8.1, we have

$$P(x_0 < X < x_0 + \Delta x) = \int_{x_0}^{x_0 + \Delta x} f(x)\,dx \qquad \textbf{(1)}$$

From the mean value theorem for definite integrals, we have

$$\int_{x_0}^{x_0 + \Delta x} f(x)\,dx = f(s)\Delta x \quad \text{where } s \text{ is such that } x_0 \le s \le x_0 + \Delta x \qquad \textbf{(2)}$$

[The quantity on the right in Eq. (2) is the area of the rectangle with width $\Delta x$ and height $f(s)$.] From Eqs. (1) and (2), we have

$$P(x_0 < X < x_0 + \Delta x) = f(s)\Delta x \qquad \text{where } x_0 \leq s \leq x_0 + \Delta x \qquad (3)$$

Now, from part (a) of Theorem 8.3, we have

$$P(x_0 < X < x_0 + \Delta x) = F(x_0 + \Delta x) - F(x_0) \qquad (4)$$

From Eqs. (3) and (4), we have

$$F(x_0 + \Delta x) - F(x_0) = f(s)\Delta x \qquad \text{where } x_0 \leq s \leq x_0 + \Delta x \qquad (5)$$

Following typical calculus terminology practice, we relabel the left side of Eq. (5) as $\Delta F$. Then, if we divide both sides of Eq. (5) by $\Delta x$, we obtain

$$\frac{\Delta F}{\Delta x} = f(s) \qquad \text{where } x_0 \leq s \leq x_0 + \Delta x \qquad (6)$$

Now, let us take the limit on both sides of Eq. (6) as $\Delta x$ approaches zero. That is, consider

$$\lim_{\Delta x \to 0} \frac{\Delta F}{\Delta x} = \lim_{\Delta x \to 0} f(s) \qquad (7)$$

From the calculus definition of derivative, the left side of Eq. (7) is the derivative $dF(x)/dx$ evaluated at $x = x_0$. The right side of Eq. (7) becomes $f(x_0)$ since $f(x)$ is continuous at $x_0$. Thus,

$$\left. \frac{dF(x)}{dx} \right|_{x=x_0} = f(x_0) \qquad (8)$$

Since $x_0$ is an arbitrary value of $x$, we have

$$\frac{dF(x)}{dx} = f(x) \qquad (9)$$

(c) Refer again to Eq. (7) of part (b). Replace $\Delta F$ on the left side of Eq. (7) by $P(x_0 < X < x_0 + \Delta x)$, to which $\Delta F$ is equal because of Eq. (4). Now, the right side term $\lim_{\Delta x \to 0} f(s)$ may be replaced by $f(x_0)$ since $f(x)$ is continuous at $x_0$. Thus, Eq. (7) may be rewritten as

$$\lim_{\Delta x \to 0} \frac{P(x_0 < X < x_0 + \Delta x)}{\Delta x} = f(x_0) \qquad (10)$$

Because of the calculus definition of the term *limit*, this means that

$$f(x_0) \approx \frac{P(x_0 < X < x_0 + \Delta x)}{\Delta x} \qquad (11)$$

if $\Delta x$ is sufficiently small. Here, "$\approx$" means *approximately equal to*. Now, Eq. (10) reveals that $f(x_0)$ is truly like a density [remember we say that $f(x)$ is the probability density function] since it is equal to the limit of the probability that $X$ is on an interval of length $\Delta x$ divided by the length $\Delta x$.

Observe that the usual mass density of science is equal to the mass of a substance divided by the volume containing that mass. Thus, as an example, the mass density of ordinary liquid water at room temperature is 62.4 pounds per cubic foot or 1.0 gram per milliliter.

---

**EXAMPLE 8-4:** Refer to Examples 8-1 and 8-2. (a) Use part (a) of Theorem 8.3 to re-evaluate the probability $P(2 < X < 3)$. (b) Use part (b) of Theorem 8.3 to verify that $f(x)$ for $1 < x < 3$ is given by $f(x) = dF(x)/dx$.

*Solution*

(a) From part (a) of Theorem 8.3, we have

$$P(2 < X < 3) = F(3) - F(2) \qquad (1)$$

From Example 8-2,

$$F(x) = (\tfrac{1}{8})[x^2 - 1] \qquad \text{for } 1 \le x \le 3 \tag{2}$$

Thus,

$$P(2 < X < 3) = (\tfrac{1}{8})[3^2 - 1] - (\tfrac{1}{8})[2^2 - 1] = \tfrac{5}{8} \tag{3}$$

as in Example 8-1.

(b) Taking the derivative of $F(x)$ as given in Eq. (2) above yields

$$F'(x) = (\tfrac{1}{8})[2x] = x/4 \qquad \text{for } 1 < X < 3 \tag{4}$$

This is as expected since from Example 8-1, we have $f(x) = x/4$ for $1 < X < 3$.

## C. Mixed probability distribution

In most situations, random variables will be either discrete or continuous. In the first case, the graph of the cumulative distribution function will have a step-like appearance as in Figures 6-1b and 6-2b. For a continuous random variable, the graph of the cdf will be continuous, as in Figure 8-4. Discontinuous cumulative distribution functions such as that graphed in Figure 8-6 occur when a random variable is *mixed*. Such a random variable is said to have a *mixed probability distribution*. A mixed random variable $X$ behaves like a discrete random variable for certain [discrete] values of $X$, and like a continuous random variable for the intervals that remain in the overall interval $(-\infty, \infty)$.

For such a random variable, $F(x)$ will be a discontinuous function [in the mathematical sense] at each value $x_0$ of $X$ where the probability that $X$ equals $x_0$ is a positive number [at such a value of $X$, the random variable behaves like a discrete random variable], and a continuous function elsewhere.

Suppose that at $X$ equals $x_0$, the random variable is discrete and that $P(X = x_0)$ is equal to the positive number $K$. At $X = x_0$, the cdf $F(x)$ will be a discontinuous function, and the change in $F(x)$ at $x_0$ will equal $K$. [In mathematical terms, where $\lim_{x \to x_0^-} F(x)$ is the symbol for the limit of $F(x)$ as $x$ approaches $x_0$ from the left, we have $F(x_0) - \lim_{x \to x_0^-} F(x) = K$.]

In summary, when $F(x)$ is discontinuous at $x_0$, then the change in $F(x)$ at $x_0$ is equal to the probability that $X$ is equal to $x_0$.

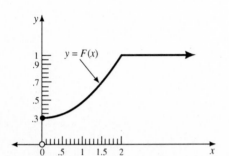

**Figure 8-6.** Graph of cumulative distribution function for mixed random variable of Example 8-5.

**EXAMPLE 8-5:** For a mixed random variable $X$, we are given that

$$P(X = 0) = .3 \tag{1}$$

Also, the probability density function for other $x$ values is given by

$$f(x) = \begin{cases} 0 & \text{for } x < 0 \\ (.35)x & \text{for } 0 < x < 2 \\ 0 & \text{for } 2 \le x \end{cases} \tag{2}$$

(a) Determine $F(x)$ in terms of $x$ for all $x$ values. (b) Sketch the graph of $F(x)$.

*Solution*

(a) Let $x_0$ be a fixed value of $X$. The general equation for $F(x_0)$ is

$$F(x_0) = P(X \le x_0) \tag{3}$$

For example, refer to Definition 8.2. Now, it is not hard to see that

$$F(x_0) = 0 \qquad \text{for } x_0 < 0 \tag{4}$$

if we make use of the first part of (2) above. For $x_0 = 0$, we have, after using Eq. (1) and the first part of (2) above,

$$F(0) = P(X \le 0) = P(X = 0) = .3 \qquad \textbf{(5)}$$

Then, we see that

$$F(x_0) = P(X \le x_0) = .3 + (.35) \int_0^{x_0} x \, dx \qquad \text{for } 0 < x_0 < 2 \qquad \textbf{(6)}$$

That is,

$$F(x_0) = .3 + (.175)(x_0)^2 \qquad \text{for } 0 < x_0 < 2 \qquad \textbf{(7)}$$

Finally,

$$F(x_0) = .3 + (.35) \int_0^2 x \, dx = 1 \qquad \text{for } 2 \le x_0 \qquad \textbf{(8)}$$

To obtain the equations for $F(x)$ in terms of $x$, we replace $x_0$ by $x$ in the equations above. Thus, we obtain:

$$F(x) = \begin{cases} 0 & \text{for } x < 0, \\ .3 & \text{for } x = 0, \\ .3 + (.175)x^2 & \text{for } 0 < x < 2, \\ 1 & \text{for } 2 \le x \end{cases} \qquad \textbf{(9)}$$

**(b)** The graph of the cdf $F(x)$ in terms of $x$ is obtained by plotting the equations of (9). This graph is shown in Figure 8-6.

*note:* Observe that $P(X = x) = 0$ if $x$ is a value of $X$ on the interval $(0, 2)$ in the previous example. The reason for this is that the probability density function $f(x) = (.35)x$ will give positive probabilities *only for intervals of nonzero length* contained within the interval $(0, 2)$.

To be specific, let us calculate $P(c < X < d)$ if the interval $(c, d)$ is contained within the interval $(0, 2)$. Using $P(c < X < d) = F(d) - F(c)$, with $F(c)$ and $F(d)$ given from the third expression in (9) above, we have

$$P(c < X < d) = (.175)[d^2 - c^2] \qquad \textbf{(10)}$$

## 8-2. Several Major Probability Density Functions

We shall now consider several major probability density functions. Recall that the corresponding random variable is called a continuous random variable. Remember that the general term probability distribution is used for either a probability function of a discrete random variable or a probability density function of a continuous random variable.

### A. Continuous uniform probability density function

**Definition 8.3:** A continuous random variable $X$ has a uniform or rectangular probability density function, and is referred to as a *continuous uniform random variable*, if the probability density function is given by the following statements, where $\alpha$ and $\beta$ denote constants:

$$f(x) = \begin{cases} \left(\dfrac{1}{\beta - \alpha}\right) & \text{for } \alpha < x < \beta \\ 0 & \text{elsewhere} \end{cases}$$

A sketch of the graph of a typical uniform probability density function is given in Figure 8-7.

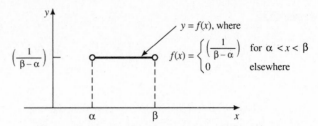

**Figure 8-7.** Uniform probability density function.

For a random variable with a uniform probability distribution, the probabilities $P(x_1 < X < x_2)$ and $P(x_3 < X < x_4)$ are equal if $(x_2 - x_1) = (x_4 - x_3)$, provided that both subintervals $(x_1, x_2)$ and $(x_3, x_4)$ are contained within the interval $(\alpha, \beta)$.

---

**EXAMPLE 8-6:** Suppose the annual rainfall in a region is a uniform random variable with rainfall strictly between 10 and 14 inches. Let $X$ denote the annual rainfall in inches. (a) Express the pdf $f(x)$ in terms of $x$ for all values of $X$. (b) Determine the expression for $F(x)$, the cumulative distribution function of $X$. (c) For a particular year, determine the probability that the annual rainfall will be between 11 and 12.5 inches, inclusive.

*Solution*

(a) From Definition 8.3, we see that $\alpha = 10$ and $\beta = 14$. Thus, the pdf is given as follows:

$$f(x) = \begin{cases} \dfrac{1}{(14 - 10)} = \dfrac{1}{4} & \text{for } 10 < x < 14 \\ 0 & \text{elsewhere} \end{cases}$$

(b) For $x_0 \le 10$, $F(x_0) = 0$. For $10 < x_0 < 14$,

$$F(x_0) = \left(\frac{1}{4}\right) \int_{10}^{x_0} dx = \left(\frac{1}{4}\right) x \Big]_{10}^{x_0} = \left(\frac{1}{4}\right)[x_0 - 10]$$

For $14 \le x_0$, $F(x_0) = 1$.

To obtain $F(x)$ in terms of $x$, merely replace $x_0$ by $x$ in the three statements for $F(x_0)$ in terms of $x_0$, immediately preceding.

(c) $$P(11 \le X \le 12.5) = \int_{11}^{12.5} f(x)\, dx = \left(\frac{1}{4}\right) \int_{11}^{12.5} dx = \left(\frac{1}{4}\right)[12.5 - 11]$$

$$= \frac{1.5}{4} = .375$$

---

The concept of parameters for a probability function was discussed in Section 6-3. The same ideas apply for a probability density function. Thus, for a uniform pdf, parameters are the endpoint numbers $\alpha$ and $\beta$.

## B. The exponential distribution

**Definition 8.4:** Given that $\theta > 0$ [here, $\theta$ is the parameter]. The *exponential probability density function* [or *exponential probability distribution*, or *expo-*

*nential distribution*] is given by

$$f(x) = \begin{cases} \dfrac{1}{\theta} e^{-x/\theta} & \text{for } x > 0 \\ \\ 0 & \text{elsewhere} \end{cases}$$

A continuous random variable $X$ is said to be an *exponential random variable* or to have an exponential distribution if its pdf is an exponential probability density function.

The exponential distribution often provides an appropriate model for calculating the probability that a piece of electrical or mechanical equipment will last for a total of $x$ time units before it fails [or dies]. This model is especially appropriate if failure is due primarily to external causes rather than internal wear.

---

**EXAMPLE 8-7:** Given that $f(x) = ke^{-x/3}$ for $x > 0$, and $f(x) = 0$, elsewhere. (**a**) Determine $k$, and sketch the $f(x)$ graph. (**b**) Determine the cdf $F(x)$ as a function of $x$. Sketch the $F(x)$ graph.

*Solution*

(**a**) Comparing the current $f(x)$ with that of Definition 8.4, we see that $k$ should equal $\frac{1}{3}$. Let us demonstrate that this is the case. First, from part (b) of Theorem 8.2, we have

$$\int_{-\infty}^{\infty} f(x)\,dx = k \int_{0}^{\infty} e^{-x/3}\,dx = 1 \tag{1}$$

Letting $I = \int_0^\infty e^{-x/3}\,dx$, we have

$$k = \frac{1}{I} \tag{2}$$

We solve for $I$ by using the fact that $\int_{x_1}^{x_2} e^u\,du = e^u\,]_{x_1}^{x_2}$, where $u$ is a function of $x$ [see Section 1-2 of Appendix B], and the method of substitution of calculus. Thus, letting $u = (-\frac{1}{3})x$, we have $du = (-\frac{1}{3})\,dx$. Now, introducing $(-3)(-\frac{1}{3}) = 1$ into the equation for $I$, we get

$$I = (-3) \int_{x=0}^{\infty} e^u \overbrace{\left(-\frac{1}{3}\right)dx}^{du} \tag{3}$$

or

$$I = -3e^u\,\Big]_{x=0}^{\infty} = -3e^{-x/3}\,\Big]_{x=0}^{\infty} \tag{4}$$

Thus,

$$I = -3 \cdot \lim_{x \to \infty} \frac{1}{e^{x/3}} + 3 \cdot e^0 = 0 + 3 = 3 \tag{5}$$

Note that the "$x = \infty$" term from Eq. (4) is evaluated by taking the limit as $x \to \infty$; observe that we have an improper integral here. Finally, substituting $I = 3$ in Eq. (2), we get

$$k = \frac{1}{I} = \frac{1}{3} \tag{6}$$

The graph of this exponential pdf is shown in Figure 8-8a.

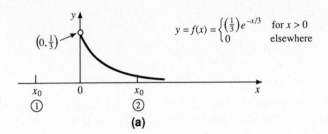

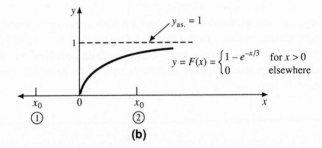

**Figure 8-8.** The pdf and cdf graphs for the exponential distribution with $\theta = 3$ [Example 8-7]. **(a)** Probability density function $f(x)$. [There is a hole at $(0, \frac{1}{3})$.] **(b)** Cumulative distribution function $F(x)$. [$y_{as.}$ indicates the horizontal asymptote.]

**(b)** Refer to Definition 8.2 and to Figure 8-8a. For the $x_0$ labeled ①, we have

$$F(x_0) = P(X \le x_0) = \int_{-\infty}^{x_0} 0 \, dx = 0 \qquad \text{for } x_0 \le 0 \tag{1}$$

For the $x_0$ labeled ②, we have

$$F(x_0) = \int_{-\infty}^{0} 0 \, dx + \left(\frac{1}{3}\right) \int_{x=0}^{x=x_0} e^{-x/3} \, dx \tag{2}$$

The first integral in Eq. (2) equals 0, and we handle the second integral by substituting $u = -x/3$ as in part (a). Thus,

$$F(x_0) = \left(\frac{1}{3}\right)(-3) \int_{x=0}^{x_0} e^u \, du = (-1)e^{-x/3} \Big]_{x=0}^{x=x_0} \tag{3}$$

Thus,

$$F(x_0) = 1 - e^{-x_0/3} \qquad \text{for } x_0 > 0 \tag{4}$$

In summary, replacing $x_0$ by $x$ in (1) and (4), we have

$$F(x) = \begin{cases} 1 - e^{-x/3} & \text{for } x > 0 \\ 0 & \text{elsewhere} \end{cases} \tag{5}$$

The graph of $F(x)$ versus $x$ is shown in Figure 8-8b. Note that $F(x)$ approaches 1 asymptotically as $x \to \infty$.

___

*notes*

(i) From Eq. (5) in part (b), we have $F'(x) = (\frac{1}{3})e^{-x/3} = f(x)$ for $x > 0$. This is consistent with part (b) of Theorem 8.3.

(ii) The graph of the cdf is continuous for all $x$ values, as expected. That is, there are no breaks in the graph. Stated differently, $F(x)$ is a continuous function of $x$ [in the mathematical sense].

It is useful to focus on $x = 0$, where $f(x)$ is discontinuous. There, we have $f(0) = 0$, but $\lim_{x \to 0^+} f(x) = \frac{1}{3}$, where $\frac{1}{3}$ is the value of the $y$

coordinate at the hole of the graph shown in part (a) of Figure 8-8. Remember that $x \to 0^+$ indicates an approach to $x = 0$ from the right.

Observe that if $X$ is a continuous random variable, then $F(x)$, the cdf, is a continuous function of $x$ [continuous in the mathematical sense]. See, for instance, Examples 8-2, 8-6, and 8-7.

---

**EXAMPLE 8-8:** A manufacturer found that the lifetime, in months, of a certain type of light bulb has an exponential pdf with $\theta = 3$ months. (**a**) Compute the probability that a light bulb of this type will have a lifetime between 1.5 and 3 months. (**b**) Determine the cdf, and make use of it to redo the computation of part (a). (**c**) Compute the probability that a light bulb will have a lifetime longer than 4.5 months.

*Solution:* First, observe that the pdf is given by

$$f(x) = \begin{cases} \frac{1}{3}e^{-x/3} & \text{for } x > 0 \\ 0 & \text{elsewhere} \end{cases}$$

This is the pdf of Example 8-7.

(**a**) From Theorem 8.1,

$$P(1.5 < X < 3) = \left(\frac{1}{3}\right) \int_{x=1.5}^{3} e^{-x/3} \, dx \qquad (1)$$

For variety, let us work out the integral by using the standard definite integral formula:

$$\int_{x_1}^{x_2} e^{ax} \, dx = \left(\frac{e^{ax}}{a}\right)\Bigg]_{x_1}^{x_2} \qquad (2)$$

Thus, since $a = -\frac{1}{3}$, $x_1 = 1.5$, and $x_2 = 3$ here, we obtain the following, after applying Eq. (2) to Eq. (1):

$$P(1.5 < X < 3) = (-1)e^{-x/3}\Bigg]_{1.5}^{3} = -e^{-1} + e^{-.5} \qquad (3)$$

Thus,

$$P(1.5 < X < 3) = -.3679 + .6065 = .2386 \qquad (4)$$

(**b**) From Example 8-7,

$$F(x) = 1 - e^{-x/3} \qquad \text{for } x > 0 \qquad (5)$$

From part (a) of Theorem 8.3,

$$P(a < X < b) = F(b) - F(a) \qquad (6)$$

Thus, applying Eq. (5) to Eq. (6), and then substituting $a = 1.5$ and $b = 3$, we get

$$P(1.5 < X < 3) = [1 - e^{-3/3}] - [1 - e^{-1.5/3}] = -e^{-1} + e^{-.5} \qquad (7)$$

as in part (a) above.

(**c**) We wish to calculate $P(X > 4.5)$. This is equal to $1 - P(\text{not } X > 4.5) = 1 - P(X \leq 4.5)$. Thus,

$$P(X > 4.5) = 1 - F(4.5) \qquad (8)$$

Applying Eq. (5) to Eq. (8) results in

$$P(X > 4.5) = 1 - [1 - e^{-4.5/3}] = e^{-1.5} = .2231 \qquad (9)$$

## C. The Poisson process

In preparation for another application of the exponential distribution, we consider the *Poisson process*. We covered the closely related Poisson probability function in Section 6-2A. [The associated Poisson random variable is, of course, discrete.] We repeat Definition 6.2.

**Definition 6.2 (Poisson Probability Function):** The Poisson probability function $p(x; \lambda)$ is given by

$$p(x; \lambda) = \frac{\lambda^x e^{-\lambda}}{x!} \quad \text{for } x = 0, 1, 2, \ldots$$

Here, the parameter $\lambda$ is a positive number.

We introduced the Poisson distribution as a distribution which approximated the binomial distribution for large $n$ and small $p$. The Poisson distribution also is a useful model in other situations. It has been found to provide a good model for the following: the number of atoms that decay [i.e., change state] in a fixed time within a radioactive material, the distribution of automobiles on a fixed length of highway, the arrival times for customers at a theater ticket office, the number of phone calls arriving at a switchboard in a one-hour time interval, the number of flaws in 1000 feet of wire, etc.

For simplicity, let us focus on a given interval of time of length $t$, though we could just as well be dealing with a spatial interval of distance or area or volume.

**Definition 8.5 (Poisson Process):** The random variable $X$ stands for the number of occurrences [or successes] in a time interval of length $t$. Let $f_t(x)$ stand for the probability of $x$ occurrences in a time interval of length $t$.

We have a Poisson process with coefficient $\alpha$ if the following three conditions are satisfied:

(i) The numbers of occurrences during nonoverlapping time intervals are independent.

(ii) The quantity $f_{\Delta t}(1)$, the probability of one occurrence in a time interval of length $\Delta t$, is approximately equal to $\alpha \Delta t$ if $\Delta t$ is sufficiently small, where $\alpha$ is a positive number. More precisely,

$$\lim_{\Delta t \to 0} \frac{f_{\Delta t}(1)}{\Delta t} = \alpha$$

This assumption says that if the length of the time interval $\Delta t$ is sufficiently small, then the probability of obtaining exactly one occurrence during that time interval is approximately proportional to the length of that interval, where the proportionality constant is $\alpha$.

(iii) The probability of $x$ occurrences in a time interval of length $\Delta t$ is approximately equal to 0 for $x = 2, 3, 4, \ldots$, if $\Delta t$ is sufficiently small. More precisely,

$$\lim_{\Delta t \to 0} \frac{f_{\Delta t}(x)}{\Delta t} = 0 \quad \text{for } x = 2, 3, 4, \ldots$$

**Theorem 8.4 (Probability Function in a Poisson Process):** Given a Poisson process with coefficient $\alpha$. Then, $f_t(x)$, the probability of $x$ occurrences [or successes] in a time interval of length $t$, is given by

$$f_t(x) = \frac{e^{-\alpha t}(\alpha t)^x}{x!} \quad \text{for } x = 0, 1, 2, 3, \ldots$$

That is, $f_t(x)$ is equal to the *Poisson* probability function of $x$ occurrences [or successes], with parameter $\lambda$ equal to $\alpha t$.

*note:* The cofficient $\alpha$ can be interpreted as the *mean* number of occurrences per unit of time. [For example, if $t$ is in minutes, then $\alpha$ would be the mean number of occurrences per minute.] The concept of mean or expected value will be explored more in Chapter 9. In particular, see Example 9-34 in Section 9-6A. For now, we envision the mean as being a type of *average quantity*.

---

**EXAMPLE 8-9:** Prove Theorem 8.4.

*Proof:* Consider a time interval of length $(t + \Delta t)$. [Remember, though, that our interval can be spatial in nature.] We can have $x_0$ occurrences in this interval in several mutually exclusive ways. Let us focus for now on the situation where $x_0 \geq 1$. [The $x_0 = 0$ case is considered in Eqs. (12) to (15) below.] Either there are $x_0$ occurrences in $t$ and none in $\Delta t$, or $(x_0 - 1)$ occurrences in $t$ and one in $\Delta t$, or $(x_0 - 2)$ occurrences in $t$ and two in $\Delta t$ [provided, of course, that $x_0 = 2$, $3, 4, \ldots$], and so on. Let us write down the equations for $f_{t+\Delta t}(x_0)$ for $x_0 = 1, 2$, and 3 to see if we can discern a pattern. Using condition (i) of Definition 8.5, and the total probability theorem [Theorem 7.2], and the multiplication rule for two independent events [Theorem 5.2], we have

$$f_{t+\Delta t}(1) = f_t(1)f_{\Delta t}(0) + f_t(0)f_{\Delta t}(1) \tag{1a}$$

$$f_{t+\Delta t}(2) = f_t(2)f_{\Delta t}(0) + f_t(1)f_{\Delta t}(1) + f_t(0)f_{\Delta t}(2) \tag{1b}$$

$$f_{t+\Delta t}(3) = f_t(3)f_{\Delta t}(0) + f_t(2)f_{\Delta t}(1) + f_t(1)f_{\Delta t}(2) + f_t(0)f_{\Delta t}(3) \tag{1c}$$

It is not hard to see that we can represent $f_{t+\Delta t}(x_0)$, for $x_0 = 1, 2, 3$, etc., as follows:

$$f_{t+\Delta t}(x_0) = f_t(x_0)f_{\Delta t}(0) + f_t(x_0 - 1)f_{\Delta t}(1) + \varepsilon \tag{2}$$

where

$$\varepsilon = 0 \qquad \text{for } x_0 = 1 \tag{3a}$$

and

$$\varepsilon = \sum_{x=0}^{x_0-2} f_t(x)f_{\Delta t}(x_0 - x) \qquad \text{for } x_0 = 2, 3, 4, \ldots \tag{3b}$$

For future reference, note that

$$\lim_{\Delta t \to 0} \frac{\varepsilon}{\Delta t} = 0 \qquad \text{for } x_0 = 1, 2, 3, \ldots \tag{4}$$

The $x_0 = 1$ result follows from Eq. (3a). Let us illustrate the correctness of Eq. (4) for the case where $x_0 = 3$. First, we determine $\varepsilon$ from Eq. (3b) for $x_0 = 3$; then, we divide both sides by $\Delta t$, and next we take the limit as $\Delta t$ approaches 0. Thus, we obtain

$$\lim_{\Delta t \to 0} \frac{\varepsilon}{\Delta t} = f_t(0) \cdot \lim_{\Delta t \to 0} \frac{f_{\Delta t}(3)}{\Delta t} + f_t(1) \cdot \lim_{\Delta t \to 0} \frac{f_{\Delta t}(2)}{\Delta t} \tag{5}$$

Now, we evaluate the limit terms on the right by making use of condition (iii) of Definition 8.5 for $x = 2$ and $x = 3$. Thus, Eq. (5) becomes

$$\lim_{\Delta t \to 0} \frac{\varepsilon}{\Delta t} = f_t(0) \cdot 0 + f_t(1) \cdot 0 = 0 \tag{6}$$

Refer again to Eq. (2), which applies for $x_0 = 1, 2, 3, \ldots$. First, subtract $f_t(x_0)$ from both sides, and then divide through by $\Delta t$. This leads to

$$\frac{f_{t+\Delta t}(x_0) - f_t(x_0)}{\Delta t} = f_t(x_0)\left[\frac{f_{\Delta t}(0) - 1}{\Delta t}\right] + f_t(x_0 - 1) \cdot \frac{f_{\Delta t}(1)}{\Delta t} + \frac{\varepsilon}{\Delta t}$$

$$\text{for } x_0 = 1, 2, 3, \ldots \tag{7}$$

[Again, note that $\varepsilon = 0$ if $x_0 = 1$.] Now, since there must be no occurrences, or one occurrence, or more than one occurrence in an interval of length $\Delta t$, we have

$$f_{\Delta t}(0) + f_{\Delta t}(1) + \sum_{x=2}^{\infty} f_{\Delta t}(x) = 1 \tag{8}$$

Let us henceforth use the symbol $f_{\Delta t}(\geq 2)$ for the last term on the left of Eq. (8). Dividing by $\Delta t$ in Eq. (8) yields

$$\frac{f_{\Delta t}(0) - 1}{\Delta t} = -\frac{f_{\Delta t}(1)}{\Delta t} - \frac{f_{\Delta t}(\geq 2)}{\Delta t} \tag{9}$$

If we take the limit in Eq. (9) as $\Delta t \to 0$, while making note of conditions (ii) and (iii) of Definition 8.5, we obtain

$$\lim_{\Delta t \to 0} \frac{f_{\Delta t}(0) - 1}{\Delta t} = -\alpha \tag{10}$$

Let us take the limit in Eq. (7) as $\Delta t \to 0$. We obtain the following differential equation, after making use of condition (ii) of Definition 8.5, and Eqs. (4) and (10):

$$\frac{df_t(x_0)}{dt} = -\alpha f_t(x_0) + \alpha f_t(x_0 - 1) \qquad \text{for } x_0 = 1, 2, 3, \ldots \tag{11}$$

Now, let us consider the special case where $x_0 = 0$. Now, the only way to have 0 occurrences in the interval of length $(t + \Delta t)$ is to have 0 occurrences in the interval of length $t$ followed by 0 occurrences in the interval of length $\Delta t$. Thus, it follows that

$$f_{t+\Delta t}(0) = f_t(0) f_{\Delta t}(0) \tag{12}$$

If we subtract $f_t(0)$ from both sides of Eq. (12), and then divide both sides by $\Delta t$, we obtain

$$\frac{f_{t+\Delta t}(0) - f_t(0)}{\Delta t} = f_t(0)\left[\frac{f_{\Delta t}(0) - 1}{\Delta t}\right] \tag{13}$$

Let us take the limit in Eq. (13) as $\Delta t \to 0$. We obtain the following differential equation, after making use of Eq. (10):

$$\frac{df_t(0)}{dt} = -\alpha f_t(0) \tag{14}$$

The general solution of Eq. (14), a separable differential equation, is

$$f_t(0) = C \cdot e^{-\alpha t} \tag{15}$$

Now, $f_t(0) = 1$ when $t = 0$ since it is certain that there will be no occurrences in an interval of zero length. Substituting this initial condition into Eq. (15) yields $C = 1$. Thus, we obtain the following solution for $f_t(0)$:

$$f_t(0) = e^{-\alpha t} \tag{15a}$$

Next, we substitute from Eq. (15a) into Eq. (11) for the case when $x_0 = 1$. Observe that $f_t(x_0 - 1)$ becomes $f_t(0)$ if $x_0 = 1$. Thus, we obtain

$$\frac{df_t(1)}{dt} = -\alpha f_t(1) + \alpha e^{-\alpha t} \tag{16}$$

The solution of this linear, first-order differential equation [after substituting the initial condition $f_t(1) = 0$ when $t = 0$] is given by

$$f_t(1) = e^{-\alpha t}(\alpha t) \tag{17}$$

Proceeding step by step [next, we would find the solution for $f_t(2)$], or by using mathematical induction, we obtain the equation of Theorem 8.4, namely

$$f_t(x) = \frac{e^{-\alpha t}(\alpha t)^x}{x!} \qquad \text{for } x = 0, 1, 2, 3, \ldots \qquad \textbf{(18)} \quad \square$$

---

**EXAMPLE 8-10:** The number of telephone calls arriving at a certain switchboard in $t$ minutes is a Poisson random variable with $\alpha = 4$ calls/minute. What is the probability of **(a)** at most four calls arriving in a two minute interval and **(b)** at least six calls arriving in a three minute interval?

*Solution*

**(a)** The Poisson parameter $\lambda$ is given by $\lambda = \alpha t$, and so $\lambda = 4 \cdot 2 = 8$ for a two minute interval. [The 8 can be interpreted as the mean number of calls in a two minute interval.] Referring to Table IV of Appendix A, we have

$$P(X \le 4) = f(0) + f(1) + \ldots + f(4)$$
$$= .0003 + .0027 + .0107 + .0286 + .0573 = .0996$$

**(b)** Here, $\lambda = 4 \cdot 3 = 12$. Now,

$$P(X \ge 6) = 1 - P(X \le 5) \qquad \textbf{(1)}$$

After using Table IV of Appendix A, we find that

$$P(X \le 5) = f(0) + f(1) + \ldots + f(5)$$
$$= .0000 + .0001 + .0004 + .0018 + .0053 + .0127 = .0203 \qquad \textbf{(2)}$$

Thus, substituting from Eq. (2) into Eq. (1), we get

$$P(X \ge 6) = 1 - .0203 = .9797 \qquad \textbf{(3)}$$

---

## D. The exponential distribution and a Poisson process

Consider a Poisson process as in Section 8-2C. Let random variable $Y$ be the waiting time between occurrences [or, successes] in a Poisson process.

**Theorem 8.5:** Given a Poisson process with coefficient $\alpha$. The pdf for $Y$, the waiting time between occurrences in a Poisson process, is given by

$$f(y) = \begin{cases} \alpha e^{-\alpha y} & \text{for } y > 0 \\ 0 & \text{elsewhere} \end{cases}$$

Thus, the pdf for $Y$ is that of an exponential distribution with parameter $\theta$ given by $\theta = 1/\alpha$. [Compare with Definition 8.4.]

*note:* Observe that since $\alpha$ indicates the mean number of occurrences per unit time, then $\theta = 1/\alpha$ indicates the mean time between occurrences.

---

**EXAMPLE 8-11:** Prove Theorem 8.5.

*Proof:* Consider the event $Y > y$, which reads that the waiting time from the most recent occurrence until the next occurrence [or from the start of the process until the first occurrence] is greater than $y$. Now, if $Y > y$, then there must have been no occurrences in the time interval of length $y$. That is,

$$P(Y > y) = P(0 \text{ occurrences in time } y) \qquad \textbf{(1)}$$

The latter probability is given from Theorem 8.4 to be

$$P(0 \text{ occurrences in time } y) = f_y(0) = \frac{e^{-\alpha y}(\alpha y)^0}{0!} = e^{-\alpha y} \qquad \textbf{(2)}$$

Thus, from Eqs. (1) and (2), we have

$$P(Y > y) = e^{-\alpha y} \tag{3}$$

Now, $P(Y \le y) = 1 - P(Y > y)$. Also, $F(y)$, the cumulative distribution function of $y$, equals $P(Y \le y)$. So,

$$F(y) = \begin{cases} 1 - e^{-\alpha y} & \text{for } y \ge 0 \\ 0 & \text{elsewhere} \end{cases} \tag{4}$$

Differentiation with respect to $y$ yields $f(y)$, the probability density function of $y$:

$$f(y) = \begin{cases} \alpha e^{-\alpha y} & \text{for } y > 0 \\ 0 & \text{elsewhere} \end{cases} \tag{5} \quad \square$$

**EXAMPLE 8-12:** Refer to the switchboard situation of Example 8-10, in which the Poisson process coefficient $\alpha$ is given by $\alpha = 4$ calls/minute. Suppose a telephone call has arrived at 2:10 P.M. **(a)** Determine the probability that the time it takes for another telephone call to arrive is more than 24 seconds. **(b)** Determine the probability that at least one telephone call will arrive in the next 30 seconds.

*Solution*

**(a)** *Method 1:* Let $Y$ be the time in minutes between arrivals of telephone calls. Thus, from Theorem 8.5, since $\alpha = 4$, we have

$$f(y) = \begin{cases} 4e^{-4y} & \text{for } y > 0 \\ 0 & \text{elsewhere} \end{cases} \tag{1}$$

Here, we wish to calculate $P(Y > .4 \text{ min.})$ since 24 seconds $= .4$ minute. Now,

$$P(Y > .4 \text{ min.}) = 4 \int_{y=.4}^{\infty} e^{-4y} \, dy \tag{2}$$

$$= -e^{-4y} \Big]_{y=.4}^{\infty} = e^{-1.6} = .2019 \tag{3}$$

*Method 2:* Consider the related Poisson process. Thus, for a time period of $t = .4$ minute, the equation for the probability of $x$ successes as given by Theorem 8.4 is

$$f_{.4}(x) = \frac{e^{-1.6}(1.6)^x}{x!} \qquad \text{for } x = 0, 1, 2, \ldots \tag{4}$$

since $\alpha = 4$ calls/minute.

Now, if the time it takes for another telephone call to arrive is more than .4 minute, then that's equivalent to a Poisson process probability of $X = 0$ phone calls in .4 minute. From Eq. (4) above, this equals

$$f_{.4}(0) = e^{-1.6} = .2019 \tag{5}$$

as before.

**(b)** Now, 30 seconds equals .5 minute. Thus, we have

$P$(at least one telephone call will arrive in next 30 seconds)

$$= P(Y < .5 \text{ min.}) = 4 \int_{y=0}^{.5} e^{-4y} \, dy = -e^{-4y} \Big]_{y=0}^{.5} = 1 - e^{-2} = .8647$$

Other major probability density functions will be considered in the Problems sections of this chapter. In Section 8-3, we will consider the gamma and beta probability density functions [distributions]. The exceedingly important normal distribution will be discussed at length in Chapter 10.

# 8-3. The Gamma Function and the Gamma Probability Distribution

## A. The gamma function

The *gamma function* occurs frequently in studies of differential equations, and its properties are explored in advanced calculus books.

**Definition 8.6 (The Gamma Function):** The gamma function with argument $\alpha$, denoted by $\Gamma(\alpha)$, is defined by the following improper integral:

$$\Gamma(\alpha) = \int_{t=0}^{\infty} t^{(\alpha-1)} e^{-t} \, dt \qquad \text{for } \alpha > 0$$

Here, "$\Gamma$" is the capital version of the Greek letter gamma. Also, $t$ is a dummy variable in the definite integral; it can be replaced, of course, by virtually any other symbol [say, $s$ or $y$, but not by $\alpha$ since that symbol is used here for the argument of the gamma function].

We see that

$$\Gamma(1) = \int_{t=0}^{\infty} e^{-t} \, dt = -e^{-t} \Big]_{t=0}^{\infty} = 1$$

This result constitutes property (a) of Theorem 8.6.

**Theorem 8.6 (Properties of the Gamma Function):**

$$\Gamma(1) = 1 \tag{a}$$

and

$$\Gamma(\alpha + 1) = \alpha \Gamma(\alpha) \qquad \text{for } \alpha > 0 \tag{b}$$

Property (b) is known as a *recursion equation* since it expresses $\Gamma(\alpha + 1)$ in terms of $\Gamma(\alpha)$. From properties (a) and (b), one can derive property (c).

$$\Gamma(\alpha + 1) = \alpha! \qquad \text{if } \alpha \text{ is a nonnegative integer} \tag{c}$$

$$\Gamma(\tfrac{1}{2}) = \sqrt{\pi} \tag{d}$$

---

**EXAMPLE 8-13:** Derive property (c) from properties (b) and (a) of Theorem 8.6.

*Solution:* Let $\alpha = 1$ in (b). Thus, we get

$$\Gamma(2) = 1 \cdot \Gamma(1) = 1 \cdot 1 = 1 \tag{1}$$

after using (a). Letting $\alpha = 2$ in (b), we now get

$$\Gamma(3) = 2 \cdot \Gamma(2) = 2 \cdot 1 = 2 \tag{2}$$

in view of Eq. (1). Now, let us use mathematical induction to prove property (c). Assume that property (c) is true for $\alpha = k$, where $k$ is a positive integer. Thus, it follows that

$$\Gamma(k + 1) = k! \qquad \text{for } k \text{ a positive integer} \tag{3}$$

Now, $\Gamma(k + 2)$ is given by

$$\Gamma(k + 2) = (k + 1)\Gamma(k + 1) \tag{4}$$

because of property (b). Now, substituting for $\Gamma(k + 1)$ from Eq. (3) into Eq. (4), we obtain

$$\Gamma(k + 2) = (k + 1) \cdot k! = (k + 1)! \tag{5}$$

Thus, property (c) is true for the positive integer $\alpha = k + 1$. By the principle of mathematical induction, it follows that

$$\Gamma(\alpha + 1) = \alpha! \tag{6}$$

for $\alpha$, a positive integer. Note that property (c) clearly holds for $\alpha = 0$ since it then reduces to $\Gamma(1) = 0! = 1$. The latter equation is the same as property (a).

---

Property (b) of Theorem 8.6 is derived in Problem 8-12, and property (d) is derived in part (b) of Problem 10-1.

## B.  The gamma probability distribution

A frequently occurring probability density function has the form

$$f(x) = \begin{cases} kx^{\alpha-1}e^{-x/\beta} & \text{for } x > 0 \\ 0 & \text{elsewhere} \end{cases} \tag{I}$$

where $\alpha > 0$ and $\beta > 0$.

---

**EXAMPLE 8-14:**  Determine $k$ of (I) in terms of $\alpha$ and $\beta$.

*Solution:*  From Theorem 8.2, we have $\int_{-\infty}^{\infty} f(x)\,dx = 1$, and for (I) above, this becomes

$$k \int_{x=0}^{\infty} x^{\alpha-1}e^{-x/\beta}\,dx = 1 \tag{1}$$

Substituting $t = x/\beta$, for $\beta > 0$, we get

$$k\beta^{\alpha} \int_{t=0}^{\infty} t^{\alpha-1}e^{-t}\,dt = 1 \tag{2}$$

[Here, observe that $dt = (1/\beta)\,dx$, and that $t = 0$ when $x = 0$, and $t \to \infty$ as $x \to \infty$.] The integral in Eq. (2) is none other than $\Gamma(\alpha)$, as we see from Definition 8.6. Thus, we have

$$k = \frac{1}{\beta^{\alpha}\Gamma(\alpha)} \tag{3}$$

---

The pdf in (I) with $k$ as given by Eq. (3) in Example 8-14 is the *gamma probability distribution* or *gamma distribution* or *gamma probability density function*.

**Definition 8.7 (Gamma Distribution):**  A random variable $X$ has a gamma distribution, and is referred to as a gamma random variable, if its pdf is given by

$$f(x) = \begin{cases} \dfrac{1}{\beta^{\alpha}\Gamma(\alpha)} x^{\alpha-1}e^{-x/\beta} & \text{for } x > 0 \\ 0 & \text{elsewhere} \end{cases}$$

Here, parameters $\alpha$ and $\beta$ are both positive numbers, and $\Gamma(\alpha)$ denotes the gamma function of $\alpha$. [See Definition 8.6.]

---

**EXAMPLE 8-15:**  Sketch the graph of the gamma distribution for $\alpha = 3$ and $\beta = 2$.

*Solution:* From property (c) of Theorem 8.6, we have $\Gamma(3) = 2! = 2$. Also, $\beta^\alpha = 2^3 = 8$. Thus, from Definition 8.7, we have

$$f(x) = \begin{cases} \dfrac{1}{16} x^2 e^{-x/2} & \text{for } x > 0 \\[2mm] 0 & \text{elsewhere} \end{cases}$$

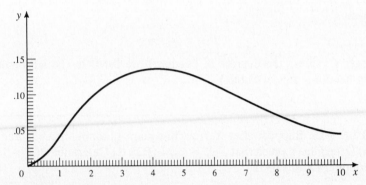

**Figure 8-9.** Gamma distribution for $\alpha = 3$ and $\beta = 2$. [Also, the chi-square distribution for $v = 6$.]

The sketch of this gamma distribution appears in Figure 8-9. The maximum of $f(x)$ occurs at $x = 4$ [to see this, set $f'(x) = 0$], and Max $f(x) = f(4) = e^{-2} \approx .1353$. Values of $f(x)$ for several values of $x$ are tabulated as follows, to 4 digits after the decimal point.

| $x$ | 0 | 2 | 4 | 6 | 8 |
|------|---|-------|-------|-------|-------|
| $f(x)$ | 0 | .0920 | .1353 | .1120 | .0733 |

Most gamma distributions have the same general shape as that of Figure 8-9.

## C. The chi-square distribution

A very important special case of the gamma distribution occurs when $\beta = 2$, and $\alpha$ is replaced by $v/2$. [Here, $v$ is the symbol for the Greek letter "nu."] This is known as the *chi-square distribution*.

**Definition 8.8 (Chi-Square Distribution):** A random variable $X$ has a *chi-square probability distribution* or *chi-square distribution* or *chi-square probability density function* if its pdf is given by the following:

$$f(x) = \begin{cases} \dfrac{1}{2^{v/2}\,\Gamma(v/2)} x^{(v-2)/2} e^{-x/2} & \text{for } x > 0 \\[2mm] 0 & \text{elsewhere} \end{cases}$$

The single parameter $v$, which is a positive number, is referred to as the *number of degrees of freedom* or, simply, *the degrees of freedom*.

*notes*

(a) Often, the chi-square random variable $X$ is replaced by the symbol $\chi^2$. [Here, $\chi$ is the Greek letter "chi."]
(b) The chi-square distribution for $v = 6$ is shown in Figure 8-9, where the

graph is also that of a gamma distribution for $\beta = 2$ and $\alpha = 3$. Most chi-square distributions have the same general shape as the graph of Figure 8-9.

Table V of Appendix A contains some probability listings for the chi-square distribution. The top margin heading contains $\alpha$ values [here, however, note that $\alpha$ stands for a *probability quantity*], while in the body of the table $x_0$ values are given, where the following equation indicates what $x_0$ and $\alpha$ mean:

$$P(X > x_0) = \alpha \tag{I}$$

Integer $\nu$ values [the degrees of freedom] are listed in the left-hand and right-hand margins of Table V.

---

**EXAMPLE 8-16:** Suppose that $X$ is a chi-square random variable, and that $\nu = 6$. **(a)** Determine $x_0$ such that $P(X > x_0) = .025$. **(b)** Determine $P(X > 1.635)$. **(c)** Determine $P(X \le 1.635)$.

*Solution*

**(a)** Here, $\alpha = .025$ in Table V. Now, refer to the extreme left- or right-hand margins for the listing of $\nu$ values. Thus, for $\nu = 6$, the $x_0$ value from the body of the table is $x_0 = 14.449$. That is,

$$P(X > 14.449) = .025$$

**(b)** Here, for the row in which $\nu = 6$, we see that 1.635 is located in the body of the table in the column labeled with $\alpha = .95$. Thus, we have

$$P(X > 1.635) = .95$$

**(c)** First, observe that

$$P(X \le 1.635) = 1 - P(X > 1.635)$$

Thus, using the result from part (b), we have

$$P(X \le 1.635) = 1 - .95 = .05$$

---

**EXAMPLE 8-17:** At the Beck department store, the annual gross revenue, in millions of dollars, is a chi-square random variable with $\nu = 4$. What is the probability that in a given year the gross revenue will exceed 9.4 million dollars?

*Solution:* From Definition 8.8, $f(x) = (\frac{1}{4})xe^{-x/2}$ since $\Gamma(\nu/2) = \Gamma(2) = 1$, $2^{\nu/2} = 2^2 = 4$, and $(\nu - 2)/2 = 1$. Thus, we wish to calculate

$$P(X > 9.4) = \left(\frac{1}{4}\right)\int_{9.4}^{\infty} xe^{-x/2}\,dx \tag{1}$$

A table of integrals [or see Section 1-2 of Appendix B] reveals that

$$\int xe^{ax}\,dx = \frac{e^{ax}}{a^2}(ax - 1) + C \tag{2}$$

Thus, since we have this integral in Eq. (1), but with $a = -\frac{1}{2}$ here, Eq. (1) becomes

$$P(X > 9.4) = -e^{-x/2}(1 + x/2)\Big]_{x=9.4}^{\infty} = e^{-4.7}(1 + 4.7) = .05184 \tag{3}$$

We could have estimated $P(X > 9.4)$ to be approximately .05 since, from Table V, we see [from the $\alpha = .05$ column] that

$$P(X > 9.488) = .05 \tag{4}$$

## D. The beta distribution

In recent years, the beta distribution has been applied extensively in *Bayesian statistical inference applications.*

**Definition 8.9 (Beta Distribution):** A random variable $X$ has a beta distribution, and is called a *beta random variable*, if its probability density function is given by the equations that follow [$\alpha$ and $\beta$ are parameters]:

$$f(x) = \begin{cases} \dfrac{\Gamma(\alpha + \beta)}{\Gamma(\alpha)\Gamma(\beta)} x^{\alpha-1}(1 - x)^{\beta-1} & \text{for } 0 < x < 1 \\[2ex] 0 & \text{elsewhere} \end{cases}$$

Here, $\alpha > 0$ and $\beta > 0$, and $\Gamma$ stands for the gamma function of Definition 8.6.

---

**EXAMPLE 8-18:** Suppose that random variable $X$ has a beta distribution with $\alpha = 2$ and $\beta = 3$. **(a)** Show that $f(x)$ can be expressed as a polynomial function for $0 < x < 1$. **(b)** Verify that $\int_{-\infty}^{\infty} f(x)\, dx = 1$.

*Solution*

**(a)**
$$f(x) = \frac{\Gamma(5)}{\Gamma(2)\Gamma(3)} x(1 - x)^2 \qquad \text{for } 0 < x < 1 \tag{1}$$

From property (c) of Theorem 8.6,

$$\Gamma(5) = 4! \qquad \Gamma(3) = 2! \qquad \text{and} \qquad \Gamma(2) = 1! \tag{2}$$

Thus,

$$\frac{\Gamma(5)}{\Gamma(2)\Gamma(3)} = 4 \cdot 3 = 12 \tag{3}$$

Next, we substitute the result from Eq. (3) into Eq. (1), and then multiply out the factors containing $x$. We thus obtain

$$f(x) = 12x(1 - 2x + x^2) = 12x - 24x^2 + 12x^3 \qquad \text{for } 0 < x < 1 \tag{4}$$

**(b)** Since $f(x)$ is given by Eq. (1) for $0 < x < 1$, and $f(x) = 0$, elsewhere,

$$\int_{-\infty}^{\infty} f(x)\, dx = \int_{0}^{1} (12x - 24x^2 + 12x^3)\, dx$$

$$= (6x^2 - 8x^3 + 3x^4)\Big]_{x=0}^{1} = 6 - 8 + 3 = 1 \tag{5}$$

---

**EXAMPLE 8-19:** Suppose the proportion of new businesses in the city of Barryville that fail within one year is a beta random variable with $\alpha = 2$ and $\beta = 3$. What is the probability that fewer than 20% of new businesses fail within a given year in Barryville?

*Solution:* Refer to Example 8-18. There, we also have a beta random variable with $\alpha = 2$ and $\beta = 3$. Thus, from the part (a) result of Example 8-18, we have

$$f(x) = 12x - 24x^2 + 12x^3 \qquad \text{for } 0 < x < 1 \tag{1}$$

Also, $f(x) = 0$, elsewhere. The probability we're interested in is $P(X < 20\%)$ $= P(X < .2)$. Thus,

$$P(X < .2) = \int_{0}^{.2} (12x - 24x^2 + 12x^3)\, dx = (6x^2 - 8x^3 + 3x^4)\Big]_{0}^{.2}$$

$$= 6(.04) - 8(.008) + 3(.0016) = .1808$$

Thus, the probability is .1808 that fewer than 20% of the businesses fail within a given year in Barryville.

# *SOLVED PROBLEMS*

**Continuous Random Variables**

**PROBLEM 8-1**   The pdf of a continuous random variable $X$ is given by

$$f(x) = \begin{cases} \dfrac{c}{x^2} & \text{for } x > 2 \\ 0 & \text{elsewhere} \end{cases}$$

**(a)** Determine $c$. **(b)** Find $P(3 < X < 5)$. **(c)** Determine the cdf $F(x)$ of the random variable $X$.

*Solution*

**(a)**  For this pdf, the requirement $\int_{-\infty}^{\infty} f(x)\,dx = 1$ becomes

$$c \int_{2}^{\infty} x^{-2}\,dx = 1 \tag{1}$$

Now,

$$\int_{2}^{\infty} x^{-2}\,dx = -\frac{1}{x}\Bigg]_{2}^{\infty} = \frac{1}{2} \tag{2}$$

Thus, substituting from Eq. (2) into Eq. (1), we get

$$c = \frac{1}{(1/2)} = 2 \tag{3}$$

**(b)**
$$P(3 < X < 5) = \int_{3}^{5} f(x)\,dx = 2\int_{3}^{5} x^{-2}\,dx = -\frac{2}{x}\Bigg]_{3}^{5} = -\frac{2}{5} + \frac{2}{3} = \frac{4}{15}$$

**(c)**
$$F(x_0) = \int_{-\infty}^{x_0} f(x)\,dx \tag{1}$$

Thus, for $x_0 < 2$, we have

$$F(x_0) = \int_{-\infty}^{x_0} 0\,dx = 0 \tag{2}$$

For $x_0 \geq 2$, we have

$$F(x_0) = \int_{-\infty}^{2} 0\,dx + 2\int_{2}^{x_0} x^{-2}\,dx \tag{3}$$

Thus, for $x_0 \geq 2$,

$$F(x_0) = 0 - \frac{2}{x}\Bigg]_{x=2}^{x_0} = 1 - \frac{2}{x_0} \tag{4}$$

Now, replacing $x_0$ by $x$ in Eqs. (2) and (4), we have

$$F(x) = \begin{cases} 0 & \text{for } x < 2 \\ 1 - \dfrac{2}{x} & \text{for } x \geq 2 \end{cases} \tag{5}$$

**PROBLEM 8-2** Let $W$ denote the actual weight of cereal boxes that are labeled 32 ounces. Let $X$ stand for the deviation from the ideal weight of 32 ounces. That is, $X = W - 32$. Suppose the pdf for $X$ is given by

$$f(x) = \begin{cases} \left(\dfrac{3}{32}\right)(4 - x^2) & \text{for } -2 < x < 2 \\ 0 & \text{elsewhere} \end{cases}$$

(a) What is the probability the actual weight of a box is between 32 and 33 ounces? (b) What is the probability that the weight of a box is between .5 ounce underweight and 1.5 ounces overweight? (c) Determine the cdf of $X$.

*Solution*

(a) Here, we wish to calculate $P(32 < W < 33)$. Since $W = X + 32$, the inequality $32 < W < 33$ is equivalent to $32 < X + 32 < 33$, and that is equivalent to $0 < X < 1$. Thus, we have

$$P(32 < W < 33) = P(0 < X < 1) = \left(\tfrac{3}{32}\right)\int_0^1 (4 - x^2)\,dx = \tfrac{11}{32} = .34375$$

(b) Since .5 ounce underweight means $W = 31.5$ or $X = -.5$, and 1.5 ounces overweight means $W = 33.5$ or $X = 1.5$, we wish to find $P(-.5 < X < 1.5)$. Now,

$$P(-.5 < X < 1.5) = \left(\tfrac{3}{32}\right)\int_{-.5}^{1.5} (4 - x^2)\,dx = \tfrac{41}{64} = .6406$$

(c) First, consider $x \leq -2$. We see that

$$F(x) = 0 \qquad \text{for } x \leq -2 \tag{1}$$

Next, consider $-2 < x < 2$. We see that

$$F(x) = \left(\tfrac{3}{32}\right)\int_{t=-2}^{t=x} (4 - t^2)\,dt = \left(\tfrac{3}{32}\right)(4t - t^2/3)\Big]_{t=-2}^{t=x}$$

$$= \left(\tfrac{3}{32}\right)(4x - x^3/3) + \left(\tfrac{1}{2}\right) \qquad \text{for } -2 < x < 2 \tag{2}$$

Finally, we have

$$F(x) = \left(\tfrac{3}{32}\right)\int_{t=-2}^{2} (4 - t^2)\,dt = 1 \qquad \text{for } x \geq 2 \tag{3}$$

**PROBLEM 8-3** For a space exploration experiment, light bulbs are turned on, and allowed to stay on until they burn out. For a typical light bulb used in the experiment, there is a .2 probability of immediate burnout when it is turned on initially. Thereafter, the pdf for the time until burnout in months, $X$, is given by $f(x) = k(4 - x)$ for $0 < x < 4$. It is certain that all light bulbs will burn out before four months have passed. (a) Determine $k$. (b) Determine the cdf, $F(x)$, and sketch its graph. (c) Determine $P(0 \leq X \leq 3)$, the probability a light bulb will burn out between 0 and 3 months, inclusive.

*Solution* Here, we have a probability distribution of the mixed type.

(a) First, observe that

$$P(X = 0) + P(0 < X < 4) = 1 \tag{1}$$

This becomes

$$P(0 < X < 4) = .8 \tag{2}$$

since $P(X = 0) = .2$. Also, using the information about the pdf for $0 < x < 4$, we have

$$P(0 < X < 4) = k\int_0^4 (4 - x)\,dx \tag{3}$$

From Eqs. (2) and (3), we derive Eq. (4), after partially working out the integral in Eq. (3).

$$k\left(4x - \frac{x^2}{2}\right)\Big]_0^4 = .8 \tag{4}$$

Thus,

$$k = \frac{.8}{8} = .1 \tag{5}$$

and, hence,

$$f(x) = k(4 - x) = (.1)(4 - x) \qquad \text{for } 0 < x < 4 \tag{6}$$

Also,

$$P(X = 0) = .2 \tag{7a}$$

where this equation describes the discrete part of random variable $X$, and

$$f(x) = 0 \qquad \text{for } x < 0 \text{ and } x \geq 4 \tag{7b}$$

**(b)** First, consider a fixed value of $x$, where $x < 0$. We have

$$F(x) = P(X \leq x) = 0 \qquad \text{for } x < 0 \tag{1}$$

Now, for the value of $x$ equal to 0, we have

$$F(0) = P(X \leq 0) = P(X = 0) = .2 \tag{2}$$

Consider a fixed value of $x$ such that $0 < x < 4$. We have

$$F(x) = P(X \leq x) = P(X = 0) + \int_{t=0}^{x} f(t)\,dt = .2 + (.1)\int_{t=0}^{x} (4 - t)\,dt$$

$$= .2 + (.1)(4x - x^2/2) \qquad \text{for } 0 < x < 4 \tag{3}$$

Lastly, consider a fixed value of $x$ such that $x \geq 4$. We have

$$F(x) = P(X = 0) + \int_{t=0}^{4} f(t)\,dt = .2 + (.1)\int_{t=0}^{4}(4 - t)\,dt = .2 + .8 = 1 \tag{4}$$

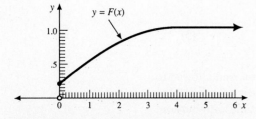

**Figure 8-10.** Graph of cumulative distribution function $F(x)$ for the mixed probability distribution of Problem 8-3.

The graph of $F(x)$ in terms of $x$ is shown in Figure 8-10. Note the discontinuity at $x = 0$.

**(c)** $P(0 \leq X \leq 3) = P(X = 0) + P(0 < X \leq 3)$

$$= .2 + \int_{x=0}^{3} f(x)\,dx$$

$$= .2 + (.1)\int_{x=0}^{3}(4 - x)\,dx$$

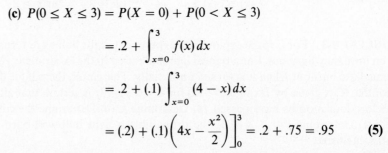

$$= (.2) + (.1)\left(4x - \frac{x^2}{2}\right)\Big]_0^3 = .2 + .75 = .95 \tag{5}$$

Alternately, $P(0 \leq X \leq 3) = P(X \leq 3) = F(3)$ since $P(X < 0) = 0$. Thus, using (3) from part (b), we get

$$P(0 \leq X \leq 3) = F(3) = (.2) + (.1)\left(4 \cdot 3 - \frac{3^2}{2}\right) = .2 + .75 = .95 \tag{6}$$

## Several Major Probability Density Functions

**PROBLEM 8-4** Suppose the annual salary of a family in Middletown in 1984 is a uniform random variable with values ranging between \$18,000 and \$30,000. **(a)** Find the probability that the salary of a family picked at random is greater than \$27,000. **(b)** Find the probability that for five families picked at random in Middletown, at least three have annual salaries greater than $27,000.

**Solution**

**(a)** Let random variable $X$ denote the annual family salary in Middletown in 1984, in thousands of dollars.
Thus,

$$f(x) = \begin{cases} \dfrac{1}{(30-18)} = \dfrac{1}{12} & \text{for } 18 < x < 30 \\ 0 & \text{elsewhere} \end{cases} \tag{1}$$

Now, the probability we wish to calculate can be expressed as $P(X > 27)$. Thus, using Eq. (1), we have

$$P(X > 27) = \left(\frac{1}{12}\right)\int_{x=27}^{30} dx = \left(\frac{1}{12}\right)x\Big]_{27}^{30} = \frac{3}{12} = .25 \tag{2}$$

**(b)** Let discrete random variable $\hat{X}$ denote the number of families [out of 5 families] which have annual salaries greater than $27,000. We see that $\hat{X}$ is a binomial random variable with $n = 5$ and $p = P(X > 27) = \frac{1}{4}$. [The latter value is from part (a). For a review of the binomial distribution parameters $n$ and $p$, refer to Section 5-3A.] Here, we wish to calculate $P(\hat{X} \geq 3)$. From Table I of Appendix A, we have

$$P(\hat{X} \geq 3) = P(\hat{X} = 3) + P(\hat{X} = 4) + P(\hat{X} = 5) = .0879 + .0146 + .0010 = .1035$$

**PROBLEM 8-5** Suppose the lifetime of an Everfine light bulb is an exponential random variable with $\theta = 4$ months. **(a)** Find the probability that an Everfine light bulb will have a lifetime greater than 4 months. **(b)** Determine the cdf, $F(x)$, for the random variable $X$, which is the lifetime of an Everfine light bulb. **(c)** Use the result from part (b) to compute the probability that the lifetime of an Everfine light bulb is between 4 months and 10 months. **(d)** Compute the lifetime $x_1$ such that $P(X \leq x_1) = .90$. Here, $x_1$ is called the 90th percentile for random variable $X$. The reason for this is that 90 percent of all lifetimes are less than or equal to $x_1$.

**Solution**

**(a)** The pdf for random variable $X$ is given by

$$f(x) = \begin{cases} \dfrac{1}{4}e^{-x/4} & \text{for } x > 0 \\ 0 & \text{elsewhere} \end{cases}$$

Here, $X$ is the lifetime of an Everfine light bulb, in months. We wish to find $P(X > 4)$.

$$P(X > 4) = \frac{1}{4}\int_{x=4}^{\infty} e^{-x/4}\,dx = (-1)e^{-x/4}\Big]_{x=4}^{\infty} \tag{1}$$

Thus,

$$P(X > 4) = e^{-1} = .3679 \tag{2}$$

**(b)** Here,

$$F(x) = 0 \quad \text{for } x < 0 \tag{3}$$

Also,

$$F(x) = \frac{1}{4}\int_{t=0}^{x} e^{-t/4}\,dt = (-1)e^{-t/4}\Big]_{t=0}^{x} = -e^{-x/4} + 1 \quad \text{for } 0 \leq x \tag{4}$$

**(c)** From part (a) of Theorem 8.3, we have

$$P(4 < X < 10) = F(10) - F(4) = (1 - e^{-10/4}) - (1 - e^{-1}) = e^{-1} - e^{-2.5} = .2858 \tag{5}$$

**(d)** Here, the unknown $x_1$ is such that $P(X \leq x_1) = .90$. We replace $P(X \leq x_1)$ by $F(x_1)$, which from part (b) leads to

$$1 - e^{-x_1/4} = .9 \qquad (6)$$

Thus,

$$e^{-x_1/4} = .1 \qquad (7)$$

Taking the natural logarithm of both sides of Eq. (7) leads to

$$(-x_1/4)\ln e = \ln .1 \qquad (8)$$

or, since $\ln e = 1$,

$$x_1 = (-4)(\ln .1) = (-4)(-2.3025) = 9.2103 \qquad (9)$$

That is, $x_1$, the 90th percentile, is equal to 9.2103 months.

**PROBLEM 8-6** The number of speeders that a radar unit observes in $t$ hours at a certain location on Route 40 is a Poisson random variable with $\alpha = 9.6$ per hour. [Here, we can interpret $\alpha = 9.6$ as being "the mean number of speeders observed per hour."] **(a)** What is the probability of 3 or fewer speeders being observed in 10 minutes? **(b)** What is the probability of a waiting time of less than 5 minutes between observations of successive speeders?

*Solution*

**(a)** From Theorem 8.4, since $t = 10$ minutes $= 1/6$ hour, it follows that $\alpha t = 9.6/6 = 1.6$, and thus the probability function for $X$, the number of speeders observed in 1/6 hour, is given by

$$f_{1/6}(x) = \frac{e^{-1.6}(1.6)^x}{x!} \qquad \text{for } x = 0, 1, 2, 3, \ldots \qquad (1)$$

Thus, we have a Poisson distribution with $\lambda = 1.6$. From Table IV of Appendix A, we have

$$P(X \le 3) = f_{1/6}(0) + \ldots + f_{1/6}(3) = .2019 + .3230 + .2584 + .1378 = .9211 \quad (2)$$

**(b)** *Method 1:* The waiting time between observations of speeders [i.e., between occurrences] $Y$ has the exponential distribution

$$f(y) = \begin{cases} \alpha e^{-\alpha y} = 9.6e^{-9.6y} & \text{for } y > 0 \\ 0 & \text{elsewhere} \end{cases} \qquad (3)$$

Here, $y$ is in hours since $\alpha y$, which appears in the exponent term, is dimensionless, and $\alpha$ is in $(\text{hr})^{-1}$. Since 5 minutes is equal to $(1/12)$ hour, our probability can be expressed as $P(Y < 1/12 \text{ hour})$. Now,

$$P(Y < 1/12 \text{ hour}) = 9.6 \int_{y=0}^{1/12} e^{-9.6y}\, dy = (-1)e^{-9.6y} \Big]_{y=0}^{1/12} = 1 - e^{-.8} = .5507 \quad (4)$$

*Method 2:* A waiting time of less than $(1/12)$ hour is equivalent to at least one speeder being observed in a time interval of length $(1/12)$ hour. Let random variable $X$ stand for the number of speeders observed in $(1/12)$ hour. Thus, we have

$$P(Y < 1/12) = P(X \ge 1) \qquad (5)$$

Now, $X$ is a Poisson random variable with parameter $\lambda$ equal to $\alpha t = (9.6)(1/12) = .8$. In the symbolism of Theorem 8.4, the probability function for $X$ is given by

$$f_{1/12}(x) = \frac{e^{-.8}(.8)^x}{x!} \qquad \text{for } x = 0, 1, 2, 3, \ldots \qquad (6)$$

Now, since $P(X \ge 1) = 1 - P(X = 0)$, we see that Eq. (5) becomes

$$P(Y < 1/12) = 1 - P(X = 0) \qquad (7)$$

Then, from Eq. (6), we have

$$P(X = 0) = f_{1/12}(0) = e^{-.8} \qquad (8)$$

Thus, from Eq. (7), we obtain

$$P(Y < 1/12) = 1 - e^{-.8} = .5507 \qquad (9)$$

as above.

**PROBLEM 8-7** A random variable has a Cauchy distribution if its pdf is given by

$$f(x) = \frac{(\beta/\pi)}{x^2 + \beta^2} \qquad \text{for } -\infty < x < \infty$$

where parameter $\beta > 0$. Verify that $\int_{-\infty}^{\infty} f(x)\,dx = 1$.

**Solution** First, note that, in general, $\int_{-\infty}^{\infty} f(x)\,dx$ exists [converges] provided that $\int_{-\infty}^{0} f(x)\,dx$ and $\int_{0}^{\infty} f(x)\,dx$ both exist [converge], in which case

$$\int_{-\infty}^{\infty} f(x)\,dx = \int_{-\infty}^{0} f(x)\,dx + \int_{0}^{\infty} f(x)\,dx \tag{1}$$

Here,

$$\int_{0}^{\infty} f(x)\,dx = \left(\frac{\beta}{\pi}\right) \int_{0}^{\infty} \frac{dx}{(x^2 + \beta^2)} = \left(\frac{\beta}{\pi}\right) \lim_{b \to \infty} \int_{0}^{b} \frac{dx}{(x^2 + \beta^2)} \tag{2}$$

Now,

$$\int_{0}^{b} \frac{dx}{(x^2 + \beta^2)} = \left(\frac{1}{\beta}\right) \tan^{-1}\left(\frac{b}{\beta}\right) \tag{3}$$

Now, substituting from Eq. (3) into Eq. (2), and then taking the limit as $b \to \infty$ leads to

$$\int_{0}^{\infty} f(x)\,dx = \left(\frac{\beta}{\pi}\right)\left(\frac{1}{\beta}\right) \lim_{b \to \infty} \tan^{-1}\left(\frac{b}{\beta}\right) = \left(\frac{\beta}{\pi}\right)\left(\frac{1}{\beta}\right)\left(\frac{\pi}{2}\right) = \frac{1}{2} \tag{4}$$

Similarly, we have

$$\int_{-\infty}^{0} f(x)\,dx = \left(\frac{\beta}{\pi}\right) \int_{-\infty}^{0} \frac{dx}{(x^2 + \beta^2)} = \left(\frac{\beta}{\pi}\right) \lim_{b \to -\infty} \int_{b}^{0} \frac{dx}{(x^2 + \beta^2)}$$
$$= -\left(\frac{\beta}{\pi}\right)\left(\frac{1}{\beta}\right) \lim_{b \to -\infty} \tan^{-1}\left(\frac{b}{\beta}\right) = -\left(\frac{\beta}{\pi}\right)\left(\frac{1}{\beta}\right)\left(-\frac{\pi}{2}\right) = \frac{1}{2} \tag{5}$$

Substituting the results of Eqs. (4) and (5) into Eq. (1), we get

$$\int_{-\infty}^{\infty} f(x)\,dx = \left(\frac{\beta}{\pi}\right) \int_{-\infty}^{\infty} \frac{dx}{(x^2 + \beta^2)} = \frac{1}{2} + \frac{1}{2} = 1 \tag{6}$$

**PROBLEM 8-8** A random variable $X$ has a *Weibull distribution* if its pdf has the form

$$f(x) = \begin{cases} kx^{\beta-1}e^{-\alpha x^\beta} & \text{for } x > 0 \\ 0 & \text{elsewhere} \end{cases}$$

Here, parameters $\alpha$ and $\beta$ are positive. This pdf arises in reliability theory, where $X$ stands for the lifetime [also called life length] of an item. Also, $X$ can be interpreted as the time until failure of an item. Show that $k = \alpha\beta$.

**Solution** From $\int_{-\infty}^{\infty} f(x)\,dx = \int_{0}^{\infty} f(x)\,dx = 1$, we have $k = 1/I$, where

$$I = \int_{0}^{\infty} x^{\beta-1}e^{-\alpha x^\beta}\,dx \tag{1}$$

If we substitute $u = \alpha x^\beta$, we have

$$du = \alpha\beta x^{\beta-1}\,dx \tag{2}$$

Also, $u = 0$ when $x = 0$ and $u \to \infty$ as $x \to \infty$. Thus, Eq. (1) may be rewritten as

$$I = \left(\frac{1}{\alpha\beta}\right) \int_{u=0}^{\infty} e^{-u}\,du = \left(\frac{-1}{\alpha\beta}\right)e^{-u}\Big]_{u=0}^{\infty} = \left(\frac{1}{\alpha\beta}\right) \tag{3}$$

Substituting the result from Eq. (3) into $k = 1/I$ leads to $k = \alpha\beta$.

**PROBLEM 8-9** Suppose random variable $T$ is the time until failure of some product, and that the pdf and cdf at time $T = t$ are $f(t)$ and $F(t)$. The failure rate at time $t$ is $Z(t)$, and it is defined as

$$Z(t) = \frac{f(t)}{1 - F(t)}$$

where it is understood that $F(t) < 1$. This concept arises in reliability theory. Observe that the failure rate at time $t$ is the pdf of failure at time $t$, on condition that failure has not occurred prior to time $t$. [Observe that $P$(failure has not occurred prior to time $t$) $= 1 - P$(failure has occurred prior to time $t$) $= 1 - F(t)$.] Show that if random variable $T$ has a Weibull distribution [Problem 8-8], then the failure rate is equal to $\alpha\beta t^{\beta-1}$ for $t > 0$.

***Solution*** From Problem 8-8, with $t$ replacing $x$, we have

$$f(t) = (\alpha\beta)t^{\beta-1}e^{-\alpha t^\beta} \qquad \text{for } t > 0 \tag{1}$$

and $f(t) = 0$ for $t \le 0$. Also, for $t_0 > 0$, we have

$$F(t_0) = (\alpha\beta) \int_{t=0}^{t_0} t^{\beta-1} e^{-\alpha t^\beta} \, dt \tag{2}$$

Let $u = \alpha t^\beta$, similar to what was done in Problem 8-8. Thus, $du = \alpha\beta t^{\beta-1} \, dt$, and Eq. (2) becomes

$$F(t_0) = \int_{u=0}^{u=u_0} e^{-u} \, du \tag{3}$$

where $u_0 = \alpha t_0^\beta$. Thus,

$$F(t_0) = 1 - e^{-\alpha t_0^\beta} \qquad \text{for } t_0 > 0 \tag{4}$$

Thus, replacing $t_0$ by $t$, we have

$$F(t) = 1 - e^{-\alpha t^\beta} \qquad \text{for } t > 0 \tag{5}$$

Substituting $f(t)$ and $F(t)$ from Eqs. (1) and (5) into the defining equation for $Z(t)$, we get

$$Z(t) = \alpha\beta t^{\beta-1} \qquad \text{for } t > 0 \tag{6}$$

***note:*** Another problem on the failure rate is Problem 8-21.

**PROBLEM 8-10** Suppose that $X$ has an exponential distribution. Show that

$$P(X \ge a + b \mid X \ge b) = P(X \ge a)$$

if $a$ and $b$ are nonnegative numbers. This so-called "memoryless property" for an exponential random variable is similar to the related property for a geometric random variable, where the latter is a discrete random variable. See Problem 6-6.

***Solution*** From the definition of conditional probability [Definition 3.1],

$$P(X \ge a + b \mid X \ge b) = \frac{P(X \ge a + b \text{ and } X \ge b)}{P(X \ge b)} \tag{1}$$

Now, the set of $x$'s [here, the $x$'s represent the numerical values of random variable $X$] which are at least equal to $a + b$ is contained within the set of $x$'s which are at least equal to $b$. [For example, if the number $x$ is at least equal to $3 + 2 = 5$, then $x$ is definitely at least equal to 3.] That means that

$$P(X \ge a + b \text{ and } X \ge b) = P(X \ge a + b) \tag{2}$$

Substitution of Eq. (2) into Eq. (1) results in

$$P(X \ge a + b \mid X \ge b) = \frac{P(X \ge a + b)}{P(X \ge b)} \tag{3}$$

Since $X$ has an exponential pdf, it follows that

$$f(x) = \begin{cases} \dfrac{1}{\theta} e^{-x/\theta} & \text{for } x > 0 \\ 0 & \text{elsewhere} \end{cases} \tag{4}$$

Thus,

$$P(X \geq b) = \frac{1}{\theta} \int_{x=b}^{\infty} e^{-x/\theta}\, dx = e^{-b/\theta} \tag{5}$$

Likewise,

$$P(X \geq a + b) = e^{-(a+b)/\theta} = e^{-a/\theta} e^{-b/\theta} \tag{6}$$

Substitution from Eqs. (5) and (6) into Eq. (3) results in

$$P(X \geq a + b \mid x \geq b) = \frac{e^{-a/\theta} e^{-b/\theta}}{e^{-b/\theta}} = e^{-a/\theta} \tag{7}$$

But, as in Eq. (5), we have $P(X \geq a) = e^{-a/\theta}$. Thus, it follows that

$$P(X \geq a + b \mid x \geq b) = P(X \geq a) \tag{8}$$

as was to be shown.

**PROBLEM 8-11**   For a type of appliance, the time in years between repairs, $T$, is an exponential random variable for which $\theta = 3$ years. Compute the probability that the time between repairs is at least 5 years, given that the time between repairs is at least 3 years.

*Solution*   Here, $f(t) = (\frac{1}{3})e^{-t/3}$ for $t > 0$, and $f(t) = 0$, elsewhere. We wish to calculate $P(T \geq 5 \mid T \geq 3)$. From the result of Problem 8-10, we can write

$$P(T \geq 5 \mid T \geq 3) = P(T \geq 2 + 3 \mid T \geq 3) = P(T \geq 2) \tag{1}$$

Thus, our answer is equal to $P(T \geq 2)$, which, from the $f(t)$ equation given above for $t > 0$, can be expressed as follows:

$$P(T \geq 2) = \left(\frac{1}{3}\right) \int_{t=2}^{\infty} e^{-t/3}\, dt = e^{-2/3} = .5134 \tag{2}$$

**The Gamma Function and the Gamma Probability Distribution**

**PROBLEM 8-12**   Prove the following recurrence property of the gamma function:

$$\Gamma(\alpha + 1) = \alpha \Gamma(\alpha) \qquad \text{for } \alpha > 0$$

This is property (b) of Theorem 8.6.

*Proof*   From Definition 8.6, we have

$$\Gamma(\alpha + 1) = \int_{t=0}^{\infty} t^{\alpha} e^{-t}\, dt \tag{1}$$

which is defined for $(\alpha + 1) > 0$, that is, for $\alpha > -1$. Let us use integration by parts on Eq. (1). Let

$$u = t^{\alpha} \qquad \text{and} \qquad dv = e^{-t}\, dt \tag{2a, 2b}$$

Thus,

$$du = \alpha t^{\alpha - 1}\, dt \qquad \text{and} \qquad v = -e^{-t} \tag{3a, 3b}$$

Thus, we may rewrite Eq. (1) as follows:

$$\Gamma(\alpha + 1) = \lim_{b \to \infty} [-e^{-t} t^{\alpha}]_{t=0}^{t=b} + \alpha \int_{t=0}^{\infty} e^{-t} t^{\alpha - 1}\, dt \tag{4}$$

Now, the integral on the right in Eq. (4) is $\Gamma(\alpha)$ for $\alpha > 0$. For $\alpha \leq 0$, the integral diverges. Thus, henceforth, we shall take $\alpha > 0$.

For $\alpha > 0$, we can show, by employing *L'Hôpital's rule* (see Section 1-3 of Appendix B), that

$$\lim_{b \to \infty} \frac{-b^{\alpha}}{e^{b}} = 0 \qquad (5)$$

Also, $e^{-t}t^{\alpha} = 0$ for $t = 0$ when $\alpha > 0$. Thus, the lead term on the right of Eq. (4) is equal to zero, and Eq. (4) becomes

$$\Gamma(\alpha + 1) = \alpha\Gamma(\alpha) \qquad \text{for } \alpha > 0 \qquad (6) \quad \square$$

**PROBLEM 8-13**   Show that the gamma probability distribution for $\alpha = 1$ and $\beta = \theta$ reduces to an exponential probability distribution.

*Solution*   The gamma probability distribution is

$$f(x) = \begin{cases} kx^{\alpha-1}e^{-x/\beta} & \text{for } x > 0 \\ 0 & \text{elsewhere} \end{cases} \qquad (1)$$

Here, $\alpha > 0$, $\beta > 0$, and

$$k = \frac{1}{\beta^{\alpha}\Gamma(\alpha)} \qquad (2)$$

Letting $\alpha = 1$ and $\beta = \theta$, we get

$$k = \frac{1}{\theta\Gamma(1)} = \frac{1}{\theta} \qquad (3)$$

and $x^{\alpha-1}e^{-x/\beta} = x^{0}e^{-x/\theta} = e^{-x/\theta}$. Thus, we see that

$$f(x) = \begin{cases} \dfrac{1}{\theta}e^{-x/\theta} & \text{for } x > 0 \\ 0 & \text{elsewhere} \end{cases} \qquad (4)$$

That is, we end up with the pdf of an exponential distribution.

**PROBLEM 8-14**   In a certain town, the daily consumption of electrical energy, in millions of kilowatt-hours, is approximately a gamma random variable with $\alpha = 2$ and $\beta = 4$. The energy sources in this town can provide a maximum of 20 million kilowatt-hours for a day. [This is known as the capacity of the energy sources.] What is the probability that the energy sources will be adequate on any given day?

*Solution*   For $\alpha = 2$ and $\beta = 4$, the value of $k$ for a gamma random variable is given by

$$k = \frac{1}{\beta^{\alpha}\,\Gamma(\alpha)} = \frac{1}{4^{2}\Gamma(2)} = \frac{1}{16} \qquad (1)$$

Let $X$ stand for the daily consumption of electric power, in millions of kilowatt-hours. Thus, for the pdf of $X$, we have

$$f(x) = kx^{\alpha-1}e^{-x/\beta} = (\tfrac{1}{16})xe^{-x/4} \qquad \text{for } x > 0 \qquad (2)$$

Thus, with $E$ being the event "energy sources are adequate on a given day," we have that $E$ is equivalent to $X \leq 20$. Thus,

$$P(E) = P(X \leq 20) = \int_{x=0}^{20} f(x)\,dx = \left(\frac{1}{16}\right)\int_{x=0}^{20} xe^{-x/4}\,dx \qquad (3)$$

From a table of integrals [or see Section 1-2 of Appendix B], we have

$$\int_{x_1}^{x_2} xe^{ax}\,dx = \frac{e^{ax}}{a^{2}}(ax - 1)\Bigg]_{x_1}^{x_2} \qquad (4)$$

Thus, applying Eq. (4) to Eq. (3), where $a = -\tfrac{1}{4}$, we obtain

$$P(E) = \frac{1}{16}\left[16e^{-x/4}\left(-\frac{x}{4}-1\right)\right]_{x=0}^{20}$$

$$= 1 - e^{-5}(5 + 1) = 1 - .04043 = .95957 \tag{5}$$

**PROBLEM 8-15** The proportion of people who vote in a mayoral election in Middletown is a beta random variable with $\alpha = 3$ and $\beta = 2$. What is the probability that more than 60% of the people will vote in a given mayoral election?

*Solution* Let $X$ stand for the proportion of people who vote in a mayoral election in Middletown. Since $X$ is a beta random variable, its pdf is given by

$$f(x) = \frac{\Gamma(\alpha + \beta)}{\Gamma(\alpha)\Gamma(\beta)}x^{\alpha-1}(1 - x)^{\beta-1} \qquad \text{for } 0 < x < 1 \tag{1}$$

and $f(x) = 0$, elsewhere. Now, using the information that $\alpha = 3$ and $\beta = 2$, we see that the lead coefficient in Eq. (1) is given by

$$\frac{\Gamma(5)}{\Gamma(3)\Gamma(2)} = \frac{4!}{2! \cdot 1!} = 12 \tag{2}$$

Thus, we see that

$$f(x) = 12(x^2 - x^3) \qquad \text{for } 0 < x < 1 \tag{3}$$

Let $E$ be the event "more than 60% vote in a mayoral election." Thus,

$$P(E) = P(X > .6) = 12\int_{.6}^{1}(x^2 - x^3)\,dx$$

$$= (4x^3 - 3x^4)]_{.6}^{1} = 1 - (.4752) = .5248 \tag{4}$$

## Supplementary Problems

**PROBLEM 8-16** Given that a probability density function is given by $f(x) = x$ for $0 < x < 1$, $f(x) = 2 - x$ for $1 \leq x < 2$, and $f(x) = 0$, elsewhere. This is a so-called *triangular distribution*. (a) Sketch the graph of $f(x)$. (b) Determine the cdf $F(x)$ as a function of $x$ for all $x$. (c) Sketch the graph of $F(x)$. [*Hint for part (b):* For $1 \leq x_0 < 2$, $F(x_0) = \int_0^1 x\,dx + \int_1^{x_0}(2 - x)\,dx$.]

*Answer* For (a) and (c), see Figure 8-11    (b) $F(x) = 0$ for $x \leq 0$; $F(x) = x^2/2$ for $0 < x < 1$; $F(x) = -x^2/2 + 2x - 1$ for $1 \leq x < 2$; $F(x) = 1$ for $2 \leq x$.

**PROBLEM 8-17** The weekly profit in thousands of dollars of Miller's Office Supply Store is a random variable $X$ whose cdf is given as follows: $F(x) = 0$ for $x < 0$; $F(x) = (\frac{3}{32})(2x^2 - x^3/3)$ for $0 \leq x \leq 4$; $F(x) = 1$ for $4 < x$. (a) Find the probability of a weekly profit of at most \$2000. That is, find $P(X \leq 2)$. (b) Find the probability of a weekly profit of at least \$3000. (c) Determine expressions for the pdf, $f(x)$.

*Answer* (a) .5    (b) 5/32    (c) $f(x) = (\frac{3}{32})(4x - x^2)$ for $0 < x < 4$; $f(x) = 0$ for $x < 0$ and $x > 4$. Each of $f(0)$ and $f(4)$ can be assigned to be equal to any finite nonnegative number.

**PROBLEM 8-18** The pdf for the lifetime $X$, in years, of a Supertuff disk drive is given as follows:

$$f(x) = \begin{cases} \dfrac{2}{x^2} & \text{for } x \geq 2 \\ 0 & \text{elsewhere} \end{cases}$$

(a) Determine the cdf, $F(x)$ for all $x$ values. (b) Find the probability $P(3 < X < 4)$. (c) Determine $x_1$ such that $P(X < x_1) = \frac{2}{3}$. [Here, $x_1$ is the 66.7th percentile.] (d) Out of a set of 10 randomly

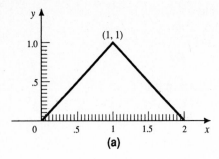

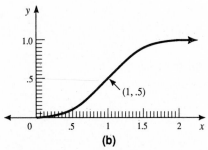

**Figure 8-11.** Graphs of pdf $f(x)$ and cdf $F(x)$ for Problem 8-16. **(a)** $y = f(x)$, where

$$f(x) = \begin{cases} x & \text{for } 0 < x < 1 \\ 2 - x & \text{for } 1 \leq x < 2 \\ 0 & \text{elsewhere} \end{cases}$$

**(b)** $y = F(x)$, where

$$F(x) = \begin{cases} 0 & \text{for } x \leq 0 \\ x^2/2 & \text{for } 0 < x < 1 \\ -x^2/2 + 2x - 1 & \text{for } 1 \leq x < 2 \\ 1 & \text{for } 2 \leq x \end{cases}$$

selected Supertuff disk drives, what is the probability that at least 4 will last longer than 5 years? [*Hint:* This is "partly" a binomial problem.]

***Answer*** **(a)** $F(x) = 2\left[ -\dfrac{1}{x} + \dfrac{1}{2} \right]$ for $x \geq 2$, and $F(x) = 0$, elsewhere    **(b)** 1/6

**(c)** $x_1 = 6$ years is the 66.7th percentile    **(d)** .6178

**PROBLEM 8-19**  In a game, at first a die is tossed. If the outcome is odd, then the player receives a number of dollars equal to the outcome of the die. If the outcome is even, then a number is selected randomly from the interval $[0, 1)$, and the player receives that portion of a dollar associated with the random number selected. [For example, if the random number turns out to be .370, then the player will receive .370 of a dollar, or 37 cents.] **(a)** Determine the combined probability function and probability density function of random variable $X$, where $X$ stands for the amount of money in dollars that the player receives. **(b)** Determine the cumulative distribution function of $X$. **(c)** Compute the probability $P(.3 < X < 1)$. **(d)** Compute the probability $P(.3 < X \leq 1)$.

***Answer*** **(a)** For the probability density function part, we have that $f(x)$ is not defined for $x = 1, x = 3$, and $x = 5$, and $f(x) = 1/2$ for $0 < x < 1$, and $f(x) = 0$ for $x \leq 0$, and $f(x) = 0$ for other values on $x > 1$ besides $x = 3$, and $x = 5$. For the probability function part, we have $P(X = 1) = P(X = 3) = P(X = 5) = 1/6$    **(b)** $F(x) = 0$ for $x \leq 0$; $F(x) = x/2$ for $0 < x < 1$, $F(x) = 4/6$ for $1 \leq x < 3$, $F(x) = 5/6$ for $3 \leq x < 5$; and $F(x) = 1$ for $5 \leq x < \infty$

**(c)** $P(.3 < X < 1) = .7/2 = .35$    **(d)** $P(.3 < X \leq 1) = .35 + 1/6 = .516\overline{6}$

**PROBLEM 8-20**  A random variable $X$ has a *Pareto distribution* if

$$f(x) = \begin{cases} \dfrac{k}{x^{\alpha+1}} & \text{for } x > 1 \\ 0 & \text{elsewhere} \end{cases}$$

Here, for the parameter $\alpha$, we have $\alpha > 0$. Determine $k$.

***Answer*** $k = \alpha$

**PROBLEM 8-21**  **(a)** Refer to Problem 8-9 for a discussion of $Z(t)$, the failure rate at time $t$, for a continuous random variable $T$. Suppose that $T$ has an exponential distribution given by $f(t) = (1/\theta)e^{-t/\theta}$ for $t > 0$ and $f(t) = 0$, elsewhere. Suppose that $T$ stands for the time until failure [of, say, a piece of equipment] here. Determine $Z(t)$ for $t > 0$. **(b)** Suppose that $X$ is a discrete random variable which stands for the trial on which a product will first fail. The failure rate at the $x$th trial, $Z(x)$, is the probability it will fail on the $x$th trial, given that it has not failed on the first $(x - 1)$ trials. Derive the equation for $Z(x)$. Suppose that $f(x)$ and $F(x)$ are the symbols, respectively, for the probability function and cumulative distribution function of $X$. **(c)** Suppose that discrete random variable $X$ has a geometric distribution [see Definition 6.4 and Theorem 6.4] where parameter $p$ stands for the probability of failure on any trial and $X$ indicates the number of the trial on which the product first fails. [Usually, the geometric random variable symbolizes the number of the trial on which the first success occurs.] Thus, the probability function for such a random variable is given by $f(x) = g(x; p) = p(1 - p)^{x-1}$ for $x = 1, 2, 3, \ldots$. Find the equation for $Z(x)$ for $x \geq 1$.

***Answer*** **(a)** $Z(t) = \dfrac{1}{\theta}$    **(b)** $Z(x) = \dfrac{f(x)}{1 - F(x - 1)}$    **(c)** $Z(x) = p$

**PROBLEM 8-22**  Given two line segments $AB$ and $CD$ where the length $a$ of $AB$ is twice the length of $CD$. A point $E$ is chosen randomly on $AB$. Let random variable $X$ stand for the distance from $A$ to $E$. Thus, $X$ is a uniform random variable where $f(x) = 1/a$ for $0 \leq x \leq a$. What is the probability that line segments *AE*, *EB*, and *CD* will form a triangle? [*Hints:* (**i**) The lengths of *AE* and *EB* are *x* and $(a - x)$, respectively. (**ii**) A triangle will be formed if the sum of the lengths of two sides exceeds the length of the third side. Here, the possible sides of the triangle are the segments *AE*, *EB*, and *CD*.]

*Answer* 1/2

# 9 THE EXPECTED VALUE AND MOMENT CONCEPTS

## THIS CHAPTER IS ABOUT

- ☑ **Expected Values of Random Variables**
- ☑ **Some Expected Value Rules**
- ☑ **Moments**
- ☑ **The Moment Generating Function**
- ☑ **Means, Variances, and Moment Generating Functions of Special Probability Distributions**
- ☑ **Practical Situations Involving Expected Values**

## 9-1. Expected Values of Random Variables

It is useful to recall the following note on terminology. When we refer to the *probability distribution* of a random variable $X$, we mean its probability function [abbreviated pf] if $X$ is discrete, or its probability density function [abbreviated pdf] if $X$ is continuous. When we refer to the cumulative distribution function [abbreviated cdf] of a random variable $X$ which is either discrete or continuous, we mean the function $F$, where $F$ is defined by $F(x) = P(X \leq x)$.

### A. Definition of $E(X)$ of a discrete random variable

A very important number associated with a random variable $X$ is $E(X)$, the expected value of the random variable. The quantity is also called the *expectation of $X$* or the *mean of $X$*.

**EXAMPLE 9-1:** Consider a binomial distribution for $n = 3$ and $p = 1/2$. A realistic model for this is the experiment of tossing a coin three times, where the random variable $X$ indicates the exact number of heads obtained. The individual values of the probability function $f(x) = b(x; n, p) = b(x; 3, 1/2)$ are presented in the second column of the table below.

| $x$ | $f(x)$ | frequency | $x \cdot$ frequency |
|-----|--------|-----------|---------------------|
| 0 | 1/8 | 100 | 0 |
| 1 | 3/8 | 300 | 300 |
| 2 | 3/8 | 300 | 600 |
| 3 | 1/8 | 100 | 300 |
| Sum | 8/8 = 1 | $T = 800$ | 1200 |

Suppose that the experiment were repeated 800 times, and that the frequencies associated with $X = 0$, 1, 2, and 3 were "ideal" values, in accord with the corresponding probabilities. Thus, $X = 0$ would occur 1/8 of the total number of times, or 100 times, $X = 1$ would occur 300 times, etc., as in the third column of the above table. Compute the average value of $X$ in this situation.

*Solution:* If we were computing an average value of $X$, we would have

$$\text{Average value of } X = \frac{0 \cdot 100 + 1 \cdot 300 + 2 \cdot 300 + 3 \cdot 100}{800} = \frac{1200}{800} = 1.5 \quad \textbf{(1)}$$

For example, the third term in the numerator indicates that the number 2 is added together 300 times, as with:

$$\overbrace{2 + 2 + 2 + \ldots + 2}^{300 \text{ times}} = 2 \cdot 300$$

Thus, from Eq. (1), the average value for $X$ would be 1.5. If we rearrange the second group of terms in Eq. (1), after dividing through by 800, we obtain the following:

$$\text{Average value of } X = 0 \cdot (\tfrac{1}{8}) + 1 \cdot (\tfrac{3}{8}) + 2 \cdot (\tfrac{3}{8}) + 3 \cdot (\tfrac{1}{8}) \quad \textbf{(2)}$$

Notice that the relative frequencies in this ideal situation, namely 1/8, 3/8, 3/8, and 1/8, are exactly equal to the probabilities $f(0)$, $f(1)$, $f(2)$, and $f(3)$. Thus, we have

$$\text{Average value of } X = \sum_{x=0}^{3} x f(x) \quad \textbf{(3)}$$

---

Equation (3) of Example 9-1 provides a motivation for the general definition of the expected value of $X$.

**Definition 9.1:** The expected value of a discrete random variable $X$ with probability function $P(X = x) = f(x)$ is given by

$$E(X) = \sum_{x} x f(x)$$

Here, $\Sigma_x$ denotes summation over all values of $X$ that are possible values of the random variable. [The set of such values of $X$ has been called $R_x$, the range space of random variable $X$. See, for example, note (b) after Definition 5.5, and the discussion following Theorem 6.1.]

In particular, observe that

$$\sum_{x} f(x) = 1$$

*notes*

(a) In the usual case, in the current book, a discrete random variable $X$ takes on only a finite number of possible values. If discrete random variable $X$ takes on a countably infinite number of possible values, then the summation in Definition 9.1 denotes the sum of an infinite series. In such a situation, the sum may not exist, in which case we also say that $E(X)$ does not exist.

(b) There is a direct analogy between $E(X)$ for a discrete distribution of probabilities along the $x$ axis, and the location of the center of mass of a discrete distribution of point masses along the $x$ axis. For a latter system of point masses $m_1, m_2, \ldots, m_n$, located, respectively, at $x_1, x_2, \ldots, x_n$, the location of the center of mass, $x_{\text{c.m.}}$, is given by

$$x_{\text{c.m.}} = (m_1 x_1 + m_2 x_2 + \ldots + m_n x_n)/(m_1 + m_2 + \ldots + m_n) \quad \textbf{(1)}$$

Suppose, for a discrete probability function, we interpret the probabilities $f(x_1), f(x_2), \ldots, f(x_n)$ that are located, respectively, at $x_1, x_2, \ldots,$

$x_n$, to be point masses. Then, the location of the center of mass is given by

$$x_{c.m.} = x_1 f(x_1) + x_2 f(x_2) + \ldots + x_n f(x_n) \tag{2}$$

in accord with Eq. (1) since $f(x_1) + f(x_2) + \ldots + f(x_n) = 1$. But the expression on the right in Eq. (2) is none other than $E(X)$, the expected value of $X$.

(c) The expected value of $X$ is also called the mean of $X$. A symbol for the mean of $X$ is $\mu_x$ or just $\mu$, if a discussion involves only one random variable. [Here, $\mu$ is the Greek letter mu.]

---

**EXAMPLE 9-2:** Suppose a pair of fair dice is tossed, and the random variable $X$ is equal to the sum of the numbers that occur on the two dice. The probability function, $f(x)$, for $X$ was presented in Example 5-16. Determine $E(X)$, the expected value of $X$.

*Solution:* From the probability function for $X$, which was tabulated in Example 5-16, we see that

$$E(X) = \sum_x xf(x) = 2(1/36) + 3(2/36) + 4(3/36) + 5(4/36) + 6(5/36) + 7(6/36)$$

$$+ 8(5/36) + 9(4/36) + 10(3/36) + 11(2/36) + 12(1/36) = 7$$

---

### B. Expected value of a function of a discrete random variable

Suppose that $X$ is a discrete random variable and that random variable $Y$ is a function of $X$, say in the form $Y = g(X)$. [In terms of values of random variables, we would have $y = g(x)$.] The random variable $Y$ will have a probability function, $h(y)$. It is meaningful to calculate $E(Y)$, the expected value of $Y$. There are two ways of doing this.

First of all, note that Definition 9.1 indicates that [replace $x$ terms by $y$ terms]:

$$E(Y) = \sum_y yh(y)$$

where the sum is over all possible values of $Y$, and $h(y)$ is equal to $P(Y = y)$.

---

**EXAMPLE 9-3:** Suppose that $Y = 3(X - 1)^2$, where the probability function for $X$ is that of Example 9-1. Calculate $E(Y)$ by using Definition 9.1. [Refer to the equation that precedes the statement of this example.]

*Solution:* Since the possible values for $X$ are 0, 1, 2, and 3, the possible values for $Y$ are $3(0 - 1)^2 = 3$, $3(1 - 1)^2 = 0$, $3(2 - 1)^2 = 3$, and $3(3 - 1)^2 = 12$. Note that $Y = 3$ occurs for both $X = 0$ and $X = 2$. This means that

$$h(3) = P(Y = 3) = P(X = 0 \text{ or } X = 2) = P(X = 0) + P(X = 2)$$

$$= \tfrac{1}{8} + \tfrac{3}{8} = \tfrac{4}{8}$$

Observe here that the events $X = 0$ and $X = 2$ are mutually exclusive. Also,

$$h(0) = P(Y = 0) = P(X = 1) = \tfrac{3}{8} \quad \text{and} \quad h(12) = P(Y = 12) = P(X = 3) = \tfrac{3}{8}$$

The calculation of $E(Y)$ is indicated in the following table [sum up values in third column]:

| $y$ | $h(y)$ | $yh(y)$ |
|-----|--------|---------|
| 0 | 3/8 | 0 |
| 3 | 4/8 | 12/8 |
| 12 | 1/8 | 12/8 |
| Sum | 1 | 3 |

Thus, we see that $E(Y) = \Sigma_y \, yh(y) = 3$.

The following theorem provides an alternate, and usually more convenient way to compute $E(Y)$.

**Theorem 9.1:** Suppose that $X$ is a discrete random variable with probability function $f(x)$, and that the random variable $Y$ is a function of $X$ given by $Y = g(X)$. Then,

$$E(Y) = \sum_x g(x)f(x) \quad \text{or, equivalently} \quad E\big(g(X)\big) = \sum_x g(x)f(x)$$

where the summation is over all possible values of $X$.

**EXAMPLE 9-4:** Refer to Example 9-3. Compute $E(Y) = E[3(X-1)^2]$ by using Theorem 9.1.

**Solution:** Here, $g(X)$ is $3(X-1)^2$. Refer to the following table. Note how we develop the numbers in the $3(x-1)^2$ column from the numbers in the $x$ column.

| $x$ | $f(x)$ | $3(x-1)^2$ | $3(x-1)^2 f(x)$ |
|-----|--------|------------|-----------------|
| 0 | 1/8 | 3 | 3/8 |
| 1 | 3/8 | 0 | 0 |
| 2 | 3/8 | 3 | 9/8 |
| 3 | 1/8 | 12 | 12/8 |
| Sum | 1 | | 3 |

The numbers in the last column are obtained by multiplying the numbers in the two previous columns. We see that $E(Y) = E[(X-1)^2] = \Sigma_x \, 3(x-1)^2 f(x) = 3$. Thus, we get the same answer as in Example 9-3.

A demonstration pertaining to Theorem 9.1 is provided by Example 9-5.

**EXAMPLE 9-5:** Given $Y = g(X)$, where the probability function $f(x)$ for discrete random variable $X$ is known, and the possible values for $X$ are $x_{11}, x_{12}, x_{13}, x_{21}, x_{22}, x_{31}, x_{32}$, and $x_{33}$. [The peculiar $x_{ij}$ labeling here is related to the discussion that follows.] Suppose that the possible values that random variable $Y$ takes on are $y_1, y_2$, and $y_3$, and that $y_1$ occurs for $X$ equal to $x_{11}, x_{12}$, or $x_{13}$. That is,

$$y_1 = g(x_{11}) = g(x_{12}) = g(x_{13}) \tag{1}$$

Suppose that $y_2$ occurs for $X$ equal to $x_{21}$ or $x_{22}$, and that $y_3$ occurs for $X$ equal to $x_{31}, x_{32}$, or $x_{33}$. That is,

$$y_2 = g(x_{21}) = g(x_{22}) \tag{2}$$

and

$$y_3 = g(x_{31}) = g(x_{32}) = g(x_{33}) \tag{3}$$

Demonstrate that $E(Y) = E\big(g(X)\big) = \Sigma_x g(x)f(x)$, where the summation is over all possible values of random variable $X$. In this example, the summation is over the eight possible $x_{ij}$ values listed above.

***Solution:*** From Definition 9.1 as applied with $Y$ as the random variable, we have

$$E(Y) = \sum_y yP(Y = y) \tag{4}$$

and here this becomes

$$E(Y) = y_1 P(Y = y_1) + y_2 P(Y = y_2) + y_3 P(Y = y_3) \tag{4a}$$

Since from Eq. (1) we see that $Y = y_1$ occurs when $X = x_{11}, x_{12},$ or $x_{13}$, and the events $X = x_{11}, X = x_{12},$ and $X = x_{13}$ are mutually exclusive, we have

$$P(Y = y_1) = P(X = x_{11}) + P(X = x_{12}) + P(X = x_{13})$$

$$= f(x_{11}) + f(x_{12}) + f(x_{13}) \tag{5}$$

Here, $f(x)$ stands for the probability function of random variable $X$. [Equation (5) is similar to the statement $P(Y = 3) = P(X = 0) + P(X = 2)$ in Example 9-3.] Similarly, we have, after using Eqs. (2) and (3) above,

$$P(Y = y_2) = f(x_{21}) + f(x_{22}) \tag{6}$$

and

$$P(Y = y_3) = f(x_{31}) + f(x_{32}) + f(x_{33}) \tag{7}$$

Note that

$$\sum_x f(x) = f(x_{11}) + f(x_{12}) + f(x_{13}) + f(x_{21}) + f(x_{22}) + f(x_{31})$$

$$+ f(x_{32}) + f(x_{33}) = 1 \tag{8}$$

Substituting from Eqs. (5), (6), and (7) into Eq. (4a) yields

$$E(Y) = y_1[f(x_{11}) + f(x_{12}) + f(x_{13})] + y_2[f(x_{21}) + f(x_{22})]$$

$$+ y_3[f(x_{31}) + f(x_{32}) + f(x_{33})] \tag{9}$$

Now, making use of Eq. (1), we can write

$$y_1[f(x_{11}) + f(x_{12}) + f(x_{13})] = g(x_{11})f(x_{11}) + g(x_{12})f(x_{12}) + g(x_{13})f(x_{13}) \tag{10}$$

Likewise, we can apply Eqs. (2) and (3) to the second and third groups of terms on the right side of Eq. (9). Doing this, and then substituting into Eq. (9) yields

$$E(Y) = g(x_{11})f(x_{11}) + g(x_{12})f(x_{12}) + g(x_{13})f(x_{13}) + g(x_{21})f(x_{21})$$

$$+ g(x_{22})f(x_{22}) + g(x_{31})f(x_{31}) + g(x_{32})f(x_{32}) + g(x_{33})f(x_{33}) \tag{11}$$

Rewriting Eq. (11), we have

$$E(Y) = \sum_x g(x)f(x) \tag{12}$$

where the summation here is with respect to the eight possible values of $X$. Replacing $Y$ on the left side by $g(X)$ yields $E\big(g(X)\big) = \Sigma_x g(x)f(x)$.

---

A generalization of Example 9-5 is given in Problem 9-2.

The main value of Theorem 9.1 is that it enables one to compute $E(Y)$, where $Y = g(X)$, without having to determine the probability function of $Y$. All one has to know is the probability function $f(x)$ of $X$.

## C. Definition of $E(X)$ for a continuous random variable

Suppose that $X$ is a continuous random variable with probability density function $f(x)$, and that we want to develop a meaningful definition for $E(X)$.

For simplicity, suppose that $f(x) \geq 0$ for $\alpha \leq x \leq \beta$, and that $f(x) = 0$ elsewhere, as shown in Figure 9-1. Suppose also that $f(x)$ is continuous for $\alpha \leq x \leq \beta$. Let us make a typical partition of the interval $[\alpha, \beta]$ into $n$ subintervals, such as is indicated in Figure 9-1. Here, we have $x_0 = \alpha$ and $x_n = \beta$, and

$$x_0 < x_1 < x_2 < \ldots < x_{n-1} < x_n \qquad \textbf{(a)}$$

Also, the length of the $i$th subinterval, $\Delta x_i$, is given by

$$\Delta x_i = x_i - x_{i-1} \qquad \text{for } i = 1, 2, \ldots, n-1, n \qquad \textbf{(b)}$$

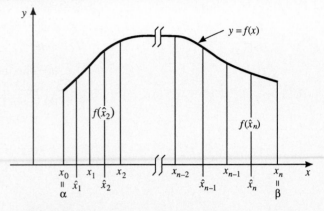

**Figure 9-1.** A partition of the interval $\alpha \leq x \leq \beta$ into $n$ subintervals.

[For example, the first subinterval goes from $x_0$ to $x_1$, the second from $x_1$ to $x_2, \ldots$, and the $n$th subinterval goes from $x_{n-1}$ to $x_n$.] Let the maximum of the $\Delta x_i$ terms be denoted as Max $\Delta x_i$. Let $\hat{x}_i$ be any intermediate $x$ value in the $i$th subinterval. That is,

$$x_{i-1} \leq \hat{x}_i \leq x_i \qquad \text{for } i = 1, 2, \ldots, n-1, n \qquad \textbf{(c)}$$

Let us suppose that each $\Delta x_i$ is fairly small. Now, let us focus on a typical subinterval $[x_{i-1}, x_i]$. Then, the probability associated with $[x_{i-1}, x_i]$ is well approximated by $f(\hat{x}_i)\Delta x_i$, that is,

$$P(x_{i-1} \leq X \leq x_i) \approx f(\hat{x}_i)\Delta x \qquad \textbf{(d)}$$

Here, and elsewhere in this book, "$\approx$" means "approximately equal to." Note that an equation similar to Eq. (d) is Eq. (11) of Example 8-3.

Now, in the definition of $E(X)$ for the discrete case, we multiplied each probability $P(X = x)$ by $x$ [where $x$ was a possible value of the random variable $X$], and then summed up all such products.

For the case of a continuous random variable, it seems plausible to multiply the probability associated with $[x_{i-1}, x_i]$ by the $x$ value $\hat{x}_i$, and then sum up all such terms for $i = 1, 2, \ldots, n-1, n$ to get an approximation to $E(X)$. Thus, as an approximation to $E(X)$, we have

$$E(X) \approx \sum_{i=1}^{n} \hat{x}_i P(x_{i-1} \leq X \leq x_i) \qquad \textbf{(e)}$$

Then, substitution of the approximation from Eq. (d) into Eq. (e) leads to

$$E(X) \approx \sum_{i=1}^{n} \hat{x}_i f(\hat{x}_i)\Delta x_i \qquad \textbf{(f)}$$

The summation on the right side of Eq. (f) is a familiar term in calculus, which is known as a *Riemann sum*. Now, the smaller we take Max $\Delta x_i$, the closer the sum in Eq. (f) will be to what we intuitively think of as the true value of $E(X)$. Thus, taking the limit on the right side of Eq. (f) as Max $\Delta x_i$ goes to 0, we obtain

$$E(X) = \lim_{\text{Max } \Delta x_i \to 0} \sum_{i=1}^{n} \hat{x}_i f(\hat{x}_i)\Delta x_i \qquad \textbf{(g)}$$

In accordance with a major definition of calculus, the right side of Eq. (g) is none other than the definite integral $\int_{\alpha}^{\beta} xf(x)\,dx$. Now, for the probability density function pictured in Figure 9-1, we can rewrite this integral as $\int_{-\infty}^{\infty} xf(x)\,dx$ since $f(x) = 0$ for $x < \alpha$ and $x > \beta$. We now take Eq. (g) with the right side replaced by this last integral as the defining equation for $E(X)$ for a continuous random variable.

**Definition 9.2:** The expected value $E(X)$, of a continuous random variable $X$ with probability density function $f(x)$, is given by

$$E(X) = \int_{-\infty}^{\infty} xf(x)\,dx$$

Here, we assume that the integral exists. Otherwise, $E(x)$ won't exist.

*notes*

(a) The improper integral in Definition 9.2 will exist, and hence $E(X)$ will exist if the definite integral $\int_{-\infty}^{\infty} |x|\,f(x)\,dx$ exists.

(b) As in the discrete random variable case, we have an analogy of $E(X)$, as given in Definition 9.2, with the center of mass concept of mechanics. Suppose that pdf $f(x)$ is thought of as the mass density at $x$ of a mass that is continuously distributed along the $x$ axis. Then, $E(X)$ represents the location of the center of mass, $x_{c.m.}$, since $\int_{-\infty}^{\infty} f(x)\,dx = 1$.

Observe that in the center of mass equation of mechanics, one has

$$x_{c.m.} = \frac{\displaystyle\int_{-\infty}^{\infty} xf(x)\,dx}{m}$$

where, since $f(x)$ is the mass density at $x$, the total mass $m$ is equal to $\int_{-\infty}^{\infty} f(x)\,dx$.

(c) Here, as in the discrete case, $E(X)$ is known as the mean of $X$. The mean is symbolized by $\mu_x$, or merely $\mu$, if the discussion involves only one random variable.

---

**EXAMPLE 9-6:** Suppose that $X$ is the lifetime of a Powermate battery, in months, and that the pdf is given by $f(x) = (\frac{3}{32})(4x - x^2)$ for $0 < x < 4$, and $f(x) = 0$, elsewhere. Calculate $E(X)$, the mean of $X$.

*Solution:* From Definition 9.2, we have

$$E(x) = \int_{-\infty}^{\infty} xf(x)\,dx = \left(\frac{3}{32}\right)\int_{0}^{4} x(4x - x^2)\,dx \qquad \textbf{(1)}$$

[Notice that the integral from $-\infty$ to $\infty$ reduces to an integral from 0 to 4 since $f(x) = 0$ for $x \le 0$ and $x \ge 4$.] Thus, from Eq. (1), we obtain

$$E(X) = \left(\frac{3}{32}\right)\left(\frac{4x^3}{3} - \frac{x^4}{4}\right)\Big]_{0}^{4} = 2 \qquad \textbf{(2)}$$

That is, the mean lifetime of a Powermate battery is two months.

---

### D. Expected value of a function of a continuous random variable

Suppose that $X$ is a continuous random variable, and that $Y = g(X)$ is a function of $X$. Then, we know that $Y$ is also a random variable, with some probability distribution. Let us suppose that $Y$ is a continuous random variable with probability density function $h(y)$.

It is meaningful to calculate $E(Y)$. First of all, note from Definition 9.2 that

$$E(Y) = \int_{-\infty}^{\infty} yh(y)\,dy$$

if $Y$ is a continuous random variable for which the pdf is $h(y)$. The following theorem, which parallels Theorem 9.1, provides an alternate, and often easier, way of computing $E(Y)$.

**Theorem 9.2:** Suppose that $X$ is a continuous random variable, with probability density function $f(x)$, and that the random variable $Y$ is a function of $X$ given by $Y = g(X)$. Then,

$$E(Y) = E\big(g(X)\big) = \int_{-\infty}^{\infty} g(x)f(x)\,dx$$

***note:*** An advantage of using Theorem 9.2 is that it is not necessary to be concerned with the probability distribution of $Y$.

---

**EXAMPLE 9-7:** For the random variable $X$, the pdf $f(x)$ is uniform and given by $f(x) = .2$ for $0 < x < 5$, and $f(x) = 0$ elsewhere. Suppose that $Y = .01X^2$. Compute $E(Y) = E(.01X^2)$ by using Theorem 9.2.

***Solution:*** From Theorem 9.2, we have

$$E(Y) = E(.01X^2) = \int_{-\infty}^{\infty} (.01x^2)f(x)\,dx \qquad (1)$$

Substituting $f(x) = .2$ for $0 < x < 5$ and $f(x) = 0$ for $x \le 0$ and $x \ge 5$ into Eq. (1), we get

$$E(Y) = E(.01X^2) = \int_{0}^{5} (.01x^2)(.2)\,dx = (.002)\left(\frac{x^3}{3}\right)\Big]_{0}^{5}$$

$$= (.002)\left(\frac{125}{3}\right) = \frac{1}{12} = .083\overline{3} \qquad (2)$$

---

It is useful to compute $E(Y)$ for the $Y$ of Example 9-7, by using $E(Y) = \int_{-\infty}^{\infty} yh(y)\,dy$, where $h(y)$ is the pdf of $Y$. Now, for $Y = .01X^2$, we will show that $Y$ is a continuous random variable, and then we will calculate $h(y)$.

---

**EXAMPLE 9-8:** Determine the probability distribution for $Y$ if $Y = .01X^2$ [thus, $g(X) = .01X^2$ also], and the pdf for $X$ is that of Example 9-7.

***Solution:*** Refer to Figure 9-2. Let us refer to an interval for which a pdf is nonnegative as the *region* or *interval of interest* for the corresponding random variable provided that the pdf is equal to zero for all points on the complement of that interval. Such an interval has also been referred to as the *range space* of the random variable. Thus, for example, the interval of interest for the random variable $X$ of Example 9-7 is $0 < x < 5$. Since $Y = .01X^2$, we see that $y \to 0$ as $x \to 0$ and $y \to .25$ as $x \to 5$. Thus, the region or interval of interest for $Y$ is $0 < y < .25$. [It is also correct to refer to the interval $0 < y < .25$ as $R_Y$, the range space of $Y$.]

If $Y = y_0$, for $y_0$ on the interval $(0, .25)$, then $X = x_0$, where $y_0$ and $x_0$ are related by $y_0 = .01(x_0)^2$ and $x_0$ is on the interval $(0, 5)$. Refer to Figure 9-2. Now,

$$P(Y \le y_0) = P(X \le x_0) \qquad (1)$$

Let us use $F$ and $H$ as the symbols for the cumulative distribution functions for $X$ and $Y$, respectively. Thus, Eq. (1) can be written as

$$H(y_0) = F(x_0) \qquad (2)$$

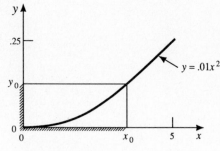

**Figure 9-2.** Graph of $y = g(x)$ for Examples 9-7, 9-8, and 9-9.

[Note that $P(X \leq x_0)$ in Eq. (1) is also equal to $P(0 < X \leq x_0)$, and similarly for $P(Y \leq y_0)$.] Differentiating both sides of Eq. (2) with respect to $y_0$, we get

$$\frac{dH(y_0)}{dy_0} = \frac{dF(x_0)}{dx_0} \cdot \frac{dx_0}{dy_0} \tag{3}$$

where we have used the chain rule on the right side. Using Part (b) of Theorem 8.3, we can rewrite this as

$$h(y_0) = f(x_0) \cdot \frac{dx_0}{dy_0} \tag{4}$$

where $h$ and $f$ indicate the pdf's of $Y$ and $X$, respectively. If we replace $x_0$ and $y_0$ by $x$ and $y$, respectively, in Eq. (4), we'll end up with

$$h(y) = f(x) \cdot \frac{dx}{dy} \tag{4a}$$

Since $y = (.01)x^2$ for $0 < x < 5$, it follows that

$$x = 10\sqrt{y} \quad \text{for } 0 < y < .25 \tag{5}$$

Thus,

$$\frac{dx}{dy} = 5y^{-1/2} \quad \text{for } 0 < y < .25 \tag{6}$$

Substitution of $f(x) = .2$ [from Example 9-7], and $dx/dy$ from Eq. (6) into Eq. (4a) yields

$$h(y) = y^{-1/2} \quad \text{for } 0 < y < .25 \tag{7}$$

Also, $h(y) = 0$ elsewhere. [It is not hard to show that $\int_0^{.25} h(y)\, dy = 1$.]

---

**EXAMPLE 9-9:** Compute $E(Y)$ for random variable $Y$ of Example 9-8 by making use of Definition 9.2.

*Solution:* From Definition 9.2, we have

$$E(Y) = \int_{y=-\infty}^{\infty} y h(y)\, dy \tag{1}$$

where $h(y)$ is the pdf of $Y$. Using $h(y)$ as determined in Example 9-8, Eq. (1) becomes

$$E(Y) = \int_{y=0}^{.25} y \cdot y^{-1/2}\, dy = \int_{y=0}^{.25} y^{1/2}\, dy = \left(\frac{2}{3}\right) y^{3/2} \Big]_0^{.25} = \frac{1}{12} \tag{2}$$

Thus, we get the same result as in Example 9-7.

---

For a calculation of $E(Y) = E(g(X))$ for a situation where $X$ is a continuous random variable, and $Y = h(X)$ is a discrete random variable, refer to Problem 9-6.

## 9-2. Some Expected Value Rules

---

**EXAMPLE 9-10:** Consider a binomial distribution for $n = 3$ and $p = 1/2$. Determine $E(X)$ and $E(X^2)$. Observe that we may give the binomial random variable here a realistic meaning by interpreting it to be the number of heads obtained in 3 tosses of a fair coin. This binomial situation was also considered in Example 9-1.

*Solution:* Here,

$$f(x) = b(x; 3, 1/2) = \frac{3!}{x!(3-x)!}\left(\frac{1}{2}\right)^3 \qquad \text{for } x = 0, 1, 2, 3 \qquad \textbf{(1)}$$

From Eq. (1), we can develop the first two columns of the following table [as in Example 9-4].

| $x$ | $f(x)$ | $xf(x)$ | $x^2f(x)$ |
|---|---|---|---|
| 0 | 1/8 | 0 | 0 |
| 1 | 3/8 | 3/8 | 3/8 |
| 2 | 3/8 | 6/8 | 12/8 |
| 3 | 1/8 | 3/8 | 9/8 |
| $\sum\limits_{x=0}^{3}$ | 1 | $12/8 = 1.5$ | $24/8 = 3$ |

Now, $E(X)$ and $E(X^2)$ are given by

$$E(X) = \sum_x xf(x) \qquad E(X^2) = \sum_x x^2f(x) \qquad \textbf{(2), (3)}$$

Thus, it is easy to compute $E(X)$ by multiplying the values in the $x$ and $f(x)$ columns together to yield the values in the third column, the $xf(x)$ column. Then, summing the values in the $xf(x)$ column yields $E(X)$. Thus, $E(X)$, also known as $\mu_x$, the mean of $X$, is given by

$$E(X) = \sum_x xf(x) = 12/8 = 1.5 \qquad \textbf{(4)}$$

To compute $E(X^2)$, we first calculate $x^2f(x)$ values in the fourth column above [by multiplying corresponding values in the $x$ and $xf(x)$ columns together, say], and then sum up these values. Thus, we find

$$E(X^2) = \sum_x x^2f(x) = 24/8 = 3 \qquad \textbf{(5)}$$

Calculations of expected values [expectations] are often simplified by using Theorem 9.3, and its corollaries, and Theorem 9.4.

**Theorem 9.3:** If $a$ and $b$ are constants, then $E(aX + b) = aE(X) + b$.

**EXAMPLE 9-11:** Prove Theorem 9.3.

*Proof:* We'll do the proof for the discrete case. [The proof for the continuous case follows a similar pattern.] From Theorem 9.1, with $g(X) = aX + b$, we have

$$E(aX + b) = \sum_x (ax + b)f(x) = a\sum_x xf(x) + b\sum_x f(x) \qquad \textbf{(1)}$$

The first sigma term after the second equals sign is equal to $E(X)$, and the second sigma term is equal to 1. Thus, we end up with

$$E(aX + b) = aE(X) + b \qquad \textbf{(2)} \quad \square$$

Setting $a = 0$, and then $b = 0$ in Theorem 9.3 yields the following corollaries:

**Corollary 1 to Theorem 9.3:** If $b$ is a constant, then

$$E(b) = b$$

**Corollary 2 to Theorem 9.3:** If $a$ is a constant, then

$$E(aX) = aE(X)$$

**Theorem 9.4 (Linearity Rule for $E$):** If $a_1, a_2, \ldots$, and $a_n$ are constants and $g_i(X)$ is a function of random variable $X$ for $i = 1, 2, 3, \ldots, n$, then

$$E[a_1 g_1(X) + a_2 g_2(X) + \ldots + a_n g_n(X)]$$
$$= a_1 E[g_1(X)] + a_2 E[g_2(X)] + \ldots + a_n E[g_n(X)] \qquad \textbf{(a)}$$

In sigma notation, Eq. (a) has the form

$$E\left[ \sum_{i=1}^{n} a_i g_i(X) \right] = \sum_{i=1}^{n} a_i E[g_i(X)] \qquad \textbf{(b)}$$

---

**EXAMPLE 9-12:** Prove Theorem 9.4.

*Proof:* We will do the proof for $n = 2$ for the case where $X$ is a discrete random variable.

Applying Theorem 9.1, where $g(X) = a_1 g_1(X) + a_2 g_2(X)$, we get

$$\underbrace{E[a_1 g_1(X) + a_2 g_2(X)]}_{g(X)} = \sum_{x} \underbrace{[a_1 g_1(x) + a_2 g_2(x)]}_{g(x)} f(x) \qquad \textbf{(1)}$$

We now multiply through by $f(x)$ on the right, and use the algebraic fact that

$$\sum_{x} [h_1(x) + h_2(x)] = \sum_{x} h_1(x) + \sum_{x} h_2(x) \qquad \textbf{(2)}$$

[Actually, for Eq. (2) to be valid, we must assume that both sigma terms on the right exist.] Next, we take the constants $a_1$ and $a_2$ to the left of the sigma symbols. This leads to

$$E[a_1 g_1(X) + a_2 g_2(X)] = a_1 \sum_{x} g_1(x) f(x) + a_2 \sum_{x} g_2(x) f(x) \qquad \textbf{(3)}$$

Applying Theorem 9.1 to each of the sigma terms on the right [assuming that both terms exist] yields the desired result, namely

$$E[a_1 g_1(X) + a_2 g_2(X)] = a_1 E[g_1(X)] + a_2 E[g_2(X)] \qquad \square$$

---

**EXAMPLE 9-13:** For the probability distribution of Example 9-10, compute $E[4X^2 - 5X + 2]$.

*Solution:* From Theorem 9.4, we have

$$E[4X^2 - 5X + 2] = 4E[X^2] - 5E[X] + 1 \cdot E[2] \qquad \textbf{(1)}$$

Here, we treated 2 as $1 \cdot 2$. Now, $E[2] = 2$ from Corollary 1 of Theorem 9.3, and the values for $E[X]$ and $E[X^2]$ are given in Example 9-10. Thus, substituting into Eq. (1), we get

$$E[4X^2 - 5X + 2] = 4(3) - 5(1.5) + 1(2) = 6.5 \qquad \textbf{(2)}$$

---

## 9-3. Moments

Equations of $E[g(X)]$ for the discrete and continuous random variable cases are given in Theorems 9.1 and 9.2. The former is

$$E[g(X)] = \sum_{x} g(x) f(x) \qquad \text{[Theorem 9.1—Discrete case]}$$

In this case, $f(x)$ is the probability function for discrete random variable $X$. Here, and in the following development, the corresponding equation for the continuous random variable case is obtained by replacing $\Sigma_x$ by $\int_x \underline{\quad} dx$, where $\int_x \underline{\quad} dx$ will stand for $\int_{-\infty}^{\infty} \underline{\quad} dx$. Also, it is important to remember that $f(x)$ symbolizes a probability density function in the continuous case. Thus, we have

$$E[g(X)] = \int_x g(x)f(x)\,dx \qquad \text{[Theorem 9.2—Continuous case]}$$

## A. Definitions of moments

Among the expected value terms that are of great significance in statistics are the *moments of a random variable*, which are also known as the *moments of the distribution of a random variable*. In the following, rv will often be used to symbolize "random variable."

**Definition 9.3:** The $r$th moment about the origin of a random variable $X$, denoted by $\mu_r'$, is equal to $E[X^r]$. This holds for $r = 0, 1, 2, 3, \ldots$. Applying the equations for $E[g(X)]$ from Theorems 9.1 and 9.2, with $g(X) = X^r$, we get

$$\mu_r' = E[X^r] = \sum_x x^r f(x) \qquad \text{[Discrete rv case]} \qquad \textbf{(a)}$$

$$\mu_r' = E[X^r] = \int_x x^r f(x)\,dx \qquad \text{[Continuous rv case]} \qquad \textbf{(b)}$$

*notes*

**(a)** The term "moment" comes from physics. For example, for the discrete-finite case, suppose we interpret the probabilities $f(x_1), f(x_2), \ldots, f(x_n)$ located, respectively, at $x_1, x_2, \ldots, x_n$ to be point masses. Then, the $r$th moment about the origin of this system of point masses is given by Eq. (a) above; the summation in this case runs from $x_1$ to $x_n$, inclusive. Of particular importance in physics are $\mu_1'$, the first moment about the origin, and $\mu_2'$, the second moment about the origin. The latter term is also known as the moment of inertia of the system of point masses with respect to the origin. Observe that $\mu_1'$ is also equal to the $x$-coordinate of the center of mass when $\Sigma_x f(x)$, the sum of the point masses, is equal to 1. [That will be the case for a discrete probability distribution. See, also, Note (b) following Definition 9.1.]

**(b)** We see that $\mu_1'$ is none other than $E(X)$, which is also known as the mean of $X$. The usual symbol for the mean [in this book] is $\mu_x$ or just $\mu$, if a discussion involves only one random variable.

For computational purposes involving probability distributions, the first and second moments about the origin, $E(X) = \mu_x$ and $E(X^2)$, are usually the most important.

**Definition 9.4:** The $r$th moment about the mean of a random variable $X$, denoted by $\mu_r$, is equal to $E[(X - \mu)^r]$. Here, $r = 0, 1, 2, 3, \ldots$ and $\mu$ is $E(X)$ [or $\mu_1'$].

Applying the equations for $E[g(X)]$ from Theorems 9.1 and 9.2, with $g(X) = (X - \mu)^r$, we get

$$\mu_r = E[(X - \mu)^r] = \sum_x (x - \mu)^r f(x) \qquad \text{[Discrete rv case]} \qquad \textbf{(a)}$$

$$\mu_r = E[(X - \mu)^r] = \int_x (x - \mu)^r f(x)\,dx \qquad \text{[Continuous rv case]} \qquad \textbf{(b)}$$

In Problem 9-9, we show that $\mu_1 = 0$. Here, $\mu_1$ is the first moment about the mean. The quantity $\mu_2 = E[(X - \mu)^2]$ is of great importance in probability and statistics.

## B.  Variance and standard deviation

**Definition 9.5**

(a) The quantity $\mu_2$ is called the variance of $X$ [or the variance of the probability distribution of $X$], and this is also symbolized by Var($X$) or $V(X)$. That is,

$$\text{Var}(X) = E[(X - \mu)^2] \quad \text{or} \quad \text{Var}(X) = \mu_2$$

(b) The standard deviation of $X$ [or the standard deviation of the probability distribution of $X$], denoted by $\sigma_x$, is equal to the positive square root of Var($X$). [Here, $\sigma$ is the lowercase version of the Greek letter sigma.] That is,

$$\sigma_x = \sqrt{\text{Var}(X)} \quad \text{or} \quad \sigma_x = \sqrt{\mu_2}$$

Often, we merely write "$\sigma$" as the symbol for the standard deviation of a random variable if a discussion involves only one random variable.

*notes*

(a) The standard deviation $\sigma_x$ is a measure of the spread, or variation, or dispersion, or scatter of the probability distribution of random variable $X$.

(b) The number Var($X$) is expressed in square units of $x$. Thus, if the units of $x$ is seconds, then the units of Var($X$) is (seconds)$^2$. The units for the standard deviation is the same as the units of $x$. Thus, if the units of $x$ is seconds, then the units of $\sigma_x$ is also seconds. This is a major reason why $\sigma_x$ is preferred to Var($X$) as a measure of spread of a probability distribution.

(c) We see from Var($X$) $= E[(X - \mu)^2]$ that the variance is the expected value [or "mean"] of the square of the deviation of $X$ with respect to $\mu$. Thus, it is reasonable to think of $\sigma_x = \sqrt{\text{Var}(X)}$ as a sort of average deviation from $\mu$.

(d) Again, let us turn to physics. If the probability distribution is interpreted as a distribution or system of point masses along the $x$ axis [for which the total mass is 1], then Var($X$) represents the moment of inertia about the point $x = \mu_x$ on the $x$ axis of this distribution.

---

**EXAMPLE 9-14:** For a binomial distribution for which $n = 3$ and $p = 1/2$, calculate (a) Var($X$), and (b) $\sigma_x$.

*Solution:* This probability distribution was considered in Example 9-10, where we calculated $E(X)$ and $E(X^2)$. Recall that $E(X) = \mu = 3/2 = 1.5$. Refer to the following table, which contains columns of Dev $= (x - \mu)$, (Dev)$^2$, and (Dev)$^2 f(x)$. Here, "Dev" stands for the deviation of an $x$ value from the mean, $\mu$.

| $x$ | $f(x)$ | Dev $=$ $(x - \mu)$ | (Dev)$^2 =$ $(x - \mu)^2$ | (Dev)$^2 f(x) =$ $(x - \mu)^2 f(x)$ |
|---|---|---|---|---|
| 0 | 1/8 | $-3/2$ | 9/4 | 9/32 |
| 1 | 3/8 | $-1/2$ | 1/4 | 3/32 |
| 2 | 3/8 | 1/2 | 1/4 | 3/32 |
| 3 | 1/8 | 3/2 | 9/4 | 9/32 |
| $\sum_{x=0}^{3}$ | 1 | | | $24/32 = .75$ |

A deviation $(x - \mu)$ for a particular $x$ value is obtained by subtracting the value of $\mu$ [here, $\mu = 3/2$] from the $x$-value. The variance is obtained by summing up values of $(x - \mu)^2 f(x)$. (See the last column of the table.) Thus, for $\text{Var}(X)$ and $\sigma_x$, we have

$$\text{Var}(X) = \sum_x (x - \mu)^2 f(x) = 24/32 = 3/4 = .75 \qquad \textbf{(1)}$$

$$\sigma_x = \sqrt{\text{Var}(X)} = \sqrt{.75} = .866 \qquad \textbf{(2)}$$

## C. Some theorems on the variance

### Theorem 9.5

$$\text{Var}(X) = E(X^2) - [E(X)]^2$$

Equivalently, $\sigma^2 = E(X^2) - \mu^2$, where $\sigma$ and $\mu$ stand for the standard deviation and mean of $X$.

**EXAMPLE 9-15:** Prove Theorem 9.5.

*Proof:* First, with $\mu$ symbolizing the mean of $X$, we have

$$\text{Var}(X) = E[(X - \mu)^2] = E[X^2 - 2\mu X + \mu^2] \qquad \textbf{(1)}$$

In Eq. (1), we obtained the last term by squaring the binomial $(X - \mu)$. Applying the linearity rule for $E$ [Theorem 9.4] to the last term of Eq. (1), we obtain

$$\text{Var}(X) = E(X^2) - 2\mu E(X) + 1 \cdot E(\mu^2) \qquad \textbf{(2)}$$

after expressing $\mu^2$ as $1 \cdot \mu^2$. Now, $E(\mu^2) = \mu^2$ by Corollary 1 of Theorem 9.3. Also, $E(X) = \mu$. Thus, Eq. (2) becomes

$$\text{Var}(X) = E(X^2) - 2\mu^2 + \mu^2 = E(X^2) - \mu^2 \qquad \textbf{(3)}$$

or

$$\text{Var}(X) = E(X^2) - [E(X)]^2 \qquad \textbf{(4)} \quad \square$$

**EXAMPLE 9-16:** Use Theorem 9.5 to compute $\text{Var}(X)$ for the binomial distribution of Examples 9-10 and 9-14.

*Solution:* From the results of Example 9-10, we have

$$E(X) = 3/2 \quad \text{and} \quad E(X^2) = 3 \qquad \textbf{(1), (2)}$$

Applying Theorem 9.5, we have

$$\text{Var}(X) = E(X^2) - [E(X)]^2 = 3 - (3/2)^2 = 3/4 = .75 \qquad \textbf{(3)}$$

This result is identical to the result of Example 9-14.

**Theorem 9.6:** If $a$ and $b$ are constants, then

$$\text{Var}(aX + b) = a^2 \text{Var}(X) \qquad \textbf{(a)}$$

Thus, taking square roots on both sides of (a), we get

$$\sigma_{ax+b} = |a|\sigma_x \qquad \textbf{(b)}$$

***note:*** This result is often used together with that of Theorem 9.3, which states that $E(ax + b) = aE(x) + b$.

A proof of Theorem 9.6 is given in Problem 9-11.

**EXAMPLE 9-17:** Suppose the average daily temperature [in degrees Fahrenheit] in July in Newtown is a random variable $T$ with mean $\mu_T = 85$ and standard deviation $\sigma_T = 7$. The daily air conditioning cost $Q$, in dollars, for a factory, is related to the average daily temperature by the equation $Q = 120T + 750$. For a typical day in July, determine the expected [or mean] air conditioning cost, and the standard deviation of the air conditioning cost. [Note that the "average" daily temperature here refers to an average with respect to time during a 24-hour period; for example, if $\theta$ denotes the time, in hours, during a day, and $t(\theta)$ denotes the temperature as a function of time, then the average daily temperature on a particular day, $t$, is given by $t = (\frac{1}{24}) \int_{\theta=0}^{24} t(\theta)\, d\theta$.]

*Solution:* We are given for the random variable $T$ that

$$\mu_T = 85°\text{F} \qquad \text{and} \qquad \sigma = 7°\text{F} \tag{1), (2}$$

Since $Q = 120T + 750$, we can compute $E(Q)$ and $\text{Var}(Q)$ from Theorems 9.3 and 9.6. Thus,

$$E(Q) = E(120T + 750) = 120E(T) + 750 = 120(85) + 750 = \$10{,}950 \tag{3}$$

$$\text{Var}(Q) = \text{Var}(120T + 750) = (120)^2\, \text{Var}(T) = (120)^2(7)^2 \tag{4}$$

As a result,

$$\sigma_Q = \sqrt{\text{Var}(Q)} = \sqrt{(120)^2(7)^2} = (120)(7) = \$840 \tag{5}$$

**Definition 9.6:** Given a random variable $X$ with mean $\mu$ and standard deviation $\sigma$. The standardized [or standard] random variable corresponding to $X$, often denoted as $Z$, is defined to be

$$Z = \frac{X - \mu}{\sigma}$$

Observe that we can also write $Z = (X - \mu)/\sigma$.

**Theorem 9.7:** For the standardized random variable $Z = (X - \mu)/\sigma$, we have

$$E(Z) = \mu_Z = 0 \qquad \text{and} \qquad \text{Var}(Z) = 1 \tag{a), (b}$$

From (b) it follows that

$$\sigma_Z = 1 \tag{c}$$

**EXAMPLE 9-18:** Prove Theorem 9.7.

*Proof:* We can express $Z = (X - \mu)/\sigma$ in the form $Z = aX + b$ by writing

$$Z = \underset{a}{\left(\frac{1}{\sigma}\right)} X + \underset{b}{\left(-\frac{\mu}{\sigma}\right)} \tag{1}$$

Now, applying Theorem 9.3 with $a = (1/\sigma)$ and $b = (-\mu/\sigma)$, we have

$$E(Z) = E\left[\left(\frac{1}{\sigma}\right)X + \left(-\frac{\mu}{\sigma}\right)\right] = \left(\frac{1}{\sigma}\right)E(X) + \left(-\frac{\mu}{\sigma}\right)$$

$$= \left(\frac{1}{\sigma}\right)\mu + \left(-\frac{\mu}{\sigma}\right) = 0 \tag{2}$$

Here, we have used the fact that $E(X) = \mu$.

Applying Theorem 9.6 to $Z$ as expressed in Eq. (1), we get

$$\text{Var}(Z) = \text{Var}\left[\left(\frac{1}{\sigma}\right)X + \left(-\frac{\mu}{\sigma}\right)\right] = \left(\frac{1}{\sigma}\right)^2 \text{Var}(X) = \left(\frac{1}{\sigma}\right)^2 \cdot \sigma^2 = 1 \qquad \textbf{(3)}$$

Here, we have used the fact that $a = (1/\sigma)$, $b = (-\mu/\sigma)$, and $\text{Var}(X) = \sigma^2$. $\qquad \square$

## D. Chebyshev's inequality theorem

Chebyshev's theorem [named after the Russian mathematician, P. L. Chebyshev, 1821–1894] demonstrates how $\sigma$ or $\sigma^2$ is an indicator of the degree of spread or deviation of a probability distribution.

**Theorem 9.8 (Chebyshev's Inequality Theorem):** Given a random variable $X$ with mean $\mu$ and standard deviation $\sigma$. Then, for any positive constant $k$, the probability is at least $1 - (1/k^2)$ that $X$ will take on a value within $k$ standard deviations of $\mu$. In symbols,

$$P(\mu - k\sigma < X < \mu + k\sigma) \geq 1 - \frac{1}{k^2} \qquad \textbf{(a)}$$

If we let $c = k\sigma$ in (a), then $1/k = \sigma/c$, and (a) becomes

$$P(\mu - c < X < \mu + c) \geq 1 - \frac{\sigma^2}{c^2} \qquad \textbf{(b)}$$

***note:*** The expressions within the probability parentheses of (a) and (b) are often written as $|X - \mu| < k\sigma$ and $|X - \mu| < c$, respectively.

---

**EXAMPLE 9-19:** Prove Chebyshev's theorem.

***Proof:*** Our proof is for the continuous case. Refer to Figure 9-3. From Definition 9.5,

$$\sigma^2 = E[(X - \mu)^2] = \int_{-\infty}^{\infty} (x - \mu)^2 f(x)\, dx \qquad \textbf{(1)}$$

Dividing the integral into three parts corresponding to the subdivision on the $x$-axis indicated in Figure 9-3, we have

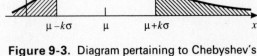

**Figure 9-3.** Diagram pertaining to Chebyshev's theorem.

$$\sigma^2 = \int_{-\infty}^{\mu - k\sigma} (x - \mu)^2 f(x)\, dx + \int_{\mu - k\sigma}^{\mu + k\sigma} (x - \mu)^2 f(x)\, dx$$

$$+ \int_{\mu + k\sigma}^{\infty} (x - \mu)^2 f(x)\, dx \qquad \textbf{(2)}$$

Since the integrand $(x - \mu)^2 f(x)$ is nonnegative, the inequality in (3) follows from Eq. (2):

$$\sigma^2 \geq \int_{-\infty}^{\mu - k\sigma} (x - \mu)^2 f(x)\, dx + \int_{\mu + k\sigma}^{\infty} (x - \mu)^2 f(x)\, dx \qquad \textbf{(3)}$$

[In going from Eq. (2) to inequality (3), we have dropped out the middle integral on the right side of Eq. (2).]

Now, in the first integral of inequality (3), the integral variable $x$ is such that $x \leq \mu - k\sigma$; in the second integral of inequality (3), the integral variable $x$ is such that $x \geq \mu + k\sigma$. In either case, we have

$$(x - \mu)^2 \geq k^2 \sigma^2 \qquad \textbf{(4)}$$

Applying inequality (4) to both integrals of inequality (3), we obtain the following double inequality:

$$\sigma^2 \geq \int_{-\infty}^{\mu-k\sigma} (x-\mu)^2 f(x)\, dx + \int_{\mu+k\sigma}^{\infty} (x-\mu)^2 f(x)\, dx$$

$$\geq \int_{-\infty}^{\mu-k\sigma} k^2\sigma^2 f(x)\, dx + \int_{\mu+k\sigma}^{\infty} k^2\sigma^2 f(x)\, dx \qquad (5)$$

From the first and last expressions of inequality (5), we obtain inequality (6), after taking $k^2\sigma^2$ to the left of the last two integrals, and then cancelling $\sigma^2$, which is assumed to be positive.

$$\frac{1}{k^2} \geq \int_{-\infty}^{\mu-k\sigma} f(x)\, dx + \int_{\mu+k\sigma}^{\infty} f(x)\, dx \qquad (6)$$

Now, since $\int_{-\infty}^{\infty} f(x)\, dx = 1$, the sum of integrals on the right of inequality (6) is equal to $1 - \int_{\mu-k\sigma}^{\mu+k\sigma} f(x)\, dx$. This last integral is also equal to the probability $P(\mu - k\sigma < X < \mu + k\sigma)$. Thus, we may rewrite inequality (6) as

$$P(\mu - k\sigma < X < \mu + k\sigma) \geq 1 - \frac{1}{k^2} \qquad (7) \quad \square$$

For example, if $k = 3/2$, then inequality (7) yields the following inequality:

$$P(\mu - k\sigma < X < \mu + k\sigma) \geq 5/9 = .555\overline{5}$$

For $k = 2$, inequality (7) becomes

$$P(\mu - 2\sigma < X < \mu + 2\sigma) \geq .75$$

In words, this last inequality says that the probability is at least .75 that random variable $X$ will take on a value within two standard deviations of the mean. To see this with respect to Figure 9-3, replace $k$ by 2 in that diagram. Then, the probability of obtaining a value for the interval on the x-axis between $\mu - 2\sigma$ and $\mu + 2\sigma$ is at least .75.

---

**EXAMPLE 9-20:** Given that $f(x) = 1$ for $0 < x < 1$, and $f(x) = 0$ elsewhere. [That is, $X$ is a uniform continuous random variable.] Calculate $P(\mu - 1.5\sigma < X < \mu + 1.5\sigma)$, and compare this probability with the lower bound provided by Chebyshev's theorem.

*Solution:* It's easy to show that $E(X) = \mu = .5$, $E(X^2) = 1/3$, and hence $\text{Var}(X) = \sigma^2 = 1/12$. Thus, $\sigma = .2887$ and

$$P(\mu - 1.5\sigma < X < \mu + 1.5\sigma) = P(.06699 < X < .93301)$$

$$= \int_{.06699}^{.93301} dx = .93301 - .06699 = .86602 \qquad (1)$$

Chebyshev's theorem predicts that

$$P(\mu - 1.5\sigma < X < \mu + 1.5\sigma) \geq \tfrac{5}{9} = .555\overline{5} \qquad (2)$$

Clearly, the statement from Eq. (1), namely, that "the probability that $X$ is within 1.5 standard deviations of the mean is equal to .86602" is stronger than the statement from Eq. (2), namely, that "the probability that $X$ is within 1.5 standard deviations of the mean is at least .555$\overline{5}$." Of course, inequality (2) is consistent with Eq. (1).

# 9-4. The Moment Generating Function

An alternative procedure for calculating moments is provided by the moment generating function. The symbol for the moment generating function for a random variable $X$ is $M_X(t)$.

## A. Definition of the moment generating function

**Definition 9.7:** The *moment generating function* for a random variable $X$, $M_X(t)$, is defined to be

$$M_X(t) = E(e^{tX}) \tag{1}$$

provided the expected value on the right exists. From the expressions for $E[g(X)]$ from Theorems 9.1 and 9.2, we have

$$M_X(t) = \sum_x e^{tx} f(x) \qquad \text{[if random variable } X \text{ is } discrete] \tag{2a}$$

$$M_X(t) = \int_x e^{tx} f(x)\, dx \qquad \text{[if random variable } X \text{ is } continuous] \tag{2b}$$

*notes*

(a) The moment generating function is abbreviated mgf. The symbol $M_X(t)$ indicates a function of the parameter $t$.
(b) We see that $M_X(t)$ is the expected value [or expectation] of $e^{tX}$. Observe that the exponent of the power of $e$ here is equal to the product of the subscript $X$ and the argument $t$ of the term $M_X(t)$.
(c) There is another function, the characteristic function, denoted by $C_X(t)$, which is often used instead of $M_X(t)$. Here, $C_X(t) = E(e^{itX})$, where $i$, which equals $\sqrt{-1}$, is the imaginary unit. There is a theoretical advantage to using $C_X(t)$. One reason is that it always exists for all values of $t$. In order to avoid calculations with complex numbers in this book, we shall restrict our attention to the moment generating function.

## B. The equation $E(X^r) = M_X^{(r)}(0)$

### Theorem 9.9 (Moments Generated from mgf Derivatives)

(a) $M_X(t)$ has the following series expansion in powers of $t$:

$$M_X(t) = 1 + E(X)t + \frac{E(X^2)}{2!}t^2 + \dots + \frac{E(X^r)}{r!}t^r + \dots$$

(b) $E(X^r) = \dfrac{d^r M_X(t)}{dt^r}\bigg|_{t=0}$ for $r = 1, 2, 3, \dots$

In words, the $r$th moment about the origin for random variable $X$ is equal to the $r$th derivative of $M_X(t)$, evaluated at $t = 0$.

*notes*

(i) The right side of Eq. (b) of Theorem 9.9 is often written as $M_X^{(r)}(0)$.
(ii) Observe that $M_X(0) = 1$. See also Problem 9-12.

---

**EXAMPLE 9-21:** Prove Theorem 9.9.

**Proof:** Let us do the proof for the case of a discrete random variable $X$. Thus,

$$M_X(t) = \sum_x e^{tx} f(x) \tag{1}$$

Recall that the Taylor series expansion of the function $e^z$ in powers of $z$ [also called a Maclaurin series expansion] is given by

$$e^z = 1 + z + \frac{z^2}{2!} + \frac{z^3}{3!} + \ldots + \frac{z^r}{r!} + \ldots = \sum_{r=0}^{\infty} \frac{z^r}{r!} \qquad (2)$$

This series converges for all values of $z$. Replacing $z$ by $tx$ in Eq. (2), we get

$$e^{tx} = 1 + tx + \frac{t^2}{2!}x^2 + \frac{t^3}{3!}x^3 + \ldots + \frac{t^r}{r!}x^r + \ldots = \sum_{r=0}^{\infty} \frac{t^r}{r!}x^r \qquad (3)$$

Substituting the expression for $e^{tx}$ from Eq. (3) into Eq. (1), we get

$$M_X(t) = \sum_x \left[ 1 + tx + \frac{t^2}{2!}x^2 + \ldots \right] f(x)$$

$$= \sum_x f(x) + t \sum_x xf(x) + \frac{t^2}{2!} \sum_x x^2 f(x) + \ldots + \frac{t^r}{r!} \sum_x x^r f(x) + \ldots \qquad (4)$$

We recognize that $\sum_x f(x) = 1$, and that the multiplier of $t^r/(r!)$ is $E(X^r)$ for $r = 1, 2, \ldots$. Thus, Eq. (4) may be written as

$$M_X(t) = 1 + E(X)t + \frac{E(X^2)}{2!}t^2 + \ldots + \frac{E(X^r)}{r!}t^r + \ldots \qquad (5)$$

Equation (5) is none other than part (a) of Theorem 9.9. Now, if a function $g(t)$ has a valid Taylor series expansion in powers of $t$ for $-r < t < r$, where $r$ is some positive number, then that Taylor series can be expressed in the form

$$g(t) = g(0) + g'(0)t + \frac{g''(0)}{2!}t^2 + \ldots + \frac{g^{(r)}(0)}{r!}t^r + \ldots \qquad (6)$$

Here, $g^{(r)}(0)$ represents the $r$th derivative of $g(t)$ evaluated at $t = 0$. Thus, for $M_X(t)$, we have the following series representation in powers of $t$:

$$M_X(t) = 1 + M_X'(0)t + \frac{M_X''(0)}{2!}t^2 + \ldots + \frac{M_X^{(r)}(0)}{r!}t^r + \ldots \qquad (7)$$

[The lead term in Eq. (7) was written as 1 since $M_X(0) = 1$. Also, observe that Eq. (7) is just a rewriting of Eq. (6), but with the function $g(t)$ being replaced by the function $M_X(t)$.]

Now, there must be a *unique* series representation of $M_X(t)$ in powers of $t$. Thus, by comparison of Eqs. (5) and (7), we see that

$$E(X^r) = M_X^{(r)}(0) \qquad \text{for } r = 1, 2, 3, \ldots \qquad (8)$$

Equation (8) is none other than part (b) of Theorem 9.9. $\qquad \square$

---

**EXAMPLE 9-22:** Show that $M_X(t) = [1 + p(e^t - 1)]^n$ for a binomial random variable $X$.

*Solution:* First, in general, for a discrete random variable $X$, we have

$$M_X(t) = \sum_x e^{tx} f(x)$$

The binomial random variable $X$ takes on integer values between 0 and $n$, inclusive. Using the equation for $f(x) = b(x; n, p)$ for a binomial distribution [Theorem 5.5], we have

$$M_X(t) = \sum_{x=0}^{n} e^{tx} \binom{n}{x} p^x (1-p)^{n-x} \qquad (1)$$

Now, from algebra,

$$e^{tx} p^x = (e^t)^x p^x = (pe^t)^x \qquad (2)$$

Using the result from Eq. (2) in Eq. (1), we get

$$M_X(t) = \sum_{x=0}^{n} \binom{n}{x} (pe^t)^x (1-p)^{n-x} \tag{3}$$

From the binomial theorem [Theorem 4.7′], we see that

$$\sum_{x=0}^{n} \binom{n}{x} a^{n-x} b^x = (a+b)^n \tag{4}$$

We obtain Eq. (4) by replacing $x$, $y$, and $r$ in Theorem 4.7′ by $a$, $b$, and $x$, respectively. Now, we can identify $pe^t$, and $(1-p)$ on the right side of Eq. (3) with $b$ and $a$, respectively, on the left side of Eq. (4). Thus, replacing $b$ by $pe^t$ and $a$ by $(1-p)$ in $(a+b)^n$ [refer to the right side of Eq. (4)], we see that $M_X(t)$ from Eq. (3) can be written as follows:

$$M_X(t) = [(1-p) + pe^t]^n = [1 + p(e^t - 1)]^n \tag{5}$$

---

**EXAMPLE 9-23:** Determine general expressions for $E(X) = \mu_x$ and $\text{Var}(X)$ for the binomial distribution.

*Solution:* From the final expression for $M_X(t)$ in Example 9-22, we obtain the following expression for $M_X'(t)$:

$$M_X'(t) = n[1 + p(e^t - 1)]^{n-1} \cdot \frac{dp(e^t - 1)}{dt}$$

$$= n[1 + p(e^t - 1)]^{n-1} \cdot pe^t \tag{1}$$

Applying the rule for finding the derivative of a product to Eq. (1), we have

$$M_X''(t) = n[1 + p(e^t - 1)]^{n-1} \cdot pe^t + n(n-1)[1 + p(e^t - 1)]^{n-2} \cdot pe^t \cdot pe^t \tag{2}$$

Thus, from part (b) of Theorem 9.9, and Eqs. (1) and (2), we have

$$E(X) = M_X'(0) = np \tag{3}$$

and

$$E(X^2) = M_X''(0) = n \cdot 1 \cdot p + n(n-1) \cdot 1 \cdot p^2 = np[1 + (n-1)p] \tag{4}$$

Now, from Theorem 9.5, we have

$$\text{Var}(X) = E(X^2) - [E(X)]^2 \tag{5}$$

Substituting the results from Eqs. (3) and (4) into Eq. (5), we get

$$\text{Var}(X) = np + np(n-1)p - n^2 p^2 = np - np^2 = np(1-p) = npq \tag{6}$$

Here, $q = 1 - p$. Thus, for the standard deviation for the binomial distribution, we have

$$\sigma_x = \sqrt{npq} \tag{7}$$

The expressions for $E(X)$ and $\text{Var}(X)$ of Eqs. (3) and (6) are very important, and are listed in Section 1-4 of Appendix B.

---

## C. An equation for $M_{aX+b}(t)$

Suppose the probability [density] function $f(x)$ of a random variable $X$ is known. Suppose further that random variable $Y$ is equal to $g(X)$, a function of $X$. Then, using Definition 9.7, but with $Y$ replacing $X$, we have

$$M_Y(t) = E(e^{tY}) = E(e^{tg(X)})$$

Now, let us regard $t$ as a constant. Thus, $e^{tg(X)}$ is a function of $X$, say $h(X)$. Thus, it follows from Theorems 9.1 and 9.2, as applied to this $h(X)$, that $M_Y(t)$, which equals $M_{g(X)}(t)$, is given as follows:

**Theorem 9.10**

$$M_{g(X)}(t) = E[e^{tg(X)}] = \begin{cases} \displaystyle\sum_x e^{tg(x)} f(x) & \text{[Discrete case]} \\ \displaystyle\int_x e^{tg(x)} f(x)\, dx & \text{[Continuous case]} \end{cases}$$

The following theorem is useful for derivation purposes.

**Theorem 9.11**

$$M_{aX+b}(t) = e^{bt} M_X(at)$$

where $a$ and $b$ are constants.

---

**EXAMPLE 9-24:** Prove Theorem 9.11.

**Proof:** From Theorem 9.10, we have $M_{g(X)}(t) = E[e^{tg(X)}]$. Let $g(X) = aX + b$. Thus,

$$M_{aX+b}(t) = E[e^{t(aX+b)}] = E[e^{(taX+tb)}] = E[e^{taX} e^{tb}] \tag{1}$$

Now, $e^{tb}$ is not a function of $X$. [For example, if we consider parameter $t$ to be a constant, the $e^{tb}$ will also be a constant.] Now, applying Theorem 9.3, we take $e^{tb}$ to the left of the $E$ symbol in the last expression of Eq. (1). Thus, we obtain

$$M_{aX+b}(t) = e^{tb} E[e^{taX}] \tag{2}$$

Now,

$$E[e^{taX}] = M_X(at) = M_{aX}(t) \tag{3}$$

[Remember the mgf "rule" for multiplying the subscript by the argument of the function. That is, $M_{\text{sub}}(\text{arg}) = E(e^{\text{sub}\cdot\text{arg}})$.] Using the middle term from Eq. (3) in Eq. (2), we get

$$M_{aX+b}(t) = e^{bt} M_X(at) \tag{4}$$

Notice that the argument of $M_X$ [see the right side] is the constant $a$ times $t$.

□

---

### D. A moment generating function uniqueness theorem

The major importance of the *moment generating function uniqueness theorem* stated below stems from the fact that it indicates that a particular moment generating function uniquely determines a probability distribution.

**Theorem 9.12 (A Moment Generating Function Uniqueness Theorem):** Given two random variables $X$ and $Y$, with moment generating functions $M_X(t)$ and $M_Y(t)$, respectively. If $M_X(t) = M_Y(t)$, that is, if $M_X(t)$ and $M_Y(t)$ are both equal to the same function of $t$, then $X$ and $Y$ have the same probability distribution. This means that if $X$ and $Y$ are both discrete random variables, then $X$ and $Y$ will have the same probability function, and if $X$ and $Y$ are both continuous random variables, then $X$ and $Y$ will have the same probability density function.

*notes*

(a) A rigorous proof of Theorem 9.12 is based on the theory of transforms of advanced calculus [or real analysis].

(b) A proof that $M_X(t)$ uniquely determines a probability function for the case of a discrete random variable $X$ which takes on a finite set of values $x_1, x_2, \ldots, x_n$ is given on pages 367–369 of Snell (1988).

(c) If $M_X(t) = M_Y(t)$, then from Theorem 9.9 that means that the moments about the origin for $r = 1, 2, 3, \ldots$ are respectively equal. That is, $E(X^r) = E(Y^r)$ for $r = 1, 2, 3, \ldots$. For that reason, it is then plausible to expect that the probability distributions for $X$ and $Y$ should be "essentially" identical.

---

**EXAMPLE 9-25:** Given that $M_V(t) = [1 + (.4)(e^t - 1)]^5$. Identify the probability distribution for the random variable $V$.

*Solution:* If we refer to Example 9-22 or to Section 1-4 of Appendix B, we see that $M_V(t)$ is the mgf of a binomial distribution with $p = .4$ and $n = 5$. To put it differently, $M_V(t) = M_X(t)$, where $X$ is the discrete random variable whose probability function is $b(x; n, p) = b(x; 5, .4)$. Thus, from Theorem 9.12, the random variable $V$ has a binomial distribution with $p = .4$ and $n = 5$.

---

## 9-5. Means, Variances, and Moment Generating Functions of Special Probability Distributions

Reviewing terminology, remember that we refer to the mean and moment generating function of a random variable as also being the mean and moment generating function of the related probability distribution. Similarly, for the variance, other moments, and other key terms [such as the standard deviation]. In Sections 1-4 and 1-5 of Appendix B, the key characteristics [means, variances, and moment generating functions] of several major probability distributions are tabulated.

### A. Means and variances for several major discrete probability distributions

In Example 9-23, we determined both $E(X)$ and $\text{Var}(X)$ for the binomial distribution by using $E(X^r) = M_X^{(r)}(0)$, where $M_X(t)$ is the moment generating function. Direct approaches for determining $E(X)$ and $\text{Var}(X)$ are presented in Examples 9-26, 9-28, and 9-29.

---

**EXAMPLE 9-26:** Show that $E(X) = np$ for the binomial distribution by making use of the probability function $b(x; n, p)$ for the binomial distribution, and the general defining equation for $E(X)$.

*Solution:* The probability function $f(x)$ or $b(x; n, p)$ for the binomial distribution is given in Theorem 5.5.

$$f(x) = b(x; n, p) = \frac{n!}{x!(n - x)!} p^x q^{n-x} \qquad \text{for } x = 0, 1, 2, \ldots, n \qquad \textbf{(1)}$$

Thus, it follows that

$$\sum_{x=0}^{n} f(x) = \sum_{x=0}^{n} \frac{n!}{x!(n - x)!} p^x q^{n-x} = 1 \qquad \textbf{(2)}$$

[This was verified in Example 5-20.] It is also true that

$$\sum_{r=0}^{\hat{n}} \frac{\hat{n}!}{r!(\hat{n} - r)!} p^r q^{\hat{n}-r} = 1 \qquad \textbf{(3)}$$

since the left side of Eq. (3) represents the sum of binomial probabilities of $r$ successes in $\hat{n}$ trials over all possible values of $r$, that is, for $r = 0, 1, 2, \ldots, \hat{n} - 1$, $\hat{n}$. Equation (3) will be useful when we simplify Eq. (9) in the development that follows.

From Definition 9.1,

$$E(X) = \sum_{x=0}^{n} xf(x) = \sum_{x=0}^{n} x \cdot \frac{n!}{x!(n-x)!} p^x q^{n-x} \qquad (4)$$

Separating out the $x = 0$ term [which is equal to zero], we get

$$E(X) = 0 \cdot q^n + \sum_{x=1}^{n} x \cdot \frac{n!}{x!(n-x)!} p^x q^{n-x} \qquad (5)$$

Now, we simplify Eq. (5) after observing that $x!/x = (x-1)!$, $n! = n \cdot (n-1)!$, and $p^x = p \cdot p^{(x-1)}$. Thus, after taking the product $np$ to the left of the $\Sigma$ sign, Eq. (5) becomes

$$E(X) = np \sum_{x=1}^{n} \frac{(n-1)!}{(x-1)!(n-x)!} p^{x-1} q^{n-x} \qquad (6)$$

Now, let us change the variable by using

$$r = x - 1 \qquad (7)$$

At the limits, $r = 0$ when $x = 1$ and $r = n - 1$ when $x = n$. Then, letting $\hat{n} = n - 1$, we see that

$$n - x = n - r - 1 = (n-1) - r = \hat{n} - r \qquad (8)$$

Using this information to replace $x$ and $n$ by $r$ and $\hat{n}$ in Eq. (6), we obtain

$$E(X) = np \sum_{r=0}^{\hat{n}} \frac{\hat{n}!}{r!(\hat{n} - r)!} p^r q^{\hat{n}-r} \qquad (9)$$

From Eq. (3), we see that the sigma term in Eq. (9) is equal to 1. Thus, Eq. (9) becomes

$$E(X) = np \cdot 1 = np \qquad (10)$$

---

**EXAMPLE 9-27:** The probability function for the hypergeometric distribution is

$$f(x) = h(x; n, a, M) = \frac{\binom{a}{x}\binom{M-a}{n-x}}{\binom{M}{n}} \qquad \text{for } x = 0, 1, 2, \ldots, n-1, n \qquad (1)$$

[See Theorem 5.7.] Observe that

$$\sum_{x=0}^{n} f(x) = \sum_{x=0}^{n} \frac{\binom{a}{x}\binom{M-a}{n-x}}{\binom{M}{n}} = 1 \qquad (1a)$$

Show that $E(X) = (na)/M$.

***Solution:*** Substituting the expression for $f(x)$ from Eq. (1) into $E(X) = \Sigma_x xf(x)$, we get

$$E(X) = \sum_{x=0}^{n} x \cdot \frac{\binom{a}{x}\binom{M-a}{n-x}}{\binom{M}{n}} \qquad (2)$$

As in the binomial case [Example 9-26], we separate out the $x = 0$ term in the summation of Eq. (2) [this term is equal to zero], and we observe that

$$x \cdot \binom{a}{x} = \frac{x \cdot a!}{x!(a-x)!} = \frac{a \cdot (a-1)!}{(x-1)!(a-x)!} = a \cdot \binom{a-1}{x-1} \tag{3}$$

Thus, after taking the constant $a$ to the left of the $\Sigma$ symbol in Eq. (2), we obtain

$$E(X) = a \sum_{x=1}^{n} \frac{\binom{a-1}{x-1}\binom{M-a}{n-x}}{\binom{M}{n}} \tag{4}$$

Now,

$$\binom{M}{n} = \frac{M}{n} \cdot \frac{(M-1)!}{(n-1)!(M-n)!} = \frac{M}{n} \cdot \binom{M-1}{n-1} \tag{5}$$

Substituting the result from Eq. (5) into Eq. (4), and then taking the fraction $n/M$ to the left of the $\Sigma$ symbol in Eq. (4), we obtain

$$E(X) = \frac{na}{M} \cdot \sum_{x=1}^{n} \frac{\binom{a-1}{x-1}\binom{M-a}{n-x}}{\binom{M-1}{n-1}} \tag{6}$$

Now, let

$$r = x - 1 \quad \text{and thus} \quad x = r + 1 \tag{7a), (7b}$$

We see that

$$r = 0 \quad \text{when} \quad x = 1 \quad \text{and} \quad r = n-1 \quad \text{when} \quad x = n \tag{8a), (8b}$$

Also, let

$$\hat{n} = n - 1 \qquad \hat{M} = M - 1 \quad \text{and} \quad \hat{a} = a - 1 \tag{9a), (9b), (9c}$$

Thus,

$$M - a = \hat{M} + 1 - (\hat{a}+1) = \hat{M} - \hat{a} \quad \text{and} \quad n - x = \hat{n} + 1 - (r+1) = \hat{n} - r \tag{10a), (10b}$$

Incorporating Eqs. (7) through (10) into Eq. (6), we get

$$E(X) = \frac{na}{M} \cdot \sum_{r=0}^{\hat{n}} \frac{\binom{\hat{a}}{r}\binom{\hat{M}-\hat{a}}{\hat{n}-r}}{\binom{\hat{M}}{\hat{n}}} \tag{11}$$

But, the summation on the right of Eq. (11) equals 1, since it is the sum of hypergeometric probabilities of $r$ successes in a draw of $\hat{n}$ objects over all possible values of $r$, that is, for $r = 0, 1, 2, \ldots, \hat{n} - 1, \hat{n}$. [See also Eq. (1a), which conveys a similar idea. There, one has a draw of $n$ objects, and the sum of hypergeometric probabilities of $x$ successes is over all possible values of $x$, that is, for $x = 0, 1, 2, \ldots, n$.] Thus, we have

$$E(X) = \frac{na}{M} \cdot 1 = \frac{na}{M} = n\left(\frac{a}{M}\right) \tag{12}$$

It is very useful to let $p = a/M$, where $p$ stands for the proportion of "successes" in the original set of $M$ objects. Thus, we have

$$E(X) = np \tag{13}$$

as with the binomial distribution. This last fact indicates that the mean number of successes obtained when drawing $n$ objects from a set of $M$ objects is the same whether the drawing is done with replacement [this leads to a binomial situation] or without replacement.

For calculations of Var(X), it is often useful to make use of Theorem 9.5, which states that

$$\text{Var}(X) = E(X^2) - [E(X)]^2$$

**EXAMPLE 9-28:** Show that $E(X^2) = E[X(X-1)] + E(X)$.

*Solution:* If we multiply out within $E[X(X-1)]$, and then use Theorem 9.4 [Linearity Rule for $E$], we get

$$E[X(X-1)] = E[X^2 - X] = E(X^2) - E(X) \qquad (1)$$

If we then solve for $E(X^2)$, we get the desired result.

Theorem 9.13 can be derived by substituting the result from Example 9-28 into Theorem 9.5.

**Theorem 9.13**

$$\text{Var}(X) = E[X(X-1)] + E(X) - [E(X)]^2$$

Theorem 9.13 is useful because it is often relatively easy to derive the expression for $E[X(X-1)]$ for a particular probability distribution. Each of the other two expressions in Theorem 9.13 are simple functions of $E(X) = \mu$.

**EXAMPLE 9-29:** **(a)** Derive the expression for $E[X(X-1)]$ for the binomial distribution. **(b)** Find the expression for $\text{Var}(X)$ for a binomial distribution.

*Solution*

**(a)** First,

$$E[X(X-1)] = \sum_{x=0}^{n} x(x-1)f(x) = \sum_{x=0}^{n} x(x-1)b(x;n,p) \qquad (1)$$

Now, both the $x=0$ and $x=1$ terms of the summation of Eq. (1) are equal to zero since the product $x(x-1)=0$ for both $x=0$ and $x=1$. Thus, we have

$$E[X(X-1)] = \sum_{x=2}^{n} x(x-1)b(x;n,p) \qquad (2)$$

Using the expression for $b(x;n,p)$ from Theorem 5.5, we get

$$E[X(X-1)] = \sum_{x=2}^{n} x(x-1)\frac{n!}{x!(n-x)!}p^x q^{n-x} \qquad (3)$$

Now, $x!/[x(x-1)] = (x-2)!$. Using this fact, and $n! = n(n-1) \times [(n-2)!]$, and $p^x = p^2 p^{x-2}$, and factoring $n(n-1)p^2$ to the left of the $\Sigma$ symbol yields

$$E[X(X-1)] = n(n-1)p^2 \sum_{x=2}^{n} \frac{(n-2)!}{(x-2)!(n-x)!}p^{x-2} q^{n-x} \qquad (4)$$

Letting $r = x-2$, and noting that $r=0$ when $x=2$, and $r=n-2$ when $x=n$, and, further, letting $\hat{n} = n-2$, we get

$$E[X(X-1)] = n(n-1)p^2 \sum_{r=0}^{\hat{n}} \frac{\hat{n}!}{r!(\hat{n}-r)!} p^r q^{\hat{n}-r} \tag{5}$$

The $\Sigma$ term is equal to 1 since this term is the same as the $\Sigma$ expression in Eq. (3) of Example 9-26. Thus,

$$E[X(X-1)] = n(n-1)p^2 \cdot 1 = n(n-1)p^2 \tag{6}$$

**(b)** Substituting the expression for $E[X(X-1)]$ from Eq. (6) and $E(X) = np$ into the equation for $\text{Var}(X)$ from Theorem 9.13, we get

$$\text{Var}(X) = n(n-1)p^2 + np - n^2 p^2 = np(1-p) = npq \tag{7}$$

We now summarize some key results for the binomial, hypergeometric, and Poisson distributions in Theorem 9.14.

**Theorem 9.14**

**(a)** For the binomial distribution, $E(X) = np$, and $\text{Var}(X) = npq$, where $q = 1 - p$.

**(b)** For the hypergeometric distribution, $E(X) = np$, and $\text{Var}(X) = npq\left(\dfrac{M-n}{M-1}\right)$. Here, $p$ is equal to $a/M$, the initial proportion of successes in the original set, which contains $M$ objects, of which $a$ are "successes." Also, $q = 1 - p$.

**(c)** For the Poisson distribution, with parameter $\lambda$, $E(X) = \lambda$ and $\text{Var}(X) = \lambda$.

The result for $\text{Var}(X)$ for the hypergeometric distribution is derived in Problem 9-15. The results for $E(X)$ and $\text{Var}(X)$ for the Poisson distribution are derived by direct approaches in Problem 9-16. In Example 9-31, we shall use a moment generating function approach for calculating $E(X)$ and $\text{Var}(X)$ for the Poisson distribution.

We see from a comparison of parts (a) and (b) of Theorem 9.14 that the variance of the hypergeometric distribution is smaller than the variance for a corresponding binomial distribution by the factor $(M-n)/(M-1) = (1 - n/M)/(1 - 1/M)$. If $n$ is much smaller than $M$ [in symbols, if $n \ll M$], then this factor will be close to 1 in value [recall that $n \geq 1$], and thus, the hypergeometric distribution variance will be close to the binomial distribution variance in value. This is yet another confirmation of the fact that the hypergeometric distribution is well approximated by the corresponding binomial distribution if $n \ll M$. [When we talk about "corresponding" here, we mean that the binomial value of $p$ is equal to the hypergeometric value of the ratio $a/M$.]

## B. Moment generating functions for key discrete probability distributions

**EXAMPLE 9-30:** Derive the mgf for a Poisson probability distribution.

*Solution:* From Definition 6.2, for a Poisson distribution with parameter $\lambda$, we have

$$f(x) = p(x; \lambda) = \frac{\lambda^x e^{-\lambda}}{x!} \qquad \text{for } x = 0, 1, 2, \ldots \tag{1}$$

Thus,

$$M_X(t) = \sum_x e^{tx} f(x) = \sum_{x=0}^{\infty} e^{tx} \frac{\lambda^x e^{-\lambda}}{x!} = e^{-\lambda} \sum_{x=0}^{\infty} \frac{(\lambda e^t)^x}{x!} \tag{2}$$

Here, we took $e^{-\lambda}$ to the left of the $\Sigma$ symbol, and we used the algebraic results that $e^{tx} = (e^t)^x$, and $(e^t)^x \cdot \lambda^x = (\lambda e^t)^x$. Now, recall that the Taylor series for $e^s$ in powers of $s$ is given by Eq. (3) below [for a similar result, see Eq. (2) of Example 9-21]:

$$e^s = 1 + s + \frac{s^2}{2!} + \dots + \frac{s^x}{x!} + \dots = \sum_{x=0}^{\infty} \frac{s^x}{x!} \qquad \text{[valid for all } s\text{]} \qquad (3)$$

Applying Eq. (3) to the final $\Sigma$ expression of Eq. (2) with $s = \lambda e^t$, we can reduce the $M_X(t)$ expression to

$$M_X(t) = e^{-\lambda} e^{\lambda e^t} = e^{(-\lambda + \lambda e^t)} = e^{\lambda(e^t - 1)} \qquad (4)$$

**EXAMPLE 9-31:** From the expression for $M_X(t)$ of Example 9-30, and Theorem 9.9, show that $E(X)$ and $Var(X)$ for the Poisson distribution are both equal to $\lambda$.

*Solution:* First,

$$M_X'(t) = \frac{de^{\lambda(e^t - 1)}}{dt} \qquad (1)$$

From the chain-rule of calculus,

$$\frac{de^u}{dt} = e^u \frac{du}{dt} \qquad (2)$$

Applying Eq. (2) to Eq. (1), with $u = \lambda(e^t - 1)$, and thus, $du/dt = \lambda e^t$, we get

$$M_X'(t) = e^{\lambda(e^t - 1)} \cdot \lambda e^t = \lambda e^{[t + \lambda(e^t - 1)]} \qquad (3)$$

Taking the derivative of $M_X'(t)$ as given in Eq. (3), we get

$$M_X''(t) = \lambda e^{[t + \lambda(e^t - 1)]} \cdot \frac{d[t + \lambda(e^t - 1)]}{dt} = \lambda e^{[t + \lambda(e^t - 1)]}(1 + \lambda e^t) \qquad (4)$$

From Eqs. (3) and (4), we get

$$E(X) = M_X'(0) = \lambda e^0 = \lambda \qquad (5)$$

and

$$E(X^2) = M_X''(0) = \lambda e^0[1 + \lambda] = \lambda + \lambda^2 \qquad (6)$$

Now, substituting the results from Eqs. (5) and (6) into Theorem 9.5, we get

$$Var(X) = E(X^2) - [E(X)]^2 = \lambda + \lambda^2 - \lambda^2 = \lambda \qquad (7)$$

## C. Means and variances for key probability density functions

We shall illustrate direct approaches for determining $E(X)$, $E(X^2)$, and $Var(X)$ for some key probability density functions.

**EXAMPLE 9-32:** For the exponential distribution [Definition 8.4], $f(x) = (1/\theta)e^{-x/\theta}$ for $x > 0$, and $f(x) = 0$ for $x \le 0$. Derive the equation for $E(X)$.

*Solution:* From Definition 9.2, and the given expressions for $f(x)$, we have

$$E(X) = \int_{-\infty}^{\infty} xf(x)\,dx = \frac{1}{\theta}\int_0^{\infty} xe^{-x/\theta}\,dx \qquad (1)$$

At this point, we can use a table of integrals or integration by parts. Let's use the latter approach here. First, let

$$u = x \qquad \text{and} \qquad dv = e^{-x/\theta}\,dx \qquad \text{(2a), (2b)}$$

Then,

$$du = dx \quad \text{and} \quad v = \int e^{-x/\theta}\,dx = -\theta e^{-x/\theta} \qquad \textbf{(3a), (3b)}$$

Thus, for the definite integral on the rightmost part of Eq. (1), we have

$$\int_{x=0}^{\infty} x e^{-x/\theta}\,dx = uv\bigg]_{x=0}^{\infty} - \int_{x=0}^{\infty} v\,du = (-\theta x e^{-x/\theta})\bigg]_{x=0}^{\infty} + \theta \int_{x=0}^{\infty} e^{-x/\theta}\,dx$$

$$= \lim_{x\to\infty} \frac{-\theta x}{e^{x/\theta}} + 0 - \theta^2 e^{-x/\theta}\bigg]_{x=0}^{\infty} \qquad \textbf{(4)}$$

From L'Hôpital's rule, we have

$$\lim_{x\to\infty} \frac{-\theta x}{e^{x/\theta}} = \lim_{x\to\infty} \frac{-\theta}{\left(\dfrac{1}{\theta}\right)e^{x/\theta}} = 0 \qquad \textbf{(5)}$$

Also,

$$-\theta^2 e^{-x/\theta}\bigg]_{x=0}^{\infty} = \lim_{x\to\infty} \frac{-\theta^2}{e^{x/\theta}} + \theta^2 \cdot 1 = 0 + \theta^2 = \theta^2 \qquad \textbf{(6)}$$

Substituting the results from Eqs. (5) and (6) into Eq. (4), we get

$$\int_{x=0}^{\infty} x e^{-x/\theta}\,dx = \theta^2 \qquad \textbf{(7)}$$

Finally, substituting the result from Eq. (7) into Eq. (1), we get

$$E(X) = \left(\frac{1}{\theta}\right)\theta^2 = \theta \qquad \textbf{(8)}$$

---

For the exponential distribution,

$$E(X^2) = \left(\frac{1}{\theta}\right)\int_{x=0}^{\infty} x^2 e^{-x/\theta}\,dx$$

and this leads to $E(X^2) = 2\theta^2$. [For example, use a table of integrals and L'Hôpital's rule.] Thus, it follows that $\text{Var}(X) = E(X^2) - [E(X)^2] = 2\theta^2 - \theta^2 = \theta^2$.

**Theorem 9.15:** For the exponential distribution as given by $f(x) = (1/\theta)e^{-x/\theta}$ for $x > 0$, and $f(x) = 0$ for $x \le 0$, we have $E(X) = \theta$ and $\text{Var}(X) = \theta^2$. Thus, it follows that $\sigma_x = \theta$.

The gamma probability density function and the related gamma function were discussed in Sections 8-3A and 8-3B.

**Theorem 9.16:** For the gamma probability distribution, the $r$th moment about the origin, $E(X^r)$, is given by

$$E(X^r) = \frac{\beta^r \Gamma(\alpha + r)}{\Gamma(\alpha)}$$

Recall that $\Gamma$ indicates the gamma function, which is defined by the equation

$$\Gamma(\alpha) = \int_{0}^{\infty} t^{(\alpha-1)}e^{-t}\,dt \qquad \text{for } \alpha > 0$$

[See Definition 8.6.]

**EXAMPLE 9-33:** Prove Theorem 9.16.

***Proof:*** From Definition 9.3 for $E(X^r)$, with $f(x)$ being that for the gamma distribution [see Definition 8.7], we have

$$E(X^r) = \int_{-\infty}^{\infty} x^r f(x)\, dx = \int_0^{\infty} x^r \frac{1}{\beta^{\alpha}\Gamma(\alpha)} x^{\alpha-1} e^{-x/\beta}\, dx \tag{1}$$

If we let $y = x/\beta$ [recall that $\beta > 0$], then we have $x^{r+\alpha-1} = \beta^{r+\alpha-1} y^{r+\alpha-1}$, $e^{-x/\beta} = e^{-y}$, and $dx = \beta\, dy$. Also, when $x = 0$, then $y = 0$, and when $x \to \infty$, then $y \to \infty$. Thus, replacing $x$ terms by $y$ terms in Eq. (1), we obtain

$$E(X^r) = \frac{\beta^r}{\Gamma(\alpha)} \int_0^{\infty} y^{\alpha+r-1} e^{-y}\, dy \tag{2}$$

From Definition 8.6, we see that the integral on the right side of Eq. (2) is $\Gamma(r + \alpha)$. Thus, Eq. (2) may be rewritten as

$$E(X^r) = \frac{\beta^r \Gamma(\alpha + r)}{\Gamma(\alpha)} \tag{3} \quad \square$$

*note:* From the definition of the gamma function as presented in Definition 8.6, we would have $\Gamma(\alpha + r) = \int_0^{\infty} t^{\alpha+r-1} e^{-t}\, dt$. But, the integration variable $t$ can be changed to virtually any symbol; in particular, $y$ was used in Eq. (2) above. This is an illustration of the fact that the value of a definite integral depends on the integrand function, and the limits of integration, and not on the variable of integration. For example, each of the definite integrals $\int_1^3 2x\, dx$, $\int_1^3 2t\, dt$, and $\int_1^3 2y\, dy$ has the value of 8.

**Theorem 9.17:** The mean and variance of the gamma distribution are given by $\mu = \alpha\beta$ and $\sigma^2 = \alpha\beta^2$. Recall that $\alpha > 0$, and $\beta > 0$.

**EXAMPLE 9-34:** Prove Theorem 9.17.

***Proof:*** Letting $r = 1$ in Theorem 9.16, we get

$$\mu = E(X) = E(X^1) = \frac{\beta^1 \Gamma(\alpha + 1)}{\Gamma(\alpha)} \tag{1}$$

From part (b) of Theorem 8.6, $\Gamma(\alpha + 1) = \alpha\Gamma(\alpha)$, and thus Eq. (1) becomes

$$\mu = \frac{\beta\alpha\Gamma(\alpha)}{\Gamma(\alpha)} = \beta\alpha \tag{2}$$

Also, from Theorem 9.16, after letting $r = 2$, we get

$$E(X^2) = \frac{\beta^2 \Gamma(\alpha + 2)}{\Gamma(\alpha)} \tag{3}$$

Applying part (b) of Theorem 8.6 twice in succession, we have

$$\Gamma(\alpha + 2) = (\alpha + 1)\Gamma(\alpha + 1) = (\alpha + 1)\alpha\Gamma(\alpha) \tag{4}$$

Substituting the expression for $\Gamma(\alpha + 2)$ from Eq. (4) into Eq. (3), we get

$$E(X^2) = \beta^2(\alpha + 1)\alpha \tag{5}$$

Substituting the results from Eqs. (2) and (5) into Theorem 9.5, we obtain

$$\mathrm{Var}(X) = E(X^2) - [E(X)]^2 = \beta^2(\alpha + 1)\alpha - \beta^2\alpha^2 = \alpha\beta^2 \tag{6} \quad \square$$

Recall that the exponential distribution is a gamma distribution with $\alpha = 1$ and $\beta = \theta$, and the chi-square distribution is a gamma distribution with $\alpha = v/2$ and $\beta = 2$. Thus, we can easily obtain the equations for the means and variances of these distributions by using Theorem 9.17.

**Theorem 9.18:** The mean and variance of the chi-square distribution with $v$ degrees of freedom are $E(X) = v$ and $\text{Var}(X) = 2v$.

## D. Moment generating functions for key probability density functions

---

**EXAMPLE 9-35:** Determine the mgf for the exponential distribution for which $f(x) = 3e^{-3x}$ for $x > 0$ and $f(x) = 0$ for $x \le 0$. For this exponential distribution, the parameter $\theta = \frac{1}{3}$.

***Solution:*** Since $M_X(t) = E(e^{tX})$, we have

$$M_X(t) = \int_{-\infty}^{\infty} e^{tx} f(x)\, dx = 3 \int_0^{\infty} e^{tx} e^{-3x}\, dx = 3 \int_0^{\infty} e^{(t-3)x}\, dx \qquad (1)$$

Let us suppose, tentatively, that $t \ne 3$, i.e., that $(t - 3) \ne 0$. Now, if $a \ne 0$,

$$\int_{x=x_1}^{x_2} e^{ax}\, dx = \frac{e^{ax}}{a} \bigg]_{x_1}^{x_2} \qquad (2)$$

Now, if we apply Eq. (2) to Eq. (1), after making note that we have an improper integral in the former, we obtain

$$M_X(t) = \frac{3e^{(t-3)x}}{(t-3)} \bigg]_{x=0}^{\infty} = \frac{3}{(t-3)} \lim_{x \to \infty} e^{(t-3)x} - \frac{3}{(t-3)} \qquad (3)$$

Now, for $t < 3$, we have $(t - 3) < 0$, and hence $\lim_{x \to \infty} e^{(t-3)x} = 0$. Also, for $t > 3$, we have $(t - 3) > 0$, and hence $\lim_{x \to \infty} e^{(t-3)x} = \infty$. Thus, we can say that

$$M_X(t) = \frac{-3}{(t-3)} = \frac{3}{(3-t)} = \frac{1}{(1 - t/3)} \qquad \text{for } t < 3 \qquad (4)$$

but $M_X(t)$ does not exist for $t > 3$.

For $t = 3$, the integrand in the rightmost expression of Eq. (1) is no longer an exponential function since $e^{(3-3)x} = e^0 = 1$. Thus, from Eq. (1), we have for $t = 3$,

$$M_X(t) = 3 \int_0^{\infty} dx = 3x \bigg]_{x=0}^{\infty} = 3 \lim_{x \to \infty} x - 0 = \infty \qquad (5)$$

Thus, $M_X(t)$ does not exist for $t = 3$.

---

In similar fashion, we can find $M_X(t)$ for the exponential distribution for any value of the parameter $\theta$ [above, $\theta = \frac{1}{3}$].

**Theorem 9.19:** The moment generating function for the exponential distribution with parameter $\theta$ [where $\theta > 0$] is given by

$$M_X(t) = \frac{1}{(1 - \theta t)} = (1 - \theta t)^{-1} \qquad \text{for } t < \frac{1}{\theta}$$

Recall that the mean, variance, and moment generating function for key probability distributions are tabulated in Sections 1-4 and 1-5 of Appendix B.

**Theorem 9.20:** The moment generating function for the gamma probability density function is

$$M_X(t) = (1 - \beta t)^{-\alpha} \qquad \text{for } t < \frac{1}{\beta}$$

Observe that Theorem 9.19 is a special case of Theorem 9.20 since the exponential distribution is a gamma distribution with $\alpha = 1$ and $\beta = \theta$. Also, since the chi-square distribution is a gamma distribution with $\alpha = v/2$ and $\beta = 2$, we have Theorem 9.21.

**Theorem 9.21:** The moment generating function for the chi-square distribution is

$$M_X(t) = (1 - 2t)^{-v/2} \qquad \text{for } t < \frac{1}{2}$$

---

**EXAMPLE 9-36:** Determine the moment generating function for the gamma distribution.

**Solution:** Substituting the equations for $f(x)$ for the gamma distribution [Definition 8.7] into the definition for $M_X(t)$ [Definition 9.7], we get

$$M_X(t) = E(e^{tX}) = \int_{-\infty}^{\infty} e^{tx} f(x)\, dx$$

$$= \int_0^{\infty} e^{tx} \frac{1}{\beta^\alpha \Gamma(\alpha)} x^{\alpha-1} e^{-x/\beta}\, dx \qquad (1)$$

Thus, we get

$$M_X(t) = \frac{1}{\beta^\alpha \Gamma(\alpha)} \int_0^{\infty} x^{\alpha-1} e^{-x(1/\beta - t)}\, dx \qquad (2)$$

Let

$$y = x\left(\frac{1}{\beta} - t\right) = \frac{x(1 - \beta t)}{\beta} \qquad (3)$$

Thus, we get

$$dy = dx\left(\frac{1}{\beta} - t\right) = dx\frac{(1 - \beta t)}{\beta} \qquad \text{and} \qquad x^{\alpha-1} = \frac{y^{\alpha-1}\beta^{\alpha-1}}{(1 - \beta t)^{\alpha-1}} \qquad \textbf{(4a), (4b)}$$

Now, let us stipulate that $[(1/\beta) - t] > 0$, or, equivalently, that $t < 1/\beta$. Thus, from Eq. (3), we see that $y = 0$ when $x = 0$, and $y \to \infty$ when $x \to \infty$. Substitution of Eqs. (3), (4a), and (4b) into Eq. (2), along with a change from $x$ limits to $y$ limits, leads to

$$M_X(t) = \frac{\beta\beta^{\alpha-1}}{\beta^\alpha \Gamma(\alpha)(1 - \beta t)^{\alpha-1}(1 - \beta t)} \int_0^{\infty} y^{\alpha-1} e^{-y}\, dy \qquad (5)$$

The integral in Eq. (5) is equal to $\Gamma(\alpha)$ for $\alpha > 0$. Using that fact, and doing further simplifications of Eq. (5) leads to

$$M_X(t) = \frac{1}{(1 - \beta t)^\alpha} = (1 - \beta t)^{-\alpha} \qquad \text{for } t < \frac{1}{\beta} \qquad (6)$$

Note that, for $t \geq 1/\beta$, it can be shown that the improper integral in Eq. (2) diverges, and thus $M_X(t)$ doesn't exist in this case.

---

**EXAMPLE 9-37:** Use the expression for $M_X(t)$ for the gamma distribution, and Theorems 9.9 and 9.5, to determine the equations for $\mu$ and $\sigma^2$ for the gamma distribution.

*Solution:* We take the derivative twice in succession of the expression for $M_X(t)$ from Theorem 9.20. Therefore,

$$M_X(t) = (1 - \beta t)^{-\alpha} \tag{1}$$

$$M'_X(t) = (-\alpha)(1 - \beta t)^{-\alpha-1}(-\beta) = (\alpha\beta)(1 - \beta t)^{-\alpha-1} \tag{2}$$

$$M''_X(t) = (\alpha\beta)(-\alpha - 1)(-\beta)(1 - \beta t)^{-\alpha-2} = (\alpha\beta^2)(\alpha + 1)(1 - \beta t)^{-\alpha-2} \tag{3}$$

Thus,

$$E(X) = M'_X(0) = \alpha\beta \quad \text{and} \quad E(X^2) = M''_X(0) = (\alpha\beta^2)(\alpha + 1) \tag{4, 5}$$

As a result, from Theorem 9.5 and Eqs. (4) and (5),

$$\text{Var}(X) = E(X^2) - [E(X)]^2 = (\alpha\beta^2)(\alpha + 1) - \alpha^2\beta^2 = \alpha\beta^2 \tag{6}$$

These results agree with those derived in Example 9-34.

# 9-6. Practical Situations Involving Expected Values

## A. The Poisson process revisited

**EXAMPLE 9-38:** Determine the expression for the mean number of occurrences in a Poisson process of length $t$ if the coefficient is $\alpha$.

*Solution:* Recall from Theorem 8.4 in Section 8-2C that in a Poisson process with coefficient $\alpha$, the probability of $x$ occurrences [successes] in a time interval of length $t$ is given by

$$f_t(x) = \frac{e^{-\alpha t}(\alpha t)^x}{x!} \quad \text{for } x = 0, 1, 2, 3, \ldots \tag{1}$$

Thus, $f_t(x)$ is the Poisson probability function with parameter $\lambda = \alpha t$. From part (c) of Theorem 9.14, the mean and variance for a Poisson distribution with parameter $\lambda$ are both equal to $\lambda$. Thus, for a Poisson process of length $t$, the mean is given by

$$E(X) = \alpha t \tag{2}$$

In particular,

$$E(X) = \alpha \quad \text{if } t = 1 \tag{3}$$

That is, the mean number of occurrences in a time interval whose length is 1 unit is $\alpha$ occurrences. Thus, $\alpha$ denotes the mean number of occurrences per time unit. For example, if $t = 1$ minute, then $\alpha$ is the mean number of occurrences per minute.

**EXAMPLE 9-39:** Given a Poisson process with coefficient $\alpha$. Show that the mean waiting between occurrences is given by $1/\alpha$.

*Solution:* From Theorem 8.5 in Section 8-2D, the pdf for $Y$, the waiting time between occurrences in a Poisson process is given by $f(y) = \alpha e^{-\alpha y}$ for $y > 0$, and $f(y) = 0$ for $y \le 0$. Thus, the pdf for $Y$ is that of an exponential distribution with parameter $\theta = 1/\alpha$.

From Theorem 9.15, the mean for an exponential distribution with parameter $\theta$ is given by $E(X) = \theta$. It follows that

$$E(Y) = \mu_Y = \frac{1}{\alpha} \tag{1}$$

**EXAMPLE 9-40:** Assume that a Poisson process applies here. The mean number of telephone calls arriving at a switchboard in a 2-minute period is 12. **(a)** Determine the coefficient $\alpha$, in occurrences per minute. **(b)** Determine the expected waiting time between telephone calls. **(c)** What is the mean number of telephone calls arriving in a 3-minute period? **(d)** What is the probability of receiving at most 14 telephone calls in a 3-minute period?

*Solution*

**(a)** Recall that $E(X) = \alpha t$. Thus, for a 2-minute period, we have

$$E(X) = \alpha 2 = 12 \tag{1}$$

From Eq. (1), we find $\alpha = 6$ occurrences/minute.

**(b)** From Example 9-39, the expected waiting time is given by $E(Y) = 1/\alpha$. Thus, here, we have $E(Y) = 1/6 = .166\overline{6}$ minute as the expected waiting time between occurrences.

**(c)** For $t = 3$ minutes,

$$E(X) = \alpha t = 6 \cdot 3 = 18 \text{ occurrences} \tag{2}$$

**(d)** From Example 9-38, for example, we see that the number of calls arriving has a Poisson distribution with $\lambda = \alpha t = 6 \cdot 3 = 18$. From Table IV of Appendix A, for $\lambda = 18$, we have

$$P(X \leq 14) = .2080 \tag{3}$$

## B.  More applications involving the expected value concept

**EXAMPLE 9-41:** Find the expected number of men on a committee of three which is randomly selected from a set containing five men and four women.

*Solution:* Let $X$ be the number of men on the committee. Thus, $X$ is a hypergeometric random variable with probability function

$$f(x) = \frac{\binom{5}{x}\binom{4}{3-x}}{\binom{9}{3}} \qquad \text{for } x = 0, 1, 2, 3 \tag{1}$$

From part (b) of Theorem 9.14, we have $E(X) = np$, where $n = 3$, and $p = \frac{5}{9}$, the proportion of men in the original set of nine persons. Thus, here, for the expected number, we have

$$E(X) = 3(\tfrac{5}{9}) = \tfrac{5}{3} = 1.66\overline{6} \tag{2}$$

**EXAMPLE 9-42:** The diameter, measured in mm, of circular disks made by a certain machine is a random variable $X$ with pdf given by $f(x) = k(x - 20)^2$ for $19 < x < 21$, and $f(x) = 0$, elsewhere. **(a)** Determine $k$. **(b)** Determine $E(X)$, the expected diameter. **(c)** Determine the expected face area of the disk.

*Solution*

**(a)** From

$$\int_{-\infty}^{\infty} f(x)\,dx = k \int_{19}^{21} (x - 20)^2 \, dx = 1 \tag{1}$$

we find that $k = 3/2$.

**(b)** For $E(X)$, the expected or mean diameter, we have

$$E(X) = \int_{-\infty}^{\infty} xf(x)\,dx = \left(\frac{3}{2}\right)\int_{19}^{21} x(x-20)^2\,dx$$

$$= \left(\frac{3}{2}\right)\int_{19}^{21} (x^3 - 40x^2 + 400x)\,dx = 20 \qquad (2)$$

Thus, the expected diameter is 20 mm.

(c) The face area, $a$, in $(mm)^2$, of a circle of diameter $x$ mm is given by $a = \pi x^2/4$. Let $A$ symbolize the face area random variable for a circular disk. We have $A = \pi X^2/4$, where $X$ is the diameter random variable. Thus, applying Theorem 9.2, we have

$$E(A) = \int_{x=19}^{21} af(x)\,dx = \left(\frac{\pi}{4}\right)\left(\frac{3}{2}\right)\int_{x=19}^{21} x^2(x-20)^2\,dx$$

$$= \left(\frac{3\pi}{8}\right)\int_{x=19}^{21} (x^4 - 40x^3 + 400x^2)\,dx = 314.6305 \qquad (3)$$

Thus, the expected face area of the circular disk is 314.6305 $(mm)^2$.

---

***note:*** Part (b) of Example 9-42 can be done by inspection. Since $f(x)$ is symmetrical with respect to the vertical line $x = 20$, this means that $E(X) = 20$. More generally, if $f(x)$, whether a pf or a pdf, is symmetrical with respect to the vertical line $x = c$, then $E(X) = c$.

# SOLVED PROBLEMS

**Expected Values of Random Variables**

**PROBLEM 9-1** Two coins are drawn from a box containing a nickel, two dimes, and a quarter. Let $X$ stand for the sum of values of the two coins. Determine the probability function and expected value of $X$.

***Solution*** We assume that two coins are drawn without replacement. Let the sample space consist of the six *equally likely combinations* obtained when drawing two objects from four objects. $\left[\text{Here,}\binom{4}{2} = 6.\right]$ We distinguish between the two dimes by writing $10_a$ and $10_b$.

| Element or Combination | $x$ | Probability |
|---|---|---|
| 5, $10_a$ | 15 | 1/6 |
| 5, $10_b$ | 15 | 1/6 |
| 5, 25 | 30 | 1/6 |
| $10_a$, $10_b$ | 20 | 1/6 |
| $10_a$, 25 | 35 | 1/6 |
| $10_b$, 25 | 35 | 1/6 |

**Table of Probability Function**

| $x$ | $f(x)$ | $xf(x)$ |
|---|---|---|
| 15 | 2/6 | 30/6 |
| 20 | 1/6 | 20/6 |
| 30 | 1/6 | 30/6 |
| 35 | 2/6 | 70/6 |
| Sum = | 1 | 150/6 = 25 |

The probability function is displayed in the table on the right in the $x$ and $f(x)$ columns. From the sum in the third column of the table on the right, we see that the expected value of $X$ is given by $E(X) = \Sigma_x\, xf(x) = 25$ cents.

**PROBLEM 9-2** Prove Theorem 9.1 for the case where $X$ is a discrete random variable that takes on a finite set of values.

***Solution*** Our approach will be a generalization of that used in Example 9-5. We are given that $Y = g(X)$, where this equation expresses random variable $Y$ in terms of discrete random variable $X$. Further, we suppose that the probability function $f(x)$ for random variable $X$ is known. Suppose that $Y$ takes on the value $y_i$ when $X$ takes on the $n_i$ values $x_{i1}, x_{i2}, \ldots, x_{in_i}$, and that the possible values for $Y$ are $y_1, y_2, \ldots, y_m$. Thus, for example, we have

$$y_i = g(x_{i1}) = g(x_{i2}) = \ldots = g(x_{in_i}) \tag{1}$$

and Eq. (1) holds for $i = 1, 2, \ldots, m$. Now,

$$E(Y) = \sum_y yP(Y = y) = \sum_{i=1}^m y_iP(Y = y_i) \tag{2}$$

Now, the event $Y = y_i$ occurs when $X = x_{i1}$, or $x_{i2}, \ldots$, or $x_{in_i}$, and the events $X = x_{i1}$, $X = x_{i2}, \ldots$, and $X = x_{in_i}$ are mutually exclusive. Thus,

$$P(Y = y_i) = P(X = x_{i1}) + P(X = x_{i2}) + \ldots + P(X = x_{in_i}) \tag{3}$$

for $i = 1, 2, \ldots, m$. Thus,

$$E(Y) = \sum_{i=1}^m y_i[P(X = x_{i1}) + P(X = x_{i2}) + \ldots + P(X = x_{in_i})]$$

$$= \sum_{i=1}^m [y_iP(X = x_{i1}) + y_iP(X = x_{i2}) + \ldots + y_iP(X = x_{in_i})] \tag{4}$$

Now, let us successively replace the $y_i$ terms in the last expression of Eq. (4) by $g(x_{i1})$, $g(x_{i2})$, $\ldots$, and $g(x_{in_i})$. Refer to Eq. (1), which states that $y_i$ is equal to each one of these $n_i$ terms. Also, let us replace each $P(X = x_{ij})$ term by $f(x_{ij})$. Thus, Eq. (4) becomes

$$E(Y) = \sum_{i=1}^m [g(x_{i1})f(x_{i1}) + g(x_{i2})f(x_{i2}) + \ldots + g(x_{in_i})f(x_{in_i})] \tag{5}$$

Expanding the summation in Eq. (5) for $i = 1, 2, \ldots, m$, we get

$$E(Y) = [g(x_{11})f(x_{11}) + g(x_{12})f(x_{12}) + \ldots + g(x_{1n_1})f(x_{1n_1})]$$
$$+ [g(x_{21})f(x_{21}) + g(x_{22})f(x_{22}) + \ldots + g(x_{2n_2})f(x_{2n_2})] + \ldots$$
$$+ [g(x_{m1})f(x_{m1}) + g(x_{m2})f(x_{m2}) + \ldots + g(x_{mn_1})f(x_{mn_1})] \tag{6}$$

But, the values $x_{11}, x_{12}, \ldots, x_{1n_1}, x_{21}, x_{22}, \ldots, x_{2n_2}, \ldots, x_{m1}, x_{m2}, \ldots, x_{mn_m}$ constitute all possible values of random variable $X$. Thus, Eq. (6) may be rewritten as

$$E(Y) = \sum_x g(x)f(x) \tag{7}$$

The left side of Eq. (7) may be replaced by $E[g(X)]$ since $Y = g(X)$.

**PROBLEM 9-3** In the gambling game "Coney Island" or "Chuck-a-Luck," a player bets a certain amount of money, say $\$K$, at a gambling house [for example, a casino], and chooses one of the numbers 1, 2, 3, 4, 5, or 6. Next, three dice are rolled. If the chosen number occurs 1, 2, or 3 times, the player wins $\$2K$, $\$3K$, or $\$4K$, respectively. If the chosen number doesn't occur, then the player loses the $\$K$ that was bet. Let random variable $G$ indicate the gain for the player. Determine $E(G)$, the expected gain for the player, if $K$ is 5.

***Solution*** The possible values for $G$, in dollars, are $-5$ [if the player loses the bet], and $+5$, $+10$, and $+15$ if the player's number comes up once, twice, or three times, respectively. The probabilities for the different values of $G$ are equal to binomial probabilities for a binomial situation in which $n = 3$ and $p = 1/6$. Let the binomial random variable $X$ stand for the number of times that the chosen number occurs in three trials, and let $b(x; 3, 1/6)$ be the corresponding binomial probability. Thus,

$$P(G = -5) = b(0; 3, 1/6) = \left(\frac{5}{6}\right)^3 = \frac{125}{216} \tag{1}$$

$$P(G = 5) = b(1; 3, 1/6) = 3\left(\frac{1}{6}\right)\left(\frac{5}{6}\right)^2 = \frac{75}{216} \tag{2}$$

Likewise, $P(G = 10) = b(2; 3, 1/6)$ and $P(G = 15) = b(3; 3, 1/6)$. Thus, we end up with the following table, where $f(g)$ denotes the probability associated with the value $g$ of the random variable $G$.

| $g$ | $f(g)$ | $gf(g)$ |
|---|---|---|
| $-5$ | 125/216 | $-625/216$ |
| 5 | 75/216 | 375/216 |
| 10 | 15/216 | 150/216 |
| 15 | 1/216 | 15/216 |
| Sum = | 216/216 = 1 | $-85/216$ |

Thus, we see that $E(G) = -85/216 = -\$.3935$. This means that if the player bets \$5 over and over again for many rolls of the three dice, then the player will lose an average of \$.3935 or 39.35 cents for each \$5 bet. Thus, on average, the house [or casino] will win \$.3935 for each \$5 bet by the player. Since .3935 is 7.870% of \$5, the "house's percentage" is $+7.870\%$ and the "player's percentage" is $-7.870\%$ for this game of "Chuck-a-Luck."

**PROBLEM 9-4** In a gambling game, for a particular bet of $B$ dollars and an associated expected gain, $E(G)$ dollars, the house percentage $H$ is defined by the equation

$$H = -\frac{100E(G)}{B} \tag{I}$$

The quantity $H$ is the average [expected] percentage of $B$ that the house will win. Use this equation to compute $H$ for the gambling game of Problem 9-3.

*Solution* There, $B = 5$, and $E(G) = -\frac{85}{216}$. Thus, using Eq. (I) above,

$$H = -\frac{100(-85/216)}{5} = 7.870\%$$

as in Problem 9-3.

**PROBLEM 9-5** Iron ingots produced in a metallurgical foundry have diameter $x$ cm, where the probability density function is $f(x) = (\frac{3}{4})(6x - 8 - x^2)$ for $2 < x < 4$ and $f(x) = 0$, elsewhere. Determine the expected value of (a) the diameter and (b) the cross-sectional area of iron ingots. [For the latter term, let $A$ denote the cross-sectional area random variable.]

*Solution*

(a)
$$E(X) = \int_x xf(x)\,dx = \left(\frac{3}{4}\right)\int_{x=2}^{4}[6x^2 - 8x - x^3]\,dx = \left(\frac{3}{4}\right)[2x^3 - 4x^2 - x^4/4]\Big|_2^4$$

$$= \left(\frac{3}{4}\right)4 = 3 \text{ cm} \tag{1}$$

(b) The cross-sectional area $a$, in (cm)$^2$, is related to diameter $x$, in cm, by $a = \pi x^2/4$. One way of finding $E(A)$ is by using $E(A) = E(\pi X^2/4) = (\pi/4)E(X^2)$, where for $E(X^2)$, we have

$$E(X^2) = \int_x x^2 f(x)\,dx = \left(\frac{3}{4}\right)\int_2^4 [6x^3 - 8x^2 - x^4]\,dx$$

$$= \left(\frac{3}{4}\right)[3x^4/2 - 8x^3/3 - x^5/5]\Big|_2^4 = 9.2 \tag{2}$$

Thus, $E(A) = (\pi/4)E(X^2) = (\pi/4)9.2 = 7.2257 \text{ (cm)}^2$.

**PROBLEM 9-6** Given two random variables $X$ and $Y$, where $Y$ is a function of $X$. For $X$, the pdf is given by $f(x) = (\frac{1}{2})e^{-x/2}$ for $x > 0$ and $f(x) = 0$ for $x \le 0$. Also, $Y = g(X)$ is given by $y = 2$ for $x < 6$ and $y = 3$ for $x \ge 6$. (a) Determine $E(Y)$ by using Theorem 9.2. (b) Determine the probability function for $Y$. (c) Determine $E(Y)$ by making use of the result from part (b), and Definition 9.1.

*Solution*

(a) Using Theorem 9.2, we have

$$E(Y) = E[g(X)] = \left(\frac{1}{2}\right) \int_{x=0}^{\infty} g(x)e^{-x/2}\, dx \tag{1}$$

Here, $y = g(x) = 2$ for $x < 6$ and $y = g(x) = 3$ for $x \geq 6$. Thus, Eq. (1) becomes

$$E(Y) = \left(\frac{1}{2}\right) \int_{0}^{6} 2e^{-x/2}\, dx + \left(\frac{1}{2}\right) \int_{6}^{\infty} 3e^{-x/2}\, dx \tag{2}$$

Working out the values for the definite integrals of Eq. (2), we get

$$E(Y) = (\tfrac{1}{2})4(1 - e^{-3}) + (\tfrac{1}{2})6e^{-3} = 2 + e^{-3} \approx 2.0498 \tag{3}$$

(b) Let $h(y)$ symbolize the probability function of $Y$. Thus,

$$h(2) = P(Y = 2) = P(X < 6) = \int_{-\infty}^{6} f(x)\, dx = \left(\frac{1}{2}\right) \int_{0}^{6} e^{-x/2}\, dx = (1 - e^{-3}) \tag{4}$$

$$h(3) = P(Y = 3) = P(X \geq 6) = \left(\frac{1}{2}\right) \int_{6}^{\infty} e^{-x/2}\, dx = e^{-3} \tag{5}$$

Note that $h(2) + h(3) = 1$, as expected.

(c) Now, from Definition 9.1 as applied to discrete random variable $Y$, we have

$$E(Y) = \sum_{y} yh(y) = 2h(2) + 3h(3) \tag{6}$$

Using the results from Eqs. (4) and (5) of part (b), we have

$$E(Y) = 2(1 - e^{-3}) + 3e^{-3} = 2 + e^{-3} \tag{7}$$

as in Eq. (3) of part (a). Note that in this problem the original random variable $X$ was continuous while $Y = g(X)$ turned out to be discrete.

## Some Expected Value Rules

**PROBLEM 9-7** Let random variable $X$ stand for the sum of numbers obtained when tossing a pair of fair dice. Determine $E(X)$ and $E(X^2)$.

*Solution* In Example 5-16, we determined the probability function $f(x)$, and it is displayed in the first two columns below.

| $x$ | $f(x)$ | $xf(x)$ | $x^2f(x)$ |
|-----|--------|---------|-----------|
| 2 | 1/36 | 2/36 | 4/36 |
| 3 | 2/36 | 6/36 | 18/36 |
| 4 | 3/36 | 12/36 | 48/36 |
| 5 | 4/36 | 20/36 | 100/36 |
| 6 | 5/36 | 30/36 | 180/36 |
| 7 | 6/36 | 42/36 | 294/36 |
| 8 | 5/36 | 40/36 | 320/36 |
| 9 | 4/36 | 36/36 | 324/36 |
| 10 | 3/36 | 30/36 | 300/36 |
| 11 | 2/36 | 22/36 | 242/36 |
| 12 | 1/36 | 12/36 | 144/36 |
| $\sum_{x}$ | 36/36 $= 1$ | 252/36 $= 7$ | 1974/36 $= 59.8\overline{3}$ |

By summing in the third and fourth columns, we find that

$$E(X) = \sum_{x} xf(x) = 7$$

and

$$E(X^2) = \sum_{x} x^2f(x) = 59.8\overline{3}$$

**PROBLEM 9-8**   For the probability function of Problem 9-7, determine $E(3 + 4X - 2X^2)$.

*Solution*  From Theorem 9.4 and the rule that $E(b) = b$, we have

$$E(3 + 4X - 2X^2) = 3 + 4E(X) - 2E(X^2) \qquad (1)$$

Next, using the values of $E(X)$ and $E(X^2)$ from Problem 9-7, we find that

$$E(3 + 4X - 2X^2) = 3 + 4(7) - 2(54.83\overline{3}) = -78.66\overline{6} \qquad (2)$$


## Moments

**PROBLEM 9-9**   Prove that $\mu_1$, the first moment about the mean, is equal to zero.

*Proof*  From Definition 9.4, the linearity rule, and the rule that $E(b) = b$, we have

$$\mu_1 = E(X - \mu) = E(X) - \mu \qquad (1)$$

But, $E(X) = \mu$, and so

$$\mu_1 = \mu - \mu = 0 \qquad (2) \quad \square$$


**PROBLEM 9-10**   Say $X$ is a random variable with finite variance. **(a)** Show that for any real number $k$, $\text{Var}(X) = E[(X - k)^2] - [E(X) - k]^2$. **(b)** Show that the quantity $E[(X - k)^2]$ is minimized when $k = \mu_x$, the mean of $X$.

*Solution*

**(a)** Let

$$A = E[(X - k)^2] \qquad \text{and} \qquad B = [E(X) - k]^2 \qquad (1), (2)$$

Thus, our goal is to show that $A - B = \text{Var}(X)$. From Eqs. (1) and (2), we have

$$A = E[X^2 - 2kX + k^2] = E(X^2) - 2kE(X) + k^2 \qquad (3)$$

and

$$B = [E(X)]^2 - 2kE(X) + k^2 \qquad (4)$$

Thus, from Eqs. (3) and (4) and Theorem 9.5, we have

$$A - B = E(X^2) - [E(X)]^2 = \text{Var}(X) \qquad (5)$$

**(b)** From the part (a) result, we have the following equation, after solving for $A$:

$$E[(X - k)^2] = \text{Var}(X) + [E(X) - k]^2 \qquad (6)$$

For a particular random variable with a definite probability distribution, the terms $\text{Var}(X)$ and $E(X) = \mu_x$ are constants. Treating real number $k$ as a variable, we see from Eq. (6) that, for any $k$ value, $E[(X - k)^2]$ exceeds $\text{Var}(X)$ by the nonnegative quantity $[E(X) - k]^2$. Now, if in Eq. (6) we let $k = E(X)$, we will cause $E[(X - k)^2]$ to be minimized at $\text{Var}(X)$.

*notes*

**(i)**  If we interpret $E(X) = \mu_x$ to be the center of mass, and $\text{Var}(X)$ to be the moment of inertia about the center of mass of a system of masses [distributed along the $x$-axis] with total mass equal to 1 unit, then the part (a) result is a statement of a version of the parallel-axis theorem of mechanics, namely that "the moment of inertia about an arbitrary point [on the $x$-axis] is equal to the moment of inertia about the center of mass plus the square of the distance of this arbitrary point from the center of mass."

**(ii)**  The part (b) result indicates that the moment of inertia of a system with total mass equal to 1 unit about an arbitrary point [on the $x$-axis] is minimized if this point is the center of mass.

**PROBLEM 9-11**   Prove Theorem 9.6 that $\text{Var}(aX + b) = a^2 \text{Var}(X)$.

*Proof* Let

$$Y = aX + b \tag{1}$$

Thus, it follows from Theorem 9.3 that

$$E(Y) = E(aX + b) = aE(X) + b \tag{2}$$

Also, since $Y$ is a random variable, it follows from Theorem 9.5 that

$$\text{Var}(Y) = E(Y^2) - [E(Y)]^2 \tag{3}$$

Now, from Eq. (1) and Theorem 9.4 [linearity rule for $E$],

$$E(Y^2) = E[(aX + b)^2] = E[a^2X^2 + 2abX + b^2] = a^2E(X^2) + 2abE(X) + b^2 \tag{4}$$

By squaring in Eq. (2), we get

$$[E(Y)]^2 = a^2[E(X)]^2 + 2abE(X) + b^2 \tag{5}$$

Substitution of the expressions for $E(Y^2)$ and $[E(Y)]^2$ from Eqs. (4) and (5) into Eq. (3) yields

$$\text{Var}(Y) = a^2E(X^2) + 2abE(X) + b^2 - a^2[E(X)]^2 - 2abE(X) - b^2$$

$$= a^2\{E(X^2) - [E(X)]^2\} = a^2 \text{Var}(X) \tag{6}$$

In the last step of Eq. (6), we used Theorem 9.5.   □

### The Moment Generating Function

**PROBLEM 9-12**   Show that $M_X(0) = 1$.

*Solution* In general, $M_X(t) = E(e^{Xt})$. Letting $t = 0$, this becomes $M_X(0) = E(e^{X0}) = E(1) = 1$. [Recall here that $E(b) = b$.]

**PROBLEM 9-13**   Given a random variable $X$ with probability function indicated by the following table:

| $x$ | 0 | 1 | 2 | 3 |
|------|------|-------|-------|------|
| $f(x)$ | 1/35 | 12/35 | 18/35 | 4/35 |

Determine $M_X(t)$.

*Solution* First, we know that $M_X(t) = E(e^{Xt}) = \sum_{x=0}^{3} e^{xt}f(x)$, and that means that

$$M_X(t) = 1(\tfrac{1}{35}) + e^t(\tfrac{12}{35}) + e^{2t}(\tfrac{18}{35}) + e^{3t}(\tfrac{4}{35})$$

after making use of the given table for $f(x)$.

*note* Here, $X$ is a hypergeometric random variable for which $M = 7$, $n = 3$, and $a = 4$.

**PROBLEM 9-14**   Given that $f(x) = (\tfrac{1}{2})e^{-|x|}$ for $-\infty < x < \infty$. **(a)** Determine the expression for $M_X(t)$. **(b)** Use the $M_X(t)$ result to find $E(X)$ and $\text{Var}(X)$.

*Solution*

**(a)** Since $|x| = x$ for $x \geq 0$, and $|x| = -x$ for $x < 0$, it follows that

$$M_X(t) = E(e^{tX}) = \int_{-\infty}^{\infty} e^{tx}f(x)\,dx = \left(\frac{1}{2}\right)\underbrace{\int_{-\infty}^{0} e^{x(t+1)}\,dx}_{I_1} + \left(\frac{1}{2}\right)\underbrace{\int_{0}^{\infty} e^{x(t-1)}\,dx}_{I_2} \tag{1}$$

Now, let us work out the improper definite integrals $I_1$ and $I_2$.

$$I_1 = \left(\frac{1}{2}\right)\frac{e^{x(t+1)}}{(t+1)}\Big]_{-\infty}^{0} = \frac{1}{2(t+1)} \qquad \text{for } t > -1 \qquad (2)$$

$$I_2 = \left(\frac{1}{2}\right)\frac{e^{x(t-1)}}{(t-1)}\Big]_{0}^{\infty} = -\frac{1}{2(t-1)} \qquad \text{for } t < 1 \qquad (3)$$

Note that the *lower* limit in Eq. (2) is $-\infty$. Observe that $I_1$ doesn't exist for $t \le -1$ and $I_2$ doesn't exist for $t \ge 1$. Substituting from Eqs. (2) and (3) back into Eq. (1), and making note of the restrictions on $t$, we obtain

$$M_X(t) = I_1 + I_2 = \frac{1}{2(t+1)} - \frac{1}{2(t-1)} = \frac{1}{(1-t^2)} \qquad \text{for } -1 < t < 1 \qquad (4)$$

**(b)** We could differentiate twice in Eq. (4) to find $M_X'(0)$ and $M_X''(0)$, but instead we will use an infinite series approach. First of all, we note that we have the following geometric series result:

$$\frac{1}{(1-t)} = 1 + t + t^2 + t^3 + t^4 + \ldots, \qquad \text{valid for } -1 < t < 1 \qquad (5)$$

Thus, for $1/(1-t^2)$, we have the following series expansion [replace $t$ by $t^2$ in Eq. (5)]:

$$\frac{1}{(1-t^2)} = 1 + t^2 + t^4 + t^6 + \ldots, \qquad \text{valid for } -1 < t < 1 \qquad (6)$$

Now, from the Taylor series for $M_X(t)$ in powers of $t$, we have

$$M_X(t) = 1 + M_X'(0)t + \frac{M_X''(0)}{2!}t^2 + \ldots + \frac{M_X^{(r)}(0)}{r!}t^r + \ldots \qquad (7)$$

as was noted in Eq. (7) of Example 9-21. Now, recall that $M_X(t) = 1/(1-t^2)$ in the current situation. Thus, comparing the series on the right sides of Eqs. (6) and (7), we see that

$$M_X'(0) = 0 \qquad \text{and} \qquad M_X''(0) = 2! = 2 \qquad \text{(8a), (8b)}$$

Now, using part (b) of Theorem 9.9, we have $E(X) = 0$ and $E(X^2) = 2$. Finally, from Theorem 9.5, we have $\text{Var}(X) = E(X^2) - [E(X)]^2 = 2$.

## Means, Variances, and Moment Generating Functions of Special Probability Distributions

**PROBLEM 9-15** For a hypergeometric distribution, determine expressions for (a) $E[X(X-1)]$ and (b) $\text{Var}(X)$.

*Solution*

**(a)** From $E[g(X)] = \Sigma_x g(x)f(x)$, with $g(X) = X(X-1)$ and $f(x)$ equal to the hypergeometric probability function [see Theorem 5.7], we obtain

$$E[X(X-1)] = \sum_{x=0}^{n} x(x-1)\frac{\binom{a}{x}\binom{M-a}{n-x}}{\binom{M}{n}} \qquad (1)$$

Now, we observe that the terms in the summation of Eq. (1) for both $x = 0$ and $x = 1$ are equal to zero. Thus, the lower index value may be replaced by $x = 2$. Next, we observe that

$$x(x-1)\binom{a}{x} = \frac{x(x-1)a!}{x!(a-x)!} = \frac{a(a-1)(a-2)!}{(x-2)!(a-x)!} = a(a-1)\binom{a-2}{x-2} \qquad (2)$$

Here, we have used the facts that $x! = x(x-1)[(x-2)!]$, and $a! = a(a-1)[(a-2)!]$. Now, if we incorporate the result from Eq. (2) into Eq. (1), and then factor $a(a-1)$ to the left of the $\Sigma$ symbol, we get

$$E[X(X-1)] = a(a-1) \sum_{x=2}^{n} \frac{\binom{a-2}{x-2}\binom{M-a}{n-x}}{\binom{M}{n}} \tag{3}$$

Next, we change the variable of summation by letting $r = x - 2$, and we express

$$\binom{M}{n} = \frac{M(M-1)}{n(n-1)}\binom{M-2}{n-2} \tag{4}$$

Then, if we factor $n(n-1)/[M(M-1)]$ to the left of the $\Sigma$ symbol, and convert from $x$ to $r$, we obtain

$$E[X(X-1)] = \frac{a(a-1)n(n-1)}{M(M-1)} \sum_{r=0}^{n-2} \frac{\binom{a-2}{r}\binom{M-a}{n-2-r}}{\binom{M-2}{n-2}} \tag{5}$$

The summation term in Eq. (5) is equal to 1 since it represents the sum of hypergeometric probabilities of $r$ successes in a drawing without replacement of $(n-2)$ objects for all possible values of $r$, that is, for $r = 0, 1, 2, \ldots, n-1, n-2$. [Here, the drawing is from a set containing a total of $(M-2)$ objects, of which $(a-2)$ are successes.] Thus, we obtain

$$E[X(X-1)] = \frac{a(a-1)n(n-1)}{M(M-1)} \tag{6}$$

(b) Using Theorem 9.13, and the result $E(X) = na/N$, we get

$$\text{Var}(X) = E[X(X-1)] + E(X) - [E(X)]^2 = \frac{a(a-1)n(n-1)}{M(M-1)} + \frac{na}{M} - \frac{n^2a^2}{M^2}$$

$$= n\left(\frac{a}{M}\right)\left(\frac{M-a}{M}\right)\left(\frac{M-n}{M-1}\right) \tag{7}$$

If we replace $a/M$ by $p$, and $(M-a)/M = 1 - p$ by $q$, we obtain

$$\text{Var}(X) = npq\left(\frac{M-n}{M-1}\right) \tag{8}$$

**PROBLEM 9-16** For the Poisson distribution, determine expressions for (a) $E(X)$, (b) $E[X(X-1)]$, and (c) $\text{Var}(X)$.

*Solution* The expression for the Poisson probability function $f(x) = p(x; \lambda)$ is given in Definition 6.2.

(a) For $E(X)$, we have

$$E(X) = \sum_x xp(x; \lambda) = \sum_{x=0}^{\infty} x\frac{\lambda^x e^{-\lambda}}{x!} \tag{1}$$

Noting that the $x = 0$ term is equal to zero, then writing $x!/x = (x-1)!$, and factoring out $\lambda$, we obtain

$$E(X) = \lambda \sum_{x=1}^{\infty} \frac{\lambda^{x-1} e^{-\lambda}}{(x-1)!} \tag{2}$$

If we let $r = x - 1$, and replace $x$ terms in Eq. (2) by $r$ terms, we get

$$E(X) = \lambda \sum_{r=0}^{\infty} \frac{\lambda^r e^{-\lambda}}{r!} = \lambda \cdot 1 = \lambda \tag{3}$$

In Eq. (3), the $\Sigma$ term was equal to 1 since it represented the sum of all possible Poisson probabilities for a Poisson probability distribution with parameter $\lambda$.

**(b)** First, for $E[X(X-1)]$, we have

$$E[X(X-1)] = \sum_{x=0}^{\infty} x(x-1)\frac{\lambda^x e^{-\lambda}}{x!} = \sum_{x=2}^{\infty} x(x-1)\frac{\lambda^x e^{-\lambda}}{x!} \qquad (4)$$

In going from the first $\Sigma$ term to the second $\Sigma$ term in Eq. (4), we observed that the $x = 0$ and $x = 1$ terms were both equal to zero [because $x(x-1) = 0$ for both $x = 0$ and $x = 1$]. Next, we observe that $x! = x(x-1)[(x-2)!]$, and factor out $\lambda^2$ to the left of the $\Sigma$ symbol, and substitute $r = x - 2$. We therefore obtain

$$E[X(X-1)] = \lambda^2 \sum_{x=2}^{\infty} \frac{\lambda^{x-2}e^{-\lambda}}{(x-2)!} = \lambda^2 \sum_{r=0}^{\infty} \frac{\lambda^r e^{-\lambda}}{r!} = \lambda^2 \cdot 1 = \lambda^2 \qquad (5)$$

Here, the second summation term was equal to 1 for the same reason as indicated in part (a).

**(c)** Using Theorem 9.13, and the results from parts (a) and (b), we obtain

$$\text{Var}(X) = E[X(X-1)] + E(X) - [E(X)]^2 = \lambda^2 + \lambda - \lambda^2 = \lambda$$

**PROBLEM 9-17**  For the uniform pdf, $f(x) = 1/(b-a)$ for $a < x < b$, and $f(x) = 0$, elsewhere. Determine expressions for **(a)** $E(X^r)$, **(b)** the mean, $E(X)$, and **(c)** the variance.

*Solution*

**(a)** From Definition 9.3,

$$E(X^r) = \int_x x^r f(x)\,dx = \left(\frac{1}{b-a}\right)\int_{x=a}^{b} x^r\,dx \qquad (1)$$

Now, since

$$\int_{x=a}^{b} x^r\,dx = \frac{x^{r+1}}{(r+1)}\Bigg]_a^b = \frac{b^{r+1} - a^{r+1}}{(r+1)} \qquad (2)$$

it follows that

$$E(X^r) = \frac{b^{r+1} - a^{r+1}}{(r+1)(b-a)} \qquad (3)$$

**(b)** If we let $r = 1$ in Eq. (3), we get

$$E(X) = \mu = \text{Mean of } X = \frac{b^2 - a^2}{2(b-a)} = \frac{a+b}{2} \qquad (4)$$

**(c)** If we let $r = 2$ in Eq. (3), we get

$$E(X^2) = \frac{b^3 - a^3}{3(b-a)} = \frac{b^2 + ab + a^2}{3} \qquad (5)$$

In going from the second to the third expression of Eq. (5), we made use of the fact that $b = a$ is a root of the polynomial $b^3 - a^3$. That means that $(b-a)$ is a factor of $b^3 - a^3$. Division of $b^3 - a^3$ by $(b-a)$ results in the second degree polynomial $b^2 + ab + a^2$.

Now, employing Theorem 9.5 and the results from Eqs. (4) and (5), we get

$$\text{Var}(X) = E(X^2) - [E(X)]^2 = \frac{(b-a)^2}{12} \qquad (6)$$

**PROBLEM 9-18**  For the uniform pdf as given in Problem 9-17, determine $M_X(t)$.

*Solution*  First

$$M_X(t) = E(e^{tX}) = \int_x e^{tx}f(x)\,dx$$

Here, this becomes

$$M_X(t) = \int_a^b \frac{e^{tx}}{(b-a)}\,dx \qquad (1)$$

For $t \neq 0$, $e^{tx}$ stays as an exponential function. Thus, we have

$$M_X(t) = \frac{1}{(b-a)} \cdot \frac{e^{tx}}{t} \Bigg]_a^b = \frac{(e^{tb} - e^{ta})}{t(b-a)} \qquad \text{for } t \neq 0 \tag{2}$$

For $t = 0$, we have $M_X(0)$, which equals 1. This was proved in general in Problem 9-12.

**PROBLEM 9-19**   For the Cauchy distribution [see Problem 8-7], show that $E(X)$ doesn't exist.

*Solution*  Here,

$$f(x) = \frac{(\beta/\pi)}{x^2 + \beta^2} \qquad \text{for } -\infty < x < \infty \tag{1}$$

Thus

$$E(X) = \int_{-\infty}^{\infty} xf(x)\,dx = \int_{-\infty}^{0} xf(x)\,dx + \int_{0}^{\infty} xf(x)\,dx \tag{2}$$

Here, we first expressed the $\int_{-\infty}^{\infty}$ improper integral as a sum of two improper integrals, as was done in Problem 8-7. Now,

$$\int_{0}^{\infty} xf(x)\,dx = (\beta/\pi) \int_{0}^{\infty} \frac{x\,dx}{x^2 + \beta^2} \tag{3}$$

Letting $v = x^2 + \beta^2$, we have $dv = 2x\,dx$, and hence

$$\int_{0}^{\infty} xf(x)\,dx = \left(\frac{\beta}{\pi}\right)\left(\frac{1}{2}\right) \int_{0}^{\infty} \frac{dv}{v} = \left(\frac{\beta}{2\pi}\right) \ln v \Bigg]_{v=0}^{\infty} \tag{4}$$

But, $\int_{0}^{\infty} (dv/v)$ diverges since $\ln(v) \to \infty$ as $v \to \infty$, and thus so does $\int_{0}^{\infty} xf(x)\,dx$. Referring back to Eq. (2), we see that this means that $E(X)$ doesn't exist.

*note*  It can also be shown for the Cauchy distribution that $\mu_r' = E(X^r)$ doesn't exist for $r$ equal to any positive integer.

**PROBLEM 9-20**   For the beta distribution [see Section 8-3D and Problem 8-15], derive the equation for $E(X)$ in terms of the parameters $\alpha$ and $\beta$. Use the fact that $\int_0^1 f(x)\,dx = 1$.

*Solution*  Since $\int_0^1 f(x)\,dx = 1$, it follows that

$$\int_0^1 \frac{\Gamma(\alpha + \beta)}{\Gamma(\alpha)\Gamma(\beta)} x^{\alpha-1}(1-x)^{\beta-1}\,dx = 1 \tag{1}$$

and hence, that

$$\int_0^1 x^{\alpha-1}(1-x)^{\beta-1}\,dx = \frac{\Gamma(\alpha)\Gamma(\beta)}{\Gamma(\alpha + \beta)} \tag{2}$$

Substitution of the expression for $f(x)$ into the $E(X)$ equation yields

$$E(X) = \int_0^1 xf(x)\,dx = \frac{\Gamma(\alpha + \beta)}{\Gamma(\alpha)\Gamma(\beta)} \int_0^1 x \cdot x^{\alpha-1}(1-x)^{\beta-1}\,dx \tag{3}$$

Now, the definite integral that appears in Eq. (3) is the same as that which appears in Eq. (2), except that $\alpha$ is replaced by $(\alpha + 1)$; here, $xx^{\alpha-1} = x^{(\alpha+1)-1}$. Thus, the definite integral that appears in Eq. (3) is equal to $\Gamma(\alpha + 1)\Gamma(\beta)/\Gamma(\alpha + \beta + 1)$. Now, if we apply the property $\Gamma(\alpha + 1) = \alpha\Gamma(\alpha)$ of Theorem 8.6 to both the $\Gamma(\alpha + 1)$ and $\Gamma(\alpha + \beta + 1)$ terms, we see that Eq. (3) becomes

$$E(X) = \frac{\Gamma(\alpha + \beta)}{\Gamma(\alpha)\Gamma(\beta)} \cdot \frac{\alpha\Gamma(\alpha)\Gamma(\beta)}{(\alpha + \beta)\Gamma(\alpha + \beta)} = \frac{\alpha}{(\alpha + \beta)} \tag{4}$$

Another key result for the beta distribution is $\text{Var}(X) = \alpha\beta/[(\alpha+\beta)^2(\alpha+\beta+1)]$.

## Practical Situations Involving Expected Values

**PROBLEM 9-21** Assume that the number of customers who arrive at a supermarket checkout counter in a $t$ minute period follows a Poisson process. Suppose that the mean number of customers who arrive in a 5 minute period is 6. **(a)** Determine the expected number who arrive during an 8 minute period. **(b)** What is the probability that at most 9 customers arrive during an 8 minute period?

*Solution*

**(a)** Since $E(X) = \alpha t$ and $E(X) = 6$ for $t = 5$ minutes, we have

$$6 = \alpha(5) \quad \text{and thus} \quad \alpha = 1.2 \text{ customers/minute} \tag{1}$$

Thus, for $t = 8$ minutes, we have

$$E(X) = \alpha t = (1.2)(8) = 9.6 \text{ customers} \tag{2}$$

So, the expected [mean] number of customers who arrive in an 8 minute period is 9.6.

**(b)** The number of customers arriving in an 8 minute period is a Poisson random variable with $\lambda = \alpha t = 9.6$. From Table IV of Appendix A, for $\lambda = 9.6$, we have

$$P(X \leq 9) = .0001 + .0007 + \dots + .1212 + .1293 = .5090 \tag{3}$$

**PROBLEM 9-22** Consider the game of American roulette. The wheel has numbers 1 through 36, and also a zero [0] and a double zero [00], marked on 38 equally spaced slots. For the wheel, 18 of the original numbers are red, and 18 are black. The wheel is spun, and a ball drops into one of the 38 slots.

In the simplest gambling situation, the player bets a certain amount of money on one of the 38 numbers. If that number "comes up" [i.e., if the ball drops into the slot labeled with that number], the player wins 35 times his original bet, and gets back the money that he originally bet. [The "35 to 1" payoff is decided on by the gambling house.] If any other number comes up, he loses the amount he has bet. **(a)** Determine the player's expected gain [i.e., the expected amount of money he wins] for this simple bet. Assume the player bets $2 on the number 9. **(b)** Determine the house percentage, $H$, for this simple bet. [*Hint:* Refer to Problems 9-3 and 9-4 of this chapter for other gambling examples.]

*Solution*

**(a)** Let $G$ be the discrete random variable which signifies the player's gain during any play of the game with respect to his particular bet. Here, the player bets $2 on the number 9. Then, the possible values of $G$ are $-2$ if the number 9 doesn't come up, and $+70$ if 9 does come up. The respective probabilities are $\frac{37}{38}$ for $G = -2$ and $\frac{1}{38}$ for $G = 70$ since all 38 numbers of the wheel are equally likely. Thus, we have the table that follows, where $f(g)$ denotes the probability associated with the value $g$ of the random variable $G$.

| $g$ | $f(g)$ | $gf(g)$ |
|---|---|---|
| $-2$ | 37/38 | $-74/38$ |
| 70 | 1/38 | 70/38 |
| Sum | $38/38 = 1$ | $-4/38$ |

In general, the expected value of $G$, which is called the expected gain for the player, is given by

$$E(G) = \sum_g gf(g) \tag{1}$$

For the current situation, the third column of the above table indicates that this becomes

$$E(G) = -4/38 = -.10526 \tag{2}$$

The interpretation of this value is as follows: if the player bets $2 on the number 9 for many spins of the roulette wheel, then the player will lose an average of 10.526 cents for each bet.

(b) Since the house percentage is given by $H = -100E(G)/B$, as we see from Eq. (I) of Problem 9-4, and $B = \$2$ here, we obtain

$$ H = -\frac{100(-4/38)}{2} = 5.263\% $$

This means that "on the average," the house wins 5.263% of all the money wagered on such bets.

**PROBLEM 9-23** A printing machine has a constant probability of .10 of being disabled on any given day. If the machine is not disabled on a particular day, then a profit of $1000 is realized. If the machine is disabled during a particular day, then a loss of $100 occurs. Let $X_i$ denote the profit associated with a period of $i$ days. (a) Compute $E(X_1)$, the expected profit for a one-day period. (b) Compute $E(X_2)$, the expected profit for a two-day period.

*Solution*

(a) Let random variable $D$ denote the number of disabled days for a one-day period. Thus, $P(D = 0) = .90$, and $P(D = 1) = .10$. Now, $X_1 = 1000$ is equivalent to $D = 0$ and $X_1 = -100$ is equivalent to $D = 1$. Thus, we have the following probability table for random variable $X_1$. [$x_1$ denotes the value of random variable $X_1$, and $f(x_1)$ denotes the probability function.]

| $x_1$ | $f(x_1)$ | $x_1 f(x_1)$ |
|---|---|---|
| 1000 | .90 | 900 |
| $-100$ | .10 | $-10$ |
| Sum | 1.00 | $890 |

From the sum in the $x_1 f(x_1)$ column, we see that the expected profit for one day, $E(X_1)$, is $890.

(b) Let random variable $D$ be the number of disabled days in a two-day period. Thus, $D$ is a binomial random variable for which $p = .10$ [the probability of the machine being disabled on a particular day], and $n = 2$ [each day represents an independent trial]. The values for $D$ are 0, 1, and 2. The corresponding values for $X_2$, and the relevant probabilities are given in the following table. Observe that $P(D = 0) = P(X_2 = 2000)$, and $P(D = 1) = P(X_2 = 1000 - 100) = P(X_2 = 900)$, and $P(D = 2) = P(X_2 = -200)$.

| $x_2$ | $f(x_2)$ | $x_2 f(x_2)$ |
|---|---|---|
| $2(1000) = 2000$ | $(.9)^2 = .81$ | 1620 |
| $1000 - 100 = 900$ | $2(.9)(.10) = .18$ | 162 |
| $2(-100) = -200$ | $(.1)^2 = .01$ | $-2$ |
| Sum = | 1.00 | 1780 |

Thus, we see that the expected profit for a two-day period is $1780, or twice that for a one-day period. As we shall see in our later development, it turns out that the expected profit for an $n$-day period is $n$ times that for a one-day period if the situation parallels what we had in this problem.

**PROBLEM 9-24** Electronic items each have a lifetime $X$, in months, for which the pdf is given by $f(x) = (\frac{1}{2})e^{-x/2}$ for $x > 0$ and $f(x) = 0$, for $x \le 0$. Suppose the cost of manufacturing each item is $3.00 and the selling price of each item is $7.00. A total refund is guaranteed if the item fails within the first 1.5 months. What is the expected profit per item?

*Solution* Let random variable $Q$ denote the profit per item, in $. Then, $Q = 7 - 3 = 4$ is equivalent to $X > 1.5$, and $Q = -3$ is equivalent to $X \le 1.5$. [Remember the total refund condition.]

Now,

$$P(X > 1.5) = \left(\frac{1}{2}\right) \int_{1.5}^{\infty} e^{-x/2}\, dx = e^{-.75} = .47237 \qquad (1)$$

Thus, $P(X \le 1.5) = 1 - .47237 = .52763$. Then, with $q$ denoting a value of $Q$,

$$E(Q) = \sum_q q P(Q = q) = (-3)(.52763) + (4)(.47237) = \$.30659 \qquad (2)$$

Thus, the expected profit per item is 30.659 cents.

## Supplementary Problems

**PROBLEM 9-25**  For the binomial distribution for which $n = 3$, and $p = 2/5$, **(a)** display the table of $f(x)$ values. Then, compute $E(X)$ and $\mathrm{Var}(X)$ **(b)** by employing the $f(x) = b(x; n, p)$ table, and **(c)** from the general equations for $E(X)$ and $\mathrm{Var}(X)$ for the binomial distribution.

*Answer*

**(a)**

| $x$ | $f(x)$ | $xf(x)$ | $x^2f(x)$ |
|---|---|---|---|
| 0 | 27/125 | 0 | 0 |
| 1 | 54/125 | 54/125 | 54/125 |
| 2 | 36/125 | 72/125 | 144/125 |
| 3 | 8/125 | 24/125 | 72/125 |
| $\sum_x$ | 125/125 <br> $= 1.0$ | 150/125 <br> $= 1.2$ | 270/125 <br> $= 2.16$ |

**(b)** From the table,

$$E(X) = \sum_x xf(x) = 1.2$$

$$E(X^2) = \sum_x x^2 f(x) = 2.16$$

$$\mathrm{Var}(X) = 2.16 - (1.2)^2 = .72$$

**(c)** $E(X) = np = 3(.4) = 1.2;$   $\mathrm{Var}(X) = npq = 3(.4)(.6) = .72$

**PROBLEM 9-26**  For the binomial distribution of the previous problem, compute **(a)** $E(Y)$ and **(b)** $\mathrm{Var}(Y)$ if $Y = 3X - 2$. Compute **(c)** $E(R)$ and **(d)** $\mathrm{Var}(R)$ if $R = 3X^{1/2}$.

*Answer* **(a)** $E(Y) = (3)(1.2) - 2 = 1.6$   **(b)** $\mathrm{Var}(Y) = 9\,\mathrm{Var}(X) = (9)(.72) = 6.48$. For (c) and (d), use $E[g(X)] = \sum_x g(x)f(x)$ with $g(x) = r = 3x^{1/2}$, and then $g(x) = r^2 = 9x$. Thus, add $3x^{1/2}f(x)$ and $9xf(x)$ columns to the table of Problem 9-25. Then, compute $\mathrm{Var}(R)$ from $E(R^2) - [E(R)]^2$.
**(c)** $E(R) = E(3X^{1/2}) = 2.8504$   **(d)** $\mathrm{Var}(R) = 2.6750$

**PROBLEM 9-27**  The age [in years] $X$ of sewing machines to be reconditioned is a random variable with the following probability distribution: $f(x) = (\frac{1}{972})x(18 - x)$ for $0 < x < 18$, and $f(x) = 0$, elsewhere. The time [in months] for reconditioning a sewing machine, $Y$, is related to $X$ by $Y = (X/9 + .4)$. **(a)** Determine $E(X)$ and $\sigma_X$. **(b)** Determine $E(Y)$ and $\sigma_Y$.

*Answer* **(a)** $E(X) = 9; \sigma_X = \sqrt{16.2} = 4.0249$   **(b)** $E(Y) = (\frac{1}{9})E(X) + .4 = 1.4; \sigma_Y = \sqrt{.20} = .44721$

**PROBLEM 9-28**  Let $X$ be a random variable with $E(X) = 3$ and $E[X(X - 1)] = 22$. **(a)** Find $\mathrm{Var}(X)$. **(b)** Find $\sigma_Y$ if $Y = 6 - 5X$.

*Answer* **(a)** $\mathrm{Var}(X) = 16$   **(b)** $\sigma_Y = 20$

**PROBLEM 9-29**  For a geometric probability distribution, $f(x) = g(x; p) = q^{x-1}p$ for $x = 1, 2, 3, \ldots$ [see Theorem 6.4; here, $q = 1 - p$]. **(a)** Determine $M_X(t)$. **(b)** Determine $E(X)$ and $\mathrm{Var}(X)$ by making use of the part (a) result. [*Hint for (a)*: Show $M_X(t) = (p/q)\sum_{x=1}^{\infty}(qe^t)^x$ and then use the geometric series result [Theorem 6.5].]

*Answer* **(a)** $M_X(t) = \dfrac{pe^t}{1 - qe^t}$, for $t < -\ln q$   **(b)** $E(X) = \dfrac{1}{p}; \mathrm{Var}(X) = \dfrac{q}{p^2}$

**PROBLEM 9-30** A random variable $X$ has a Rayleigh distribution if its pdf is given by $f(x) = 2\alpha x e^{-\alpha x^2}$ for $x > 0$ and $f(x) = 0$ for $x \le 0$. Here, parameter $\alpha > 0$. Determine the expression for $\mu$. [*Hint:* Use integration by parts [or a table of integrals], and then use the fact that $\int_0^\infty e^{-t^2/2} dt = \sqrt{\pi/2}$.]

*Answer* $\mu = E(X) = \left(\dfrac{1}{2}\right)\sqrt{\dfrac{\pi}{\alpha}}$

**PROBLEM 9-31** A random variable $X$ has a Weibull distribution if its pdf is given by $f(x) = kx^{\beta-1}e^{-\alpha x^\beta}$ for $x > 0$ and $f(x) = 0$ for $x \le 0$. Also, $\alpha > 0$ and $\beta > 0$. [In Problem 8-8, we showed that $k = \alpha\beta$.] Determine the expression for $\mu$. [*Hint:* Make use of the gamma function; it is defined in Definition 8.6.]

*Answer* $\mu = \alpha^{-1/\beta}\Gamma(1 + 1/\beta)$

**PROBLEM 9-32** For the exponential distribution, $f(x) = (1/\theta)e^{-x/\theta}$ for $x > 0$, and $f(x) = 0$ for $x \le 0$. **(a)** Determine a lower bound for the probability $P(0 < X < 3\theta)$ by using Chebyshev's theorem [Theorem 9.8]. **(b)** Determine the exact value for the probability $P(0 < X < 3\theta)$.

*Answer* **(a)** 3/4    **(b)** $1 - e^{-3} = .9502$

**PROBLEM 9-33** Prove the following alternate forms of Chebyshev's theorem [Theorem 9.8]. Here, $k$ and $c$ are positive numbers. **(a)** $P(|X - \mu| \ge k\sigma) \le 1/k^2$. **(b)** $P(|X - \mu| \ge c) \le \sigma^2/c^2$. [*Hints for* (a): Refer to Theorem 9.8, part (a), and observe that $P(|X - \mu| \ge k\sigma) = 1 - P(\mu - k\sigma < X < \mu + k\sigma)$. Here, we use the complementary event idea: $P(\text{not } A) = 1 - P(A)$. Observe that the event $|X - \mu| \ge k\sigma$ is equivalent to $X \ge \mu + k\sigma$ or $X \le \mu - k\sigma$, where the latter two events are mutually exclusive.]

**PROBLEM 9-34** Show that $M_{X-\mu}^{(r)}(0)$ is equal to $\mu_r$, the $r$th moment about the mean. Here, $M_{X-\mu}^{(r)}(t)$ denotes the $r$th derivative of $M_{X-\mu}(t)$ with respect to $t$. [*Hint:* Start with $M_{X-\mu}(t) = E[e^{(X-\mu)t}]$, and then proceed as in Example 9-21.]

**PROBLEM 9-35** Let $Q_X(t) = \ln M_X(t)$. **(a)** Show that $Q_X'(0) = \mu$ and $Q_X''(0) = \sigma^2$. **(b)** Use the part (a) results to determine $\mu$ and $\sigma^2$ for a random variable $X$ for which $M_X(t) = e^{(3t+8t^2)}$. [*Hint for* (a): First, show that $Q_X'(t) = M_X'(t)/M_X(t)$.]

*Answer* **(b)** $\mu = 3$ and $\sigma^2 = 16$

**PROBLEM 9-36** A player keeps tossing a fair die until he gets the result of his first toss a second time. If this occurs in three or fewer tosses [after the first toss], the player wins \$2. Otherwise, the gambling house wins \$2. **(a)** What is the probability the player wins \$2? **(b)** What is the expected gain to the player? **(c)** What is the house percentage? **(d)** What is the expected number of tosses after the first toss to obtain the same result as on the first toss?

*Answer* **(a)** 91/216    **(b)** $(-2)(34/216) = -\$.3148$    **(c)** $(34/216)(100) = 15.74\%$    **(d)** $E(X) = 6$ tosses

**PROBLEM 9-37** Suppose the life, in hours, of a particular computer component is a random variable $X$ with a Weibull distribution [see Problems 8-8 and 9-31] with $\alpha = .040$ and $\beta = .500$. **(a)** Calculate the expected life of that computer component. **(b)** What is the probability that the component will last longer than 2500 hours?

*Answer* **(a)** 1250 hours    **(b)** $e^{-2} = .135335$

**PROBLEM 9-38** The proportion of new health food stores that fail annually in Chicago is a random variable with a beta distribution for which $\alpha = 3$ and $\beta = 2$. [For a review of the beta distribution, see Section 8-3D, and Problems 8-15 and 9-20.] **(a)** Find the mean of this distribution [the mean represents the mean proportion of new health food stores that fail in a given year]. **(b)** Find the probability that at least 50% of all new health food stores in Chicago will fail in a given year.

*Answer* **(a)** 3/5    **(b)** 11/16

**PROBLEM 9-39** A lot of 12 lamps must either be totally rejected or sold, depending on the results of the following testing operation. Two lamps are randomly chosen and tested. If one or more are defective, the whole lot is rejected. Otherwise, the lot is accepted, and then sold. Say each lamp costs $60 and is sold for $90. What is the expected profit per lot if the lot contains two defective lamps?

*Answer* $2160/132 = \$16.\overline{36}$

# 10 THE NORMAL DISTRIBUTION AND RELATED TOPICS

## THIS CHAPTER IS ABOUT

- ☑ **Introduction to the Normal Distribution**
- ☑ **Calculations Involving a Normal Distribution**
- ☑ **Skewness of a Random Variable**
- ☑ **Normal Approximation to the Binomial Distribution**
- ☑ **A Law of Large Numbers**
- ☑ **More Normal Curve Problems**

## 10-1. Introduction to the Normal Distribution

One of the most important probability distributions in the continuous random variable category is the *normal distribution* [or *Gaussian distribution* or *bell-shaped distribution*].

### A. Definition of normal distribution

A normal pdf is given as follows:

$$f(x) = ce^{-(1/2)[(x-a)/b]^2} \qquad \text{for } -\infty < x < \infty \qquad \textbf{(I)}$$

where $a$, $b$, and $c$ are constants.

> ***note:*** In order to make our notation less awkward with respect to the exponential function, we shall sometimes use $\exp(s)$ in place of $e^s$. Thus, for example, Eq. (Ia) below is an alternate version of Eq. (I).

$$f(x) = c \exp\left[-\frac{1}{2}\left(\frac{x-a}{b}\right)^2\right] \qquad \text{for } -\infty < x < \infty \qquad \textbf{(Ia)}$$

---

**EXAMPLE 10-1:** Determine the constant $c$. Then, express $a$ and $b$ in terms of $\mu$ and $\sigma$, the mean and standard deviation of the normal pdf [as given by Eq. (I) or (Ia)].

***Solution:*** *Preliminaries:* Here, we shall merely summarize the key results. The details of the computations and manipulations can be found in the solutions of Problems 10-1, 10-2, and 10-3 of the Solved Problems section.

From the requirement $\int_{-\infty}^{\infty} f(x)\,dx = 1$, we show that $c = 1/[b\sqrt{2\pi}]$ in Problem 10-1. In general, the mean $E(X)$, or $\mu_x$, or $\mu$ is given by

$$\mu = \int_{-\infty}^{\infty} x f(x)\,dx \qquad \textbf{(1)}$$

From this equation, with $f(x)$ as given by Eq. (I), we show, in Problem 10-2, that

$$\mu = a \qquad \textbf{(2)}$$

In general, the variance of $X$, $\text{Var}(X)$, is given by

$$\text{Var}(X) = E[(X - \mu)^2] \qquad (3)$$

as we know from Definition 9.5, or by $\text{Var}(X) = E(X^2) - [E(X)]^2$. (See Theorem 9.5.) Recall that $\text{Var}(X) = \sigma^2$, where the nonnegative [and usually positive] quantity $\sigma$ is called the standard deviation. In Problem 10-3, we show that $\text{Var}(X) = b^2$, or, equivalently, $\sigma^2 = b^2$, from which, if we choose $b$ to be nonnegative, we get

$$\sigma = b \qquad (4)$$

Substituting from Eq. (4) into $c = 1/[b\sqrt{2\pi}]$, we get

$$c = \frac{1}{\sigma\sqrt{2\pi}} \qquad (5)$$

---

*note:* If $\text{Var}(X)$ exists, it is nonnegative, and it is most often the case that $\text{Var}(X)$ is positive.

Substituting the results from Eqs. (2), (4), and (5) into Eq. (I), we obtain Definition 10.1.

**Definition 10.1 (Normal Distribution):** A random variable $X$ has a normal distribution, and is referred to as a normal random variable, if its pdf is given by

$$f(x) = \frac{1}{\sigma\sqrt{2\pi}} e^{-(1/2)[(x-\mu)/\sigma]^2} \qquad \text{for } -\infty < x < \infty$$

Here, $\mu$ and $\sigma$ are the mean and standard deviation of $X$. Frequently used symbols for the normal pdf are $n(x; \mu, \sigma)$, $N(x; \mu, \sigma)$, and $N(\mu, \sigma^2)$.

A diagram illustrating the symmetry and some key probability properties of the normal distribution is Figure 10-1. Some key features of the normal distribution are the following:

(a) It has a vertical axis of symmetry, namely the vertical line $x = \mu$.
(b) It has two points of inflection, at $x = \mu - \sigma$ and at $x = \mu + \sigma$. [At such points, the concavity of the graph changes in sign.]
(c) The numbers in the body of the diagram are probabilities [or areas, since area equals probability for a continuous random variable] to two significant digits. Thus, for example,

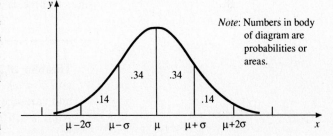

**Figure 10-1.** Key properties of the normal distribution.

$$P(\mu - \sigma < X < \mu) = P(\mu < X < \mu + \sigma) \approx .34$$

and

$$P(\mu - 2\sigma < X < \mu) = P(\mu < X < \mu + 2\sigma) \approx .34 + .14 = .48$$

## B. Moment generating function for a normal distribution

**Theorem 10.1:** The moment generating function for a normal random variable is given by

$$M_X(t) = e^{(\mu t + \sigma^2 t^2 / 2)} = e^{\mu t} e^{\sigma^2 t^2 / 2}$$

**EXAMPLE 10-2:** Prove Theorem 10.1.

*Proof:* In general, $M_X(t) = E(e^{tX})$, as we know from Definition 9.7. Thus, for $f(x)$ as given in Definition 10.1, we have

$$M_X(t) = \frac{1}{\sigma\sqrt{2\pi}} \int_{-\infty}^{\infty} \exp(tx) \exp\left[-\frac{1}{2}\left(\frac{x-\mu}{\sigma}\right)^2\right] dx \qquad (1)$$

Let $s = (x - \mu)/\sigma$. Thus, $x = \sigma s + \mu$, and $dx = \sigma\, ds$. Applying these results to Eq. (1), we get

$$M_X(t) = \frac{1}{\sqrt{2\pi}} \int_{-\infty}^{\infty} e^{t(\sigma s + \mu)} e^{-s^2/2}\, ds$$

$$= \frac{e^{\mu t}}{\sqrt{2\pi}} \int_{-\infty}^{\infty} e^{-(s^2 - 2\sigma ts)/2}\, ds \qquad (2)$$

Now, we complete the square in the exponent of the $e$ term inside the integral to get $s^2 - 2\sigma ts = (s - \sigma t)^2 - \sigma^2 t^2$. Thus, Eq. (2) becomes

$$M_X(t) = \frac{e^{(\mu t + \sigma^2 t^2/2)}}{\sqrt{2\pi}} \int_{-\infty}^{\infty} e^{-(s-\sigma t)^2/2}\, ds \qquad (3)$$

[Note that we took $e^{\mu t}$ and $e^{\sigma^2 t^2/2}$ to the left of the integral symbol in the development of Eqs. (2) and (3). This was allowable since the variable of integration is $s$.]

Next, let us label the integral in Eq. (3) as $K$. Also, in Eq. (3), let $u = s - \sigma t$. Thus, $du = ds$, and we get

$$K = \int_{-\infty}^{\infty} e^{-u^2/2}\, du \qquad (4)$$

But from Problem 10-1 [in the Solved Problems section], we see that $K = \sqrt{2\pi}$ since $K$, here, is equal to $J$ of Problem 10-1. Thus, Eq. (3) reduces to

$$M_X(t) = e^{(\mu t + \sigma^2 t^2/2)} \qquad (5) \quad \square$$

---

**Theorem 10.2:** If $X$ has a normal distribution with mean $\mu$ and variance $\sigma^2$, and $Y = aX + b$, then $Y$ has a normal distribution with mean $\mu_Y = a\mu + b$, and variance $\sigma_Y^2 = a^2\sigma^2$. Briefly, we say that if $X$ is $N(\mu, \sigma^2)$, then $aX + b$ is $N(a\mu + b, a^2\sigma^2)$.

A proof is given in Problem 10-4.

## C. The standardized normal random variable

In Definition 9.6, we used the symbol $Z$ for the *standardized* [or *standard*] *random variable* corresponding to a random variable $X$ with mean $\mu$ and standard deviation $\sigma$:

$$Z = \frac{(X - \mu)}{\sigma} = \left(\frac{1}{\sigma}\right)X + \left(-\frac{\mu}{\sigma}\right)$$

**Theorem 10.3:** If $X$ has a normal distribution with mean $\mu$ and variance $\sigma^2$, then $Z = (X - \mu)/\sigma$, the standard random variable corresponding to $X$, has a normal distribution with mean equal to 0, and variance equal to 1. In this case, $Z$ is referred to as a standardized [or standard] normal random variable. Briefly, we say that if $X$ is $N(\mu, \sigma^2)$, then $Z$ is $N(0, 1)$.

**EXAMPLE 10-3:** Prove Theorem 10.3.

*Proof:* Since

$$Z = \frac{(X - \mu)}{\sigma} = \left(\frac{1}{\sigma}\right)X + \left(-\frac{\mu}{\sigma}\right) \qquad (1)$$

we see that $Z$ is of the form $aX + b$ with $a = 1/\sigma$ and $b = -\mu/\sigma$. Thus, applying Theorem 10.2, we see that $Z$ has a normal distribution [equivalently, $Z$ is a normal random variable] with mean $a\mu + b = \mu/\sigma - \mu/\sigma = 0$ and variance $a^2\sigma^2 = \sigma^2/\sigma^2 = 1$.

Actually, the information that is new here is that $Z$ is a normal random variable. We established, in general, in Theorem 9.7 that $\mu_Z = 0$ and $\text{Var}(Z) = 1$.

Now, from Definition 10.1 as applied to the standard normal random variable $Z$, it follows that the pdf of $Z$ is given by

$$f_N(z) = \frac{1}{\sqrt{2\pi}}e^{-z^2/2} \qquad \text{for } -\infty < z < \infty \qquad (2)$$

[Here, and elsewhere, the subscript $N$ denotes "normal pdf."] What we did here was replace the mean and standard deviation of Definition 10.1 by 0 and 1, respectively, and the symbol $x$ by $z$. $\qquad \square$

---

**Theorem 10.4:** Given a random variable $X$ [which is not necessarily normal] with mean $\mu$ and standard deviation $\sigma$, and the corresponding standard random variable $Z = (X - \mu)/\sigma$. Then, we have

$$P(x_0 < X < x_1) = P(z_0 < Z < z_1) \qquad (a)$$

where $z_0 = (x_0 - \mu)/\sigma$ and $z_1 = (x_1 - \mu)/\sigma$. [We assume that $x_0 < x_1$; hence, $z_0 < z_1$.]

If we let $x_0 = \mu$, then Eq. (a) becomes

$$P(\mu < X < x_1) = P(0 < Z < z_1) \qquad (b)$$

where $z_1$ is positive. If we let $x_1 = \mu$, then Eq. (a) becomes

$$P(x_0 < X < \mu) = P(z_0 < Z < 0) \qquad (c)$$

where $z_0$ is negative.

A proof of Theorem 10.4 is given in Problem 10-5. Now, in Table VI of Appendix A, there are tabulated values of the probability $P(0 < Z < z_1)$ for a standardized normal random variable.

**Theorem 10.5:** If $Z$ is a standard normal random variable, then

$$P(0 < Z < z_1) = \frac{1}{\sqrt{2\pi}}\int_0^{z_1} e^{-z^2/2}\,dz \qquad (a)$$

Also, from symmetry, we have that

$$P(-z_1 < Z < 0) = P(0 < Z < z_1) \qquad (b)$$

In (a) and (b), $z_1$ is positive; hence, $-z_1$ is negative.

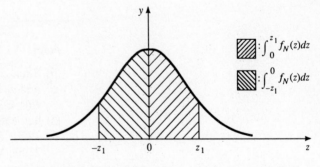

**Figure 10-2.** Probabilities from standard normal distribution table.

---

**EXAMPLE 10-4:** Prove Theorem 10.5.

*Proof:* Since $Z$ is a standard normal random variable, its pdf is as given in the proof of Theorem 10.3 [see Eq. (2) of Example 10-3]. Thus,

$$P(0 < Z < z_1) = \int_0^{z_1} f_N(z)\,dz = \frac{1}{\sqrt{2\pi}} \int_0^{z_1} e^{-z^2/2}\,dz \qquad (1)$$

So, part (a) is proved. Now,

$$P(-z_1 < Z < 0) = \frac{1}{\sqrt{2\pi}} \int_{-z_1}^0 e^{-z^2/2}\,dz \qquad (2)$$

The function $e^{-z^2/2}$ is an even function since $e^{-(-z)^2/2} = e^{-z^2/2}$. Hence, it follows that $\int_{-z_1}^0 f_N(z)\,dz = \int_0^{z_1} f_N(z)\,dz$. [An even function $g(s)$ is one for which $g(-s) = g(s)$. As a consequence, $\int_{-c}^0 g(s)\,ds = \int_0^c g(s)\,ds$. Here, the letter of integration, $s$, is arbitrary.] This means that

$$P(-z_1 < Z < 0) = P(0 < Z < z_1) \qquad (3)$$

and hence part (b) is proved. $\qquad\qquad\qquad\qquad\qquad\qquad\qquad\square$

Pertaining to Theorem 10.5, the diagram that applies is Figure 10-2.

**EXAMPLE 10-5:** Use Table VI of Appendix A to determine **(a)** $P(0 < Z < .6)$, **(b)** $P(-.9 < Z < 0)$, **(c)** $P(0 < Z < 1)$, **(d)** $P(0 < Z < 1.4)$, **(e)** $P(0 < Z < 1.96)$, and **(f)** $P(-2 < Z < 0)$.

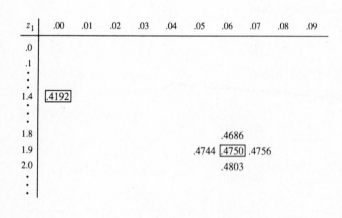

**Solution: (a)** to **(f)**: The subtable indicates how to read off $P(0 < Z < 1.4)$ and $P(0 < Z < 1.96)$ from the body of Table VI. [These probabilities are surrounded by rectangular boxes in the subtable.]

The top margin contains the hundredths parts of $z_1$, namely .00, .01, .02, ..., .08, and .09. In the left margin, we have $z_1$ values to the nearest tenth.

Thus, we read off $P(0 < Z < 1.4) = .4192$ and $P(0 < Z < 1.96) = .4750$ as the answers to parts (d) and (e). Observe that any value in the body of the table can be interpreted either as $P(0 < Z < z_1)$ or $P(-z_1 < Z < 0)$.

Thus, for example, we have $P(-1.96 < Z < 0) = .4750$, also. The other probabilities are read off from Table VI as follows: **(a)** $P(0 < Z < .6) = .2257$, **(b)** $P(-.90 < Z < 0) = .3159$, **(c)** $P(0 < Z < 1) = .3413$, and **(f)** $P(-2 < Z < 0) = .4772$.

**notes**

(i) Since $Z$ is a continuous random variable, we can replace $<$ by $\leq$, and $>$ by $\geq$, wherever those symbols appear. Thus, for example, we have $P(0 < Z < 1.96) = P(0 < Z \leq 1.96) = P(0 \leq Z \leq 1.96) = .4750$.

(ii) To calculate the probability $P(1 < Z < 2)$, we use $P(1 < Z < 2) = P(0 < Z < 2) - P(0 < Z < 1)$, where we look up the latter two values in Table VI. Thus, we have $P(1 < Z < 2) = .4772 - .3413 = .1359$. Then, rounding off to two significant digits, we recover the numbers .34 and .14 listed in the body of Figure 10-1. [We'll have more to say about this in Section 10-2A.] Thus, we have $P(0 < Z < 1) = P(-1 < Z < 0) = .3413 \approx .34$, and $P(1 < Z < 2) = P(-2 < Z < -1) = .1359 \approx .14$.

## 10-2. Calculations Involving a Normal Distribution

### A. Calculating $P(a < X < b)$ for a normal random variable

For the calculation of $P(a < X < b)$, where $X$ is a normal random variable, we can make use of Theorems 10.4, 10.5, and Table VI. It is especially useful to recall that $P(a < X < b)$ represents the area between the graph of the pdf $y = f(x)$ and the $x$-axis for the interval going from $x = a$ to $x = b$. This, of course, is the case for any continuous random variable $X$.

*note:* In much of the following development, we'll refer to either of $Z = (X - \mu)/\sigma$ or $X = \sigma Z + \mu$ as being the *correspondence equation* between random variables $X$ and $Z$ [with similar equations applying to the values $x$ and $z$ of these random variables].

---

**EXAMPLE 10-6:** Suppose the weights of men at Billings College are approximately normally distributed with a mean of 160 pounds and a standard deviation of 15 pounds. Let $X$ symbolize the weight random variable. (**a**) Calculate $P(160 < X < 181)$, which represents the probability that a man picked at random at Billings College will have a weight between 160 and 181 pounds. Calculate (**b**) $P(151 < X < 160)$, (**c**) $P(151 < X < 181)$, and (**d**) $P(X \geq 181)$. (**e**) Interpret part (c).

*Solution: Preliminaries:* Here, we have $\mu = 160$ pounds, and $\sigma = 15$ pounds. Thus, the correspondence equation between $x$ and $z$ is

$$z = (x - 160)/15 \tag{1}$$

The correspondence is indicated in Figure 10-3 by the parallel $z$- and $x$-axes, where directly below any $x$ value is the corresponding $z$ value. Thus, for example, the $x$ value 181 and the $z$ value 1.4 correspond to one another. Refer to Figure 10-3 for parts (a) through (d).

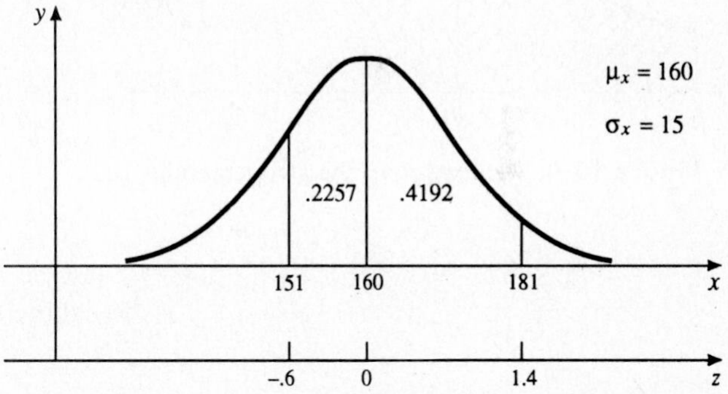

(**a**) Using Theorem 10.4, we have

$$P(160 < X < 181) = P(0 < Z < 1.4) \tag{2}$$

since for $x = \mu = 160$, we have $z = (160 - 160)/15 = 0$, and for $x = 181$, we have $z = (181 - 160)/15 = 1.4$. From Table VI, we have

$$P(0 < Z < 1.4) = .4192 \tag{3}$$

and thus from Eqs. (2) and (3), we have

$$P(160 < X < 181) = .4192 \tag{4}$$

**Figure 10-3.** Normal curve diagram for Example 10-6.

(**b**) Here,

$$P(151 < X < 160) = P(-.6 < Z < 0) \tag{5}$$

since for $x = 151$, we have $z = (151 - 16)/15 = -.6$. Now from part (b) of Theorem 10.5, and Table VI, we have

$$P(-.6 < Z < 0) = P(0 < Z < .6) = .2257 \tag{6}$$

Thus, from Eqs. (5) and (6),

$$P(151 < X < 160) = .2257 \tag{7}$$

(**c**) Using the area property of $P(a < X < b)$ for a continuous random variable $X$, we have that

$$P(151 < X < 181) = P(151 < X < 160) + P(160 < X < 181) \tag{8}$$

Thus, using the results from parts (b) and (c), we have

$$P(151 < X < 181) = .2257 + .4192 = .6449 \qquad (9)$$

The summing of probabilities [or areas] is illustrated nicely in Figure 10-3.

(d) Again thinking in terms of areas, we have

$$P(X \geq 181) = P(X > 160) - P(160 < X < 181) \qquad (10)$$

Now, $P(X > 160) = .50$ since the mean of $X$ is 160 pounds, and each normal curve is symmetrical with respect to its mean. Thus, using the result from part (a) in Eq. (10), we get

$$P(X \geq 181) = .5000 - .4192 = .0808 \qquad (11)$$

(e) The probability $P(151 < X < 181)$ can be interpreted in either of the following ways:

(i) The probability that a male drawn at random [from among the male students at Billings College] weighs between 151 and 181 pounds is .6449.

(ii) The fraction of male students at Billings College who weigh between 151 and 181 pounds is .6449 [or 64.49%].

---

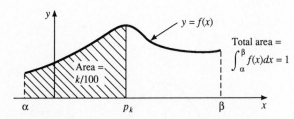

**Figure 10-4.** Illustration of the *k*th percentile, $p_k$.

## B.   Percentiles for a probability distribution of a continuous random variable

Refer to Figure 10-4, in which we have the graph of a pdf of a typical continuous random variable. The $k$th percentile $p_k$ is a value of random variable $X$ such that the area to the left of $p_k$ is equal to $k/100$. For example, the 95th percentile $p_{95}$ is such that $P(X \leq p_{95}) = 95/100 = .95$.

**Definition 10.2:** The $k$th percentile $p_k$ is a value of random variable $X$ such that

$$P(X \leq p_k) = \frac{k}{100} \qquad (a)$$

Recalling that $P(X \leq p_k)$ is the cdf $F$ evaluated at $p_k$ it is also correct to say that

$$F(p_k) = \frac{k}{100} \qquad (b)$$

*notes*

(i) The median $Md$ is an often-used measure of average value for a random variable $X$. The median is the same as the 50th percentile; that is, $MD = p_{50}$. [The other popular measure of average for a random variable $X$ is the mean $E(X)$ or $\mu$.] Thus, the median $Md$ is such that the area or probability to the left of $Md$ is equal to 50/100 or 1/2 of the total area under the graph of the probability distribution.

(ii) The three quartiles $Q_1$, $Q_2$, and $Q_3$ divide the probability distribution into four equal probability [or area] portions. In fact, $Q_1 = p_{25}$, $Q_2 = Md = p_{50}$, and $Q_3 = p_{75}$.

(iii) A commonly used measure of spread or deviation for a random variable $X$ is the interquartile range, which equals $Q_3 - Q_1$ or $p_{75} - p_{25}$. [Another major measure of spread or deviation for a random variable $X$ is, of course, the standard deviation, $\sigma_x$.]

**EXAMPLE 10-7:** Given a random variable $X$ for which $f(x) = (\frac{2}{9})x$ for $0 < x < 3$, and $f(x) = 0$ elsewhere. Determine (a) the 25th percentile, (b) the 50th percentile, and (c) the 75th percentile. (d) Compute the median and the interquartile range.

*Solution:* The cumulative distribution function evaluated at $x_0$, $F(x_0)$, is given by

$$F(X_0) = P(X \le x_0) = \left(\frac{2}{9}\right) \int_0^{x_0} x\, dx = (x_0)^2/9 \quad \textbf{(1)}$$

for $0 \le x_0 \le 3$. Refer to Figure 10-5.

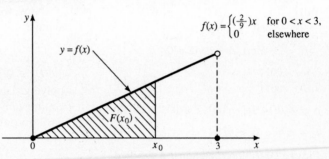

(a) If $x_0 = p_{25}$, the 25th percentile, we have $F(X_0) = 25/100$. Thus, Eq. (1) becomes

$$\frac{(x_0)^2}{9} = \frac{25}{100} = \frac{1}{4} \quad \text{or} \quad x_0 = \sqrt{9/4} = 1.5 \quad \textbf{(2)}$$

**Figure 10-5.** Calculation of percentiles in Example 10-7.

Thus, $p_{25} = 1.5$.

(b) If $x_0 = p_{50}$, the 50th percentile [or median], we have $F(X_0) = 50/100$. Thus, Eq. (1) becomes

$$\frac{(x_0)^2}{9} = \frac{1}{2} \quad \text{or} \quad x_0 = \sqrt{4.5} = 2.121 \quad \textbf{(3)}$$

Thus, $p_{50} = 2.121$.

(c) If $x_0 = p_{75}$, the 75th percentile, we have $F(X_0) = 75/100$. Thus, Eq. (1) becomes

$$\frac{(x_0)^2}{9} = \frac{75}{100} = \frac{3}{4} \quad \text{or} \quad x_0 = \sqrt{6.75} = 2.598 \quad \textbf{(4)}$$

Thus, $p_{75} = 2.598$.

(d) Thus, using the results from parts (a), (b), and (c), we have

$$\text{Median} = Md = Q_2 = p_{50} = 2.121 \quad \textbf{(5)}$$

and

$$\text{Interquartile range} = Q_3 - Q_1 = 2.598 - 1.5 = 1.098 \quad \textbf{(6)}$$

*note:* The mean for the random variable $X$ is $E(X) = (\frac{2}{9}) \int_0^3 x^2\, dx = 2$.

**EXAMPLE 10-8:** Refer to the normal curve data of Example 10-6. (a) Determine the weight $x_1$, such that 97.5% of all male weights lie below the weight $x_1$. [This $x_1$ is $p_{97.5}$, the 97.5th percentile.] Determine those weights that are (b) the median and (c) the 75th percentile.

*Solution*

(a) Refer to Figure 10-6. Ultimately, we wish to calculate $x_1$, the 97.5th percentile. The quantity $z_1$ corresponds to $x_1$. (See the parallel $x$- and $z$-axes.) The numbers in the body of the graph are probabilities [or areas]. Working with the $z$-axis, we see that $z_1$ is such that $P(Z \le z_1) = .975$. Thus, $P(0 < Z \le z_1) = .475$, and Table VI indicates that $z_1 = 1.96$. Turning to the correspondence equation relating $z_1$ and $x_1$, we have

$$x_1 = \sigma z_1 + \mu = 15(1.96) + 160 = 189.4 \text{ pounds} \quad \textbf{(1)}$$

Thus, the 97.5th percentile weight is 189.4 pounds [i.e., $p_{97.5} = 189.4$ lb]. Observe that $z_1 = 1.96$ is the 97.5th percentile $z$ score.

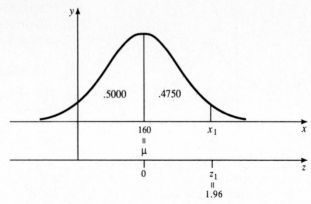

**Figure 10-6.** Normal curve diagram for part (a) of Example 10-8.

**(b)** Here, $Md = \mu = 160$ pounds since $P(X \leq 160) = .50$.

**(c)** The 75th percentile weight is a weight $x_1$ such that $P(X \leq x_1) = F(X_1) = .75$. Thus,

$$P(\mu < X \leq x_1) = .25 \qquad (2)$$

Now, working with the corresponding standardized random variable $Z$, we have

$$P(0 < Z \leq z_1) = .25 \qquad (3)$$

where $z = 0$ corresponds to $x = \mu$, and $z = z_1$ corresponds to $x = x_1$. Thus, applying Table VI to Eq. (3), we see that $z_1 \approx .675$, as the interpolation subtable of Figure 10-7 reveals. Therefore, from the correspondence equation relating $x_1$ and $z_1$, we have

$$x_1 = \sigma z_1 + \mu = 15(.675) + 160 = 170.125 \approx 170.1 \text{ lb} \qquad (4)$$

and $p_{75} \approx 170.1$ lb.

| $z_1$ | $P(0 < Z < z_1)$ |
|---|---|

$$.010 \begin{bmatrix} \Delta \begin{bmatrix} .670 \\ z_1 \\ .680 \end{bmatrix} & \begin{bmatrix} .2486 \\ .2500 \\ .2517 \end{bmatrix} .0014 \end{bmatrix} .0031$$

Here, $\Delta = z_1 - .670$

$$\frac{\Delta}{.010} = \frac{.0014}{.0031} \qquad (a)$$

$$\Delta = .010(14/31) = .00452 \approx .005 \qquad (b)$$

$$z_1 = .670 + \Delta \approx .675 \qquad (c)$$

**Figure 10-7.** Interpolation in a standard normal table for Example 10-8.

*note:* A normal distribution is completely determined from a probability point of view if the parameters $\mu$ and $\sigma$ are specified.

## 10-3. Skewness of a Random Variable

If a probability distribution has a vertical axis of symmetry [as with a normal distribution or a binomial distribution with $p = q = .5$], it is said to be *symmetrical* or *nonskewed*. The quantity $\mu_3 = E[(X - \mu)^3]$, the third moment about the mean, is a measure of skewness.

**Definition 10.3 (Coefficient of Skewness):** The *coefficient of skewness*, $\alpha_3$, is defined as follows:

$$\alpha_3 = \frac{E[(X - \mu)^3]}{\sigma^3}$$

The quantity $\alpha_3$ is a dimensionless quantity since both the numerator and denominator of $\alpha_3$ have the same dimension [namely that of $x$, cubed].

Sketches of typical pdf's having positive, negative, and zero skewness are shown in Figure 10-8. If there is a tendency for a longer "tail" to the right, the distribution has a positive skewness; it is also then said to be "skewed to the right." Likewise, a negative skewness is referred to as a skewness to the left.

*note:* If a distribution has zero skewness, it is symmetrical with respect to a vertical axis of symmetry, and conversely. In this case, the equation for the vertical axis of symmetry is $x = \mu$, where $\mu$ is the mean of the probability distribution. In such a symmetrical case, it is also true that the mean is equal to the median, that is, $\mu = Md$.

### A. Calculation of $\alpha_3$, the coefficient of skewness

The following relationship, similar to that contained in Theorem 9.5, expresses $E[(X - \mu)^3]$ in terms of moments of the form $E(X^r)$.

**Theorem 10.6**

$$E[(X - \mu)^3] = E(X^3) - 3\mu E(X^2) + 2\mu^3$$

Here, $\mu = E(X)$.

A proof of Theorem 10.6 is given in Problem 10-8.

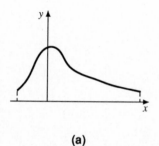

**(a)**

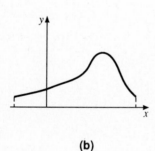

**(b)**

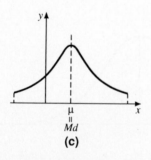

**(c)**

**Figure 10-8.** Illustrations of positive, negative, and zero skewness.
**(a)** Positive skewness ($\alpha_3 > 0$).
**(b)** Negative skewness ($\alpha_3 < 0$).
**(c)** Zero skewness ($\alpha_3 = 0$).

**EXAMPLE 10-9:** For the binomial distribution in which $n = 3$ and $p = 2/5$, calculate **(a)** $\mu$, **(b)** $\sigma$, and **(c)** the coefficient of skewness, $\alpha_3$. **(d)** Sketch a graph of the probability function.

*Solution:* **(a)**, **(b)**, **(c)**: Here,

$$f(x) = b(x; 3, 2/5) = \binom{3}{x}\left(\frac{2}{5}\right)^x \left(\frac{3}{5}\right)^{3-x} \qquad \text{for } x = 0, 1, 2, 3 \qquad \textbf{(1)}$$

The table of $f(x)$ values, which can also be read off from Table I of Appendix A [in decimal form], is displayed below. We observe that

$$E(X) = \mu = np = 3(2/5) = 6/5 \qquad \textbf{(2)}$$

$$\sigma = \sqrt{npq} = \sqrt{3(2/5)(3/5)} = \frac{\sqrt{18}}{5} = .8485 \qquad \textbf{(3)}$$

[See also Problem 9-25 for data on $\mu$ and $\sigma^2$.]

| $x$ | $f(x)$ | $x^2 f(x)$ | $x^3 f(x)$ |
|-----|--------|-----------|-----------|
| 0 | 27/125 | 0 | 0 |
| 1 | 54/125 | 54/125 | 54/125 |
| 2 | 36/125 | 144/125 | 288/125 |
| 3 | 8/125 | 72/125 | 216/125 |
| $\Sigma_x$ | 125/125 | 270/125 | 558/125 |

$$E(X^2) = \sum_x x^2 f(x) = 270/125 \qquad \textbf{(4)}$$

$$E(X^3) = \sum_x x^3 f(x) = 558/125 \qquad \textbf{(5)}$$

From Theorem 10.6, and the above results for $\mu$, $E(X^2)$, and $E(X^3)$, we have

$$\mu_3 = \frac{558 - 972 + 432}{125} = \frac{18}{125} \qquad \textbf{(6)}$$

Then,

$$\alpha_3 = \frac{18/125}{18\sqrt{18/125}} = \frac{1}{\sqrt{18}} = .2357 \qquad \textbf{(7)}$$

**(d)** The bar chart graph of $f(x)$ [see Figure 10-9] reveals a slight rightward skewness.

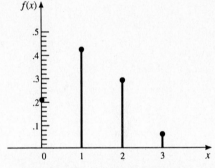

**Figure 10-9.** Bar chart of binomial distribution for $n = 3$, $p = .4$.

The coefficient of skewness is calculated for a continuous random variable in Problem 10-9.

## B. Skewness for binomial and normal distributions

**Theorem 10.7:** The coefficients of skewness for binomial and normal distributions are as follows:

$$\alpha_3 = \frac{(q - p)}{\sigma} = \frac{(q - p)}{\sqrt{npq}} \qquad \text{(binomial)}$$

$$\alpha_3 = 0 \qquad \text{(normal)}$$

Since any normal distribution has vertical symmetry [where, naturally, the vertical axis of symmetry has the equation $x = \mu$], it follows that all odd moments about the mean equal zero. That is,

$$\mu_k = E[(X - \mu)^k] = 0 \qquad \text{for } k = 1, 3, 5, \ldots$$

In particular, $\mu_3 = 0$, and thus, it follows from Definition 10.3 that $\alpha_3 = 0$ for a normal distribution.

**EXAMPLE 10-10:** Verify the calculation for $\alpha_3$ of Example 10-9 by making use of Theorem 10.7.

*Solution:* Substituting $n = 3$, $p = 2/5$, and $q = 3/5$ into the binomial equation for $\alpha_3$ in Theorem 10.7, we get

$$\alpha_3 = \frac{(3/5 - 2/5)}{\sqrt{18/5}} = \frac{1}{\sqrt{18}}$$

## 10-4. Normal Approximation to the Binomial Distribution

The normal distribution provides a close approximation to the binomial distribution when $n$, the number of binomial trials, is large.

### A. Normal approximation to the binomial distribution theorem

Recall that $b(x; n, p)$, the binomial probability of exactly $x$ successes in $n$ trials, is given by

$$b(x; n, p) = \binom{n}{x} p^x (1 - p)^{n-x} \qquad \text{for } x = 0, 1, 2, \ldots, n - 1, n \qquad \textbf{(I)}$$

For the normal distribution, the pdf is given by [Definition 10.1]:

$$f(x) = \frac{1}{\sigma\sqrt{2\pi}} e^{-(1/2)[(x-\mu)/\sigma]^2} \qquad \text{for } -\infty < x < \infty \qquad \textbf{(II)}$$

### Theorem 10.8 (Limit Theorem of DeMoivre and LaPlace)

(A) [*Ordinate Approximation*]: For sufficiently large $n$, the binomial probability $b(x; n, p)$ is well approximated by the normal pdf $f(x)$ at $x = 0, 1, 2, \ldots, n - 1, n$. For $\mu$ and $\sigma$ [as they appear in the normal $f(x)$ equation], the appropriate binomial expressions are used, namely $\mu_x = np$ and $\sigma_x = \sqrt{npq}$. Thus, we have the approximation

$$b(x; n, p) \approx \frac{1}{\sigma\sqrt{2\pi}} e^{-(1/2)[(x-\mu)/\sigma]^2}$$

for $x = 0, 1, 2, \ldots, n - 1, n$, where $\mu$ and $\sigma$ are replaced by $np$ and $\sqrt{npq}$, respectively.

(B) [*Area Approximation, with Continuity Correction*]: Let $P(a \le X \le b)$ denote the binomial probability of obtaining between $a$ and $b$ successes, inclusive, where each of $a$ and $b$ is an integer between 0 and $n$, inclusive. Then, for $n$ sufficiently large,

$$P(a \le X \le b) \approx \int_{a-.5}^{b+.5} f(x)\, dx$$

where, in the equation for the normal pdf $f(x)$, $\mu$ and $\sigma$ are taken to be the binomial expressions, namely $np$ and $\sqrt{npq}$, respectively.

*notes*

(i) For practical purposes, $n$ is said to be sufficiently large if both $np \ge 5$ and $nq \ge 5$.

(ii) Direct proofs of parts (A) and (B) are given in Appendix 1 of Kreysig (1970).

(iii) In part (B), the "continuity correction" relates to the limits $a - .5$ and

$b + .5$ in the integral. In Example 10-12, we will illustrate the continuity correction idea. The key point is that the probability distribution of a discrete random variable [here, a binomial rv] is being approximated by the probability distribution of a continuous random variable [here, a normal rv].

In a normal area approximation without a continuity correction, we would have

$$P(a \le X \le b) \approx \int_a^b f(x)\,dx$$

in place of the equation of part (B) of Theorem 10.8. This approximation is not, as a rule, as good as that given by part (B) of Theorem 10.8.

(iv) From the result of part (B), it follows that an area approximation for the binomial probability $b(k; n, p) = P(X = k)$ is given by $\int_{k-.5}^{k+.5} f(x)\,dx$.

---

**EXAMPLE 10-11:** A fair coin is tossed 16 times. [Equivalently, 16 fair coins are tossed, each once.] Let random variable $X$ denote the number of heads obtained. **(a)** Calculate the ordinate and area approximations of $P(X = 10)$, the probability of getting exactly 10 heads. Also, calculate the exact value. **(b)** Calculate the area approximation of $P(X \ge 10)$, and the exact value for this probability.

*Solution: Preliminaries:* Since $n = 16$ and $p = 1/2$, we have

$$\mu_x = np = 8 \quad \text{and} \quad nq = 8 \tag{1}$$

Thus, $n$ is sufficiently large for a normal approximation to be valid. Also,

$$\sigma_x = \sqrt{8(.5)} = \sqrt{4} = 2 \tag{2}$$

**(a)** *Ordinate Approximation for $P(X = 10)$:* Using part (a) of Theorem 10.8, we have

$$P(X = 10) \approx \frac{1}{2\sqrt{2\pi}} \exp\left[-\frac{1}{2}\left(\frac{10 - 8}{2}\right)^2\right] = .1210 \tag{3}$$

*Exact Value for $P(X = 10)$:* The exact value for this binomial probability can easily be calculated, or looked up in Table I of Appendix A.

$$P(X = 10) = b(10; 16, .5) = \frac{16!}{10!6!}\left(\frac{1}{2}\right)^{16} = .1222 \tag{4}$$

*Area Approximation for $P(X = 10)$:* Refer to Figure 10-10. Let us use the $B$ and $N$ subscripts to refer to the binomial and normal distributions, respectively. The rectangles shown are parts of the binomial histogram that applies for $n = 16$ and $p = 1/2$. The rectangle heights [indicated by $b(8)$, $b(9)$, and $b(10)$] are the binomial probabilities $P_B(X = 8)$, $P_B(X = 9)$, and $P_B(X = 10)$. For each rectangle, the height equals the area since the width is 1. For example, $P_B(X = 10) = A_B(9.5 \to 10.5)$, where the latter term refers to the rectangle area. Now, the area under the normal curve between 9.5 and 10.5, denoted by $A_N(9.5 \to 10.5)$, is approximately equal to $A_B(9.5 \to 10.5)$ since the normal ordinate at $x = 10$ is approximately equal to $P_B(X = 10)$, and the graph of $f_N(x)$ from 9.5 to 10.5 is roughly that of a straight line [with negative slope], as we see from Figure 10-10. Thus,

$$P_B(X = 10) = A_B(9.5 \to 10.5) \approx A_N(9.5 \to 10.5) = P_N(9.5 < X < 10.5) \tag{5}$$

Now, for $P_N(9.5 < X < 10.5)$, which is equal to $A_N(9.5 \to 10.5)$, we proceed as in Example 10-6. That is, we employ Theorem 10.4. Thus,

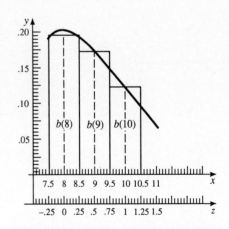

**Figure 10-10.** Normal approximation to binomial: ordinate and area approximations [Examples 10-11 and 10-12].

$$P_N(9.5 < X < 10.5) = P_N(.75 < Z < 1.25) = P_N(0 < Z < 1.25)$$

$$- P_N(0 < Z < .75) = .3944 - .2734 = .1210 \qquad \textbf{(6)}$$

Here, we made use of Table VI of Appendix A and the fact that $z = (9.5 - 8)/2 = .75$ and $z = (10.5 - 8)/2 = 1.25$ correspond to $x = 9.5$ and $10.5$, respectively. Thus, we have

$$P_B(X = 10) \approx .1210 \qquad \textbf{(7)}$$

It's a coincidence that the ordinate and area approximations are equal; we would expect them to be close, however.

**(b)** *Area Approximation for $P(X \geq 10)$:* We wish to approximate

$$P_B(X \geq 10) = P_B(X = 10) + P_B(X = 11) + \ldots + P_B(X = 16) \qquad \textbf{(8)}$$

Now, as in part (a), we have the following normal area approximation for $P_B(X = k)$, where $k$ is a nonnegative integer not greater than $n$:

$$P_B(X = k) \approx P_N(k - .5 < X < k + .5) \qquad \textbf{(9)}$$

If we now apply (9) for $k = 10, 11, \ldots, 16$, and substitute into (8), we get

$$P_B(X \geq 10) \approx P_N(9.5 < X < 10.5) + P_N(10.5 < X < 11.5) + \ldots$$

$$+ P_N(14.5 < X < 15.5) + P_N(15.5 < X < 16.5) \qquad \textbf{(10)}$$

Now, the total right side of (10) is equal to the area under the normal curve between $x = 9.5$ and $x = 16.5$ since each individual term is equivalent to an area under the normal curve for an $x$ interval of width 1. That is,

$$P_B(X \geq 10) \approx P_N(9.5 < X < 16.5) \qquad \textbf{(11)}$$

Here, Eq. (11) yields the result predicted by part (B) of Theorem 10.8. Now, for the right side of Eq. (11), we employ Theorem 10.4. Thus, we obtain

$$P_N(9.5 < X < 16.5) = P_N(.75 < Z < 4.25)$$

$$= P_N(0 < Z < 4.25) - P_N(0 < Z < .75) \qquad \textbf{(12)}$$

since $z = (9.5 - 8)/2 = .75$ and $z = (16.5 - 8)/2 = 4.25$ correspond to $x = 9.5$ and $x = 16.5$, respectively. [Note that Table VI goes up to $z_1 = 3.09$ in the main part of the table, but observe that values are listed at the bottom of the table for $z_1 = 4, 5,$ and $6$. Our convention will usually be to take $P(0 < Z < z_1) \approx .5000$, to four significant digits, for $z_1$ greater than 3.09.] Thus, substituting (12) in (11), and then using Table VI, we obtain

$$P_B(X \geq 10) \approx P_N(0 < Z < 4.25) - P_N(0 < Z < .75) = .5000 - .2734 = .2266$$
$$\textbf{(13)}$$

*Exact Value for $P(X \geq 10)$:* From Table I, we have

$$P_B(X \geq 10) = \sum_{x=10}^{16} b(x; 16, .5)$$

$$= .1222 + .0667 + \ldots + .0002 + .0000 = .2272 \qquad \textbf{(14)}$$

---

**note:** For the area approximation of part (b), we could have used

$$P_B(X \geq 10) \approx P_N(X > 9.5) = P_N(9.5 < X < \infty)$$

Actually, for $n$ sufficiently large [that is, for $np$ and $nq$ both equal to at least 5], we can consider the areas to the left of $x = -.5$, and to the right of $x = n + .5$ to be approximately equal to zero, to 4 decimal places. That is, for $n$ sufficiently large, we can take $P(X < -.5) = P(X > n + .5) \approx .0000$.

## B. More normal approximation of the binomial distribution examples

**EXAMPLE 10-12:** For the situation of Example 10-11, (a) compute the normal area approximation of $P_B(9 \leq X \leq 10)$, and the exact value. (b) Determine the normal area approximation if the continuity correction is neglected.

*Solution*

(a) *Area Approximation with Continuity Correction:*

$$P_B(9 \leq X \leq 10) = P_B(X = 9 \text{ or } X = 10) \approx P_N(8.5 < X < 10.5)$$

$$= P_N(.25 < Z < 1.25) = P_N(0 < Z < 1.25)$$

$$- P_N(0 < Z < .25) = .3944 - .0987 = .2957 \quad \textbf{(1)}$$

Here, $z = .25$ and $z = 1.25$ correspond to $x = 8.5$ and $x = 10.5$, respectively, and the two $P_N(0 < Z < z_1)$ probabilities were looked up in Table VI. Refer also to Figure 10-10.

    *Exact Value:*

$$P_B(9 \leq X \leq 10) = b(9; 16, 8) + b(10; 16, 8) = .1746 + .1222 = .2968 \quad \textbf{(2)}$$

Here, we used Table I of Appendix A.

(b) *Area Approximation Without Continuity Correction:* See note (iii) after Theorem 10.8. Here, we use

$$P_B(9 \leq X \leq 10) \approx P_N(9 < X < 10) = P_N(.5 < Z < 1)$$

$$= P_N(0 < Z < 1) - P_N(0 < Z < .5)$$

$$= .3413 - .1915 = .1498 \quad \textbf{(3)}$$

Observe that $z = .5$ and $z = 1$ correspond to $x = 9$ and $x = 10$, respectively. We see that the resulting probability estimate is seriously in error.

*note:* The accuracy of a normal area approximation without a continuity correction improves as the width of the interval increases. Refer to Example 10-13.

**EXAMPLE 10-13:** A manufacturer of cotter pins knows from experience that 20% of his product is defective. If a box contains 400 cotter pins, what is the approximate probability that at least 70 will be defective?

*Solution:* We have approximately a binomial situation here with $n = 400$ and $p = .20$. [Actually, we have a hypergeometric situation here, but the unknown population size is presumed to be extremely large with respect to $n$. Thus, as a first step, the binomial approximation applies. Refer to the comments after Example 5-28.] Since $\mu_x = np = 80$ and $nq = n - np = 320$, the normal approximation to the binomial should be accurate. Here, $\sigma_x = \sqrt{npq} = \sqrt{80(.8)} = 8$. Now, we wish to calculate $P_B(X \geq 70)$. From part (B) of Theorem 10.8, we have

$$P_B(X \geq 70) = P_B(70 \leq X \leq 400) \approx P_N(69.5 \leq X \leq 400.5) \quad \textbf{(1)}$$

[Here, we could just as well use $P_B(X \geq 70) \approx P_N(X \geq 69.5) = P_N(69.5 \leq X < \infty)$.]

    Refer to Figure 10-11.

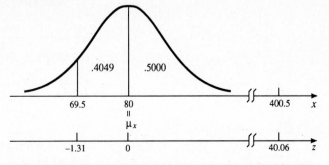

**Figure 10-11.** Normal area diagram for Example 10-13.

$$\text{For} \quad x = 69.5, \qquad z = (69.5 - 80)/8 \approx -1.31 \qquad (2)$$

$$\text{For} \quad x = 400.5, \qquad z = (400.5 - 80)/8 \approx 40.06 \qquad (3)$$

Thus,

$$P_B(X \geq 70) \approx P_N(-1.31 \leq Z \leq 40.06)$$
$$= P_N(-1.31 \leq Z \leq 0) + .5000$$
$$= .4049 + .5000 = .9049 \qquad (4)$$

*note:* If we ignore the continuity correction, we end up with

$$P_B(X \geq 70) \approx P_N(70 \leq X \leq 400) = P_N(-1.25 \leq Z \leq 40)$$
$$= P_N(-1.25 \leq Z \leq 0) + .5000 = .3944 + .5000 = .8944$$
$$(5)$$

The difference between the approximations given in Eqs. (4) and (5) is slight because the width of the interval of interest for the binomial random variable—here, $70 \leq X \leq 400$—is fairly large.

## C. Another moment generating function uniqueness theorem

In Theorem 9.12 of Section 9-4D, we indicated a useful moment generating function uniqueness theorem. A similar uniqueness theorem is the following, whose proof is also beyond the scope of this book.

**Theorem 10.9:** Suppose $V$ is a random variable with mgf $M_V(t)$ and cumulative distribution function $F$. Suppose that $V_1, V_2, V_3, \ldots$, constitute a sequence of random variables with mgf's $M_{V_1}(t), M_{V_2}(t), M_{V_3}(t), \ldots$, and cdf's $F_1, F_2, F_3, \ldots$.

Also, suppose that $\lim_{n \to \infty} M_{V_n}(t) = M_V(t)$ for all values of $t$ in some interval containing $t = 0$. Then, $\lim_{n \to \infty} F_n(v) = F(v)$ for all real values of random variable $V$.

Theorem 10.9 will prove useful in the proof of Theorem 10.12 [see Example 10-15 of Section 10-4D]. We will gradually lead up to that theorem. In particular, Theorems 10.10 and 10.11 will provide a convenient route.

Suppose that $X$ is a normal random variable with mean $\mu$ and standard deviation $\sigma$. We know from Theorem 10.3 that $Z = (X - \mu)/\sigma$, the standardized random variable corresponding to $X$, is also a normal random variable.

**Theorem 10.10:** The standardized random variable $Z = (X - \mu)/\sigma$ corresponding to a normal random variable $X$ has the following moment generating function: $M_Z(t) = e^{t^2/2}$.

Recall that we refer to the $Z$ of Theorem 10.10 as a standard [or standardized] normal random variable.

**EXAMPLE 10-14:** Prove Theorem 10.10.

**Proof:** It follows from Theorem 10.1 that the mgf for a normal random variable $V$ with mean $\mu_V$ and standard deviation $\sigma_V$ is given by

$$M_V(t) = \exp(\mu_V t) \exp(\sigma_V^2 t^2/2) \qquad (1)$$

Since $Z$ is a normal random variable with mean $\mu_Z = 0$ and variance $\sigma_Z^2 = 1$, it follows from Eq. (1) that

$$M_Z(t) = \exp(0t) \exp(1 \cdot t^2/2) = \exp(t^2/2) = e^{t^2/2} \qquad (2) \quad \square$$

**Theorem 10.11:** Suppose that $X_n$ is a binomial random variable for a binomial situation with $n$ trials, and probability of success on a single trial equal to $p$. [Thus, the mean and standard deviation of $X_n$ are $\mu = np$ and $\sigma = \sqrt{npq}$.] Consider $Z_n = (X_n - \mu)/\sigma = (X_n - np)/\sqrt{npq}$, where $Z_n$ symbolizes the standardized random variable corresponding to binomial random variable $X_n$. Then the moment generating function of $Z_n$ approaches that of a standard normal random variable as $n \to \infty$. That is, $\lim_{n\to\infty} M_{Z_n}(t) = e^{t^2/2}$.

*note:* In previous discussions of random variables pertaining to a binomial distribution, we used the symbols $X$ and $Z$, and not $X_n$ and $Z_n$, as was done in Theorem 10.11. Our purpose for using $X_n$ and $Z_n$ in Theorem 10.11 was to stress the dependence on $n$, the number of binomial trials.

A proof of Theorem 10.11 is given in Problem 10-11.

## D. A proof of the normal approximation to the binomial theorem

**Theorem 10.12:** The cdf of the standardized binomial random variable $Z_n$ approaches the cdf of a standardized normal random variable as $n \to \infty$.

---

**EXAMPLE 10-15:** Prove Theorem 10.12.

*Proof:* We have from Theorem 10.11 that

$$\lim_{n\to\infty} M_{Z_n}(t) = e^{t^2/2} \tag{1}$$

Here, $Z_n$ is the standardized binomial random variable pertaining to $n$ binomial trials. We also know that

$$M_Z(t) = e^{t^2/2} \tag{2}$$

for a standardized normal random variable [Theorem 10.10]. Thus, we see that

$$\lim_{n\to\infty} M_{Z_n}(t) = M_Z(t) \tag{3}$$

Now, let $F_n$ symbolize the cdf of a standard binomial random variable for $n$ trials, and let $F$ symbolize the cdf of a standardized normal random variable. Thus, it follows from Theorem 10.9 [replace $V_n$ and $V$ by $Z_n$ and $Z$, respectively] that

$$\lim_{n\to\infty} F_n(z) = F(z) \tag{4}$$

Here, $F(z)$ denotes the cdf of a standardized normal random variable evaluated at $z$. [Remember that $F(z) = (1/\sqrt{2\pi})\int_{t=-\infty}^{z} e^{-t^2/2}\, dt$ since the pdf is given by $f(t) = (1/\sqrt{2\pi})e^{-t^2/2}$ for a standardized normal random variable.] □

---

Theorem 10.12 implies that the cdf of a binomial random variable $X_n$ [for $n$ trials, and with $p$ equal to the probability of success in a single trial] will be approximately equal to the cdf of a normal random variable $X$ with mean $np$ and variance $npq$ provided $n$ is sufficiently large.

---

**EXAMPLE 10-16:** Demonstrate the validity of part (B) of Theorem 10.8.

*Solution:* For this example, let $F(\ )$ and $F_N(\ )$, respectively, indicate the cdf's for a binomial rv and normal rv. That is, capital $N$ indicates a "normal" distribution.

From the statement immediately preceding this example, it follows that

$$F(b + .5) - F(a - .5) \approx F_N(b + .5) - F_N(a - .5) \tag{1}$$

On the left side, we have a difference of cdf's for a binomial rv, and on the right side, we have a difference of cdf's for a normal rv. [For both the binomial rv and the approximating normal rv, the mean and standard deviation are given by $\mu_x = np$ and $\sigma_x = \sqrt{npq}$.] Also, in Eq. (1), both $a$ and $b$ denote nonnegative integers, such that neither exceeds $n$.

Now, the left side of Eq. (1) can be expressed as follows:

$$F(b + .5) - F(a - .5) = P(X = a) + P(X = a + 1) + \ldots + P(X = b)$$

$$= \sum_{x=a}^{b} P(X = x) = P(a \leq X \leq b) \tag{2}$$

The two summation expressions in Eq. (2) containing the $P$ terms represent the sum of binomial probabilities between $X = a$ and $X = b$; inclusive.

Now, the right side of Eq. (1) can be written as

$$P(a - .5 < X < b + .5) = \int_{a-.5}^{b+.5} f(x)\,dx \tag{3}$$

where either expression represents the normal curve probability [or area] between $x = a - .5$ and $x = b + .5$. Also, either $<$ can be replaced by $\leq$ since the normal rv is continuous.

Replacing the left side of Eq. (1) by the rightmost expression of Eq. (2), and the right side of Eq. (1) by the right side expression of Eq. (3), we obtain

$$P(a \leq X \leq b) \approx \int_{a-.5}^{b+.5} f(x)\,dx \tag{4}$$

Equation (4) is a repeat of part (B) of Theorem 10.8. [Remember that $P(a \leq X \leq b)$ represents the binomial probability of obtaining between $a$ and $b$ successes, inclusive, where $a$ and $b$ are nonnegative integers, such that neither exceeds $n$.]

---

### E.  The binomial proportion of successes random variable

Recall that $X$ and $X_n$ are symbols we have used for the binomial random variable for the number of successes in $n$ trials. [Another frequently used symbol is $S_n$.] Let us use $X$ in the current discussion. The possible values for $X$ are 0, 1, 2, ..., $n$, and

$$P(X = x) = f(x) = b(x; n, p) = \binom{n}{x} p^x q^{n-x} \tag{I}$$

Also,

$$E(X) = np \quad \text{and} \quad \text{Var}(X) = npq = np(1 - p) \tag{IIa, IIb}$$

**Definition 10.4:** The binomial proportion of successes [in $n$ trials] random variable, $\hat{P}$, is given by

$$\hat{P} = \frac{X}{n}$$

where $X$ denotes the binomial random variable for the number of successes in $n$ trials. Here, the possible values for $\hat{P}$ are 0, $1/n$, $2/n$, ..., $(n - 1)/n$, and $n/n = 1$.

**Theorem 10.13**

$$E(\hat{P}) = \mu_{\hat{p}} = p \tag{a}$$

and

$$\text{Var}(\hat{P}) = \sigma_{\hat{P}}^2 = \frac{p(1-p)}{n} \qquad \textbf{(b)}$$

Thus,

$$\sigma_{\hat{P}} = \sqrt{pq/n} = \sqrt{p(1-p)/n} \qquad \textbf{(c)}$$

A proof of Theorem 10.13 is given in Problem 10-12.

---

**EXAMPLE 10-17:** Consider the binomial situation for which $n = 3$ and $p = 2/5 = .4$. **(a)** Display the probability function for the random variable $\hat{P}$ in table form. **(b)** Use the table of part (a) to compute $E(\hat{P})$. [Note that we consider this binomial situation in Problem 9-25 and Example 10-9.]

*Solution*

**(a)** Here, the probability function for $X$ is given by

$$P(X = x) = \frac{3!}{x!(3-x)!}(2/5)^x(3/5)^{3-x} \qquad \text{for } x = 0, 1, 2, 3 \qquad \textbf{(1)}$$

Now, the random variable $\hat{P}$ is related to $X$ by $\hat{P} = X/3$. Thus, it follows that

$$P(\hat{P} = x/3) = P(X = x) \qquad \textbf{(2)}$$

since the event $X = x$ is equivalent to the event $\hat{P} = x/3$. [To see this, merely divide both sides of $X = x$ by 3.] From Eqs. (1) and (2), it follows that

$$P\left(\hat{P} = \frac{x}{3}\right) = \frac{3!}{x!(3-x)!}(2/5)^x(3/5)^{3-x} \qquad \text{for } x = 0, 1, 2, 3 \qquad \textbf{(3)}$$

In the tables below, we shall denote $P(X = x)$ as $f(x)$ and $P(\hat{P} = x/3)$ as $\hat{f}(\hat{p})$, where $\hat{p} = x/3$. Observe that we are using the symbol $\hat{p}$ to denote a value of the random variable $\hat{P}$. [Clearly, in general, $\hat{p} = x/n$, where $x$ denotes a value of random variable $X$.]

| $x$ | $f(x)$ | $\hat{p}$ | $\hat{f}(\hat{p})$ | $\hat{p}\hat{f}(\hat{p})$ |
|---|---|---|---|---|
| 0 | $(3/5)^3 = 27/125$ | 0 | 27/125 | 0 |
| 1 | $3(3/5)^2(2/5) = 54/125$ | 1/3 | 54/125 | 54/375 |
| 2 | $3(3/5)(2/5)^2 = 36/125$ | 2/3 | 36/125 | 72/375 |
| 3 | $(2/5)^3 = 8/125$ | 1 | 8/125 | 24/375 |
| $\sum_{x=0}^{3}$ | $125/125 = 1$ | $\sum_{\hat{p}=0}^{1}$ | 1 | $150/375 = 2/5$ |

**(b)** In the third column of the table on the right, we have a $\hat{p}\hat{f}(\hat{p})$ column. The sum of the entries in that column yields the value of $E(\hat{P})$. That is,

$$E(\hat{P}) = \sum_{\hat{p}} \hat{p}\hat{f}(\hat{p}) = 150/375 = 2/5 \qquad \textbf{(4)}$$

Thus, part (a) of Theorem 10.13 is verified since $E(\hat{P})$ is equal to $p$ here.

---

Theorems similar to Theorems 10.8 and 10.12 apply to the binomial proportion random variable, $\hat{P}$. The following theorem is analogous to part (B) of Theorem 10.8.

**Theorem 10.14 (Normal Approximation for $\hat{P}$, with Continuity Correction):** Suppose that $P(a/n \leq \hat{P} \leq b/n)$ denotes the probability that the binomial random variable $\hat{P}$ is between $a/n$ and $b/n$, inclusive. [Here, $a$ and $b$ are nonnegative integers such that each does not exceed $n$.] Then, for sufficiently large $n$,

$$P(a/n \le \hat{P} \le b/n) \approx \int_{z_a'}^{z_b'} f(z)\, dz \qquad \text{(a)}$$

where

$$z_a' = \frac{a - .5 - np}{\sqrt{npq}} = \frac{(a - .5)/n - p}{\sqrt{pq/n}} \qquad \text{(b)}$$

and

$$z_b' = \frac{b + .5 - np}{\sqrt{npq}} = \frac{(b + .5)/n - p}{\sqrt{pq/n}} \qquad \text{(c)}$$

On the right side of Eq. (a), we have the integral of the pdf for a standard normal random variable.

***notes***

(i) Practically speaking, $n$ is said to be sufficiently large if both $np \ge 5$ and $nq \ge 5$. [Compare with note (i) that follows Theorem 10.8.]

(ii) The approximate equation in Theorem 10.14 is said to contain a "continuity correction." Often, in practice, we will ignore the continuity correction, and use

$$P(a/n \le \hat{P} \le b/n) \approx \int_{z_a}^{z_b} f(z)\, dz$$

where

$$z_a = \frac{a - np}{\sqrt{npq}} = \frac{a/n - p}{\sqrt{pq/n}} \quad \text{and} \quad z_b = \frac{b - np}{\sqrt{npq}} = \frac{b/n - p}{\sqrt{pq/n}}$$

This is fairly accurate provided that the difference between $b$ and $a$ is fairly big.

(iii) Observe that the $z$ corresponding to binomial random variable $X$ [where $X$ stands for the number of successes in $n$ trials] is identical to the $z$ corresponding to the proportion of successes random variable, $\hat{P}$. Here is a demonstration of this fact. First of all,

$$z_{\text{for } X} = \frac{x - np}{\sqrt{npq}} \qquad \text{(1)}$$

since $E(X) = np$ and $\sigma_x = \sqrt{npq}$. Also,

$$z_{\text{for } \hat{P}} = \frac{\hat{p} - p}{\sqrt{pq/n}} \qquad \text{(2)}$$

since $E(\hat{P}) = p$ and $\sigma_{\hat{p}} = \sqrt{pq/n}$.

Dividing numerator and denominator on the right side of Eq. (1) by $n$ yields

$$z_{\text{for } X} = \frac{x/n - p}{\dfrac{\sqrt{npq}}{n}} = \frac{x/n - p}{\sqrt{\dfrac{npq}{n^2}}} = \frac{x/n - p}{\sqrt{\dfrac{npq}{n^2}}} = \frac{\hat{p} - p}{\sqrt{\dfrac{pq}{n}}} = z_{\text{for } \hat{P}} \qquad \text{(3)}$$

[Here, we have used the algebraic result that $\sqrt{u}/\sqrt{v} = \sqrt{u/v}$.] Thus, comparing the extreme left and right expressions of Eq. (3), we see that we have shown that the $z$ corresponding to random variable $X$ is equal to the $z$ corresponding to random variable $\hat{P}$.

## F. Examples involving the normal approximation for the binomial proportion random variable

**EXAMPLE 10-18:** Refer to Example 10-13. A manufacturer knows that 20% of the cotter pins he produces are defective. (a) In a box containing 400 cotter pins, what is the probability that at least 17.5% will be defective? Use the continuity correction here. (b) Repeat part (a) if the continuity correction is neglected.

*Solution*

(a) Here, 17.5% is the same as a proportion of $17.5/100 = .175$. Thus, we wish to calculate the probability $P(\hat{P} \geq .175)$, or, equivalently,

$$P(.175 \leq \hat{P} \leq 1) \tag{1}$$

Now, $\hat{p} = .175$ means $x/400 = .175$, and thus, $x = (400)(.175) = 70$. Also, $\hat{p} = 1$ means $x = (400)(1) = 400$. Thus, we have $a = 70$ and $b = 400$ in an application of Theorem 10.14 to this example.

Also, $E(\hat{P}) = p = .2$, $\text{Var}(\hat{P}) = pq/n = .16/400$, and $\sigma_{\hat{p}} = .4/20 = .02$. Thus, we have

$$P(.175 \leq \hat{P} \leq 1) \approx \int_{z'_a}^{z'_b} f(z)\,dz \tag{2}$$

where

$$z'_a = \frac{(70 - .5)/400 - .2}{.02} \approx -1.31 \tag{3}$$

and

$$z'_b = \frac{(400 + .5)/400 - .2}{.02} \approx 40.06 \tag{4}$$

Thus, we have

$$P(.175 \leq \hat{P} \leq 1) \approx \int_{-1.31}^{40.06} f(z)\,dz \tag{5}$$

We therefore end up with the same integral [for a standard normal rv] as in Example 10-13. This should be no surprise as the extreme values of $Z$, namely $-1.31$ and $40.06$, are the same in both examples. [Refer to the horizontal $z$ scale in Figure 10-11. Recall that the $z$ for random variable $X$ is the same as the $z$ for random variable $\hat{P}$.] Let us now use $P_N(r < Z < s)$ as the symbol for $\int_r^s f(z)\,dz$, the integral of a standard normal rv from $z = r$ to $z = s$. Thus, for $P(\hat{P} \geq .175)$, or $P(.175 \leq \hat{P} \leq 1)$, we have

$$P(\hat{P} \geq .175) \approx P_N(-1.31 < Z < 40.06) = P_N(-1.31 < Z < 0)$$
$$+ P_N(0 < Z < 40.06) = .4049 + .5000 = .9049 \tag{6}$$

Observe that we obtain the same answer as in Example 10-13. This is not surprising since $P(\hat{P} \geq .175)$ is equivalent to $P(X \geq 70)$.

(b) Here, for $P(\hat{P} \geq .175)$ or $P(.175 \leq \hat{P} \leq 1)$, we use

$$P(.175 \leq \hat{P} \leq 1) \approx \int_{z_a}^{z_b} f(z)\,dz = P_N(z_a < Z < z_b) \tag{1}$$

with

$$z_a = \frac{70/400 - .2}{.02} = \frac{.175 - .2}{.02} = -1.25 \tag{2}$$

and

$$z_b = \frac{1 - .2}{.02} = 40 \qquad (3)$$

Refer to note (ii) after Theorem 10.14 for the general expressions for $z_a$ and $z_b$. Thus, we have

$$P(.175 \leq \hat{P} \leq 1) \approx P_N(-1.25 < Z < 0) + P_N(0 < Z < 40)$$

$$= .3944 + .5000 + .8944 \qquad (4)$$

This agrees with the answer found in the note at the end of Example 10-13. It is quite close to the answer found in part (a), thereby indicating that it is not serious in this example to neglect the continuity correction.

## 10-5.  A Law of Large Numbers

### A.  A statement and proof of the law

**Theorem 10.15 (A Law of Large Numbers—Bernoulli's Form):**  Let $\mathscr{E}$ denote any experiment and let $A$ denote an event associated with $\mathscr{E}$. Let $P(A) = p$, where $p$ is considered to be a constant. Consider $n$ independent repetitions [or trials] of $\mathscr{E}$, and let random variable $X$ denote the number of times that event $A$ occurs among the $n$ trials. Let random variable $\hat{P}$ stand for $X/n$. [Thus, $X$ and $\hat{P}$ are *binomial* random variables.]

Then, for any positive constant $c$, the probability is at least $1 - p(1-p)/[nc^2]$ that $\hat{P}$, the *proportion* of occurrences of $A$, falls between $p - c$ and $p + c$. That is, for any $c > 0$,

$$P(p - c < \hat{P} < p + c) \geq 1 - \frac{p(1-p)}{nc^2} \qquad (a)$$

It follows from (a) that as $n \to \infty$, the probability approaches 1 that $\hat{P}$ will differ from $p$ by less than any arbitrary positive constant $c$, no matter how shall $c$ is. That is,

$$\lim_{n \to \infty} P(p - c < \hat{P} < p + c) = 1 \qquad \text{for any } c > 0 \qquad (b)$$

*notes*

(i)  Theorem 10.15 is called a *weak* law of large numbers.
(ii)  Looking at (b) of Theorem 10.15, we see that the proportion of occurrences, or *relative frequency*, for any event $A$ [namely, $\hat{P}$] "converges" in some sense to $P(A)$ as $n$ increases without bound. Observe that in Chapter 1, we used the symbol $\hat{f}$ for relative frequency. Note that in (b) of Theorem 10.15, we are not saying that $\lim_{n \to \infty} x/n = p$, or that $\lim_{n \to \infty} x = np$. It is fallacious to think that the number of successes, namely $x$, will be close to $np$ when $n$ is large.

---

**EXAMPLE 10-19:**  Prove Theorem 10.15.

***Proof:***  Given a random variable $Y$ for which $E(Y) = \mu$ and $\text{Var}(Y) = \sigma^2$. From part (b) of Theorem 9.8 [Chebyshev's theorem], we have that

$$P(\mu - c < Y < \mu + c) \geq 1 - \frac{\sigma^2}{c^2} \qquad (1)$$

Here, $c$ is any positive number. Suppose that $Y = \hat{P}$, the binomial proportion random variable. We know from Theorem 10.13 that $E(Y) = E(\hat{P}) = p$ and $\text{Var}(\hat{P}) = p(1 - p)/n$. Thus, replacing $Y$ terms by $\hat{P}$ terms in (1), we obtain

$$P(p - c < \hat{P} < p + c) \geq 1 - \frac{p(1 - p)}{nc^2} \qquad (2)$$

Thus, part (a) of Theorem 10.15 is established. Now, if we hold $c$ constant and let $n$ increase without bound in (2), we see that $\lim_{n \to \infty} p(1 - p)/[nc^2] = 0$. Hence, part (b) is established. $\qquad \square$

## B. Illustrating the law of large numbers

**EXAMPLE 10-20:** **(a)** Suppose a fair coin is tossed 100 times. Use Theorem 10.15 to determine a lower bound for the probability $P(.4 < \hat{P} < .6)$. Here, $\hat{P}$ denotes the proportion of heads in 100 tosses. **(b)** Use Theorem 10.14 to estimate the value for $P(.4 < \hat{P} < .6)$.

*Solution*

**(a)** Here, $n = 100$, $p = P(\text{Head}) = .50$, and $c = .10$ [from $2c = (p + c) - (p - c) = .6 - .4$]. Thus, from part (a) of Theorem 10.15, we obtain

$$P(.4 < \hat{P} < .6) \geq 1 - \frac{(.5)^2}{100(.1)^2} = .75 \qquad (1)$$

Thus, a lower bound for $P(.4 < \hat{P} < .6)$ is .75.

**(b)** Let us neglect the continuity correction here. Now, $E(\hat{P}) = p = .5$, and $\sigma_{\hat{P}} = \sqrt{pq/n} = \sqrt{(.5)^2/100} = .05$. Corresponding to $\hat{P}_2 = .6$, we have

$$z_2 = (\hat{P}_2 - p)/\sigma_{\hat{P}} = \frac{(.6 - .5)}{.05} = 2 \qquad (2)$$

and for $\hat{P}_1 = .4$, we have $z_1 = -2$. Thus, we have

$$P(.4 < \hat{P} < .6) = P(-2 < Z < 2) \qquad (3)$$

The normal curve area approximation applies since $np = nq = 50 \gg 5$. From Table VI of Appendix A, we have $P(0 < Z < 2) = .4772$, and so

$$P(.4 < \hat{P} < .6) \approx 2(.4772) = .9544 \qquad (4)$$

This result is consistent with the lower bound of .75 predicted from part (a).

# 10-6. More Normal Curve Problems

**EXAMPLE 10-21:** Given a community with several hundred thousand voters in which the true fraction of Republicans is .4 [that is, $p_{\text{Rep.}} = .4$]. How large a sample should be taken so that the probability is .9544 that the sample proportion of Republicans is within .02 of $p_{\text{Rep.}}$?

*Solution:* Let the sample size be $n$. Assume that $n$ is small with respect to the number of voters in the community, $M$, so that we approximately have a binomial situation. [Here, we suppose that the sample is drawn without replacement—if the sample were drawn with replacement, then $n$ could have any value. See also the text in Section 5-4B right before Example 5-29. There, it is indicated that a hypergeometric distribution is well approximated by a binomial distribution if $n/M \ll .05$.] Also, we assume that $n$ is sufficiently large so that the normal approximation to the binomial is valid. Let us work with $\hat{P}$, the binomial

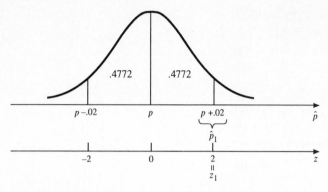

**Figure 10-12.** Normal curve approximation in Example 10-21.

proportion random variable, though we can just as well work with the binomial random variable $X$. Also, let us neglect the continuity correction. Refer to Figure 10-12.

We have

$$P(p - .02 < \hat{P} < p + .02) = .9544 \qquad (1)$$

and thus, by symmetry,

$$P(p < \hat{P} < p + .02) = .4772 \qquad (2)$$

Let $\hat{p}_1 = p + .02$, and let the corresponding $z$ value be $z_1$. We see from Table VI of Appendix A that

$$P(0 < Z < z_1) = .4772 \qquad \text{implies } z_1 = 2 \qquad (3)$$

Thus, $z_1 = 2$ corresponds to $\hat{p}_1 = p + .02 = .42$ since $p = .40$. From

$$z_1 = \frac{\hat{p}_1 - p}{\sqrt{\dfrac{p(1-p)}{n}}} \qquad (4)$$

we have

$$2 = \frac{.42 - .40}{\sqrt{\dfrac{(.4)(.6)}{n}}} = \frac{.02\sqrt{n}}{\sqrt{.24}} \qquad \text{and thus} \qquad \sqrt{n} = \frac{2\sqrt{.24}}{.02} \qquad (5)$$

Then, $n = (100)^2(.24) = 2400$. That is, a sample of size 2400 should be drawn. See Problem 10-14 for a related problem. (There, the proportion of Republicans, $p$, is unknown.)

---

**EXAMPLE 10-22:** A person is firing a gun at a rectangular target that is 10 feet wide and sufficiently high so that if a miss occurs it will be in the horizontal direction. The horizontal distance a person's shot hits from the centerline is normally distributed with $\sigma = 4$ feet. See Figure 10-13a. (a) What is the probability that a person will hit the target on a single shot? (b) If 200 shots are fired at the target, how many would be expected to hit the target? (c) What is the minimum number of shots to be fired so as to be certain with a probability of at least .90 that the target will be hit at least 40 times?

*Solution: Preliminaries:* Here, we take random variable $X$ to symbolize the distance in the horizontal direction [in feet] where the shot hits the target relative to the center of the target. The value $x$ of $X$ is negative to the left of the centerline. From the given information, $\mu_x = 0$ since we take the centerline to be at $x = 0$. Also, $X$ is given to be a normal rv.

**(a)** For a single shot,

$$P(\text{hit target}) = P(-5 \leq X \leq 5) \qquad (1)$$

Transforming to a standard normal rv $Z$ [see Figure 10-13b], we have

$$P(\text{hit target}) = P(-1.25 \leq Z \leq 1.25) = 2P(0 \leq Z \leq 1.25)$$

$$= 2(.3944) = .7888 \qquad (2)$$

after using Table VI. Recall that here $\sigma = 4$.

**(b)** Let random variable $Y$ represent the number of shots that hit the target out of 200 shots. Thus, $Y$ is a binomial rv with $p = .7888$ and

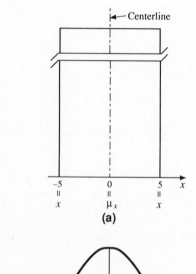

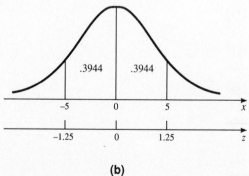

**Figure 10-13.** Diagrams for Example 10-22. **(a)** The target [$x$ is distance from centerline]. **(b)** Normal curve diagram for part (a).

$n = 200$. It follows that the expected number of shots that hit the target, $E(Y)$, is given by

$$E(Y) = np = 200(.7888) = 157.76 \qquad (3)$$

as we see from Theorem 9.14.

(c) Suppose the minimum number of shots that are fired is denoted by $m$ — this is our unknown. Let $V$, a binomial rv, signify the number of times the target is hit. As in part (b), the probability of hitting the target on a single shot is given by $p = .7888$. For $\mu_V$ and $\sigma_V$, we have [Theorem 9.14] that

$$\mu_V = m(.7888) \qquad \text{and} \qquad \sigma_V = \sqrt{m(.7888)(.2112)} \qquad (4), (5)$$

Refer to Figure 10-14. To find $m$ such that $P(V \geq 40)$ is at least .90, we consider

$$P(V \geq 40) = .90 \qquad (6)$$

Thus, treating $V$ as a normal random variable [assuming $m$ is sufficiently large], and employing the continuity correction, we have

$$P_N(V \geq 39.5) = .90 \qquad (7)$$

where the subscript $N$ indicates a normal probability distribution. Transforming to $Z$, the standard normal random variable corresponding to $V$, we get

$$P_N(Z \geq z_0) = .90 \qquad (8)$$

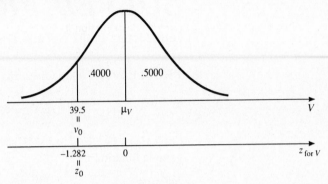

**Figure 10-14.** Normal curve diagram for part (c) of Example 10-22.

From Table VI of Appendix A, we see that $z_0 = -1.282$, after interpolating. Now $z_0 = -1.282$ corresponds to $v_0 = 39.5$ [refer to the parallel $z$ and $v$ axes in Figure 10-14]. Thus, using the correspondence equation $z_0 = (v_0 - \mu_V)/\sigma_V$, with the above values for $z_0$ and $v_0$, and the expressions for $\mu_V$ and $\sigma_V$ from Eqs. (4) and (5), we get

$$\frac{39.5 - m(.7888)}{\sqrt{m}\sqrt{(.7888)(.2112)}} = -1.282 \qquad (9)$$

After simplifying, we get

$$(.7888)m - (.52326)\sqrt{m} - 39.5 = 0 \qquad (10)$$

Letting $t = \sqrt{m}$, we get a quadratic equation involving $t^2$ [for $m$] and $t$. From the quadratic formula as applied to $t$, we get two solutions for $t$ as given by

$$t = \frac{.52326 \pm \sqrt{124.9042}}{2(.7888)} \qquad (11)$$

This yields $t_1 = \sqrt{m_1} = 7.41589$ and $t_2 = \sqrt{m_2} = -6.75253$. The correct $m$ value is $m_1$, where

$$m_1 = (7.41589)^2 = 54.9954 \approx 55 \qquad (12)$$

Thus, the minimum number of shots to be fired is 55.

---

*notes*

(i) For $m_1 = 54.9954$, we find that $\mu_V = E(V) = m_1 p \approx 43.38$. This conforms with Figure 10-14 since $39.5 < 43.38$. From $m_2 = (-6.75253)^2 = 45.5967$, we calculate $E(V) = 35.9667 \approx 35.97$. This is unreasonable since $35.97 < 39.5$.

(ii) One way of checking our calculation for $m$ is to substitute the value obtained into the left side of Eq. (9). Substitution of $m_1$ into the left side of Eq. (9)

yields $-1.282$, which is what we expect since the right side is $-1.282$. Substitution of $m_2$ into the left side of Eq. (9) yields $+1.282$, and this indicates an inconsistency.

(iii) The value $m_1 = 55$ is sufficiently large to ensure that the normal approximation to the binomial holds [$m_1 p = 43.38$ and $m_1 q = 11.62$].

# SOLVED PROBLEMS

## Introduction to the Normal Distribution

**PROBLEM 10-1** (a) Given the normal pdf in the form

$$f(x) = c \exp\left(-\frac{1}{2}\left(\frac{x-a}{b}\right)^2\right) \qquad \text{for } -\infty < x < \infty$$

Show that $c = 1/[b\sqrt{2\pi}]$. (b) Use the result from part (a) to show that $\Gamma(\frac{1}{2}) = \sqrt{\pi}$, where $\Gamma(\alpha)$ is the gamma function.

### Solution

(a) First, let $I = \int_{-\infty}^{\infty} f(x)\,dx$, with $f(x)$ as given above. [Later, we'll use the fact that $I = 1$. Note that it should be clear that the constant $c$ in the above equation for $f(x)$ has to be positive since $e^t > 0$ for any real number $t$. Also, we will suppose that $b$ is positive.] Let $s = (x-a)/b$. Thus, $dx = b\,ds$, and the integral expression for $I$ becomes

$$I = (bc) \int_{-\infty}^{\infty} e^{-s^2/2}\,ds \tag{1}$$

Now, $s$ is just a dummy variable of integration in Eq. (1). Thus, if we replace $s$ by $t$, we get the following equation for $I$:

$$I = (bc) \int_{-\infty}^{\infty} e^{-t^2/2}\,dt \tag{2}$$

The "trick" used to evaluate $(bc)$ is to consider $I^2$. Now, $I^2 = 1$ since $I = 1$. In (3) below, we rewrite the product of two equal single integrals as a double integral. If we let each of the single integrals be denoted as $J$, then the double integral equals $J^2$.

$$I^2 = 1 = (bc)^2 \left( \int_{-\infty}^{\infty} e^{-s^2/2}\,ds \right) \left( \int_{-\infty}^{\infty} e^{-t^2/2}\,dt \right)$$

$$= (bc)^2 \underbrace{\int_{s=-\infty}^{s=\infty} \int_{t=-\infty}^{t=\infty} e^{-(s^2+t^2)/2}\,dt\,ds}_{J^2} \tag{3}$$

Thus, from (3) we obtain

$$(bc)^2 = 1/J^2 \tag{4}$$

Let us introduce polar coordinates $r$ and $\theta$ in order to evaluate the double integral $J^2$ of (3). With $s = r\cos(\theta)$ and $t = r\sin(\theta)$, the element of area $dt\,ds$ transforms to $r\,dr\,d\theta$. Also, $s^2 + t^2 = r^2$. The region of integration given by $-\infty < t < \infty$, $-\infty < s < \infty$ in Eq. (3) is the entire $ts$ plane; in polar coordinates, it is given by $0 < r < \infty$, $0 < \theta < 2\pi$. Thus,

$$J^2 = \int_{\theta=0}^{2\pi} \left( \int_{r=0}^{\infty} r e^{-r^2/2}\,dr \right) d\theta \tag{5}$$

and

$$J^2 = \int_{\theta=0}^{2\pi} \left( -e^{-r^2/2} \Big]_{r=0}^{\infty} \right) d\theta = \int_{\theta=0}^{2\pi} 1 \cdot d\theta = 2\pi \tag{6}$$

Hence, $J = \sqrt{2\pi}$. Note that the positive square root applies since $J$ is equal to the integral $\int_{-\infty}^{\infty} e^{-s^2/2}\, ds$, which is positive. Now, from Eq. (4), we have

$$c = \frac{1}{bJ} = \frac{1}{b\sqrt{2\pi}} \tag{7}$$

b) First, observe from Definition 8.6 that

$$\Gamma(\alpha) = \int_{t=0}^{\infty} t^{\alpha-1} e^{-t}\, dt \tag{1}$$

Let $t = z^2/2$ with $z$ positive. Thus, $dt = z\, dz$ and $0 < t < \infty$ corresponds to $0 < z < \infty$. Also,

$$t^{\alpha-1} = (z^2/2)^{\alpha-1} = (\tfrac{1}{2})^{\alpha-1} z^{2\alpha-2} = 2^{1-\alpha} z^{2\alpha-2} \tag{2}$$

Thus, if we replace the $t$ terms by $z$ terms in Eq. (1), we obtain the following alternate equation for $\Gamma(\alpha)$:

$$\Gamma(\alpha) = 2^{1-\alpha} \int_{z=0}^{\infty} z^{2\alpha-1} e^{-z^2/2}\, dz \tag{3}$$

Now, if we let $\alpha = 1/2$ in Eq. (3), we obtain

$$\Gamma(1/2) = 2^{1/2} \int_{z=0}^{\infty} e^{-z^2/2}\, dz \tag{4}$$

Now, a definite integral for $J$ of part (a) is $J = \int_{-\infty}^{\infty} e^{-z^2/2}\, dz$. But since $e^{-z^2/2}$ is an *even function*, we have

$$J = 2 \int_{0}^{\infty} e^{-z^2/2}\, dz \tag{5}$$

[An *even function* $g(t)$ is one for which $g(-t) = g(t)$; consequently, $\int_{-c}^{0} g(t)\, dt = \int_{0}^{c} g(t)\, dt$, and hence $\int_{-c}^{c} g(t)\, dt = 2 \int_{0}^{c} g(t)\, dt$.] Thus, from Eqs. (4) and (5), and the result $J = \sqrt{2\pi}$ from part (a), we obtain

$$\Gamma(1/2) = (2^{1/2})(J/2) = J/(2^{1/2}) = \sqrt{2\pi}/\sqrt{2} = \sqrt{\pi} \tag{6}$$

**PROBLEM 10-2**   For the normal distribution pdf in the form

$$f(x) = c \exp\left( -\frac{1}{2}\left(\frac{x-a}{b}\right)^2 \right) \quad \text{for } -\infty < x < \infty$$

show that $E(X) = \mu = a$.

**Solution**   In general, we have $\mu = \int_x x f(x)\, dx$, for a continuous random variable $X$. Thus, for the $f(x)$ as given above, we have

$$\mu = c \int_{-\infty}^{\infty} x \exp\left( -\frac{1}{2}\left(\frac{x-a}{b}\right)^2 \right) dx \tag{1}$$

Let $t = (x - a)/b$. It follows that $dx = b\, dt$, and $x = bt + a$. Also, if we assume that $b$ is positive, we see that for the interval of integration for $t$, we have $t \to \infty$ when $x \to \infty$ and $t \to -\infty$ when $x \to -\infty$. Thus, Eq. (1) becomes

$$\mu = cb^2 \int_{-\infty}^{\infty} t e^{-t^2/2}\, dt + cab \int_{-\infty}^{\infty} e^{-t^2/2}\, dt \tag{2}$$

The first integral equals zero since $t e^{-t^2/2}$ is an odd function. [A function $h(t)$ is an *odd function* if $h(-t) = -h(t)$; as a consequence, $\int_{-c}^{0} h(t)\, dt = -\int_{0}^{c} h(t)\, dt$, and hence $\int_{-c}^{c} h(t)\, dt = 0$.] From

Problem 10-1, we see that the second integral in Eq. (2) equals $J = \sqrt{2\pi}$, and also that $cb = 1/J$. Thus, Eq. (2) may be rewritten as

$$\mu = aJ/J = a \tag{3}$$

**PROBLEM 10-3**   For the normal pdf in the form

$$f(x) = c\exp\left(-\frac{1}{2}\left(\frac{x-a}{b}\right)^2\right) \quad \text{for } -\infty < x < \infty$$

show that $\text{Var}(X) = b^2$.

**Solution**   In general, we have $E(X^2) = \int_x x^2 f(x)\,dx$, for a continuous random variable $X$. Thus, for the $f(x)$ as given above, we have

$$E(X^2) = c\int_{-\infty}^{\infty} x^2\exp\left(-\frac{1}{2}\left(\frac{x-a}{b}\right)^2\right)dx \tag{1}$$

As in Problem 10-2, let $t = (x-a)/b$. Thus, $dx = b\,dt$, $x = bt + a$, and $x^2 = b^2t^2 + 2abt + a^2$. Thus, Eq. (1) becomes

$$E(X^2) = cb^3\int_{-\infty}^{\infty} t^2 e^{-t^2/2}\,dt + 2ab^2c\int_{-\infty}^{\infty} te^{-t^2/2}\,dt + a^2bc\int_{-\infty}^{\infty} e^{-t^2/2}\,dt \tag{2}$$

Now, the second integral on the right equals zero [see Problem 10-2], and the third integral equals $J = \sqrt{2\pi} = 1/(bc)$ as we learned in Problem 10-1. Thus, we may rewrite Eq. (2) as

$$E(X^2) = cb^3\int_{-\infty}^{\infty} t^2 e^{-t^2/2}\,dt + a^2 \tag{3}$$

Integrating by parts in the integral of Eq. (3) [let $u = t$ and $dv = te^{-t^2/2}\,dt$; thus, $du = dt$ and $v = -e^{-t^2/2}$], we get

$$\int_{-\infty}^{\infty} t^2 e^{-t^2/2}\,dt = -te^{-t^2/2}\Big]_{-\infty}^{\infty} + \int_{-\infty}^{\infty} e^{-t^2/2}\,dt = \int_{-\infty}^{\infty} e^{-t^2/2}\,dt \tag{4}$$

Thus, the integral in Eq. (3) is equal to $J = \sqrt{2\pi} = 1/(bc)$. Thus, Eq. (3) becomes

$$E(X^2) = b^2 + a^2 \tag{5}$$

Now, using the fact that $\text{Var}(X) = E(X^2) - [E(X)]^2$, and the result $\mu = a$ from Problem 10-2, we get

$$\text{Var}(X) = b^2 + a^2 - a^2 = b^2 \tag{6}$$

**PROBLEM 10-4**   Prove Theorem 10.2, namely that if $X$ has a normal distribution with mean $\mu$ and variance $\sigma^2$, and $Y = aX + b$, then $Y$ has a normal distribution with mean $a\mu + b$ and variance $a^2\sigma^2$.

**Proof**   From Theorem 10.1,

$$M_X(t) = e^{\mu t}e^{\sigma^2 t^2/2} \tag{1}$$

From Theorem 9.11,

$$M_{aX+b}(t) = e^{bt}M_X(at) \tag{2}$$

Evaluating $M_X(at)$ in the current case from Eq. (1), after first remembering to replace $t$ by $at$, and then substituting into Eq. (2), we get

$$M_Y(t) = M_{aX+b}(t) = e^{bt}e^{a\mu t}e^{\sigma^2 a^2 t^2/2} = e^{(a\mu + b)t}e^{\sigma^2 a^2 t^2/2} \tag{3}$$

Now, Theorem 9.12 [a moment generating function uniqueness theorem] indicates that if $M_W(t) = e^{ct}e^{d^2 t^2/2}$, with $c$ and $d$ constant, then $W$ has a normal distribution with mean $c$ and variance $d^2$. Thus, it follows from Eq. (3) that $Y = aX + b$ has a normal distribution with mean $(a\mu + b)$ and variance $a^2\sigma^2$.   □

**PROBLEM 10-5**   Prove Theorem 10.4.

***Proof***  First of all, we see that

$$P(x_0 < X < x_1) = P(x_0 - \mu < X - \mu < x_1 + \mu) \tag{1}$$

since the condition $x_0 < X < x_1$ is equivalent to $x_0 - \mu < X - \mu < x_1 + \mu$ from the simple algebra of inequalities. Likewise,

$$P(x_0 - \mu < X - \mu < x_1 - \mu) = P\left(\frac{x_0 - \mu}{\sigma} < \frac{X - \mu}{\sigma} < \frac{x_1 - \mu}{\sigma}\right) \tag{2}$$

Here, we divided through all terms in the left probability expression by the positive constant $\sigma$. From Eqs. (1) and (2), we have

$$P(x_0 < X < x_1) = P(z_0 < Z < z_1) \tag{3}$$

since $Z = (X - \mu)/\sigma, z_0 = (x_0 - \mu)/\sigma$, and $z_1 = (x_1 - \mu)/\sigma$. Thus, Eq. (3) is part (a) of Theorem 10.4.

For part (b), choose $x_0 = \mu$ in Eq. (3). Then, we have $z_0 = (x_0 - \mu)/\sigma = (\mu - \mu)/\sigma = 0$, and Eq. (3) becomes

$$P(\mu < X < x_1) = P(0 < Z < z_1) \tag{4}$$

The proof of part (c) of Theorem 10.4 is similar to that of part (b).   □

## Calculations Involving a Normal Distribution

**PROBLEM 10-6**   In Example 9-17, the average daily temperature $T$ [in degrees Fahrenheit] in July in Newtown was a random variable with mean $\mu_T = 85$ and standard deviation $\sigma_T = 7$. The daily air conditioning cost $Q$, in dollars, for a factory, is related to $T$ by $Q = 120T + 750$. Suppose that $T$ is a normal random variable. Compute the probability that the daily air conditioning cost on a typical July day for the factory will exceed \$12,210.

***Solution***  In Example 9-17, we found that

$$\mu_Q = E(Q) = \$10,950 \quad \text{and} \quad \sigma_Q = \$840 \tag{1a), (1b}$$

If $T$ is a normal random variable, then so is $Q$ since it is of the form $Q = aT + b$ [Theorem 10.2]. Here, we wish to calculate $P(Q > \$12,210)$. From Theorem 10.4, we have

$$P(Q > 12,210) = P(Z > z_1) \tag{2}$$

where $Z$ is the standard normal random variable corresponding to $Q$, and $z_1$ corresponds to $q_1 = 12,210$. Thus,

$$z_1 = (q_1 - \mu_Q)/\sigma_Q = (12,210 - 10,950)/840 = 1.50 \tag{3}$$

Then, after substituting from Eq. (3) into Eq. (2), and using Table VI of Appendix A, we have

$$P(Q > 12,210) = P(Z > 1.50) = .5 - .4332 = .0668 \tag{4}$$

**PROBLEM 10-7**   Refer to Problem 10-6. (a) What percentile is the cost $q = \$12,210$? (b) Determine the cost that is the 40th percentile.

***Solution***  Here, we have a normal random variable $Q$ for which $\mu_Q = \$10,950$ and $\sigma_Q = \$840$.

(a) In Problem 10-6, we found that $P(Q > 12,210) = .0668$. Thus, $P(Q \le 12,210) = .9332$, which means that \$12,210 is $p_{93.32}$, the 93.32nd percentile.

(b) Let $q_1$ symbolize the 40th percentile, $p_{40}$. Thus,

$$P(Q \le q_1) = .40 \tag{1}$$

The 40th percentile for $Z$, namely $z_1$, say, is such that

$$P(Z \le z_1) = .40 \quad \text{or} \quad P(z_1 < Z < 0) = .10 \tag{2a), (2b}$$

From Eq. (2b), and from Table VI of Appendix A, we see that $z_1 = -.253$, after interpolating. From the correspondence equation relating $q_1$ and $z_1$, we find that

$$q_1 = \sigma_Q z_1 + \mu_Q = 840(-.253) + 10{,}950 = 10{,}737.48 \tag{3}$$

Thus, $p_{40} = \$10{,}737$, to five significant digits.

### Skewness of a Random Variable

**PROBLEM 10-8**   Prove Theorem 10.6. Here, we wish to express the third moment about the mean in terms of moments about the origin.

**Proof**   First, expanding $(X - \mu)^3$, we have

$$\mu_3 = E[(X - \mu)^3] = E[(X^3 - 3\mu X^2 + 3\mu^2 x - \mu^3)] \tag{1}$$

Next, we use Theorem 9.4 [linearity rule for $E$], the rule that $E(c) = c$, and the fact that $E(X) = \mu$. Thus, from Eq. (1), we get

$$E[(X - \mu)^3] = E(X^3) - 3\mu E(X^2) + 3\mu^2 E(X) - \mu^3 = E(X^3) - 3\mu E(X^2) + 3\mu^3 - \mu^3$$

$$= E(X^3) - 3\mu E(X^2) + 2\mu^3 \tag{2} \quad \square$$

**PROBLEM 10-9**   Calculate the coefficient of skewness $[\alpha_3]$ for the pdf given by $f(x) = (\frac{2}{9})x$ for $0 < x < 3$, and $f(x) = 0$, elsewhere.

**Solution**   Let us calculate $\mu_i' = E(X^i)$ for $i = 1, 2$, and 3.

$$E(X) = \mu = \left(\frac{2}{9}\right) \int_0^3 x^2 \, dx = 2 \tag{1}$$

$$E(X^2) = \left(\frac{2}{9}\right) \int_0^3 x^3 \, dx = 4.5 \tag{2}$$

$$E(X^3) = \left(\frac{2}{9}\right) \int x^4 \, dx = 10.8 \tag{3}$$

Thus, from Theorem 10.6,

$$\mu_3 = 10.8 - 3(2)(4.5) + 2(2)^3 = -.2 \tag{4}$$

Now, for $\mathrm{Var}(X) = \sigma^2$, we have

$$\sigma^2 = E(X^2) - [E(X)]^2 = 4.5 - (2)^2 = .5 \tag{5}$$

and thus, $\sigma = \sqrt{.5}$. So, for $\alpha_3$, we have

$$\alpha_3 = \frac{\mu_3}{\sigma^3} = \frac{(-.2)}{(.5)\sqrt{.5}} = -.56569 \tag{6}$$

### Normal Approximation to the Binomial Distribution

**PROBLEM 10-10**   Weather records over many years show that the average number [or expected number] of rainy days in Dampsville in September is 7.5. Compute the probability of at most 4 rainy days next September in Dampsville.

**Solution**   Assume that the number of rainy days in Dampsville in September is a binomial random variable $X$. Thus, with $n = 30$, and $E(X) = 7.5$, and the general result $E(X) = np$, we have

$$p = E(X)/n = 7.5/30 = .25 \tag{1}$$

Let us use the normal approximation to the binomial. This should be acceptable here since $np = 7.5$ and $nq = n - np = 22.5$. For the variance and standard deviation of random variable $X$, we have

$$\text{Var}(X) = npq = (7.5)(.75) = 5.625; \qquad \sigma_x = (5.625)^{1/2} = 2.372 \qquad \textbf{(2), (3)}$$

In the following, the subscripts $B$ and $N$ will refer to binomial and approximating normal, respectively. We wish to calculate $P_B(X \le 4)$, and in the normal approximation by area, we have [Theorem 10.8, part (B)]:

$$P_B(X \le 4) = P_B(X = 0 \text{ or } 1 \text{ or } 2 \text{ or } 3 \text{ or } 4) \approx P_N(-.5 < X < 4.5) \qquad \textbf{(4)}$$

[Here, we are using the continuity correction.] Now, the standard rv $Z$ corresponds to $X$, where

$$Z = (X - np)/\sqrt{npq} = (X - 7.5)/2.372 \qquad \textbf{(5)}$$

Thus, in the rightmost expression of (4), $x_1 = -.5$ corresponds to $z_1 = (-.5 - 7.5)/2.372 = -3.37$, and $x_2 = 4.5$ corresponds to $z_2 = -1.26$. Thus, from Table VI and Eq. (4), we find that

$$P_B(X \le 4) \approx P_N(-.5 < X < 4.5) = P_N(-3.37 < Z < -1.26) = P_N(-3.37 < Z < 0)$$

$$- P_N(-1.26 < Z < 0) = .5000 - .3962 = .1038 \qquad \textbf{(6)}$$

**PROBLEM 10-11**  Prove Theorem 10.11.

**Proof**  For a binomial random variable $X_n$, where $X_n$ stands for the number of successes in $n$ Bernoulli trials, we saw in Example 9-22 that

$$M_{X_n}(t) = [1 + p(e^t - 1)]^n \qquad \textbf{(1)}$$

We can find an expression for $M_{Z_n}(t)$ by making use of Theorem 9.11, that is, of

$$M_{(aX+b)}(t) = e^{bt} M_X(at) \qquad \textbf{(2)}$$

with

$$Z_n = (X_n - \mu)/\sigma = \left(\frac{1}{\sigma}\right) X_n + \left(-\frac{\mu}{\sigma}\right) \qquad \textbf{(3)}$$

Thus, comparing the right side of Eq. (3) with the $aX + b$ expression of Eq. (2), we have $a = (1/\sigma)$, $b = (-\mu/\sigma)$, and $X$ replaced by $X_n$, here. [It is important to note that $\mu = E(X_n) = np$, and $\sigma^2 = \text{Var}(X_n) = np(1 - p)$ for binomial random variable $X_n$.] Thus,

$$M_{Z_n}(t) = M_{(X_n - \mu)/\sigma}(t) = e^{-\mu t/\sigma}[1 + p(e^{t/\sigma} - 1)]^n \qquad \textbf{(4)}$$

Taking logarithms to the base $e$, we get

$$\ln M_{Z_n}(t) = -\left(\frac{\mu t}{\sigma}\right) + n \ln[1 + p(e^{t/\sigma} - 1)] \qquad \textbf{(5)}$$

Now, we shall use Maclaurin's series for $e^s$, namely

$$e^s = 1 + s + \frac{s^2}{2!} + \sum_{k=3}^{\infty} \frac{s^k}{k!}$$

to express $e^{t/\sigma}$ in Eq. (5). Thus, replacing $s$ by $t/\sigma$, we obtain Eq. (6):

$$\ln M_{Z_n}(t) = -\left(\frac{\mu t}{\sigma}\right) + n \ln\left[1 + p\left\{\frac{t}{\sigma} + \frac{1}{2}\left(\frac{t}{\sigma}\right)^2 + \frac{1}{6}\left(\frac{t}{\sigma}\right)^3 + \dots\right\}\right] \qquad \textbf{(6)}$$

Next, we use the Maclaurin series $\ln[1 + s] = s - (s^2/2) + (s^3/3) - \dots$ on the right-hand "ln" term [this series converges for $-1 < s < 1$, while the $e^s$ series converges for all $s$]. Here, the $s$ term is

$$p\left\{\frac{t}{\sigma} + \frac{1}{2}\left(\frac{t}{\sigma}\right)^2 + \frac{1}{6}\left(\frac{t}{\sigma}\right)^3 + \dots\right\}$$

[In the current situation, this term will be "small" for large $n$.] Thus, we get

$$\ln M_{Z_n}(t) = -\left(\frac{\mu t}{\sigma}\right) + np\left[\frac{t}{\sigma} + \frac{1}{2}\left(\frac{t}{\sigma}\right)^2 + \frac{1}{6}\left(\frac{t}{\sigma}\right)^3 + \cdots\right]$$

$$-\frac{np^2}{2}\left[\frac{t}{\sigma} + \frac{1}{2}\left(\frac{t}{\sigma}\right)^2 + \frac{1}{6}\left(\frac{t}{\sigma}\right)^3 + \cdots\right]^2$$

$$+\frac{np^3}{3}\left[\frac{t}{\sigma} + \frac{1}{2}\left(\frac{t}{\sigma}\right)^2 + \frac{1}{6}\left(\frac{t}{\sigma}\right)^3 + \cdots\right]^3 - \cdots \qquad (7)$$

Collecting like powers of $t$, we get

$$\ln M_{Z_n}(t) = \left(-\frac{\mu}{\sigma} + \frac{np}{\sigma}\right)t + \left(\frac{1}{\sigma^2}\right)\left(\frac{np}{2} - \frac{np^2}{2}\right)t^2 + \left(\frac{n}{\sigma^3}\right)\left(\frac{p - 3p^2 + 2p^3}{6}\right)t^3 + \cdots \qquad (8)$$

Using the fact that $\mu = np$ [this causes the coefficient of $t$ to equal zero], then substituting $\sigma^2 = np(1 - p)$ in the coefficient of $t^2$, and simplifying, we get

$$\ln M_{Z_n}(t) = \frac{t^2}{2} + \left(\frac{n}{\sigma^3}\right)\left(\frac{p - 3p^2 + 2p^3}{6}\right)t^3 + \cdots \qquad (9)$$

Now, for $r \geq 3$, the coefficient of $t^r$ is a constant times $n/\sigma^r$, that is, times $n^{(1-r/2)}/(pq)^{r/2}$ since $\sigma = (npq)^{1/2}$. Thus, for $r \geq 3$, each coefficient of $t^r$ in (9) approaches zero as $n \to \infty$. Thus, from (9), it follows that

$$\lim_{n \to \infty} \ln M_{Z_n}(t) = \frac{t^2}{2} \qquad (10)$$

Now, since the limit of a logarithm equals the logarithm of a limit, it follows from Eq. (10) that

$$\ln \lim_{n \to \infty} M_{Z_n}(t) = \frac{t^2}{2} \qquad (11)$$

Taking antilogarithms in Eq. (11), we get

$$\lim_{n \to \infty} M_{Z_n}(t) = e^{t^2/2} \qquad (12) \quad \square$$

**PROBLEM 10-12**   Prove Theorem 10.13.

*Proof*  Recall from Theorems 9.3 and 9.6 the results indicated in Eqs. (1) and (2).

$$E(aX + b) = aE(X) + b \qquad \text{Var}(aX + b) = a^2\,\text{Var}(X) \qquad \textbf{(1), (2)}$$

Also,

$$\hat{P} = X/n \qquad (3)$$

In addition,

$$E(X) = np \quad \text{and} \quad \text{Var}(X) = np(1 - p) \qquad \textbf{(4), (5)}$$

are true for the binomial random variable $X$. Observe from Eq. (3) that $\hat{P}$ is of the form $aX + b$ with $a = 1/n$ and $b = 0$. Thus, applying Eqs. (1) and (2) to $\hat{P}$, we obtain

$$E(\hat{P}) = E\left(\frac{1}{n}X\right) = \frac{1}{n}E(X) \qquad (6)$$

and

$$\text{Var}(\hat{P}) = \text{Var}\left(\frac{1}{n}X\right) = \left(\frac{1}{n}\right)^2 \text{Var}(X) \qquad (7)$$

Now, applying Eqs. (4) and (5) to the right sides of Eqs. (6) and (7), we get

$$E(\hat{P}) = \frac{1}{n}np = p \qquad (8)$$

and

$$\text{Var}(\hat{P}) = \left(\frac{1}{n}\right)^2 np(1-p) = \frac{p(1-p)}{n} \qquad (9) \quad \square$$

## A Law of Large Numbers

**PROBLEM 10-13** The probability of drawing a spade from a usual deck of 52 cards is $p = .25$. (a) Suppose that 100,000 cards are drawn with replacement from such a deck of cards. Thus, $n = 100,000$ in this binomial situation. Let $\hat{P}$ denote the proportion of spades obtained in 100,000 such draws. Compute a lower bound for the probability $P(.24 < \hat{P} < .26)$ by using Theorem 10.15. (b) Repeat part (a) if $n = 100,000,000$ and the probability for $\hat{P}$ is $P(.249 < \hat{P} < .251)$.

*Solution*

(a) Here, $n = 100,000 = 10^5$, $p = .25$, and $c = (.26 - .24)/2 = .01$. Now, the right side of the inequality

$$P(p - c < \hat{P} < p + c) \geq 1 - \frac{p(1-p)}{nc^2} \qquad (1)$$

for the above values of $n$, $p$, and $c$ is the lower bound we seek. Thus,

$$\text{lower bound} = 1 - \frac{(.25)(.75)}{10^5(.01)^2} = .98125 \qquad (2)$$

and

$$P(.24 < \hat{P} < .26) \geq .98125 \qquad (3)$$

(b) Here, $n = 10^8$, $p = .25$, and $c = (.251 - .249)/2 = .001$. Thus, here we have

$$\text{lower bound} = 1 - \frac{(.25)(.75)}{10^8(.001)^3} = .998125 \qquad (4)$$

and

$$P(.249 < \hat{P} < .251) \geq .998125 \qquad (5)$$

*note:* Thus, we see that it is highly probable that the proportion [or relative frequency] for success [here, spade on a draw of a card] will be extremely close to $p$, provided that $n$ is large enough.

## More Normal Curve Problems

**PROBLEM 10-14** Rework Example 10-21 if the true fraction of Republicans, $p$, is unknown. Here, find a *minimum* value for $n$ such that the probability $P(p - .02 < \hat{P} < p + .02)$ is *at least* .9544.

*Solution* Let $n_0$ be the minimum $n$ value. At the outset, we assume, as in Example 10-21, that $n$ is sufficiently large so that the binomial proportion rv $\hat{P}$ is approximately normally distributed. [Also, here, as in Example 10-21, we shall neglect the continuity correction.] Here, we want

$$P(p - .02 < \hat{P} < p + .02) \geq .9544 \qquad (1)$$

for $n \geq n_0$. In Figure 10-15, we have

$$P(p - .02 < \hat{P} < p + .02) = .9544 \qquad (2)$$

Working with reference to Figure 10-15, we find from $P(0 < Z < z) = .9544/2 = .4722$ that $z = 2$. Now, in Figure 10-15, $\hat{p} = p + .02$ corresponds to $z = 2$.

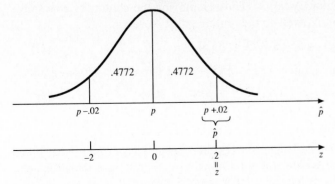

**Figure 10-15.** Normal curve approximation for Problem 10-14.

More generally, let $z_1$ correspond to $\hat{p}_1 = p + .02$. This means that

$$z_1 = \frac{\hat{p}_1 - p}{\sigma_{\hat{p}}} = \frac{.02\sqrt{n}}{\sqrt{p(1-p)}} \tag{3}$$

Thus, in order for inequality (1) to hold, we must have $z_1 \geq 2$, which means [after using (3)] that

$$\frac{.02\sqrt{n}}{\sqrt{p(1-p)}} \geq 2 \qquad \text{for any } p \text{ on } 0 < p < 1 \tag{4}$$

Now, it can be shown that

$$p(1-p) \leq .25 \qquad \text{for any } p \text{ on } 0 < p < 1 \tag{5}$$

[Let us digress to demonstrate (5). Consider $h = p(1-p)$ for $0 < p < 1$. From $h' = 1 - 2p$, we see that $h' = 0$ for $p = .5$. From $h'' = -2$, we see that $h$ is *maximized* at $p = .5$. Recalling what $h$ is, this means that $p(1-p) \leq (.5)^2 = .25$ for all $p$ on the interval $0 < p < 1$.] From (5), it follows that

$$\frac{1}{\sqrt{p(1-p)}} \geq \frac{1}{\sqrt{.25}} = 2 \qquad \text{for any } p \text{ on } 0 < p < 1 \tag{6}$$

If we multiply both sides of (6) by $.02\sqrt{n}$, we get

$$\frac{.02\sqrt{n}}{\sqrt{p(1-p)}} \geq .04\sqrt{n} \qquad \text{for any } p \text{ on } 0 < p < 1 \tag{7}$$

Now, compare (4) and (7). If we choose $.04\sqrt{n} \geq 2$, it will follow from (7) that $.02\sqrt{n}/\sqrt{p(1-p)} \geq 2$ for any $p$ on $0 < p < 1$. Thus, choosing $.04\sqrt{n} \geq 2$, we get

$$\sqrt{n} \geq \frac{2}{.04} = 50 \qquad \text{and} \qquad n \geq 2500 \tag{8}$$

Thus, $n_0$ is equal to 2500. In summary, if we choose $n \geq 2500$, that will cause (4) to hold. In turn, it follows, upon examining (3), that $z_1$ will certainly be at least equal to 2 for any $p$ on $0 < p < 1$. But, if $z_1 \geq 2$, that means that (1) will hold.

In Example 10-21, where $p$ was equal to .4, we found that the $n$ value was 2400.

**PROBLEM 10-15**   A restaurant can accommodate 55 people. The restaurant takes reservations for 60 people for Sunday dinner. The probability a person making a reservation will not show up is .15. What is the probability that all the people who show up for Sunday dinner can be accommodated?

*Solution*   Let random variable $X$ denote the number of people who show up out of the 60 people who make reservations. Thus, $X$ is a binomial rv for which $n = 60$, and $p = 1 - .15 = .85$. We see that

$$\mu_X = np = 60(.85) = 51 \qquad \text{and} \qquad \sigma_X = \sqrt{60(.85)(.15)} = \sqrt{7.65} = 2.766 \quad \textbf{(1a), (1b)}$$

The probability that all the people who show up can be accommodated is the same as $P_B(X \leq 55)$, where subscript $B$ stands for *binomial*. Since $np = 51$ and $nq = 9$, we can approximate $P_B(X \leq 55)$ by using a normal area approximation. Thus, employing the continuity correction, we have the following, where subscript $N$ refers to the normal rv:

$$P_B(X \leq 55) \approx P_N(X \leq 55.5) \tag{2}$$

For $x_1 = 55.5$, the corresponding $z_1$ value is $z_1 = (55.5 - 51)/2.766 \approx 1.63$. Thus, we find

$$P_B(X \leq 55) \approx P_N(Z \leq 1.63) = .5 + .4484 = .9484 \tag{3}$$

after using Table VI.

**PROBLEM 10-16**   The heights of female students at Gates College are normally distributed with $\mu = 64$ inches, and $\sigma = 2.5$ inches. What is the conditional probability that a female student will be taller than 68 inches if it is known that she is taller than 65 inches?

***Solution*** Let random variable $X$ stand for a female student's height, in inches, at Gates College. Here, we wish to compute the conditional probability $P(X > 68 \mid X > 65)$. From Definition 3.1, this becomes

$$P(X > 68 \mid X > 65) = \frac{P(X > 65 \text{ and } X > 68)}{P(X > 65)} \tag{1}$$

Now, for a female student to be both taller than 65 inches and 68 inches, she must be taller than 68 inches. Thus,

$$P(X > 65 \text{ and } X > 68) = P(X > 68) \tag{2}$$

Substituting Eq. (2) in Eq. (1), and then transforming to the standard random variable $Z$, we have

$$P(X > 68 \mid X > 65) = \frac{P(X > 68)}{P(X > 65)} = \frac{P(Z > 1.60)}{P(Z > .40)} \tag{3}$$

For example, for $x_1 = 68$, we have $z_1 = (x_1 - \mu)/\sigma = (68 - 64)/2.5 = 1.60$. Determining the probabilities $P(Z > 1.60)$ and $P(Z > .40)$ by making use of Table VI, we get

$$P(X > 68 \mid X > 65) = \frac{.5 - P(0 < Z < 1.60)}{.5 - P(0 < Z < .40)} = \frac{.5 - .4452}{.5 - .1554} = .1590 \tag{4}$$

Thus, our answer is .1590 or 15.9%.

## Supplementary Problems

**PROBLEM 10-17**  Verify that the inflection points for a standard normal random variable occur at $z = -1$ and $z = 1$.

***Answer***  From $f''(z) = (1/\sqrt{2\pi})(z^2 - 1)e^{-z^2/2}$, we see that $f''(z) = 0$ for $z_1 = -1$ and $z_2 = +1$, and that $f''(z) > 0$ for $z < -1$ and $z > 1$, and $f''(z) < 0$ for $-1 < z < 1$.

**PROBLEM 10-18**  For a normal random variable $X$ with mean $\mu$ and variance $\sigma^2$, show that the points of inflection occur at $x_1 = \mu - \sigma$ and $x_2 = \mu + \sigma$. [*Hint:* Use the answer to Problem 10-17, and the correspondence equation $z = (x - \mu)/\sigma$ relating $x$ and $z$.]

**PROBLEM 10-19**  For the exponential distribution given by $f(x) = (\frac{1}{3})e^{-x/3}$ for $x > 0$, and $f(x) = 0$ for $x \leq 0$, determine (a) $Q_1$, the first quartile; (b) $Q_2 = Md$, the median; (c) $Q_3$, the third quartile; and (d) the interquartile range. [*Hint:* First, find the equation for the cdf $F(x_0)$ in terms of $x_0$, and then solve for $x_0$ in terms of $F(x_0)$.]

***Answer***  (a) $Q_1 = .8630$    (b) $Md = 2.0794$    (c) $Q_3 = 4.1589$
(d) interquartile range $= Q_3 - Q_1 = 3.2959$

**PROBLEM 10-20**  For the exponential distribution of Problem 10-19, determine what percentiles the following $x$ values are: (a) $x_1 = .7$; (b) $x_2 = 2$; (c) $x_3 = 7.5$.

***Answer***  (a) $x_1 = p_{20.81}$    (b) $x_2 = p_{48.66}$    (c) $x_3 = p_{91.79}$

**PROBLEM 10-21**  For the beta distribution with $\alpha = 1$ and $\beta = 3$, determine (a) the mean, $\mu$; (b) $\mu'_2$, or $E(X^2)$; (c) $\mu'_3$, or $E(X^3)$; (d) $\sigma^2$; (e) $E[(X - \mu)^3]$; (f) $\alpha_3$, the coefficient of skewness. [*Hint:* For a review of the beta distribution, see Section 8-3D and Problems 8-15 and 9-20.]

***Answer***  (a) .25    (b) .10    (c) .05    (d) .0375    (e) .00625    (f) .86067

**PROBLEM 10-22**  If $Z$ is a standard normal random variable, find the probability that $Z$ will take on a value (a) greater than 1.58, (b) between 1.58 and 2.86, (c) less than $-1.35$, (d) between $-.64$ and 1.45.

*Answer* (a) .0571    (b) .0550    (c) .0885    (d) .6654

**PROBLEM 10-23**  Say $Z$ is a standard normal random variable. Find the value $z_0$ if (a) $P(Z < z_0) = .9236$, (b) $P(Z > z_0) = .2946$, (c) $P(Z > z_0) = .6480$, (d) $P(-z_0 < Z < z_0) = .8740$.

*Answer* (a) $z_0 = 1.43$    (b) $z_0 = .54$    (c) $z_0 = -.38$    (d) $z_0 = 1.53$

**PROBLEM 10-24**  The intelligence quotient (I. Q.) of students at Wells College is normally distributed with a mean of 110 and a standard deviation of 15. What is the probability that a student selected at random will have an I. Q. (a) less than 99.5; (b) greater than 131; (c) between 92 and 128? (d) What I. Q. is the 95th percentile?

*Answer* (a) .242    (b) .0808    (c) .7698    (d) $134.675 \approx 134.7$ [Here, use $z_1 = 1.645$.]

**PROBLEM 10-25**  A pair of fair dice is tossed 180 times. Find the probability of getting a sum equal to seven exactly 30 times by using (a) the binomial probability equation and (b) a normal curve approximation by area. (c) Find the approximate probability of getting at least 29 sevens. [Use a normal curve approximation by area.]

*Answer* (a) .07956    (b) .0796    (c) .6179

**PROBLEM 10-26**  Given a container of marbles of which 60% [or .60] are colored red. How large a random sample should be drawn [that is, drawn with replacement] so that the probability is .99 that the sample proportion of red marbles is within .08 of .60? [*Hints:* The random variable that applies here is $\hat{P}$, the binomial proportion rv. Use normal approximation to the binomial here [by area], and neglect the continuity correction.]

*Answer* $n = 249$ after rounding up from 248.65. This $n$ value is sufficiently large to justify using the normal approximation to the binomial distribution.

**PROBLEM 10-27**  Assume that the lifetime in hours of a computer chip is normally distributed with mean 5000 hours. If a producer wants at least 80% of the chips to have lifetimes exceeding 4200 hours, what is the largest value the standard deviation $\sigma$ can have? [*Hint:* Find the cut-off value $\sigma_c$ which corresponds to exactly 80% having lifetimes exceeding 4200 hours.]

*Answer* 950.1 hours

**PROBLEM 10-28**  Cereal packages are labeled as containing 16 ounces of cereal. The actual amount of cereal which a filling machine puts into a box labeled "16 ounces" is a normal random variable with $\sigma = .16$ ounce. (a) If the machine is set to fill a package with 16.2 ounces, what fraction of packages will be short [i.e., contain less than 16 ounces]? (b) What weight should the machine be set for so that only 1% will contain less than 16 ounces? [*Hint:* For part (a), consider the mean of the pertinent normal rv to be 16.2 ounces, and use a similar idea for part (b).]

*Answer* (a) .1056 or 10.56%    (b) $16.3723 \approx 16.37$ ounces

# 11 DESCRIPTIVE STATISTICS

## THIS CHAPTER IS ABOUT

☑ **Frequency Distributions**
☑ **Histograms**
☑ **The Empirical Cumulative Distribution Function**
☑ **Percentiles for a Frequency Distribution of Continuous Data**
☑ **Moments for a Sample of Data**
☑ **Moments for a Sample of Grouped Data**

*Preliminary Note:* In this chapter, we shall present a brief treatment of descriptive statistics. A much more extensive treatment may be found in Tanis (1987a), Freund (1988), and many other books. A key portion of descriptive statistics is concerned with methods for estimating *parameters* such as the mean, variance, and various percentiles for a given probability distribution. To do this, one works with an empirically [experimentally] obtained *sample of data values* pertaining to the probability distribution under consideration.

It is from such a sample of data that we shall, in the current chapter, calculate *statistics* such as the mean, variance, and various percentiles of the sample. Statistics such as these are the entities that provide one with estimates of comparable parameters for the related probability distribution. For example, the mean for a sample of data will often serve as an estimate of the mean of the associated probability distribution.

An important point is that this chapter will focus on some of the main methods for determining those statistics [such as the mean, variance, and various percentiles of a sample of data], which will have a bearing on our later work involving statistical analyses.

In particular, we shall be concerned later, to a great degree, with problems of estimating parameters and testing hypotheses with respect to theoretical probability distributions [or populations]. Our method of attack will often involve using statistics calculated from samples of data pertaining to those probability distributions.

Other concepts pertaining to descriptive statistics will be developed as the need arises.

## 11-1. Frequency Distributions

### A. Data

A set of data obtained by performing a random experiment a number of times is called a *sample of data*, or just *sample*. [Note that in many situations a sample of data will not be obtained in such an idealized way as this, however.]

## B.  Discrete data

Suppose we have a random [or probability] experiment with which there is associated a discrete random variable. If we repeat the experiment over and over again, and observe values of the random variable for each performance of the experiment, then that collection of values is said to be a *sample of discrete data.*

For example, consider the binomial experiment of tossing a particular coin three times. In this case, a typical binomial random variable $X$ pertains to the number of heads obtained. [Thus, in a performance of this experiment, $X$ will equal either 0 or 1 or 2 or 3.] A collection of values of $X$ resulting from performing the experiment [say, the collection of five values 2, 1, 2, 3, 0] then constitutes a sample of discrete data pertaining to the binomial experiment.

If we agree to perform an experiment [discrete or otherwise] $N$ times, then $N$ is known as the *total frequency* for the related sample of data.

A *frequency distribution* for a sample of discrete data consists of a tallying and recording of the number of times each value of the discrete random variable occurs within the sample.

---

**EXAMPLE 11-1:** Given a coin for which the probability of head on a single toss is .6. Consider the binomial experiment of tossing the coin three times. Thus, for the binomial random variable $X$, which stands for the number of heads obtained, we have $n = 3$ and $p = .6$. [See Section 5-3 for a review.] Suppose the experiment is repeated 40 times, and the following discrete data [i.e., values of $X$] are obtained.

$$1 \quad 0 \quad 2 \quad 2 \quad 3 \quad 1 \quad 2 \quad 1 \quad 2 \quad 2 \qquad 2 \quad 1 \quad 2 \quad 0 \quad 2 \quad 1 \quad 3 \quad 2 \quad 1 \quad 2$$

$$2 \quad 1 \quad 2 \quad 1 \quad 3 \quad 3 \quad 2 \quad 1 \quad 2 \quad 3 \qquad 1 \quad 2 \quad 2 \quad 0 \quad 2 \quad 1 \quad 3 \quad 2 \quad 3 \quad 2$$

Generate a frequency distribution for this sample of discrete data.

*Solution:* Observe that $N$, the total frequency, is equal to 40. To generate the frequency distribution, we count the number of times each value of $X$ occurs. The number of times a particular value $x$ of $X$ occurs is called the *frequency* for $x$.

| Number of heads; $x$ | Tally | Frequency, $f$ |
|:---:|:---:|:---:|
| 0 | ||| | 3 |
| 1 | ⫴⫴ | 11 |
| 2 | ⫴⫴⫴ |||| | 19 |
| 3 | ⫴ || | 7 |
| | | Total = 40 |

---

## C.  Continuous data

Given a random experiment with which we have associated a continuous random variable. If we repeat the experiment over and over again, and observe the values of the continuous random variable, then the collection of those values is said to constitute a *sample of continuous data.*

Examples of continuous random variables [see, for example, Chapters 8, 9, and 10] are the height of a male person, the weight of a box of cereal, and the lifetime of an electrical component.

Because very often we have to use imprecise measuring devices to determine measurements of continuous data [such as using a wooden ruler to record a length measurement], we often have no choice but to record continuous data in a discrete fashion. Suppose, for example, we are measuring the diameters of nails with a measuring device [say, a type of ruler] which is accurate to the nearest thousandth of an inch. Thus, typical diameters in inches would be recorded as .109, .113, .107, etc. Thus, we would expect that a recorded diameter of .113 inch would actually indicate a true diameter between .1125 and .1135 inch since .1135 − .1125 = .001, that is, a thousandth of an inch. Our convention, in a case like this, will be to assume that the true diameter $d$ is in the following interval: $.1125 \le d < .1135$.

Each recorded diameter measurement is a *multiple* of the unit, that is, a whole number times the unit. Thus, in the above example, the diameter recorded as .113 inch is 113 times the value of the unit, which is .001 inch here.

**Definition 11.1:** Given a unit of measurement $u$. Then, any recorded score [or recorded measurement] $r$ is a multiple of the unit. That is, $r = ku$, where $k$ is a whole number [which is usually positive]. The true value $\tau$ of a measurement recorded as $r$ is on the interval: $r - u/2 \le \tau < r + u/2$.

If we wish to make a frequency distribution for a set of continuous data [usually, consisting of 20 or more data values], we first group the data into classes. The number of data values that are placed into a class is known as the *class frequency*.

---

**EXAMPLE 11-2:** The heights of a sample of 50 women at Gates College, as recorded to the nearest inch [i.e., $u = 1$ inch, here], are as follows:

61  60  64  63  69  $\boxed{72}$  65  67  66  66     66  69  59  64  65  60  67  65  60  64

66  63  $\boxed{59}$  67  65  67  64  71  61  69     65  70  66  $\boxed{72}$  60  69  62  69  64  69

65  66  68  70  62  64  69  63  64  65

(a) Group the height data into five classes. (b) Generate the frequency distribution corresponding to these five classes.

*Solution*

(a) First, we find the *range*, which is the difference between the largest and smallest data values. Note that we have "boxed in" the smallest and largest values in the above data list, as with the largest value, $\boxed{72}$. Thus, here, we have

$$\text{range} = 72 - 59 = 13 \text{ inches}$$

Next, we decide upon a width for each of the five classes. Here, let us choose the width to be the odd number [of units] which is "just greater than" the range divided by 5. [The 5 pertains to the fact that we want to have five classes.] Since here we have range/5 = 2.6, we choose the width of each class to equal 3. With this width, we have the classes as listed in the left-hand column of the table of part (b). Note that each of the 50 data values in the original list will be placed in one of the five classes. Now, the so-called "class limits" designation of the first class as 58–60 means that the recorded scores in the first class are 58, 59, and 60.

Observe that there is more than one correct solution for choosing the five classes. For example, we could just as well have used the following five classes since they also will contain all of the 50 original data values: 59–61; 62–64; 65–67; 68–70; 71–73. The important thing is that we want the lowest score [here, 59] to be in the lowest class, and the highest score to be in the highest class.

(b) The frequency distribution follows:

| Class label | Class limits | Tally | Class frequency |
|---|---|---|---|
| 1 | 58–60 | ||| | 3 |
| 2 | 61–63 | 卌 卌 | 10 |
| 3 | 64–66 | 卌 卌 卌 卌 | 20 |
| 4 | 67–69 | 卌 卌 || | 12 |
| 5 | 70–72 | 卌 | 5 |

Total = 50

Let us focus on the first class, which is designated by 58–60. The recorded scores 58 and 60 are the class limits for the class, where 58 is the lower class limit (LCL) and 60 is the upper class limit (UCL). The middle recorded score, 59, is called the class mark (CM) for the class. Now, let us generalize.

**Definition 11.2:** Given a frequency distribution consisting of a collection of classes [or class intervals], and associated class frequencies. Each class contains at least one recorded score [and usually an odd number of recorded scores]. The following terminology applies to a particular class.

(a) The class limits are the smallest and largest recorded scores in a given class. They are often denoted as the *lower class limit* [abbreviated LCL] and the *upper class limit* [abbreviated UCL].

(b) The *class boundaries* are the smallest and largest limiting values for a given class. The lower class boundary (LCB) for a class is half a unit less than the lower class limit. That is, $LCB = LCL - u/2$. The upper class boundary (UCB) for a class is half a unit greater than the upper class limit. That is, $UCB = UCL + u/2$.

(c) The width of a class, which is denoted by $w$ [or $c$], is the difference between the upper and lower class boundaries of a class. In symbols, $w = UCB - LCB$. It is referred to as the *class width, class interval width, class length,* or *class size.*

(d) The *class mark* or *class midpoint* [abbreviated as CM] is the midpoint of a class interval. Thus, we have that $CM = (LCL + UCL)/2$ or $CM = (LCB + UCB)/2$.

The class mark will be a recorded score if there is an odd number of recorded scores in a class. Note that it is advantageous in statistical work involving frequency distributions for continuous-type data to have the class mark equal to a recorded score.

*notes*

(i) It is useful if all class intervals have the same width. The number of class intervals is usually taken to be between 5 and 20. To obtain a preliminary estimate for the [constant] class width, divide the range by the desired number of class intervals. Using $R$ as the symbol for the range, and $k$ for the desired number of classes, then $w \approx R/k$.

(ii) It is advantageous to have a class width such that there is an odd number of recorded scores in a class. [This will be the case if $w$ is equal to an odd number multiple of the unit, $u$; for example, if $w = 3u$ or $w = 5u$.] If this is the case, then the class mark will be a recorded score. The grouping error involved in mathematical approximation calculations [such as approximating various moments for a frequency distribution] is lessened if the class marks are recorded scores. Calculations of moments, and terms derived from moments such as the mean and standard deviation, for situations in which data are grouped into classes, will be covered in Section 11-6C.

(iii) In our simplified and limited presentation of descriptive statistics, we shall not deal with class intervals that are unbounded from above or from below. Also, we shall not deal with any situation in which there are class intervals of different widths.

---

**EXAMPLE 11-3:** For the frequency distribution of Example 11-2, determine **(a)** the class width, $w$, **(b)** the class limits for the second class, **(c)** the class boundaries for the second class, **(d)** the recorded scores of the second class, and **(e)** the class marks for all the classes.

*Solution*

(a) If all the class intervals have the same width, as is the case here, then $w$ is equal to the difference between lower [or upper] class limits for a pair of adjacent classes. Thus,

$$w = 61 - 58 = 3 \quad \text{or} \quad w = 72 - 69 = 3$$

(b) For the class designated by 61–63, the class limits are 61 [lower] and 63 [upper].

(c) The lower class boundary is half a unit less than the LCL, and the upper class boundary is half a unit greater than the UCL. Thus,

$$\text{LCB} = \text{LCL} - u/2 = 61 - .5 = 60.5 \quad \text{and}$$

$$\text{UCB} = \text{UCL} + u/2 = 63 + .5 = 63.5$$

As expected, we see that $w = \text{UCB} - \text{LCB} = 63.5 - 60.5 = 3$.

(d) The recorded scores are 61, 62, and 63.

(e) The recorded score 62 is the class mark for the second class. Successive class marks are obtained by adding or subtracting the width $w$ [here, 3] from any class mark. Thus, the five class marks are 59 [from $62 - 3$], 62, 65 [from $62 + 3$], 68, and 71.

---

*note:* Let us again review what possible actual scores a recorded score represents. For instance, in Examples 11-2 and 11-3, where $u = 1$, the recorded score 61 represents any actual score $s$ on the interval $61 - u/2 \leq s < 61 + u/2$, that is, on the interval $60.5 \leq s < 61.5$. Thus, for example, if a score is actually 60.73 inches, it will be recorded as 61 inches.

Likewise, the recorded score 62 represents any actual score $s$ on $61.5 \leq s < 62.5$, and the recorded score 63 represents any actual score $s$ on $62.5 \leq s < 63.5$. But, since 61, 62, and 63 are the recorded scores of the second class, this means that any actual score $s$ in the second class is such that $60.5 \leq s < 63.5$. The limits of this interval, 60.5 and 63.5, are, of course, the class boundaries of the second class.

---

**EXAMPLE 11-4:** The diameters of 100 nails go from .106 to .132 inch. The unit of measurement is .001 inch. Suppose one wants to construct a frequency distribution with approximately six classes of equal width such that each class contains an odd number of recorded scores. **(a)** Determine a suitable collection of class intervals. **(b)** Determine the class boundaries, and class marks for these class intervals.

*Solution*

(a) For the range, $R = .132 - .106 = .026$ inch. Thus, the range consists of 26 units since each unit is .001 inch. Since we want approximately six class intervals, the approximate width, $w$, of each class interval is given by $w = .026/6 = .0043\overline{3}$ inch. Thus, the approximate width, $w'$, in units, is given by $w' = 26/6 = 4.3\overline{3}$ units.

Since we want an odd number of recorded scores per class, or, equivalently, a width equal to an odd number of units, we round up as follows:

$$w = .005 \text{ inch} \qquad \text{or} \qquad w' = 5 \text{ units}$$

Choosing the lower class limit of the first class to be .105 inch [we could just as well use .106 or .104 inch, say], we have the following designation of six classes by class limits:

.105–.109; .110–.114; .115–.119; .120–.124; .125–.129; .130–.134

For the first class, the five recorded scores are .105, .106, .107, .108, and .109. Observe that the lowest recorded score, .106, for the set of 100 nails, is a recorded score in the first class, while the highest recorded score, .132, is in the last class.

(b) For the first class, we have $\text{LCB} = \text{LCL} - u/2 = .105 - .0005 = .1045$. The designation by class boundaries for the six classes, where we use the "LCB-UCB" for each class, is as follows:.

.1045–.1095; .1095–.1145; .1145–.1195; .1195–.1245;

.1245–.1295; .1295–.1345

The class marks for the six classes are .107, .112, .117, .122, .127, and .132.

# 11-2. Histograms

A *histogram* is one type of graphical representation for a frequency distribution. We previously learned about histograms in our discussion of graphical representations for probability functions of discrete random variables [Section 5.2].

## A. Histograms for discrete data

For discrete data [i.e., data pertaining to values of a discrete random variable], the histogram consists of a collection of rectangles, one for each random variable value that occurs.

The *center* of the base of a rectangle is at the value of the random variable. Also, the base extends an equal distance to the left and right of the center. The *areas* for the rectangles are proportional to the frequencies. If the values of the discrete random variable happened to be spaced apart by 1 [as with the random variable of Example 11-1, where the values of $X$ were 0, 1, 2, and 3], then a usual procedure for each rectangle is to take the width of each base equal to one, and the height equal to the frequency associated with the observed random variable value.

**EXAMPLE 11-5:** Construct a histogram that graphically displays the frequency table data of Example 11-1.

*Solution:* We plot the $x$ values on the horizontal axis and the frequencies on the vertical axis. The base of the rectangle for $X = x$ extends from $x - .5$ to $x + .5$ in the horizontal direction. The height of such a rectangle is equal to the frequency for that value of *x*. Thus, the total area for the histogram is equal to the total frequency [here, 40] since each rectangle has a width equal to 1.

It is often useful to make a modification in the construction of a histogram so that the total area is equal to 1. If one does this, then graphical comparisons with the histogram of a related theoretical discrete probability distribution can more easily be made. [See Example 11-6.]

Henceforth, we shall often use the symbol *N* for the total frequency of a sample of data.

**EXAMPLE 11-6:** For the frequency distribution of Examples 11-1 and 11-5, develop a histogram for which the total area is equal to 1.

*Solution:* One easy approach is to keep the width of each rectangle the same, and to modify the heights of the rectangles. In Figure 11-1, we have

$$\text{total area} = 1 \cdot 3 + 1 \cdot 11 + 1 \cdot 19 + 1 \cdot 7 = 40 = N \quad (1)$$

Thus, for a new area equal to 1, we take each rectangle height equal to one fortieth of its original value. Since originally, the height was equal to *f*, the frequency for a value of *x*, the new height is equal to $f/N = f/40$. The quantity $f/N$, known as the *relative frequency*, and symbolized by $\hat{f}$, was introduced back in Section 1-2. In Table 11-1, we have a tabulation of both the relative frequencies—for the data of Example 11-1—and the theoretical probabilities [i.e., values of $f(x)$] for a binomial distribution for which $n = 3$ and $p = .6$.

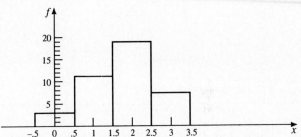

**Figure 11-1.** Histogram for Example 11-5.

**TABLE 11-1: Relative Frequencies for Sample, and Probabilities for Binomial Distribution in which n = 3 and p = .6**

| *x* | $\hat{f}$, relative frequency for sample of data | $f(x)$, probability for *x* |
|---|---|---|
| 0 | 3/40 = .075 | .064 |
| 1 | .275 | .288 |
| 2 | .475 | .432 |
| 3 | .175 | .216 |
| Sum | 1.000 | 1.000 |

In Figure 11-2a, we have a histogram for the frequency distribution, where the ordinate is $\hat{f}$, the relative frequency. In Figure 11-2b, we have a histogram for the probability function $f(x) = b(x; 3, .6)$. These binomial probabilities can be looked up in Table I of Appendix A.

As we would expect, the histogram for the frequency distribution of our sample data is very close in shape to the histogram of the related theoretical probability distribution. Restricting the total histogram area to be equal to one for a frequency distribution simplifies such visual comparisons.

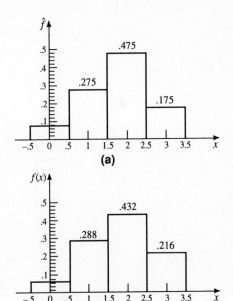

**Figure 11-2.** Histograms for binomial distribution in which total area equals 1. **(a)** Histogram for frequency distribution; $\hat{f}$ is relative frequency. **(b)** Histogram for binomial probability function $f(x)$ [$n = 3$; $p = .6$].

## B. Histograms for continuous data

A histogram for a frequency distribution for continuous data consists of a set of rectangles, one for each class [interval]. The base of a rectangle for a particular class extends from the lower to the upper class boundary of the class. Thus, the width of each rectangle is equal to *w*, the class width.

The areas of the rectangles are to be proportional to class frequencies. If the class width $w$ is constant [as will be the case in our work, unless otherwise noted], this means that the heights of the rectangles are to be proportional to the class frequencies. A usual procedure in this case is to take the heights of the rectangles equal to the corresponding class frequencies.

Along the horizontal axis, it is customary to label class boundaries and/or class marks.

---

**EXAMPLE 11-7:** Construct a histogram for the frequency distribution of Example 11-2 [heights of 50 women at Gates College].

*Solution:* In Figure 11-3, the height of each rectangle equals the frequency for the corresponding class. The class boundaries and class marks are displayed on the horizontal axis. The base of each rectangle extends from the lower class boundary to the upper class boundary of the corresponding class.

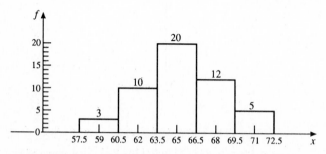

**Figure 11-3.** Histogram for frequency distribution of Example 11-2.

---

### C. Histogram variations for continuous data

The relative frequency of a class, $\hat{f}$, is the frequency of a class, $f$, divided by $N$, the total frequency. If the relative frequency is used instead of the frequency as the ordinate for a histogram, then the resulting pictorial representation is known as a *relative frequency histogram*. The resulting shape is similar [in a geometrical sense] to that of a histogram in which the ordinate is the class frequency. The sum of the relative frequencies for all classes is, of course, equal to 1.

Often, it is desirable to compare a histogram for continuous-type data with the graph of a probability distribution for some continuous random variable. For example, we may believe that a sample of continuous-type data [from which we have subsequently developed a frequency table and histogram] comes from some definite continuous probability distribution, such as, say, a normal distribution or exponential distribution. But, recall that our convention has been to take the total area under the graph of a probability distribution for a continuous random variable [the latter is also known as a probability density function, abbreviated pdf] to be equal to 1. That is, we have $\int_{-\infty}^{\infty} f(x)\,dx = 1$. Also, the probability for the interval from $a$ to $b$ is the area under the graph of $y = f(x)$ between $x = a$ and $x = b$. [See Definition 8.1 and Theorems 8.1 and 8.2.]

Suppose in our original histogram that the ordinate is class frequency, and the base of each rectangle extends between class boundaries of the corresponding class. Then, we can construct a histogram with total area equal to 1 by using a new ordinate $f^*$, which is equal to the class frequency divided by the total area of the original histogram, namely the total frequency times the class width. That is, we choose $f^*$ such that

$$f^* = \frac{f}{Nw} \qquad \textbf{(I)}$$

[The total area of the original histogram is given by $A = f_1 w + f_2 w + \ldots + f_k w = (f_1 + f_2 + \ldots + f_k)w = Nw.$]

---

**EXAMPLE 11-8:** Consider the frequency distribution of Examples 11-2 and 11-7. Construct histograms for which the ordinate is **(a)** the relative frequency, $\hat{f}$, and **(b)** $f^* = f/(Nw)$.

## Solution

(a) The ordinate for class $i$, $\hat{f}_i$, is given by $\hat{f}_i = f_i/N$, where $f_i$ denotes the frequency of class $i$, and the $N$ is the total frequency. Here, $N = 50$. Refer to Figure 11-4a.

(b) Here, the ordinate for class $i$, $f_i^*$, is given by $f_i^* = f_i/(Nw) = f_i/150$. See Figure 11-4b. The total area for this histogram is 1.

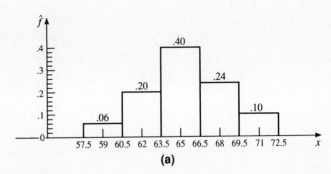

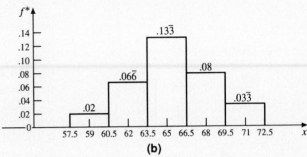

**Figure 11-4.** Histograms for Example 11-8. (a) Ordinate is $\hat{f} = f/N$. (b) Ordinate is $f^* = f/(Nw)$.

Naturally, the shapes of the histograms of Figures 11-3, 11-4a, and 11-4b are similar to one another. This is so because the ordinates are proportional to one another.

*note:* Some practitioners express relative frequency in terms of a percentage. Such a relative frequency is equal to our relative frequency, $\hat{f}$, times 100.

## 11-3. The Empirical Cumulative Distribution Function

### A. Discrete data: The empirical cumulative distribution function

Given a value $x$ of a discrete random variable $X$. The empirical cumulative distribution function at $x_0$, $\hat{F}(x_0)$, is equal to the relative frequency accumulated up to and including $x_0$. In the following, $f_x$ and $\hat{f}(x)$ will respectively stand for the frequency and relative frequency at $x$.

**Definition 11.3:** The *empirical cumulative distribution function* [or *empirical distribution function* or *cumulative relative frequency distribution*] at $x_0$ is defined to be

$$\hat{F}(x_0) = \sum_{x \le x_0} \hat{f}(x) = \frac{1}{N} \sum_{x \le x_0} f_x$$

We can develop a table for $\hat{F}(x)$ by adding two new columns to a frequency distribution table. The first column is for the cumulative frequency, where the cumulative frequency at $x_0$ is given by $\sum_{x \le x_0} f_x$.

The second column is for the cumulative relative frequency, and this is obtained by dividing the cumulative frequency at a value $x_0$ by the total frequency. That is, cumulative relative frequency at $x_0 = \hat{F}(x_0) = (1/N) \sum_{x \le x_0} f_x$.

It is useful to graph the empirical cumulative distribution function. Observe that this function is a step function, just as was the case for the cdf of a discrete random variable.

---

**EXAMPLE 11-9:** Refer to the frequency distribution of Examples 11-1, 11-5, and 11-6. (a) Develop a table for the cumulative relative frequency in terms of $x$. (b) Draw a graph of the empirical cumulative distribution function.

## Solution

(a) First, we repeat the $x$ and frequency columns of the frequency distribution table of Example 11-1. Then, we add a third column which has values for the cumulative frequencies at $x = 0$, 1, 2, and 3. The fourth column contains values for the cumulative relative frequencies for $x = 0$, 1, 2, and 3. These values are obtained by dividing each cumulative frequency by the total frequency [here, $N = 40$].

| No. of heads; $x$ | Frequency, $f$ | Cumulative frequency | Cumulative relative frequency |
|---|---|---|---|
| 0 | 3 | 3 | $3/40 = .075$ |
| 1 | 11 | $3 + 11 = 14$ | $14/40 = .350$ |
| 2 | 19 | $14 + 19 = 33$ | $33/40 = .825$ |
| 3 | 7 | $33 + 7 = 40$ | $40/40 = 1.000$ |

In the preceding table, we have the cumulative relative frequency tabulated for the discrete values $x = 0, 1, 2,$ and 3. But the cumulative relative frequency $\hat{F}(x)$ is defined for *all* values of $x$. What happens here and for frequency distributions for other discrete random variables is similar to what happens with respect to the theoretical cdf for discrete random variables. Refer to Section 6-1, for example.

Thus, the complete table for the cumulative relative frequency as a function of $x$ for all $x$ is as follows: $\hat{F}(x) = 0$ for $x < 0$; $\hat{F}(x) = 3/40 = .075$ for $0 \le x < 1$; $\hat{F}(x) = 14/40 = .350$ for $1 \le x < 2$; $\hat{F}(x) = 33/40 = .825$ for $2 \le x < 3$; $\hat{F}(x) = 40/40 = 1.000$ for $3 \le x$.

**(b)** The graph of the empirical cdf [or cumulative relative frequency] in terms of $x$ is shown in Figure 11-5. This graph clearly shows why the function $\hat{F}(x)$ is called a step function [essential discontinuities occur at $x = 0, 1, 2,$ and 3].

The "holes" [o] at $x = 0, 1, 2,$ and 3 indicate points that are not on the graph, while the "filled-in dots" [•] at $x = 0, 1, 2,$ and 3 indicate points that are on the graph of $\hat{F}(x)$. For example, the point $(1, .075)$ is not on the graph [hole at that point] while the point $(1, .350)$ is on the graph [filled-in dot at that point].

Recall that $f_0, \ldots, f_3$ denote the respective frequencies for $x = 0, \ldots, 3$. Thus, we have

$$\hat{F}(1) = (f_0 + f_1)/N = (3 + 11)/40 = 14/40 = .350$$

while

$$\hat{F}(x) = f_0/N = 3/40 = .075 \qquad \text{for all } x \text{ on the interval } 0 \le x < 1$$

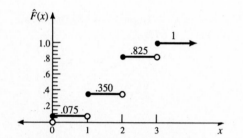

**Figure 11-5.** Graph of empirical cdf for a discrete random variable [Example 11-9].

## B.  Continuous data: the empirical cumulative distribution function

We wish to generate a piecewise-linear function that represents the empirical cumulative distribution function for a sample of continuous data. [By "piecewise linear," we mean that the function will consist of *straight line* pieces or portions. As we shall see, the straight line pieces will be joined in a continuous fashion.] Supposedly, the empirical cdf will serve as an approximation to the cdf for some continuous random variable. See Section 8-1B for a review of the properties of $F(x)$, the cdf for a continuous random variable.

Our first task will be to generate a table for the empirical cdf $\hat{F}(x)$ at the class boundaries of a frequency distribution. [Recall that for a frequency distribution for a sample of continuous data, we have a tabulation of frequencies for a set of classes, where, in our work, all classes have the same width, $w$.] If $x_l$ denotes the lowest class boundary for a frequency distribution, then clearly

$$\hat{F}(x) = 0 \qquad \text{for } x \le x_l \tag{I}$$

Also, if $x_h$ denotes the highest class boundary for a frequency distribution, then clearly

$$\hat{F}(x) = 1 \qquad \text{for } x_h \le x \tag{II}$$

In the following development, we shall use $f_i$ to indicate the frequency for class $i$, where $i = 1, 2, \ldots, k$.

Now, at the upper class boundary of the first class [the class with the lowest class mark], the empirical cdf is equal to $f_1/N$. [Here, $f_1$ is the frequency for the first class.] At the upper class boundary of the second class, the empirical cdf is equal to $(f_1 + f_2)/N$. Thus, we observe that we have accumulated the relative frequency $(f_1 + f_2)/N$ at the upper class boundary of the second class. This pattern continues at the remaining class boundaries for a frequency table.

*note:* We shall often refer to the empirical cumulative distribution function as the cumulative relative frequency or cumulative relative frequency function.

The technique for obtaining $\hat{F}(x)$ at the class boundaries of a frequency distribution is illustrated in Example 11-10.

---

**EXAMPLE 11-10:** Refer to the frequency distribution pertaining to Examples 11-2, 11-3, 11-7, and 11-8. Those examples deal with the height data for a sample of 50 women at Gates College. Develop a table which contains columns displaying the cumulative frequency and cumulative relative frequency [or empirical cdf] at the class boundaries.

*Solution:* In Table 11-2 below, the first two columns are for the class marks and class frequencies. The third column contains values for the six class boundaries, running from 57.5 [lowest] to 72.5 [highest]. Notice that the class boundary values are not on the same horizontal lines as the class mark values. In fact, for each class mark, the lower class boundary occurs on the line physically above the class mark, and the upper class boundary occurs on the line physically below the class mark.

**TABLE 11-2: Generating Cumulative Relative Frequencies at the Class Boundaries for a Frequency Distribution**

| Class mark | Class frequency | Class boundary | Cumulative frequency | Cumulative relative frequency |
|---|---|---|---|---|
| | | 57.5 | 0 | 0 |
| 59 | 3 | | | |
| | | 60.5 | $0 + 3 = 3$ | $3/50 = .06$ |
| 62 | 10 | | | |
| | | 63.5 | $3 + 10 = 13$ | $13/50 = .26$ |
| 65 | 20 | | | |
| | | 66.5 | $13 + 20 = 33$ | $33/50 = .66$ |
| 68 | 12 | | | |
| | | 69.5 | 45 | $45/50 = .90$ |
| 71 | 5 | | | |
| | | 72.5 | 50 | $50/50 = 1.00$ |

The values for the cumulative frequency [fourth column] and cumulative relative frequency [fifth column] are on the same horizontal lines as are the class boundary values. Each cumulative relative frequency is equal to the corresponding cumulative frequency divided by the total frequency.

At the lowest class boundary, the cumulative frequency is 0. At any other class boundary, the cumulative frequency is equal to the cumulative frequency at the preceding class boundary plus the class frequency for the class enclosed by those class boundaries.

## C. Continuous data: cumulative relative frequency polygon

Once we have a tabulation of cumulative relative frequencies in terms of class boundaries, we can construct the *cumulative relative frequency polygon* [or *cumulative relative frequency graph* or *cumulative relative frequency ogive* or *graph of the empirical cumulative distribution function*]. This polygon is supposedly an approximation to the graph of the cdf of a related continuous random variable. Let us now discuss the construction of the polygon. First, we agree to plot cumulative relative frequencies vertically and score values horizontally [as with the related histogram diagram].

Consider two class boundaries $x$ and $y$, where $x < y$, such that they are *adjacent* [that is, $y - x = w$, the class width]. The points $(x, \hat{F}(x))$ and $(y, \hat{F}(y))$ are plotted, and then joined by a straight line segment. Similarly, straight line segments are drawn between every pair of such adjacent points. The resulting figure is known as a cumulative relative frequency polygon.

Observe that if we start with a frequency distribution comprised of $k$ classes, then there will be $(k + 1)$ such original plotted points since there are $(k + 1)$ class boundaries. Clearly, there will be $k$ line segments connecting these $(k + 1)$ points.

Again, let $x_l$ and $x_h$ stand for the lowest and highest class boundaries. On the cumulative relative frequency polygon, we construct a horizontal line segment which extends, in unlimited fashion, through and to the left of the point $(x_l, 0)$, and another horizontal line segment which extends, in unlimited fashion, through and to the right of the point $(x_h, 1)$.

---

**EXAMPLE 11-11:** From the tabulated data in Example 11-10, construct a cumulative relative frequency polygon.

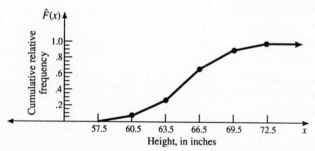

**Figure 11-6.** Cumulative relative frequency polygon for a frequency distribution.

*Solution:* Refer to Figure 11-6. First, we plot the points $(x, \hat{F}(x))$, where $x$ is a class boundary value, and $\hat{F}(x)$ is the corresponding cumulative relative frequency. The coordinates for these six points are obtained from the third and fifth columns of Table 11-2. [See Example 11-10.]

Next, we connect the point $(57.5, 0)$ to the point $(60.5, .06)$ with a straight line segment. We continue connecting points that have adjacent class boundaries with straight line segments until we have constructed five such segments. Finally, we draw horizontal line segments of unlimited extent, one to the left from $(57.5, 0)$, and the other to the right from $(72.5, 1)$. The former line segment coincides with the $x$-axis, and hence is not clearly shown.

---

It is useful to compare the cumulative relative frequency polygon of Figure 11-6 with graphs of cdf's of continuous random variables. See, for example, Figures 8-4, 8-8, and 8-11. Remember that Figure 11-6 is also known as a *graph of an empirical cdf*.

## 11-4. Percentiles for a Frequency Distribution of Continuous Data

In Section 10-2B, we defined the $k$th percentile $p_k$ for the probability distribution of a continuous random variable. In particular, in Definition 10.2, we said that the $k$th percentile $p_k$ was a value of the continuous random variable $X$ such that

$$P(X \le p_k) = k/100 \qquad \text{or} \qquad F(p_k) = k/100 \qquad \textbf{(a), (b)}$$

## A. Definition of a percentile

The definition of the $k$th percentile $p_k$ for a frequency distribution of a sample of continuous data is similar, except that we replace the cdf $F$ by the empirical cdf [or cumulative relative frequency] $\hat{F}$.

**Definition 11.4:** The $k$th percentile $p_k$ is a value of $x$ [where $x$ denotes a score] such that

$$\hat{F}(p_k) = k/100$$

We can easily compute a percentile if we know how the cumulative relative frequency is related to $x$. We can work from a cumulative relative frequency polygon [for a particular example, see Figure 11-6], or, to be more precise in our calculations, we may proceed algebraically from a tabulation of the cumulative relative frequency in terms of class boundary. For a particular example of such a tabulation, refer to the third and fifth columns of Table 11-2.

Our algebraic approach will involve the technique of linear interpolation [also known as "interpolation by proportional parts"]. After all, points corresponding to adjacent class boundaries are connected by straight line segments on the cumulative relative frequency polygon.

Henceforth, let us agree that the pair of columns consisting of the class boundary column and the corresponding cumulative relative frequency column constitutes a *cumulative relative frequency table* [or *empirical cdf table*].

## B. Illustrating how to calculate percentiles

**EXAMPLE 11-12:** Refer to Table 11-2 in Example 11-10. [Recall that the original data consisted of heights of a sample of 50 women at Gates College. These data were then grouped into five classes.] **(a)** Calculate $p_{25}$, the 25th percentile. **(b)** Calculate the median and the third quartile. **(c)** Demonstrate how the median and third quartile can be calculated by using a cumulative relative frequency polygon.

*Solution: Preliminaries:* Refer to Table 11-2 of Example 11-10. Let us reproduce the cumulative relative frequency table, which consists of the third and fifth columns of Table 11-2.

**TABLE 11-3: Cumulative Relative Frequency Table for Sample of Heights of Women at Gates College**

| Class boundary, $x$ | Cumulative relative frequency, $\hat{F}(x)$ |
|---|---|
| 57.5 | .00 |
| 60.5 | .06 |
| 63.5 | .26 |
| 66.5 | .66 |
| 69.5 | .90 |
| 72.5 | 1.00 |

**(a)** Observe that the 25th percentile, $p_{25}$, is also the first quartile, $Q_1$. Corresponding to $p_{25}$ is a cumulative relative frequency equal to $25/100 = .25$.

To find $p_{25}$, we do linear interpolation with respect to the points in Table 11-3 for which $\hat{F}(x) = .06$ and $\hat{F}(x) = .26$ since these $\hat{F}(x)$ values are *just less than* and *just greater than* $\hat{F}(x) = .25$. Refer to the subtable below. We form proportions corresponding to the numbers connected by the braces. [The number next to a brace is the difference obtained by subtracting the numbers adjacent to the ends of the brace. For example, $63.5 - 60.5 = 3.0, .25 - .06 = .19$, and $.26 - .06 = .20$. The difference $\Delta$ is given by $p_{25} - 60.5$, where $p_{25}$ is our unknown.]

| $x$ | $\hat{F}(x)$ |
|---|---|
| 60.5 | .06 |
| $p_{25}$ | .25 |
| 63.5 | .26 |

$$\frac{\Delta}{3.0} = \frac{.19}{.20} \qquad \textbf{(1)}$$

Since

$$\Delta = p_{25} - 60.5 \qquad \textbf{(2a)}$$

$$p_{25} = 60.5 + \Delta \qquad \textbf{(2b)}$$

From Eq. (1), we have $\Delta = 3(19/20) = 2.85$. Thus,

$$p_{25} = 60.5 + \Delta = 63.35 \approx 63.4 \qquad \textbf{(3)}$$

Here, we carry our calculation to one decimal place since our original $x$ values [which are class boundaries] are accurate to one decimal place.

**(b)** Let us symbolize the median by $Md$ as in Section 10-2B. Recall that $Md$ is also $p_{50}$. Now, corresponding to $p_{50}$, we have $\hat{F}(x) = 50/100 = .50$. Doing linear interpolation with respect to $\hat{F}(x) = .50$, we find that

$$Md = p_{50} = 63.5 + \overbrace{\frac{\overbrace{(.50 - .26)}^{.24}}{\underbrace{(.66 - .26)}_{.40}}}(3.0) = 63.5 + 1.8 = 65.3 \qquad \textbf{(4)}$$

The third quartile $Q_3$ is the same as $p_{75}$. Corresponding to $p_{75}$, we have $\hat{F}(x) = 75/100 = .75$. Doing linear interpolation with respect to $\hat{F}(x) = .75$, we find that

$$Q_3 = p_{75} = 66.5 + \frac{(.75 - .66)}{(.90 - .66)}(3.0) = 66.5 + 1.125$$

$$= 67.625 \approx 67.6 \qquad \textbf{(5)}$$

**(c)** Refer to Figure 11-7, which, at the outset, is a reproduction of Figure 11-6. For the median, first we draw a horizontal line from .50 on the vertical axis to its intersection with the cumulative relative frequency polygon. Then, we draw a vertical line from this intersection point to the horizontal axis. At the intersection point of the vertical line with the horizontal axis, we read off $Md \approx 65.3$.

The third quartile is the horizontal coordinate [abscissa] of that point on the cumulative relative frequency polygon whose vertical coordinate [ordinate] is .75. We read off $Q_3 \approx 67.6$.

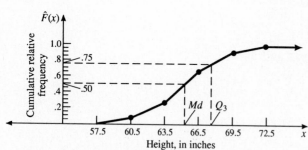

**Figure 11-7.** Using a cumulative relative frequency polygon to calculate percentiles.

*note:* It should be clear that an algebraic approach, such as that used in parts (a) and (b), is much more reliable than a graphical approach [such as that used in part (c)] for calculating percentiles.

# 11-5. Moments for a Sample of Data

## A. Review of moments for a discrete random variable

We discussed the moment concept in Section 9-3. It is useful to repeat the equation for the $r$th moment about the origin that applied for the case of a discrete random variable. That was given in Definition 9.3, as follows:

$$\mu'_r = \sum_x x^r f(x) \qquad \text{for } r = 0, 1, 2, 3, \dots \qquad \textbf{(I)}$$

Here, $\mu$ is a Greek letter [read as "mu"], which is comparable to the letter $m$ of the Roman alphabet. The symbol $\mu'_r$ is read as "mu sub $r$, prime."

A moment of great importance is $\mu'_1 = \mu = \Sigma_x \, xf(x)$, the mean of random variable $X$. Recall that the mean $\mu$ is a primary measure of the average value of random variable $X$. The moments $\mu'_2$ and $\mu'_3$ are also fairly important.

The $r$th moment about the mean of random variable $X$, denoted as $\mu_r$, was given in Definition 9.4 as follows, for a discrete random variable:

$$\mu_r = \sum_x (x - \mu)^r f(x) \qquad \text{for } r = 0, 1, 2, 3, \dots \qquad \textbf{(II)}$$

The moment $\mu_2$ is of great importance, and it is also known as the variance of $X$ [this is often symbolized as $\text{Var}(X)$]. The standard deviation of $X$, denoted by $\sigma_x$, is given by

$$\sigma_x = \sqrt{\mu_2} \qquad \text{[Definition 9.5]} \qquad \textbf{(III)}$$

Recall that the standard deviation is a primary measure of *spread* or *dispersion* of a random variable.

## B. Moments for a sample of ungrouped data

Suppose we are collecting data from the repeat performance of a probability [or random] experiment, or just collecting data. Usually, in a probability situation, we have specified some random variable $X$, and we are interested in the values of $X$ that occur during repeat performances of the probability experiment. Suppose that $N$ observed values of random variable $X$ are denoted as $x_1, x_2, \dots, x_N$.

The collection of values $x_1, x_2, \dots, x_N$ is often called a sample [or set] of data, or a sample of ungrouped data. We would like to use Eqs. (I) and (II) of Section 11-5A to develop equations for moments of a sample of ungrouped data.

*notes*

(a) We could have a grouping of data if certain values in a sample of data are repeated. So far in this chapter, we dealt with two situations in which data were grouped. One was for the sample data for a discrete random variable [Section 11-1B]. The other was for data for a continuous random variable, which were grouped into classes [Section 11-1C]. We will deal with calculations of moments pertaining to grouped data in Section 11-6.

One of the main reasons for the use of the word "ungrouped" in the phrase "ungrouped data" in the current discussion is to distinguish our equations from those that will arise in the discussion of grouped data.

(b) Another name for a sample of data values $x_1, x_2, \dots, x_N$ is *empirical distribution*.

We shall use the letter $m$ instead of $\mu$ in equations dealing with moments for samples of data [both ungrouped and grouped]. Thus, instead of $\mu'_r$ and $\mu_r$, we will have $m'_r$ and $m_r$ as symbols for moments about the origin and mean, respectively.

Let us now set up a hypothetical model in order to determine a reasonable way to express $\Sigma_x x^r f(x)$ in Eq. (I) of Section 11-5A. We assume that the $N$ measurements $x_1, x_2, \ldots, x_N$ are written one to a card, and that the cards are placed in a container. The probability experiment is then to draw a single card from the container. Thus, it is clear that we have the following probability distribution associated with the values $x_1, x_2, \ldots, x_N$:

$$f(x_i) = \frac{1}{N} \quad \text{for } i = 1, 2, \ldots, N, \quad \text{that is, for } x_1, x_2, \ldots, x_N \quad \textbf{(α)}$$

Substitution of $f(x)$ from Eq. (α) into Eqs. (I) and (II) of Section 11-5A yields the following results, after taking the constant $1/N$ to the left of the $\Sigma$ sign.

**Definition 11.5:** The equation for $m'_r$, the $r$th moment about the origin for a sample of ungrouped data, is

$$m'_r = \frac{1}{N} \sum_{i=1}^{N} x_i^r \quad \text{for } r = 0, 1, 2, 3, \ldots \quad \textbf{(A)}$$

For example, for $r = 1$ and 2, Eq. (A) yields

$$m'_1 = \frac{1}{N} \sum_{i=1}^{N} x_i \quad \text{and} \quad m'_2 = \frac{1}{N} \sum_{i=1}^{N} x_i^2$$

The term $m'_1$ is the mean for a sample of ungrouped data; it is usually symbolized by $\bar{x}$.

$$\bar{x} = \frac{1}{N} \sum_{i=1}^{N} x_i \quad \text{[sample mean—ungrouped data]} \quad \textbf{(B)}$$

Substitution of $f(x_i) = 1/N$ for $i = 1, 2, \ldots, N$, and the use of the symbol $\bar{x}$ for the mean in Eq. (II) of Section 11-5A yields the following equation for $m_r$, the $r$th moment about the mean for a sample of ungrouped data:

$$m_r = \frac{1}{N} \sum_{i=1}^{N} (x_i - \bar{x})^r \quad \text{for } r = 0, 1, 2, 3, \ldots \quad \textbf{(C)}$$

For $r = 2$, we get the equation for $m_2$, which is a *type of sample variance*. This will also be denoted as $\tilde{s}^2$, where $\tilde{s}$ is a *type of standard deviation* for a sample of ungrouped data.

$$\text{variance} = \tilde{s}^2 = m_2 = \frac{1}{N} \sum_{i=1}^{N} (x_i - \bar{x})^2 \quad \textbf{(D)}$$

*notes*

(i) Observe that we replaced the general summation symbol $\Sigma_x$—see Eqs. (I) and (II) of Section 11-5A—by $\Sigma_{i=1}^{N}$. The index $i$ pertains to the typical sample measurement $x_i$.

(ii) The mean $\bar{x}$ is a measure of average or central tendency for a sample of data. The standard deviation $\tilde{s}$ is a measure of deviation or spread for a sample of data.

The following example pertains to a sample of data for a continuous random variable.

---

**EXAMPLE 11-13:** Krunchee cereal comes in boxes that are labeled 32 ounces. The actual weights, in ounces, for a sample of five boxes are listed as follows: 30.68, 32.52, 33.24, 31.83, 31.47. Find **(a)** the sample mean, **(b)** the sample variance (i.e., $\tilde{s}^2$), and **(c)** the sample standard deviation, $\tilde{s}$.

### Solution

(a)

$$\bar{x} = \frac{1}{N} \sum_{i=1}^{N} x_i = \frac{30.68 + 32.52 + 33.24 + 31.83 + 31.47}{5}$$

$$= 31.948 \approx 31.95 \text{ oz} \tag{1}$$

(b), (c) It is useful to set up a table to assist us with the calculations.

| $x$ | $(x - \bar{x})$ | $(x - \bar{x})^2$ |
|---|---|---|
| 30.68 | $-1.268$ | 1.607824 |
| 32.52 | .572 | .327184 |
| 33.24 | 1.292 | 1.669264 |
| 31.83 | $-.118$ | .013924 |
| 31.47 | $-.478$ | .228484 |
| $\sum_{i=1}^{5} = 159.74$ | .000 | 3.846680 |

Thus,

$$\tilde{s}^2 = \frac{\sum_{i=1}^{5} (x_i - \bar{x})^2}{5} = \frac{3.846680}{5} = .769336 \tag{2}$$

and hence,

$$\tilde{s} = \sqrt{.769336} = .877118 \approx .877 \text{ oz} \tag{3}$$

#### notes

(i) Observe from Eq. (1) that the mean is given by an equation that we would naturally expect, namely

$$\bar{x} = \frac{\text{sum of data values}}{\text{number of data values}} = \frac{x_1 + x_2 + \dots + x_5}{5}$$

(ii) Observe from the calculation in part (b) that the variance is the expected [or mean or average] square deviation of $x_i$ from $\bar{x}$. That is, for general $N$, we can write $\tilde{s}^2$ as follows:

$$\tilde{s}^2 = \frac{(x_1 - \bar{x})^2 + (x_2 - \bar{x})^2 + \dots + (x_N - \bar{x})^2}{N}$$

(iii) Very often, we shall abbreviate $\sum_{i=1}^{N}$ by $\sum$.

## C. More information on the variance and standard deviation for ungrouped data

Another type of sample variance that is quite popular is the following:

$$\hat{s}^2 = \frac{1}{(N-1)} \sum_{i=1}^{N} (x_i - \bar{x})^2 \tag{I}$$

If we compare this equation with that for $\tilde{s}^2$, we see that $\hat{s}^2$ and $\tilde{s}^2$ differ solely in the denominator term. For $\tilde{s}^2$, which appears more "natural" at this point in the development, we divide by $N$, while for $\hat{s}^2$, we divide by $(N-1)$. Associated with $\hat{s}^2$ is a second type of standard deviation, namely $\hat{s}$. It, of course, is equal to $\sqrt{\hat{s}^2}$.

**EXAMPLE 11-14:** For the sample of data for the weights of five Krunchee cereal packages of Example 11-13, compute the variance $\hat{s}^2$ and the standard deviation $\hat{s}$.

*Solution:* From the table in Example 11-13, we have

$$\hat{s}^2 = \frac{\sum(x-\bar{x})^2}{(N-1)} = \frac{3.846680}{4} = .96167 \tag{1}$$

Thus,

$$\hat{s} = \sqrt{.96167} = .980648 \approx .981 \text{ oz} \tag{2}$$

For large $N$, it is clear that $\hat{s}^2$ and $\tilde{s}^2$ will be close in value, and likewise for $\hat{s}$ and $\tilde{s}$.

note: The reason for the importance of $\hat{s}^2$ has to do with point [or single value] estimation theory. We will focus on estimation theory in more detail later in the development. [Briefly, we can say that estimation theory is concerned with estimating parameters of a probability distribution by using statistical quantities obtained from a sample of data.]

For now, we shall merely outline some key ideas. An equation similar to our equation for $\hat{s}^2$ is the following equation for the random variable $\hat{S}^2$:

$$\hat{S}^2 = \frac{1}{(N-1)} \sum_{i=1}^{N} (X_i - \bar{X})^2 \tag{Ia}$$

Here, $X_1, X_2, \ldots, X_N$, and $\bar{X}$ and $\hat{S}^2$ are all *random variables*. [The situation is different from what we had in Eq. (I) where, in particular, $x_1, x_2, \ldots$, and $x_N$ were all *data values* in a sample.] It can be shown for a particular kind of sampling of data from a probability distribution—known as *random sampling*—that the expected value of the random variable $\hat{S}^2$ is equal to $\sigma^2$, the variance for the probability distribution. That is,

$$E(\hat{S}^2) = \sigma^2 \tag{II}$$

Now, a random variable $V$ is said to be an *unbiased estimator* of a parameter $\theta$ if $E(V) = \theta$. Thus, if random sampling exists, then $\hat{S}^2$ is an unbiased estimator of $\sigma^2$. Now, a particular value of $\hat{S}^2$ is $\hat{s}^2$, where the latter is obtained from a particular sample of data.

The above discussion and, in particular, Eq. (II) indicate why $\hat{s}$ has become a very popular type of sample standard deviation.

In Theorem 9.5, we listed the "shortcut" equation for calculating Var($X$). We repeat it here:

$$\text{Var}(X) = E(X^2) - [E(X)]^2$$

The following comparable equation holds if we are dealing with a sample of ungrouped data:

**Theorem 11.1:** For a sample of ungrouped data,

$$m_2 = m_2' - (\bar{x})^2 \tag{a}$$

Here, of course, $m_2$ is also known as $\tilde{s}^2$. Also, $m_2' = (1/N)\Sigma x^2$ and $\bar{x} = (1/N)\Sigma x$.

The following example deals with ungrouped data pertaining to a sample for a discrete random variable.

**EXAMPLE 11-15:** Given a coin for which the probability of head on a single toss is .6 [i.e., $p = .6$]. Suppose the binomial experiment of tossing the coin three times is repeated 10 times. The 10 values for the random variable $X$, which stands for the number of heads obtained, are as follows: $x_1 = 2, x_2 = 2, x_3 = 0, x_4 = 1, x_5 = 2, x_6 = 2, x_7 = 1, x_8 = 2, x_9 = 3, x_{10} = 1$. Here, the values of the random variable $X$ are symbolized by $x_1, x_2, \ldots, x_{10}$. For this sample, find **(a)** the mean, **(b)** the variance $m_2 = \tilde{s}^2$, and **(c)** the variance $\hat{s}^2$. Make use of Theorem 11.1 to compute $\tilde{s}^2$.

**Solution:** *Preliminaries:* The following tabulation is useful.

| | | | | | | | | | | | $\sum$ | $\frac{1}{10}\sum$ |
|---|---|---|---|---|---|---|---|---|---|---|---|---|
| $x$ | 2 | 2 | 0 | 1 | 2 | 2 | 1 | 2 | 3 | 1 | 16 | 1.6 |
| $x^2$ | 4 | 4 | 0 | 1 | 4 | 4 | 1 | 4 | 9 | 1 | 32 | 3.2 |

**(a)**, **(b)**, and **(c)**: Thus,

$$\bar{x} = \frac{1}{10}\sum x = \frac{16}{10} = 1.6 \tag{1}$$

$$m_2' = \frac{1}{10}\sum x^2 = \frac{32}{10} = 3.2 \tag{2}$$

From Theorem 11.1, we have

$$\tilde{s}^2 = m_2' - (\bar{x})^2 = 3.2 - (1.6)^2 = .64 \tag{3}$$

Now,

$$\hat{s}^2 = \left(\frac{N}{N-1}\right)\tilde{s}^2 = \left(\frac{10}{9}\right)(.64) = .711\bar{1} \approx .711 \tag{4}$$

---

**notes**

**(i)** Examples dealing with a data sample of size 40 for the same binomial random variable were considered in Examples 11-1, 11-5, 11-6, etc.

**(ii)** The values for $\bar{x}$ and $\tilde{s}^2$ [or $\hat{s}^2$] are supposed to serve, respectively, as estimates for the parameters $\mu$ [mean] and $\sigma^2$ [variance] of the related random variable. For this binomial random variable with $n = 3$ and $p = .6$, we have $\mu = np = 1.8$ and $\sigma^2 = npq = .72$.

# 11-6. Moments for a Sample of Grouped Data

## A. Introduction to moments for a sample of grouped data

Suppose one has a sample of data, usually such that the total frequency $N$ is fairly large, for which the data values are $x_1, x_2, \ldots, x_k$, and $x_1$ occurs with frequency $f_1$, $x_2$ occurs with frequency $f_2, \ldots$, and $x_k$ occurs with frequency $f_k$. Naturally, $N = f_1 + f_2 + \ldots + f_k$. We saw this type of situation in both Sections 11-1B and 11-1C. In Section 11-1B, we dealt with a frequency distribution of discrete data. In Section 11-1C, we dealt with a frequency distribution of continuous data in which the data were first grouped into classes.

Let us now repeat the equations of Section 11-5A for the $r$th moments about the origin and about the mean, only now let us replace $\mu_r'$ by $m_r'$, $\mu_r$ by $m_r$, and $\mu$ [the symbol for the mean] by $\bar{x}$.

$$m_r' = \sum_x x^r f(x) \qquad \text{for } r = 0, 1, 2, 3, \ldots \qquad \text{[Mom. about origin]} \tag{I}$$

$$m_r = \sum_x (x - \bar{x})^r f(x) \qquad \text{for } r = 0, 1, 2, 3, \ldots \qquad \text{[Mom. about mean]} \tag{II}$$

Here, $m_1' = \bar{x}$, and $m_2 = \tilde{s}^2$ as with a sample of ungrouped data.

We shall now set up a hypothetical model in order to determine a reasonable way to express $m'_r$ and $m_r$ in Eqs. (I) and (II) above. We assume that the $N$ data values, of which $f_1$ values are $x_1$, $f_2$ values are $x_2$, ..., and $f_k$ values are $x_k$, are written one to a card, and that the cards are placed in a container. The probability experiment is then to draw a single card from the container. Thus, it is clear that we have the following probability distribution associated with the values $x_1, x_2, ...,$ and $x_k$:

$$f(x_i) = \frac{f_i}{N} \quad \text{for } i = 1, 2, ..., k, \quad \text{that is, for } x_1, x_2, ..., x_k \quad \text{(}\alpha\text{)}$$

Substitution of ($\alpha$) into Eqs. (I) and (II) above yields the following results, after taking the constant $1/N$ to the left of the $\Sigma$ sign.

**Definition 11.6:** The equation for $m'_r$, the $r$th moment about the origin for a sample of grouped data, is

$$m'_r = \frac{1}{N} \sum_{i=1}^{k} x_i^r f_i \quad \text{for } r = 0, 1, 2, 3, ... \quad \text{[Mom. about origin]} \quad \textbf{(A)}$$

Observe that the index $i$ runs from 1 to $k$, and that $N = \Sigma_{i=1}^{k} f_i$. For $r = 1$ and 2, Eq. (A) becomes

$$m'_1 = \frac{1}{N} \sum_{i=1}^{k} x_i f_i \quad \text{and} \quad m'_2 = \frac{1}{N} \sum_{i=1}^{k} x_i^2 f_i$$

Here, the term $m'_1$ is known as the mean for a sample of grouped data; it is usually symbolized by $\bar{x}$. Thus,

$$\bar{x} = \frac{1}{N} \sum_{i=1}^{k} x_i f_i \quad \textbf{(B)}$$

Substitution of $f(x_i) = f_i/N$ for $i = 1, 2, ..., k$, into Eq. (II) for $m_r$ leads to (C):

$$m_r = \frac{1}{N} \sum_{i=1}^{k} (x_i - \bar{x})^r f_i \quad \text{for } r = 0, 1, 2, 3, ... \quad \text{[Mom. about mean]}$$

$$\textbf{(C)}$$

For $r = 2$, we get the following equation for $\tilde{s}^2$, a particular type of sample variance:

$$\text{variance} = \tilde{s}^2 = m_2 = \frac{1}{N} \sum_{i=1}^{k} (x_i - \bar{x})^2 f_i \quad \textbf{(D)}$$

The term $\tilde{s} = \sqrt{\tilde{s}^2}$, with $\tilde{s}^2$ given by (D), is a type of standard deviation for a sample of grouped data.

*notes*

(i) As with a sample of ungrouped data, the mean $\bar{x}$ for a sample of grouped data [given above by (B)] is a statistic which is a measure of average or central tendency for the sample. Likewise, the standard deviation $\tilde{s}$ for a sample of grouped data [with $\tilde{s}^2$ given above by (D)] is a statistic which is a measure of spread or deviation for a sample of grouped data.

(ii) Another type of variance, $\hat{s}^2$, was introduced in Section 11-2. Here, too, we have a division by $(N - 1)$ instead of by $N$. Thus, for a sample of grouped data,

$$\hat{s}^2 = \frac{1}{(N - 1)} \sum_{i=1}^{k} (x_i - \bar{x})^2 f_i$$

(iii) Very often, we will abbreviate $\Sigma_{i=1}^{k}$ by $\Sigma$, and omit the subscript $i$ from terms inside the summation symbol.

## B. Computing moments for a sample of grouped discrete data

**EXAMPLE 11-16:** Refer to the grouped data of Examples 11-1, 11-5, 11-6, etc. There, we had a sample of 40 values for a discrete random variable, which happened to be a binomial random variable with $n = 3$ and $p = .6$. Compute **(a)** the mean $\bar{x}$, **(b)** the variance $\tilde{s}^2$, and **(c)** the standard deviation $\tilde{s}$.

**Solution:** *Preliminaries:* It is useful to set up a computational table containing columns for $x$, $f$ [the frequency], $xf$, $(x - \bar{x})$, $(x - \bar{x})^2$, and $(x - \bar{x})^2 f$.

| $x$ | $f$, freq. | $xf$ | $(x - \bar{x})$ | $(x - \bar{x})^2$ | $(x - \bar{x})^2 f$ |
|---|---|---|---|---|---|
| 0 | 3 | 0 | $-1.75$ | 3.0625 | 9.1875 |
| 1 | 11 | 11 | $-.75$ | .5625 | 6.1875 |
| 2 | 19 | 38 | .25 | .0625 | 1.1875 |
| 3 | 7 | 21 | 1.25 | 1.5625 | 10.9375 |
| $\sum$ | 40 | 70 | | | 27.5000 |

**(a)** First, from the sum in the third column,

$$\bar{x} = \frac{1}{N} \sum xf = \frac{70}{40} = 1.75$$

**(b), (c)** Once $\bar{x}$ is known, the numbers in the $(x - \bar{x})$ column, as well as the other columns, can be computed. We thus obtain the following values for $\tilde{s}^2$ and $\tilde{s}$:

$$\tilde{s}^2 = \frac{1}{N} \sum (x - \bar{x})^2 f = \frac{27.5}{40} = .6875$$

$$\tilde{s} = \sqrt{\tilde{s}^2} = \sqrt{.6875} = .8292 \approx .829$$

*note:* As in Example 11-15, we appreciate that our values for $\bar{x}$ and $\tilde{s}^2$ are supposed to serve, respectively, as estimates for the mean $\mu$ and variance $\sigma^2$ of the related random variable. For this binomial random variable with $n = 3$ and $p = .6$, we have $\mu = 1.8$ and $\sigma^2 = .72$. Thus, we see that our estimates $\bar{x} = 1.75$ and $\tilde{s}^2 = .6875$ are fairly accurate.

## C. Computing moments for a sample of grouped continuous data

Suppose we have a frequency distribution for a sample of data for a continuous random variable [as in Section 11-1C]. Recall that the data were first grouped into classes.

The equations given in Definition 11.6 for $m'_r$, $m_r$, $\bar{x}$, and $\tilde{s}^2$ will be applicable provided we make a grouping assumption.

In this grouping assumption, we assume that all the data values in a particular class are equal to the class mark of that class. [Thus, for example, we assume that all $f_1$ data values in the first class are equal to the class mark of the first class.] Our policy will be to denote the class marks for the $k$ classes as $x_1, x_2, \ldots, x_k$. Thus, the equations of Section 11-6A are directly applicable, where the $x_i$ that appears in those equations now means the class mark of the $i$th class.

Because the grouping assumption does not necessarily match with reality, our values for $\bar{x}$, $\tilde{s}^2$, and other quantities based on moments will only be approximate.

Naturally, if we have the original sample of data available, that is, the collection of values that exists before the data are grouped into classes, we can obtain accurate values for $\bar{x}$, $\tilde{s}^2$, etc., by using the equations that apply for ungrouped data. [See Definition 11.5 in Section 11-5B.]

Two points are important to remember here:

One is that it is appropriate to group sample data for a continuous random variable [i.e., continuous-type data] if the total frequency $N$ is fairly large. The definition of what constitutes "large" is vague, and really depends on the sampling situation. To be specific, if not overly precise, we shall take large to mean that $N$ is at least equal to approximately 20.

Another important point is that very often we won't have the original continuous-type data available; all we'll have is a frequency distribution table containing class marks, and associated class frequencies. In such a situation, the best we can do is use the moment equations for grouped data, and thereby obtain approximate values for $\bar{x}$, $\tilde{s}^2$, etc.

When calculating $\tilde{s}^2$, it is often useful to make use of the result of Theorem 11.1, as applied to grouped data. For convenience, we now state that result for grouped data, in Theorem 11.2.

**Theorem 11.2:** For a sample of grouped data, the result of Theorem 11.1 holds. That is,

$$m_2 = m_2' - (\bar{x})^2 \qquad \text{(a)}$$

Of course, $m_2 = \tilde{s}^2$. Here, $\bar{x} = (1/N)\Sigma xf$ and $m_2' = (1/N)\Sigma x^2 f$.

*note:* Theorem 11.2 also applies to grouped discrete data.

---

**EXAMPLE 11-17:** Refer to the frequency distribution data of Examples 11-2, 11-3, 11-7, etc. Use the equations of Definition 11.6 for grouped data to calculate (a) the mean $\bar{x}$; (b) the variance $\tilde{s}^2$, and the standard deviation $\tilde{s}$; and (c) the variance $\hat{s}^2$, and the standard deviation $\hat{s}$. In computing $\tilde{s}^2$, make use of Theorem 11.2.

*Solution:* *Preliminaries:* First, we form a table similar to that of Example 11-16. Here, however, we have an $x^2 f$ column instead of the three last columns used in Example 11-16. Also, here, we make use of the index $i$ for purposes of illustration.

| $i$ | Class mark $x_i$ | Class frequency $f_i$ | $x_i f_i$ | $x_i^2 f_i$ |
|---|---|---|---|---|
| 1 | 59 | 3 | 177 | 10,443 |
| 2 | 62 | 10 | 620 | 38,440 |
| 3 | 65 | 20 | 1,300 | 84,500 |
| 4 | 68 | 12 | 816 | 55,488 |
| 5 ($=k$) | 71 | 5 | 355 | 25,205 |
| $\sum_{i=1}^{5}$ | | 50 | 3,268 | 214,076 |

(a)
$$\bar{x} = \frac{\Sigma xf}{N} = \frac{3,268}{50} = 65.36$$

(b)
$$m_2' = \frac{\Sigma x^2 f}{N} = \frac{214,076}{50} = 4,281.52$$

From Theorem 11.2,

$$\tilde{s}^2 = m_2' - (\bar{x})^2 = 4,281.52 - (65.36)^2 = 9.5904$$

$$\tilde{s} = \sqrt{\tilde{s}^2} = 3.096837 \approx 3.10$$

(c) Finally,

$$\hat{s}^2 = \left(\frac{N}{N-1}\right)\tilde{s}^2 = \left(\frac{50}{49}\right)(9.5904) = 9.78612$$

$$\hat{s} = \sqrt{\hat{s}^2} = 3.128278 \approx 3.13$$

*notes*

(i) Recall that two statistics which measure central tendency for a sample of data are the median ($Md$) and the mean ($\bar{x}$). In Example 11-12, we found that $Md = 65.30$ for the same sample of data that was considered in Example 11-17. It is a fact that if a frequency distribution is symmetrical [with respect to a vertical axis of symmetry], then $Md = \bar{x}$. Here, we have a frequency distribution which is highly symmetrical; this is illustrated, for example, by the histogram of Figure 11-3 [see Example 11-7]. Thus, it comes as no surprise that $Md \approx \bar{x}$.

(ii) For probability distributions of random variables, it is also true that the mean is equal to the median if the distribution is symmetrical.

(iii) In Problem 11-10, we show that the exact value for $\bar{x}$ is 65.32. To do this, we use the original height data [see Example 11-2], that is, the height data that existed before grouping into classes, and the ungrouped data equation for $\bar{x}$, namely $\bar{x} = \Sigma x/N$.

Usually, the grouped data equation $\bar{x} = \Sigma xf/N$ will give a result that is fairly close to the exact value, particularly for a large value of $N$.

# SOLVED PROBLEMS

## Frequency Distributions; Histograms; The Empirical Cumulative Distribution Function

**PROBLEM 11-1** The number of children per residence for 25 residences in a middle-class community were observed, yielding the following data:

$$1\ 0\ 3\ 4\ 1 \quad 2\ 1\ 0\ 4\ 1 \quad 2\ 3\ 1\ 2\ 1 \quad 0\ 3\ 1\ 2\ 0 \quad 3\ 1\ 2\ 0\ 1$$

(a) Group the sample data into a frequency distribution. [Here, the related random variable $X$ takes on the values 0, 1, 2, 3, 4 for this sample.] (b) Draw the histogram for the frequency distribution. (c) Determine the cumulative relative frequency in terms of $x$. (d) Draw the graph of the empirical cumulative distribution function.

*Solution*

(a) The frequency distribution table follows:

| Number of children; $x$ | Tally | Frequency, $f$ |
|:---:|:---:|:---:|
| 0 | ︲︲︲︲︲ | 5 |
| 1 | ︲︲︲︲︲ ︲︲︲︲ | 9 |
| 2 | ︲︲︲︲︲ | 5 |
| 3 | ︲︲︲︲ | 4 |
| 4 | ︲︲ | 2 |
| | | Sum = 25 |

(b) We plot the number of children [$x$] on the horizontal axis, and the frequency

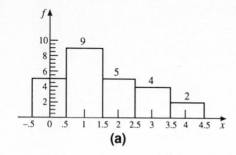

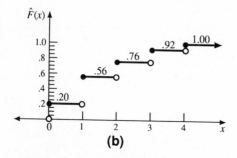

**Figure 11-8.** Histogram and graph of empirical cdf for Problem 11-1. (a) Histogram. (b) Graph of empirical cdf.

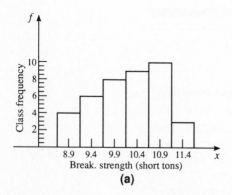

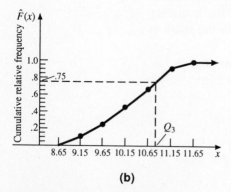

**Figure 11-9.** Histogram and cumulative relative frequency polygon for sample of continuous data. (a) Histogram (Problem 11-2). (b) Cumulative relative frequency polygon. [Dashed lines pertain to part (d) of Problem 11-3.]

[$f$] on the vertical axis. For each $x$ value, we plot a rectangle with base width equal to one, and height equal to the frequency. See Figure 11-8a.

(c) The table below has the same $x$ and $f$ columns that appear above. In addition, the third column has cumulative frequencies at $x = 0, 1, \ldots, 4$, and the fourth column has cumulative relative frequencies at $x = 0, 1, \ldots, 4$.

| $x$ | $f$ | Cumulative frequency | Cumulative relative frequency, $\hat{F}(x)$ |
|---|---|---|---|
| 0 | 5 | 5 | .20 |
| 1 | 9 | 14 | .56 |
| 2 | 5 | 19 | .76 |
| 3 | 4 | 23 | .92 |
| 4 | 2 | 25 | 1.00 |

There follows a tabulation of $\hat{F}(x)$ in terms of $x$ for all $x$: $\hat{F}(x) = 0$ for $x < 0$; $\hat{F}(x) = .20$ for $0 \le x < 1$; $\hat{F}(x) = .56$ for $1 \le x < 2$; $\hat{F}(x) = .76$ for $2 \le x < 3$; $\hat{F}(x) = .92$ for $3 \le x < 4$; $\hat{F}(x) = 1.000$ for $4 \le x$.

(d) See Figure 11-8b for the graph of $\hat{F}(x)$ in terms of $x$.

**PROBLEM 11-2**  The breaking strengths in short tons [one short ton is 2,000 pounds] for a sample of 40 cables are listed below. [Note that the unit of measurement is .1 ton; each recorded score in the list below is a multiple of .1.]

*The list:*  10.0  10.5  10.7  9.3  10.9  10.1  9.7  8.8  9.5  10.2

11.0  10.1  8.7  10.9  11.2  9.7  10.8  9.4  10.8  11.3

9.8  10.6  10.2  10.3  9.6  11.1  9.0  9.9  11.4  10.8

10.5  10.2  9.4  11.0  9.1  10.0  10.4  10.7  10.6  9.5

(a) Group these data into a frequency distribution of 6 classes, each having a class width of .5 [which equals 5 units]. (b) Draw a histogram for this frequency distribution. Label the class marks on the horizontal axis.

*Solution*

(a) Range $= 11.4 - 8.7 = 2.7$ tons or 27 units since $u = .1$. Now, $27/6 = 4.5$. Since it is desirable to have an odd number of recorded scores per class, we choose the class width $w$ to equal 5 units, or, equivalently, .5 ton. We choose the lower and upper class limits for the first class to be 8.7 and 9.1, respectively. [Other choices are possible; for example, the choice of 8.6 and 9.0 for lower and upper class limits is also acceptable.] Now, for the first class, $LCB_1 = LCL_1 - u/2 = 8.7 - .05 = 8.65$, and $UCB_1 = UCL_1 + u/2 = 9.1 + .05 = 9.15$.

The table for the frequency distribution is as follows:

| Class label, $i$ | Class limits | Class boundaries | Class mark | Tally | Class frequency |
|---|---|---|---|---|---|
| 1 | 8.7– 9.1 | 8.65– 9.15 | 8.9 | IIII | 4 |
| 2 | 9.2– 9.6 | 9.15– 9.65 | 9.4 | ЖHI I | 6 |
| 3 | 9.7–10.1 | 9.65–10.15 | 9.9 | ЖHI III | 8 |
| 4 | 10.2–10.6 | 10.15–10.65 | 10.4 | ЖHI IIII | 9 |
| 5 | 10.7–11.1 | 10.65–11.15 | 10.9 | ЖHI ЖHI | 10 |
| 6 (= $k$) | 11.2–11.6 | 11.15–11.65 | 11.4 | III | 3 |

Sum = 40

(b) The histogram, consisting of six rectangles, one for each class, is shown in Figure 11-9a. For each rectangle, the height is the class frequency, and the base extends between the class boundaries.

**PROBLEM 11-3**  From the tabulated data in the solution of Problem 11-2, (**a**) develop a table that lists values of the cumulative relative frequency at the class boundaries. (**b**) Construct the cumulative relative frequency polygon.

*Solution*

(**a**) In Table 11-4 below, observe how the class boundaries for a class are positioned; one class boundary is placed physically above, and the other is placed physically below the class mark for the class.

   The cumulative frequency equals zero at the lowest class boundary. At any other class boundary, the cumulative frequency is equal to the cumulative frequency at the preceding class boundary, plus the class frequency for the class enclosed by these class boundaries.

   Each cumulative relative frequency is equal to the corresponding cumulative frequency divided by the total frequency, $N$. Here, $N = 40$.

**TABLE 11-4: Generating Cumulative Relative Frequencies at Class Boundaries of a Frequency Distribution [Problem 11-3]**

| Class mark | Class frequency | Class boundary | Cumulative frequency | Cumulative relative freq. |
|---|---|---|---|---|
| | | 8.65 | 0 | 0 |
| 8.9 | 4 | | | |
| | | 9.15 | $0 + 4 = 4$ | $4/40 = .10$ |
| 9.4 | 6 | | | |
| | | 9.65 | $4 + 6 = 10$ | $10/40 = .25$ |
| 9.9 | 8 | | | |
| | | 10.15 | 18 | .45 |
| 10.4 | 9 | | | |
| | | 10.65 | 27 | .675 |
| 10.9 | 10 | | | |
| | | 11.15 | 37 | .925 |
| 11.4 | 3 | | | |
| | | 11.65 | 40 | 1.000 |

(**b**) Refer to Figure 11-9b. First, we plot the points $(x, \hat{F}(x))$, where $x$ is a class boundary, and $\hat{F}(x)$ is the cumulative relative frequency. The coordinates for these seven points are obtained from the third and fifth columns of Table 11-4. Next, starting with the point $(8.65, 0)$, and going to the right, we draw a line segment from each point to the point on the right whose class boundary is greater by one class width. Finally, we draw unlimited horizontal line segments, one extending to the left from $(8.65, 0)$, and the other to the right from $(11.65, 1.00)$.

## Percentiles for a Frequency Distribution of Continuous Data

**PROBLEM 11-4**  Refer to Table 11-4 of Problem 11-3. Calculate (**a**) the 40th percentile, (**b**) the first quartile, (**c**) the median, and (**d**) the third quartile. Indicate how to estimate the third quartile by using the cumulative relative frequency polygon of Figure 11-9b.

*Solution*

(**a**) For $p_{40}$, the corresponding cumulative relative frequency is given by $\hat{F}(x) = 40/100 = .40$. Doing a linear interpolation with respect to the nearest $\hat{F}(x)$ values above and below .40 in Table 11-4, we find

$$p_{40} = 9.65 + \frac{(.40 - .25)}{(.45 - .25)}(10.15 - 9.65) = 9.65 + .375 \approx 10.03 \qquad (1)$$

(**b**) For $Q_1 = p_{25}$, we have $\hat{F}(x) = 25/100 = .25$. Thus, reading directly from Table 11-4 [no interpolation is needed], we have $Q_1 = 9.65$.

(c) For $Md = p_{50}$, we have $\hat{F}(x) = 50/100 = .50$. Doing a linear interpolation relative to Table 11-4, we find

$$Md = 10.15 + \frac{(.50 - .45)}{(.675 - .45)}(.50) = 10.15 + .11\overline{1} \approx 10.26 \qquad (2)$$

(d) For $Q_3 = p_{75}$, we have $\hat{F}(x) = 75/100 = .75$. Doing a linear interpolation relative to Table 11-4, we find

$$Q_3 = 10.65 + \frac{(.75 - .675)}{(.925 - .675)}(.50) = 10.65 + .15 = 10.80 \qquad (3)$$

The graphical determination of $Q_3$ from a cumulative relative frequency polygon is shown in Figure 11-9b. The term $Q_3$ is the horizontal coordinate of the point on the polygon whose vertical coordinate is .75. We see that $Q_3 \approx 10.8$.

**PROBLEM 11-5**   In studies of data, a statistic commonly used to measure deviation [or spread or variation] in a sample of data is the *interquartile range*:

$$\text{Interquartile Range} = \text{I. R.} = (Q_3 - Q_1)$$

A statistic commonly used to measure skewness [which is equivalent to antisymmetry] in a sample of data is the following:

$$\text{Quartile Coefficient of Skewness} = \text{Q. C. S.} = \frac{(Q_3 - Q_2) - (Q_2 - Q_1)}{(Q_3 - Q_1)} = \frac{(Q_3 - 2Q_2 + Q_1)}{(Q_3 - Q_1)}$$

The Q. C. S. is dimensionless.

For the sample of continuous data pertaining to Problems 11-3 and 11-4, determine (a) the interquartile range and (b) the quartile coefficient of skewness.

*Solution* (a) and (b): Using the results of Problem 11-4, we have

$$\text{I. R.} = (Q_3 - Q_1) = 10.80 - 9.65 = 1.15$$

$$\text{Q. C. S.} = \frac{(Q_3 - 2Q_2 + Q_1)}{(Q_3 - Q_1)} = \frac{10.80 - 2(10.26) + 9.65}{1.15} = \frac{-.07}{1.15} \approx -.0609$$

This last calculation confirms that the sample data are "skewed to the left." See Figure 11-9a for a pictorial demonstration of this. Refer to Section 10-3, where a general discussion of skewness is given.

**PROBLEM 11-6**   Refer to Table 11-4 of Problem 11-3. What percentile is the score $x = 9.75$?

*Solution*   We see that $x = 9.75$ is between the class boundaries 9.65 and 10.15 in Table 11-4. We use linear interpolation relative to 9.65 and 10.15 to find the value of the cumulative relative frequency $\hat{F}(x)$ that corresponds to $x = 9.75$. We obtain

$$\hat{F}(x) = .25 + \frac{(9.75 - 9.65)}{(10.15 - 9.65)}(.45 - .25) = .25 + .04 = .29 = \frac{29}{100}$$

This means that the score 9.75 is $p_{29}$, the 29th percentile.

## Moments for a Sample of Data

**PROBLEM 11-7**   Refer to Example 11-13. For another sample of five boxes of Krunchee cereal, labeled 32 ounces, the weights are as follows: 31.16, 32.22, 32.42, 32.30, 31.80. Find (a) the sample mean, (b) the sample variance $\tilde{s}^2$, and (c) the sample standard deviation $\tilde{s}$. (d) By using the statistics $\bar{x}$ and $\tilde{s}$, make a simplified comparison of the sample of data of Example 11-13 and the current sample.

*Solution* *Preliminaries:* First, we set up a table with columns for $x$, $(x - \bar{x})$, and $(x - \bar{x})^2$. We fill in the values in the second and third columns after calculating $\bar{x}$.

| $x$ | $(x - \bar{x})$ | $(x - \bar{x})^2$ |
|---|---|---|
| 31.16 | $-.82$ | .6724 |
| 32.22 | .24 | .0576 |
| 32.42 | .44 | .1936 |
| 32.30 | .32 | .1024 |
| 31.80 | $-.18$ | .0324 |
| Sum = 159.90 | 0 | 1.0584 |

**(a)** From the first column, $\bar{x} = \Sigma x/5 = 159.90/5 = 31.98$ oz.
**(b)** From the third column, $\tilde{s}^2 = \Sigma(x - \bar{x})^2/5 = 1.0584/5 = .21168$.
**(c)** Using the result from part (b), $\tilde{s} = \sqrt{.21168} = .460087 \approx .460$ oz.

> ***note:*** Observe that the sum in the second column equals 0. This will always be the case for the reason that the moment $m_1 = 0$ for any sample of data [ungrouped or grouped]. That result follows, for example, from the more general result that $\mu_1 = 0$ for the case of a random variable.
> Next, the result $\Sigma(x - \bar{x}) = 0$ follows from the fact that $\Sigma(x - \bar{x}) = Nm_1$.

**(d)** Labeling the sample from Example 11-13 with the letter $a$, and the current sample with $b$, we have the following statistics:

$$\bar{x}_a = 31.95 \text{ oz} \qquad \tilde{s}_a = .877 \text{ oz} \qquad \bar{x}_b = 31.98 \text{ oz} \qquad \tilde{s}_b = .460 \text{ oz}$$

Thus, from the respective values for $\bar{x}$, we see that the average values are approximately equal [i.e., roughly the same central tendency]. From the respective values for $\tilde{s}$, we see that the data for the sample of Example 11-13 are much more spread out than the current data.

**PROBLEM 11-8** The data for the number of automobiles per residence for 8 homes in a middle-class community are as follows: 1, 2, 0, 1, 3, 0, 2, 1.
Compute $\bar{x}$, $\tilde{s}$, and $\hat{s}$ by using equations appropriate for ungrouped data.

*Solution* We tabulate values for $x$ and $x^2$ in two rows, and then find the sum for each row.

| | | | | | | | | | $\Sigma$ | $\frac{1}{8}\Sigma$ |
|---|---|---|---|---|---|---|---|---|---|---|
| $x$ | 1 | 2 | 0 | 1 | 3 | 0 | 2 | 1 | 10 | 1.25 |
| $x^2$ | 1 | 4 | 0 | 1 | 9 | 0 | 4 | 1 | 20 | 2.50 |

$$\bar{x} = \sum x/N = 10/8 = 1.25 \tag{1}$$

$$m_2' = \sum x^2/N = 20/8 = 2.50 \tag{2}$$

$$\tilde{s}^2 = m_2' - (\bar{x})^2 = 2.50 - (1.25)^2 = .9375 \tag{3}$$

$$\tilde{s} = \sqrt{\tilde{s}^2} = \sqrt{.9375} = .968246 \approx .968 \tag{4}$$

$$\hat{s}^2 = \left(\frac{N}{N-1}\right)\tilde{s}^2 = \left(\frac{8}{7}\right)(.9375) = 1.07143 \tag{5}$$

$$\hat{s} = \sqrt{\hat{s}^2} = \sqrt{1.07143} = 1.035098 \approx 1.035 \tag{6}$$

## Moments for a Sample of Grouped Data

**PROBLEM 11-9** Refer to the grouped discrete data of Problem 11-1. Compute **(a)** the mean, $\bar{x}$, **(b)** the variance, $\tilde{s}^2$, and **(c)** the standard deviation, $\tilde{s}$.

*Solution* We shall use the following table:

| x | f | xf | $x^2f$ |
|---|---|----|--------|
| 0 | 5 | 0 | 0 |
| 1 | 9 | 9 | 9 |
| 2 | 5 | 10 | 20 |
| 3 | 4 | 12 | 36 |
| 4 | 2 | 8 | 32 |
| $\Sigma =$ | 25 | 39 | 97 |

(a)
$$\bar{x} = \frac{\sum xf}{N} = \frac{39}{25} = 1.56 \tag{1}$$

(b)
$$\tilde{s}^2 = \frac{\sum x^2f}{N} - \bar{x}^2 = \frac{97}{25} - (1.56)^2 = 1.4464 \tag{2}$$

(c)
$$\tilde{s} = \sqrt{1.4464} \approx 1.203 \tag{3}$$

**PROBLEM 11-10**   Refer to the frequency distribution data of Problem 11-2 [continuous-type data]. Use the equations of Definition 11.6 for grouped data to calculate (a) the mean, $\bar{x}$, (b) the variance, $\tilde{s}^2$, and (c) the standard deviation, $\tilde{s}$.

*Solution* We shall make use of the following table.

| Class mark, x | Class frequency, f | xf | $x^2f$ |
|---------------|--------------------|----|--------|
| 8.9 | 4 | 35.6 | 316.84 |
| 9.4 | 6 | 56.4 | 530.16 |
| 9.9 | 8 | 79.2 | 784.08 |
| 10.4 | 9 | 93.6 | 973.44 |
| 10.9 | 10 | 109.0 | 1,188.10 |
| 11.4 | 3 | 34.2 | 389.88 |
| $\Sigma$ | 40 | 408.0 | 4,182.50 |

(a)
$$\bar{x} = \frac{\sum xf}{N} = \frac{408.0}{40} = 10.2$$

(b)
$$\tilde{s}^2 = \frac{\sum x^2f}{N} - \bar{x}^2 = \frac{4,182.50}{40} - (10.2)^2 = .5225$$

(c)
$$\tilde{s} = \sqrt{.5225} = .722842 \approx .723$$

**PROBLEM 11-11**   Refer to Examples 11-2 and 11-17. Make use of the equation that applies to ungrouped data to compute the exact value of $\bar{x}$.

*Solution* We refer back to the original height data for 50 women at Gates College, as presented in Example 11-2. The appropriate equation for $\bar{x}$ is $\bar{x} = \sum_{i=1}^{50} x_i/50$ since we have a sample of ungrouped data. Summing in groups of 10, we get

$$50\bar{x} = \sum_{i=1}^{10} x_i + \sum_{i=11}^{20} x_i + \ldots + \sum_{i=41}^{50} x_i \tag{1}$$

Thus, for the current data, we have

$$50\bar{x} = 653 + 639 + 652 + 666 + 656 = 3,266 \tag{2}$$

which leads to

$$\bar{x} = 3,266/50 = 65.32 \qquad \text{[exact value]} \tag{3}$$

This value is very close to the approximate value $\bar{x} = 65.36$ that was calculated in Example 11-17. Recall that the latter value is based on the "grouping assumption" that all data values in a class of a frequency distribution are equal to the class mark of the class.

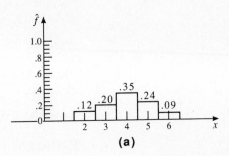

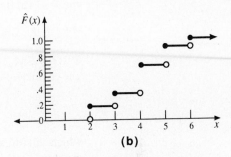

(a)

(b)

## Supplementary Problems

**PROBLEM 11-12** Equal numbers of balls in a container are marked with the digits 1, 2, and 3, respectively. The experiment of drawing two balls with replacement is repeated 100 times. Let random variable $X$ stand for the sum of the numbers obtained on a draw of two balls. A tabulation of frequencies for different values of $x$ follows:

| Sum, $x$ | 2 | 3 | 4 | 5 | 6 |
|---|---|---|---|---|---|
| Frequency, $f$ | 12 | 20 | 35 | 24 | 9 |

$N = \sum f = 100$

**Figure 11-10.** Histogram and graph of empirical cumulative distribution function for Problem 11-12. (**a**) Histogram, where relative frequency is ordinate. (**b**) Graph of empirical cdf.

(**a**) Construct a histogram with relative frequency plotted on the vertical axis. (**b**) Determine the probability function for $X$ for the related theoretical random variable. (**c**) Determine the cumulative relative frequency $\hat{F}(x)$ in terms of $x$. (**d**) Sketch the graph of the empirical cumulative distribution function.

*Answer* (**a**) See Figure 11-10a      (**b**) A tabulation for the probability function follows:

| Sum, $x$ | 2 | 3 | 4 | 5 | 6 |
|---|---|---|---|---|---|
| Probability function, $f(x)$ | 1/9 | 2/9 | 3/9 | 2/9 | 1/9 |

(**c**) $\hat{F}(x) = 0$ for $x < 2$; $\hat{F}(x) = .12$ for $2 \le x < 3$; $\hat{F}(x) = .32$ for $3 \le x < 4$; $\hat{F}(x) = .67$ for $4 \le x < 5$;
$\hat{F}(x) = .91$ for $5 \le x < 6$; $\hat{F}(x) = 1.00$ for $6 \le x$      (**d**) See Figure 11-10b

**PROBLEM 11-13** A frequency distribution for the weights, in lb, of 80 male college seniors, is given as follows:

| Class mark, lb | 140 | 155 | 170 | 185 | 200 | 215 |
|---|---|---|---|---|---|---|
| Class frequency, $f$ | 17 | 27 | 18 | 12 | 4 | 2 |

$N = \sum f = 80$

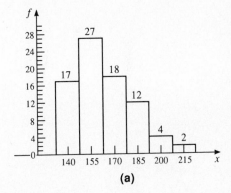

(a)

(**a**) Construct a histogram for these data. (**b**) Develop a table listing values of the cumulative relative frequency $\hat{F}(x)$ at the class boundaries. (**c**) Sketch the cumulative relative frequency polygon. (**d**) Determine the first and third quartiles.

*Answer* (**a**) See Figure 11-11a      (**b**) The table is as follows:

| $x$, class boundary | 132.5 | 147.5 | 162.5 | 177.5 | 192.5 | 207.5 | 222.5 |
|---|---|---|---|---|---|---|---|
| $\hat{F}(x)$, cum. rel. freq. | .0000 | .2125 | .550 | .775 | .925 | .975 | 1.000 |

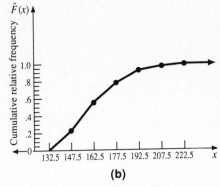

(b)

**Figure 11-11.** Histogram and cumulative relative frequency polygon for Problem 11-13. (**a**) Histogram. (**b**) Cumulative relative frequency polygon.

(**c**) See Figure 11-11b      (**d**) $Q_1 \approx 149.2$, $Q_3 \approx 175.8$

**PROBLEM 11-14** For 100-watt light bulbs made by the Flashfirm Company, the average lifetime listed on the package is 750 hours. For a sample of five of these light bulbs, the actual lifetimes attained, in hours, are as follows: 725, 742, 755, 760, 753. For this sample, determine (a) the mean, $\bar{x}$, (b) the standard deviation, $\tilde{s}$, and (c) the standard deviation, $\hat{s}$.

*Answer* (a) $\bar{x} = 747.0$ hr    (b) $\tilde{s} = 12.474 \approx 12.5$ hr    (c) $\hat{s} = 13.946 \approx 13.9$ hr

**PROBLEM 11-15**  For the data of Problem 11-12, determine (a) the mean and (b) the standard deviation, $\tilde{s}$.

*Answer* (a) $\bar{x} = 3.98$    (b) $\tilde{s} = 1.13119 \approx 1.131$

**PROBLEM 11-16**  For the grouped continuous-type data of Problem 11-13, determine (a) the mean, $\bar{x}$, (b) the standard deviation, $\tilde{s}$, and (c) the standard deviation, $\hat{s}$.

*Answer* (a) $\bar{x} = 163.4375 \approx 163.44$    (b) $\tilde{s} = 18.80149 \approx 18.801$    (c) $\hat{s} = 18.92012 \approx 18.920$

**PROBLEM 11-17**  The *moment coefficient of skewness* for a sample of data, $a_3$, is analogous to the coefficient of skewness for a random variable as defined in Section 10-3. In fact, we have

$$a_3 = \frac{m_3}{\tilde{s}^3}$$

which is similar to what appears in Definition 10.3. Here, the term $m_3$ can be expressed as $m_3 = m'_3 - 3m'_2\bar{x} + 2\bar{x}^3$, and this equation is comparable to that which appears in Theorem 10.6. Use both of these equations to compute $a_3$ for the discrete sample data that occurs in Problems 11-1 and 11-9.

*Answer* Before rounding off, we have $\tilde{s} = 1.20266$, $m'_3 = 11.4$, and $m_3 = .834432$. Thus, we calculate $a_3 = .47969179 \approx .4797$.

# 12 MULTIVARIATE PROBABILITY DISTRIBUTIONS

## THIS CHAPTER IS ABOUT

☑ **Introduction to Multivariate Probability Distributions**
☑ **The Multivariate Cumulative Distribution Function**
☑ **Independent Random Variables and Conditional Probability Distributions**
☑ **Extensions to the *n* Random Variables Case**
☑ **The Multinomial and Multivariate Hypergeometric Distributions**

## 12-1. Introduction to Multivariate Probability Distributions

Previously, we considered the case of a sample space for a probability experiment, with a random variable defined with respect to the sample space. In Chapters 5, 6, and 7, we focused on the case of a single discrete random variable, and in Chapter 8, we focused on the case of a single continuous random variable.

The related concepts of expected value, moments, and moment generating functions for both discrete and continuous random variables were discussed in Chapter 9. Always in these discussions, however, the main focus was on a single random variable with respect to some underlying sample space. [The sample space in question was somewhat obscured in the continuous random variable case].

In this chapter, we will be concerned with two or more random variables with respect to a given sample space, and the related probability distribution.

### A. Discrete random variables case: an introduction

First, consider the case of two discrete random variables defined with respect to the same sample space.

**Definition 12.1:** Given two discrete random variables $X$ and $Y$ defined with respect to a sample space $S$. The probability $P(X = x$ and $Y = y)$, which is sometimes written as $P(X = x, Y = y)$, is the joint probability function [or multivariate probability function or joint probability distribution] of $X$ and $Y$, and it is denoted by $f(x, y)$.

*notes*

**(a)** Some authors use $f_{X,Y}(x, y)$ instead of $f(x, y)$, and this is really more precise since this notation stresses that we have a function pertaining to the two random variables $X$ and $Y$. See, for example, Larsen and Marx (1986).

**(b)** Later, we'll often label our random variables $X_1$ and $X_2$. The corresponding joint probability function in this case is $f(x_1, x_2)$.

**EXAMPLE 12-1:** Given a deck with five cards, three of which are spades, namely the ace, two, and three, and two hearts, namely the ace and two. [We shall abbreviate these as $A_S, 2_S, 3_S, A_H, 2_H$.] The experiment is to draw two cards without replacement. Let random variable $X$, the "first card" random variable, be defined as follows: $X = 0$ if first card drawn is heart, and $X = 1$ if first card drawn is spade. The random variable $Y$, the "second card" random variable, is defined as follows: $Y = 0$ if second card drawn is heart, and $Y = 1$ if second card drawn is spade. Determine the joint probability function $f(x, y)$ for all relevant values of $X$ and $Y$, and display it in a table.

*Solution:* Here, $f(x, y) = P(X = x, Y = y)$. Since the possible values for $X$ are 0 and 1, and likewise for $Y$, it follows that the possible values for $(x, y)$ are given by the following set: $\{(0, 0), (0, 1), (1, 0), (1, 1)\}$. This set is said to be the range space for the random variables $X$ and $Y$. [Some authors describe this set as the range space for the two-dimensional random variable $(X, Y)$.] Observe that the range spaces for $X$ and $Y$, individually, are given by $R_X = R_Y = \{0, 1\}$.

As an example of a calculation of an $f(x, y)$, we have

$$f(0, 1) = P(X = 0, Y = 1) = P(H_1 \text{ and } S_2) \tag{1}$$

Here, $H_1$ means heart on the first card, and $S_2$ means spade on the second card. Clearly,

$$f(0, 1) = P(H_1)P(S_2 \mid H_1) = (\tfrac{2}{5})(\tfrac{3}{4}) = \tfrac{3}{10} \tag{2}$$

Similarly,

$$f(0, 0) = P(H_1)P(H_2 \mid H_1) = (\tfrac{2}{5})(\tfrac{1}{4}) = \tfrac{1}{10} \tag{3}$$

Then, $f(1, 0)$ and $f(1, 1)$ are computed in similar fashion, and we can fill in the four values for $f(x, y)$ in the body of Table 12-1.

**TABLE 12-1: Table for $f(x, y)$ for Example 12-1**

| $\downarrow y \quad \overset{\rightarrow}{x}$ | 0 | 1 | $f_Y(y)$ |
|---|---|---|---|
| 0 | 1/10 | 3/10 | 2/5 |
| 1 | 3/10 | 3/10 | 3/5 |
| $f_X(x)$ | 2/5 | 3/5 | 1 |

For now, we will ignore the symbols and numbers in the right margin [$f_Y(y)$, 2/5, etc.] and bottom margin [$f_X(x)$, 2/5, etc.]. Notice that the sum of all four $f(x, y)$ values equals 1.

**notes**

(a) We will continue to abbreviate probability function by pf.

(b) A joint pf of two random variables is also called a bivariate pf. Some authors refer to the probability function of two or more discrete random variables as being a probability density function. Likewise, they refer to the pf of a single random variable as being a probability density function.

(c) Suppose that $R_X$ and $R_Y$ are the range spaces for $X$ and $Y$. Then, the range space associated with the pair of random variables $X$ and $Y$ is taken by some to be the Cartesian product set $R_X \times R_Y$. This set is the set of pairs $(x, y)$ such that $x$ is in $R_X$ and $y$ is in $R_Y$. However, it may turn out that $f(x, y) = 0$ for some of the $(x, y)$ pairs in $R_X \times R_Y$. Thus, some practitioners consider the range space for $X$ and $Y$ to be the set that remains when such $(x, y)$ pairs are deleted from $R_X \times R_Y$.

Some practitioners consider the domain of the function $f(x, y)$ to be identical to the range space of $X$ and $Y$.

**Theorem 12.1:** Suppose that $f(x, y)$ is a joint [multivariate] probability function of discrete random variables $X$ and $Y$. Then, $f(x, y)$ satisfies the following properties:

(a) $0 \le f(x, y) \le 1$ for each possible pair $(x, y)$.
(b) $\Sigma_x \Sigma_y f(x, y) = 1$.

Here, the double summation may be regarded as extending over all possible pairs $(x, y)$ in the domain of $f(x, y)$. Also, we may replace $\Sigma_x \Sigma_y f(x, y)$ by $\Sigma_y \Sigma_x f(x, y)$.

## B. Marginal probability functions

Suppose we have two discrete random variables $X$ and $Y$ defined on a sample space $S$ for some probability experiment as, for example, the experiment of Example 12-1. In addition to the $f(x, y)$ entries in the body of the table [four such entries in Table 12-1], we can compute "marginal" totals; that is, sums for the columns and rows of $f(x, y)$ entries of the table [two such columns and two such rows for Table 12-1].

The summation numbers in the bottom row and right column margins are values for the probability functions of $X$ and $Y$, respectively. For example, with reference to Table 12-1 of Example 12-1, we have $P(X = 0) = P(X = 0$ and $Y = 0$, or $X = 0$ and $Y = 1) = P(X = 0$ and $Y = 0) + P(X = 0$ and $Y = 1) = f(0,0) + f(0,1) = \frac{1}{10} + \frac{3}{10} = \frac{2}{5}$. We denote this term, which represents the probability of heart on the first draw, as $f_X(0)$ or $f_1(0)$. It appears as the item in the bottom row margin of the table for which $X = 0$. Also, $P(Y = 1) = P(X = 0$ and $Y = 1$, or $X = 1$ and $Y = 1) = f(0,1) + f(1,1) = \frac{3}{10} + \frac{3}{10} = \frac{3}{5}$. We denote this term, which represents the probability of spade on the second draw, as $f_Y(1)$ or $f_2(1)$. It appears as the item in the right column margin of the table for which $Y = 1$.

**EXAMPLE 12.2:** For the $f(x, y)$ data of Table 12-1 of Example 12-1, determine the marginal probabilities $f_X(1)$ and $f_Y(0)$.

*Solution:* Summing in the $x = 1$ column of Table 12-1, we have

$$f_X(1) = f(1,0) + f(1,1) = \tfrac{3}{10} + \tfrac{3}{10} = \tfrac{3}{5} \tag{1}$$

That is, the probability of spade on the first card is 3/5. This number appears as the item in the bottom row margin of the table for which $X = 1$.

Summing in the $Y = 0$ row, we have

$$f_Y(0) = f(0,0) + f(1,0) = \tfrac{1}{10} + \tfrac{3}{10} = \tfrac{2}{5} \tag{2}$$

That is, the probability of heart on the second card is 2/5. This number appears as the item in the right column margin of the table for which $Y = 0$.

**Definition 12.2:** Given two discrete random variables $X$ and $Y$ defined on the same sample space, for which $f(x, y)$ is the related joint pf. Then, the marginal probability function of $X$, denoted by $f_X(x)$ or $f_1(x)$, is given by

$$f_X(x) = \sum_y f(x, y) \tag{a}$$

In (a), $x$ is constant and the sum is over all possible values of $y$. The marginal probability function of $Y$, denoted by $f_Y(y)$ or $f_2(y)$, is given by

$$f_Y(y) = \sum_x f(x, y) \qquad \textbf{(b)}$$

In (b), $y$ is constant and the sum is over all possible values of $x$.

#### notes

(i) Some authors consider Definition 12.2 to be a theorem. For example, $f_X(x_0)$ can properly be considered as $P(X = x_0)$ for $x_0$ in the range space $R_X$. It can be proven [see page 127 of Larsen and Marx (1986)] that $f_X(x_0) = \sum_y f(x_0, y)$ by using Axiom 1.3, which states that the probability of a union of mutually exclusive events is equal to the sum of the probabilities of those mutually exclusive events.

(ii) It should be clear that $\sum_x f_X(x) = 1$ and $\sum_y f_Y(y) = 1$. To prove the former, we substitute (a) of Definition 12.2 and obtain

$$\sum_x f_X(x) = \sum_x \left( \sum_y f(x, y) \right)$$

But, from part (b) of Theorem 12.1, we see that the latter sum is equal to 1.

---

**EXAMPLE 12-3:** A box contains two red, three green, and five blue chips. Two chips are selected from the box. Let $X_1$ and $X_2$ denote the number of red and green chips obtained. **(a)** Find the probabilities associated with all possible pairs of values $(x_1, x_2)$. **(b)** Determine the marginal probability functions associated with $X_1$ and $X_2$.

#### Solution

**(a)** It is reasonable to assume that the chips are drawn without replacement. The total number of equally likely ways of selecting two chips when drawing from ten chips is $\binom{10}{2} = 45$. Now, consider $f(1, 1)$, which is the probability of one red and one green chip.

$$f(1, 1) = P(X_1 = 1 \text{ and } X_2 = 1) = \frac{\binom{2}{1}\binom{3}{1}}{45} = \frac{6}{45} \qquad \textbf{(1)}$$

For $f(0, 1)$, we have

$$f(0, 1) = P(X_1 = 0 \text{ and } X_2 = 1) = P(0 \text{ red and } 1 \text{ green})$$

$$= P(1 \text{ green and } 1 \text{ blue}) = \frac{\binom{3}{1}\binom{5}{1}}{45} = \frac{15}{45} \qquad \textbf{(2)}$$

For other examples,

$$f(1, 2) = P(1 \text{ red and } 2 \text{ green}) = 0 \qquad \textbf{(3)}$$

$$f(0, 0) = P(2 \text{ blue}) = \frac{\binom{5}{2}}{45} = \frac{10}{45} \qquad \textbf{(4)}$$

Continuing in this way, we obtain the values shown in the body of Table 12-2.

**TABLE 12-2: Table for pf $f(x_1, x_2)$ and Associated Marginal pf's**

| $\downarrow x_2 \quad \overrightarrow{x_1}$ | 0 | 1 | 2 | $f_2(x_2)$ |
|---|---|---|---|---|
| 0 | 10/45 | 10/45 | 1/45 | 21/45 |
| 1 | 15/45 | 6/45 | 0 | 21/45 |
| 2 | 3/45 | 0 | 0 | 3/45 |
| $f_1(x_1)$ | 28/45 | 16/45 | 1/45 | 1 |

**(b)** If we sum probabilities within the $x_1 = 0$, 1, and 2 columns, we obtain the marginal probability function $f_1(x_1)$. For example,

$$f_1(1) = f(1,0) + f(1,1) + f(1,2) = (10 + 6 + 0)/45 = 16/45$$

Of course, $f_1(1)$ is the hypergeometric probability of getting exactly one red chip, and this can also be calculated from $f_1(1) = \binom{2}{1}\binom{8}{1}\Big/45$. Recall that there are eight non-red chips.

If we sum probabilities within the $x_2 = 0$, 1, and 2 rows, we obtain the marginal probability function $f_2(x_2)$. For example,

$$f_2(0) = f(0,0) + f(1,0) + f(2,0) = (10 + 10 + 1)/45 = 21/45$$

Of course, $f_2(0)$ is the hypergeometric probability of getting exactly zero green chips, and this can also be calculated from $f_2(0) = \binom{3}{0}\binom{7}{2}\Big/45 = 21/45$ since there are seven non-green chips.

---

*notes*

**(i)** The range spaces for $X_1$ and $X_2$ are given by $R_{X_1} = R_{X_2} = \{0, 1, 2\}$. Thus, we may take as the range space for the pair $X_1$ and $X_2$, the Cartesian product set $R_{X_1} \times R_{X_2}$. The latter set consists of the nine points $(x_1, x_2)$ such that $x_1 = 0, 1, 2$, and $x_2 = 0, 1, 2$. But, we can delete the points $(1, 2)$, $(2, 1)$, and $(2, 2)$ since $f(x_1, x_2) = 0$ for each of these points. Thus, we can just as well take the range space for the pair $X_1$ and $X_2$, and, also, the domain of $f(x_1, x_2)$, to consist of the six remaining points. The latter six points can be described symbolically by writing $x_1 = 0, 1, 2$, and $0 \leq x_2 \leq 2 - x_1$, where $x_2$ is an integer.

**(ii)** The pf of $x_1$ and $x_2$ in Example 12-3 is called a multivariate [or joint] hypergeometric pf of $x_1$ and $x_2$. In fact, we see that we can write

$$f(x_1, x_2) = \frac{\binom{2}{x_1}\binom{3}{x_2}\binom{5}{2 - x_1 - x_2}}{\binom{10}{2}} \qquad \begin{array}{l} \text{for } x_1 = 0, 1, 2; \\ 0 \leq x_2 \leq 2 - x_1 \end{array}$$

This has the form of a natural extension of the equation for a simple hypergeometric pf. [See Theorem 5.7 in Section 5-4; also, refer ahead to Section 12-5B.]

## C. Continuous random variables case: an introduction

Let us now extend our ideas to the continuous random variables case.

**Definition 12.3:** Given two continuous random variables $X$ and $Y$ defined on the same sample space $S$. The joint [or multivariate] probability density function [abbreviated pdf] of $X$ and $Y$ is a function $f(x, y)$ such that

$$P[(X, Y) \text{ is somewhere in region } A] = \iint\limits_{A} f(x, y)\, dx\, dy \qquad \textbf{(a)}$$

Here, $A$ denotes some region in the $xy$-plane. For the special case where the region $A$ is the rectangle given by $a_1 \le x \le a_2$ and $b_1 \le y \le b_2$ the probability on the left side of (a) becomes $P(a_1 \le X \le a_2 \text{ and } b_1 \le Y \le b_2)$. Thus, replacing the "and" by a comma, we have

$$P(a_1 \le X \le a_2, b_1 \le Y \le b_2) = \int_{b_1}^{b_2} \int_{a_1}^{a_2} f(x, y)\, dx\, dy \qquad \textbf{(b)}$$

Here, any of the $\le$ symbols may be replaced by $<$.

***notes***

(i) We assume here and elsewhere, unless otherwise noted, that $f(x, y)$ is sufficiently well behaved so that we may freely interchange the order of integration. Thus, the double integral on the right side of (b) can also be expressed as $\int_{a_1}^{a_2} \int_{b_1}^{b_2} f(x, y)\, dy\, dx$.

(ii) The symbols $x_1$ and $x_2$ are sometimes used in place of $x$ and $y$.

(iii) Often, $f(x, y)$ is referred to as being a bivariate pdf.

(iv) As in the single random variable case, $f(x, y)$ does not, by itself, represent a probability. However, for $\Delta x$ and $\Delta y$ sufficiently small, the probability $P(x_0 \le X \le x_0 + \Delta x, y_0 \le Y \le y_0 + \Delta y)$ is approximately equal to $f(x_0, y_0)\Delta x \Delta y$. Thus, we see that it is appropriate to regard $f(x, y)$ as a probability density, per area of the $xy$-plane.

---

**EXAMPLE 12-4:** Given the joint pdf of $X$ and $Y$ given by $f(x, y) = (\frac{1}{2})x^2 y + (\frac{1}{3})y$ for $0 < x < 2; 0 < y < 1$ and $f(x, y) = 0$, elsewhere. **(a)** Determine the probability $P(0 < X < 1; 0 < Y < 1)$. **(b)** Determine the probability $P(Y \le X)$.

***Solution:*** *Preliminaries:* The region of interest for $f(x, y)$ is shown in Figures 12-1a and 12-1b as the rectangle with corners at $(0, 0)$, $(2, 0)$, $(0, 1)$, and $(2, 1)$. In general, in situations like this, the *region of interest* is the region for which $f(x, y)$ is positive.

**(a)** The region for integration is shown doubly cross-hatched in Figure 12-1a. From Definition 12.3, we have

$$P(0 < X < 1, 0 < Y < 1) = \int_{x=0}^{1} \int_{y=0}^{1} f(x, y)\, dy\, dx$$

$$= \int_{x=0}^{1} \left( \int_{y=0}^{1} \left[ \left(\frac{1}{2}\right)x^2 y + \left(\frac{1}{3}\right)y \right] dy \right) dx \qquad \textbf{(1)}$$

Let us call the inner integral [the $y$ integral] $I$. Thus, we have

$$I = \left[ \left(\frac{1}{4}\right)x^2 y^2 + \left(\frac{1}{6}\right)y^2 \right]_{y=0}^{1} = \left(\frac{1}{4}\right)x^2 + \frac{1}{6} \qquad \textbf{(2)}$$

Substituting from Eq. (2) into Eq. (1), and then integrating with respect to $x$, we get

$$P(0 < X < 1, 0 < Y < 1) = \int_{x=0}^{1} \left[ \left(\frac{1}{4}\right)x^2 + \frac{1}{6} \right] dx$$

$$= \left(\frac{1}{12}\right)x^3 + \left(\frac{1}{6}\right)x \Big]_{x=0}^{1} = \frac{1}{4} \qquad \textbf{(3)}$$

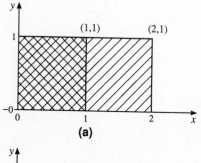

**(a)**

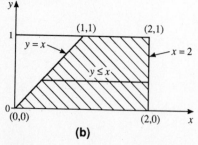

**(b)**

**Figure 12-1.** Region of interest and regions for probability calculations in Example 12-4.

Note that one would get the same result if one integrated with respect to $x$ first.

**(b)** The region for integration is shown cross-hatched in Figure 12-1b. Observe that $y < x$ for points below and to the right of the line with equation $y = x$. Integrating with respect to $x$ first [see horizontal line segment through the cross-hatched region], we have

$$P(Y \le X) = P(y \le X \le 2, 0 \le Y \le 1)$$

$$= \int_{y=0}^{1} \left( \int_{x=y}^{2} \left[ \left(\frac{1}{2}\right)x^2 y + \left(\frac{1}{3}\right)y \right] dx \right) dy \qquad \textbf{(4)}$$

Working out the inner integral [call it $I$], we get

$$I = \left(\frac{1}{6}\right)x^3 y + \left(\frac{1}{3}\right)yx \Big]_{x=y}^{2} = 2y - \left(\frac{1}{6}\right)y^4 - \left(\frac{1}{3}\right)y^2 \qquad \textbf{(5)}$$

Substituting from Eq. (5) into Eq. (4), and then integrating with respect to $y$, we get

$$P(Y \le X) = \int_{y=0}^{1} I\, dy = \int_{y=0}^{1} \left[ 2y - \left(\frac{1}{6}\right)y^4 - \left(\frac{1}{3}\right)y^2 \right] dy$$

$$= \left[ y^2 - \left(\frac{1}{30}\right)y^5 - \left(\frac{1}{9}\right)y^3 \right] \Big]_{y=0}^{1} = \frac{77}{90} \qquad \textbf{(6)}$$

---

**Theorem 12.2:** Suppose that $f(x, y)$ is a joint [or multivariate] pdf of continuous random variables $X$ and $Y$. Then, $f(x, y)$ satisfies the following properties:

**(a)** $0 \le f(x, y)$ for each point $(x, y)$ in the $xy$-plane.
**(b)** $\int_{-\infty}^{\infty} \int_{-\infty}^{\infty} f(x, y)\, dx\, dy = 1$, or $\int_{-\infty}^{\infty} \int_{-\infty}^{\infty} f(x, y)\, dy\, dx = 1$.

Either double integral in (b) is equivalent to the following probability statement:

$$P[(X, Y) \text{ is somewhere in the } xy\text{-plane}]$$

$$= P(-\infty < X < \infty, -\infty < Y < \infty) = 1$$

Also, part (b) is the equivalent of Axiom 1.2 [Section 1-5B], which says that $P(\text{sample space}) = 1$.

## D. Marginal probability density functions

Consider Definition 12.2 for the marginal probability functions of discrete random variables $X$ and $Y$. When $X$ and $Y$ are continuous random variables, the term "probability function" is replaced by "probability density function" (pdf), the summations are replaced by integrals, and we thereby obtain Definition 12.4.

**Definition 12.4:** Given that $X$ and $Y$ are continuous random variables, and that $f(x, y)$ is the joint pdf at the point $(x, y)$. The marginal pdf of $X$, which is denoted by $f_X(x)$ or $f_1(x)$, is given by

$$f_X(x) = \int_{-\infty}^{\infty} f(x, y)\, dy \qquad \textbf{(a)}$$

In Eq. (a), $x$ is constant. The marginal pdf of $Y$, which is denoted by $f_Y(y)$ or $f_2(y)$, is given by

$$f_Y(y) = \int_{-\infty}^{\infty} f(x, y)\, dx \qquad \textbf{(b)}$$

In Eq. (b), $y$ is constant.

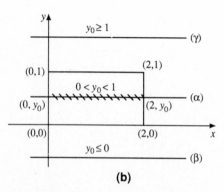

**Figure 12-2.** Determining $f_X(x)$ and $f_Y(y)$ in Example 12-5. (a) $f_X(x)$. (b) $f_Y(y)$.

**EXAMPLE 12-5:** For the joint pdf of Example 12-4, determine the marginal pdf's (a) $f_X(x)$, and (b) $f_Y(y)$.

**Solution**

(a) Refer to Figure 12-2a. Consider the vertical line $x = x_0$ for which $0 < x_0 < 2$. Such a line, which is labeled with ($\alpha$), cuts the region of interest along the line segment which runs between $(x_0, 0)$ and $(x_0, 1)$. Thus,

$$f_X(x_0) = \int_{-\infty}^{\infty} f(x_0, y)\, dy = \int_{y=0}^{1} \left[ \left(\frac{1}{2}\right) x_0^2 y + \left(\frac{1}{3}\right) y \right] dy$$

$$= \left[ \left(\frac{1}{4}\right) x_0^2 y^2 + \left(\frac{1}{6}\right) y^2 \right]_{y=0}^{1} = \left(\frac{1}{4}\right) x_0^2 + \frac{1}{6} \qquad \text{for } 0 < x_0 < 2 \quad (1)$$

For $x_0 \le 0$ and $x_0 \ge 2$ [see vertical lines labeled ($\beta$) and ($\gamma$) in Figure 12-2a], the region of interest is not intersected by a vertical line $x = x_0$, and we have

$$f_X(x_0) = \int_{-\infty}^{\infty} f(x_0, y)\, dy = \int_{-\infty}^{\infty} 0\, dy = 0 \qquad \text{for } x_0 \le 0 \text{ and } x_0 \ge 2 \quad (2)$$

Replacing $x_0$ by $x$, we get

$$f_X(x) = (\tfrac{1}{4})x^2 + \tfrac{1}{6} \qquad \text{for } 0 < x < 2, \qquad \text{and } f_X(x) = 0 \text{ elsewhere} \quad (3)$$

(b) Refer to Figure 12-2b. Consider the horizontal line $y = y_0$ for which $0 < y_0 < 1$. Such a line, which is labeled with ($\alpha$), cuts the region of interest along the line segment which runs between $(0, y_0)$ and $(2, y_0)$. Thus,

$$f_Y(y_0) = \int_{-\infty}^{\infty} f(x, y_0)\, dx = \int_{x=0}^{2} \left[ \left(\frac{1}{2}\right) x^2 y_0 + \left(\frac{1}{3}\right) y_0 \right] dx$$

$$= \left[ \left(\frac{1}{6}\right) x^3 y_0 + \left(\frac{1}{3}\right) y_0 x \right]_{x=0}^{2}$$

$$= \left(\frac{8}{6}\right) y_0 + \left(\frac{2}{3}\right) y_0 = 2y_0 \qquad \text{for } 0 < y_0 < 1 \quad (4)$$

For $y_0 \le 0$ and $y_0 \ge 1$ [see horizontal lines labeled ($\beta$) and ($\gamma$) in Figure 12-2b], the region of interest is not intersected by a horizontal line $y = y_0$, and we have

$$f_Y(y_0) = \int_{-\infty}^{\infty} f(x, y_0)\, dx = \int_{-\infty}^{\infty} 0\, dx = 0 \qquad \text{for } y_0 \le 0 \text{ and } y_0 \ge 1 \quad (5)$$

Replacing $y_0$ by $y$, we get

$$f_Y(y) = 2y \qquad \text{for } 0 < y < 1, \text{ and } f_Y(y) = 0 \text{ elsewhere} \quad (6)$$

## 12-2. The Multivariate Cumulative Distribution Function

As in the single variable case, we are interested in a function which measures accumulation of probability up to and including some point. This quantity is known as the *multivariate* [or *joint*] *cumulative distribution function*, and it is symbolized by $F$, as in the single variable case. At the fixed point $(x_0, y_0)$, it is given by

$$F(x_0, y_0) = P(X \le x_0, Y \le y_0) \quad (I)$$

where the comma on the right is read as "and." As previously, we shall abbreviate cumulative distribution function by cdf.

## A. Joint cumulative distribution function: discrete random variables case

**Definition 12.5:** Given that $X$ and $Y$ are discrete random variables, where $x_0$ and $y_0$ are fixed values of $X$ and $Y$. The joint cdf at $(x_0, y_0)$ is given by

$$F(x_0, y_0) = \sum_{x \le x_0} \sum_{y \le y_0} f(x, y)$$

Here, $-\infty < x_0 < \infty$ and $-\infty < y_0 < \infty$. Also, the order of summation may be reversed.

*note:* Some authors refer to $F(x, y)$ as being the joint [or multivariate] distribution function—both for the discrete and continuous random variable cases.

---

**EXAMPLE 12-6:** Refer to the joint pf of Example 12-1. [See, for example, Table 12-1.] Determine **(a)** $F(1.5, .6)$, and **(b)** $F(1, 1)$.

*Solution*

**(a)** Refer to Figure 12-3a. Consider the cross-hatched region which extends up to and including the point $(1.5, .6)$. To find $F(1.5, .6)$, we sum up ["accumulate"] probabilities for all the points within the cross-hatched region at which positive probabilities occur. There are only two such points, namely $(0, 0)$ and $(1, 0)$. Thus,

$$F(1.5, .6) = P(X \le 1.5, Y \le .6) = f(0, 0) + f(1, 0) = \tfrac{1}{10} + \tfrac{3}{10} = \tfrac{4}{10} \quad \textbf{(1)}$$

In fact, it is true that

$$F(x, y) = \tfrac{4}{10} \quad \text{for any point } (x, y) \text{ such that } x \ge 1 \text{ and } 0 \le y < 1 \quad \textbf{(2)}$$

**(b)** Refer to Figure 12-3b. Consider the cross-hatched region which extends up to and including the point $(1, 1)$. To calculate $F(1, 1)$, we sum up probabilities for all points within the cross-hatched region at which positive probabilities occur. There are four such points, namely $(0, 0)$, $(1, 0)$, $(0, 1)$, and $(1, 1)$, and these constitute the only points for which $f(x, y)$ is positive. Thus,

$$F(1, 1) = P(X \le 1, Y \le 1) = f(0, 0) + f(1, 0) + f(0, 1) + f(1, 1) = 1 \quad \textbf{(3)}$$

In fact, it is true that

$$F(x, y) = 1 \quad \text{for any point } (x, y) \text{ such that } x \ge 1 \text{ and } y \ge 1 \quad \textbf{(4)}$$

---

## B. Joint cumulative distribution function: continuous random variables case

**Definition 12.6:** Given that $X$ and $Y$ are continuous random variables where $x_0$ and $y_0$ are fixed values of $X$ and $Y$. The joint cdf evaluated at $(x_0, y_0)$ is given by

$$F(x_0, y_0) = \int_{-\infty}^{x_0} \int_{-\infty}^{y_0} f(x, y)\, dy\, dx$$

or by

$$F(x_0, y_0) = \int_{-\infty}^{y_0} \int_{-\infty}^{x_0} f(x, y)\, dx\, dy$$

Here, $(x_0, y_0)$ is any point in the $xy$-plane.

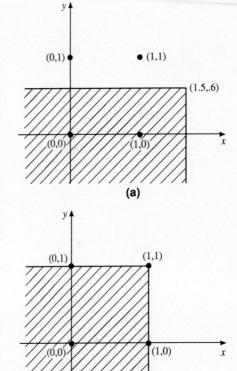

**Figure 12-3.** Diagrams for $F(x, y)$ calculations in Example 12-6.
**(a)** Diagram for $F(1.5, .6)$ calculation.
**(b)** Diagram for $F(1, 1)$ calculation.

---

**EXAMPLE 12-7:** For the joint pdf of Examples 12-4 and 12-5, determine $F(x, y)$ for all possible points $(x, y)$ in the $xy$-plane.

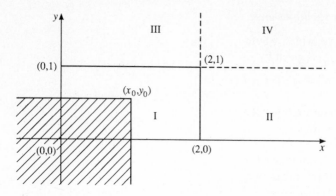

**Figure 12-4.** Diagram for determining $F(x, y)$ in Example 12-7.

**Solution:** First of all, it should be clear that $F(x, y) = 0$ for either $x \leq 0$ or $y \leq 0$, or both $x \leq 0$ and $y \leq 0$. Now, let us focus on the four remaining regions in the $xy$-plane, as indicated by I, II, III, and IV in Figure 12-4.

*Region* I: $0 < x_0 < 2$ and $0 < y_0 < 1$. We integrate over the cross-hatched region shown in Figure 12-4, and obtain

$$
\begin{aligned}
F(x_0, y_0) &= \int_{x=0}^{x_0} \int_{y=0}^{y_0} \left[ \left(\frac{1}{2}\right)x^2 y + \left(\frac{1}{3}\right)y \right] dy \, dx \\
&= \int_{x=0}^{x_0} \left[ \left(\frac{1}{4}\right)x^2 y_0^2 + \left(\frac{1}{6}\right)y_0^2 \right] dx \\
&= \left(\frac{1}{12}\right)x_0^3 y_0^2 + \left(\frac{1}{6}\right)x_0 y_0^2
\end{aligned}
\tag{1}
$$

*Region* II: $x_0 \geq 2$ and $0 < y_0 < 1$. We obtain

$$
\begin{aligned}
F(x_0, y_0) &= \int_{x=0}^{2} \int_{y=0}^{y_0} \left[ \left(\frac{1}{2}\right)x^2 y + \left(\frac{1}{3}\right)y \right] dy \, dx \\
&= \int_{x=0}^{2} \left[ \left(\frac{1}{4}\right)x^2 y_0^2 + \left(\frac{1}{6}\right)y_0^2 \right] dx = \left(\frac{2}{3}\right)y_0^2 + \left(\frac{1}{3}\right)y_0^2 = y_0^2
\end{aligned}
\tag{2}
$$

*Region* III: $0 < x_0 < 2$ and $y_0 \geq 1$. We obtain

$$
\begin{aligned}
F(x_0, y_0) &= \int_{x=0}^{x_0} \int_{y=0}^{1} \left[ \left(\frac{1}{2}\right)x^2 y + \left(\frac{1}{3}\right)y \right] dy \, dx \\
&= \int_{x=0}^{x_0} \left[ \left(\frac{1}{4}\right)x^2 \cdot 1 + \left(\frac{1}{6}\right)\cdot 1 \right] dx = \left(\frac{1}{12}\right)x_0^3 + \left(\frac{1}{6}\right)x_0
\end{aligned}
\tag{3}
$$

*Region* IV: $x_0 \geq 2$ and $y_0 \geq 1$. We obtain

$$
\begin{aligned}
F(x_0, y_0) &= \int_{x=0}^{2} \int_{y=0}^{1} \left[ \left(\frac{1}{2}\right)x^2 y + \left(\frac{1}{3}\right)y \right] dy \, dx \\
&= \int_{x=0}^{2} \left[ \left(\frac{1}{4}\right)x^2 \cdot 1 + \left(\frac{1}{6}\right)\cdot 1 \right] dx = 1
\end{aligned}
\tag{4}
$$

## 12-3. Independent Random Variables and Conditional Probability Distributions

The concept of independent events, which was introduced in Section 3-2 [Definition 3.2], leads us to a similar definition for independent random variables.

**Definition 12.7:** Given a typical interval $a_1 \leq X \leq a_2$ for random variable $X$, and a typical interval $b_1 \leq Y \leq b_2$ for random variable $Y$ [here, any of the $\leq$ symbols may be replaced by $<$]. Random variables $X$ and $Y$ are said to be independent if for any intervals $a_1 \leq X \leq a_2$ and $b_1 \leq Y \leq b_2$,

$$
P(a_1 \leq X \leq a_2, b_1 \leq Y \leq b_2) = P(a_1 \leq X \leq a_2) \cdot P(b_1 \leq Y \leq b_2)
$$

The following theorem, which is much more useful in establishing whether $X$ and $Y$ are independent, is proved in Larsen and Marx (1986).

**Theorem 12.3:** Two random variables $X$ and $Y$ are independent if and only if

$$
f(x, y) = f_X(x) \cdot f_Y(y)
$$

for all $x$ and all $y$.

*notes*

(i) Theorem 12.3 is often stated as a definition of independence of $X$ and $Y$.

(ii) Some authors state that if two random variables are not independent, then they are dependent. Unless otherwise noted, we will not make use of such terminology in this book.

(iii) To establish that $X$ are $Y$ are not independent, one only has to find a single point $(x_0, y_0)$ for which $f(x_0, y_0)$ is not equal to the product $f_X(x_0)f_Y(y_0)$.

## A. Independent random variables: the discrete case

**EXAMPLE 12-8:** Determine if the discrete random variables $X$ and $Y$ of Examples 12-1 and 12-2 are independent.

*Solution:* Refer to Table 12-1 of Example 12-1, in which the values for $f(x, y)$ for $X = 0, 1$ and $Y = 0, 1$, and $f_X(x)$ for $X = 0, 1$, and $f_Y(y)$ for $Y = 0, 1$, are presented. We see that $f(0, 1) = \frac{3}{10}$ and $f_X(0)f_Y(1) = (\frac{2}{5})(\frac{3}{5}) = \frac{6}{25}$, and hence that $f(0, 1) \neq f_X(0)f_Y(1)$. Thus, random variables $X$ and $Y$ are not independent.

Note that here we don't expect $X$ and $Y$ to be independent, since $X$ refers to the first draw and $Y$ to the second draw for a situation in which we have drawing without replacement. See part (d) of Problem 12-5 for a situation in which two discrete random variables are independent.

## B. Independent random variables: the continuous case

**EXAMPLE 12-9:** Refer to the continuous random variables $X$ and $Y$ of Examples 12-4 and 12-5. Determine if $X$ and $Y$ are independent.

*Solution:* First of all, observe that

$$f(x, y) = (\tfrac{1}{2})x^2 y + (\tfrac{1}{3})y \qquad \text{for } 0 < x < 2, 0 < y < 1, \text{ and } f(x, y) = 0 \text{ elsewhere} \tag{1}$$

$$f_X(x) = (\tfrac{1}{4})x^2 + \tfrac{1}{6} \qquad \text{for } 0 < x < 2, \text{ and } f_X(x) = 0 \text{ elsewhere} \tag{2}$$

$$f_Y(y) = 2y \qquad \text{for } 0 < y < 1, \text{ and } f_Y(y) = 0 \text{ elsewhere} \tag{3}$$

Now, observe that we can factor $f(x, y)$ as follows:

$$f(x, y) = [(\tfrac{1}{2})x^2 + (\tfrac{1}{3})]y \qquad \text{for } 0 < x < 2, 0 < y < 1 \tag{4}$$

The ability to factor leads us to suspect that $X$ and $Y$ are independent. Moreover, if, in Eq. (4), we multiply the first factor by $\frac{1}{2}$ and the second factor by 2 [thus, the net effect is to multiply $f(x, y)$ by 1], we obtain

$$f(x, y) = f_X(x)f_Y(y) \qquad \text{for } 0 < x < 2, 0 < y < 1 \tag{5}$$

Thus, we see that random variables $X$ and $Y$ are independent.

**EXAMPLE 12-10:** In Problem 12-6 [see Solved Problems section], the following conditions hold:

$$f(x, y) = 2(x + y) \qquad \text{for } 0 < y < x, 0 < x < 1, \text{ and } f(x, y) = 0 \text{ elsewhere} \tag{1}$$

$$f_X(x) = 3x^2 \qquad \text{for } 0 < x < 1, \text{ and } f_X(x) = 0 \text{ elsewhere} \tag{2}$$

$$f_Y(y) = 1 + 2y - 3y^2 \qquad \text{for } 0 < y < 1, \text{ and } f_Y(y) = 0 \text{ elsewhere} \tag{3}$$

The region of interest [or range space] for random variables $X$ and $Y$ is shown cross-hatched in Figure 12-5. Determine if random variables $X$ and $Y$ are independent.

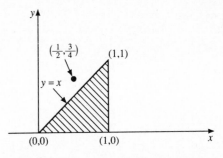

**Figure 12-5.** Test of independence of $X$ and $Y$ in Example 12-10.

**Solution:** Consider the point $(\frac{1}{2}, \frac{3}{4})$. This point is in the square such that $0 < x < 1$ and $0 < y < 1$, but not in the region of interest. See Figure 12-5. Clearly, $f(\frac{1}{2}, \frac{3}{4}) = 0$. But, $f_X(\frac{1}{2}) = \frac{3}{4}$, and $f_Y(\frac{3}{4}) = \frac{13}{16}$, and thus $f_X(\frac{1}{2})f_Y(\frac{3}{4})$ is positive. Thus, $f(\frac{1}{2}, \frac{3}{4}) \neq f_X(\frac{1}{2})f_Y(\frac{3}{4})$, and that means that $X$ and $Y$ are not independent.

*note:* Similarly, whenever the region of interest is not rectangular, the continuous random variables $X$ and $Y$ are not independent. Recall that the region of interest [also called the *range space* or the *support* for $X$ and $Y$] is that region in the $xy$-plane such that for $(x_0, y_0)$ in the region, $f(x_0, y_0) > 0$, and for $(x_1, y_1)$ not in the region, $f(x_1, y_1) = 0$.

## C. Conditional probability distributions

In Section 3-1, we encountered the conditional probability of event $B$, given event $A$. This was symbolized by $P(B \mid A)$, and defined as follows:

$$P(B \mid A) = \frac{P(A \cap B)}{P(A)}, \qquad \text{provided that } P(A) > 0 \qquad \textbf{(I)}$$

Suppose we have two discrete random variables $X$ and $Y$, and that $A$ and $B$ are the events $X = x$ and $Y = y$, respectively. Then, Eq. (I) leads directly to the following equation:

$$P(Y = y \mid X = x) = \frac{P(X = x \text{ and } Y = y)}{P(X = x)} = \frac{f(x, y)}{f_X(x)}, \qquad \text{provided } f_X(x) > 0$$

$$\textbf{(II)}$$

We shall use the symbol $f_Y(y \mid x)$ for $P(Y = y \mid X = x) = f(x, y)/f_X(x)$, and likewise, we shall use $f_X(x \mid y)$ to denote $P(X = x \mid Y = y) = f(x, y)/f_Y(y)$.

**Definition 12.8:** Say $f(x, y)$ is the joint pf of discrete random variables $X$ and $Y$ at the point $(x, y)$, and $f_X(x)$ is the value of the marginal pf at $X = x$. Then, the equation for $f_Y(y \mid x)$, the conditional pf of $Y$, given that $X = x$, is

$$f_Y(y \mid x) = f(x, y)/f_X(x), \qquad \text{provided } f_X(x) > 0 \qquad \textbf{(a)}$$

Suppose that $f_Y(y)$ is the value of the marginal pf at $Y = y$. Then, the equation for $f_X(x \mid y)$, the conditional pf of $X$, given that $Y = y$, is

$$f_X(x \mid y) = f(x, y)/f_Y(y), \qquad \text{provided } f_Y(y) > 0 \qquad \textbf{(b)}$$

*notes*

(i) Other symbols commonly used for $f_Y(y \mid x)$ are $g(y \mid x)$—other letters besides $g$ are employed—and $f_{Y|x}(y)$. Similarly, other symbols are used in place of $f_X(x \mid y)$.

(ii) If we use the symbols $X_1$, $X_2$, $x_1$, and $x_2$ in place of $X$, $Y$, $x$, and $y$, then parts (a) and (b) of Definition 12.8 take on the following forms:

$$f_{X_2}(x_2 \mid x_1) = f(x_1, x_2)/f_1(x_1) \quad \text{and} \quad f_{X_1}(x_1 \mid x_2) = f(x_1, x_2)/f_2(x_2)$$

(iii) If $f_X(x) = 0$, then $f_Y(y \mid x)$ is undefined. Likewise, if $f_Y(y) = 0$, then $f_X(x \mid y)$ is undefined.

---

**EXAMPLE 12-11:** Refer to Example 12-3. Determine **(a)** $f_{X_2}(1 \mid 0)$, and **(b)** $f_{X_1}(2 \mid 0)$.

**Solution:** *Preliminaries:* Remember that $x_1$ and $x_2$ indicate the number of red and green chips in a drawing without replacement of two chips from a box containing two red, three green, and five blue chips. The probability functions $f(x_1, x_2)$, $f_1(x_1)$, and $f_2(x_2)$ are displayed in Table 12-2 of Example 12-3.

(a) For $f_{X_2}(1 \mid 0)$, we want to compute the probability that $X_2 = 1$ given that $X_1 = 0$. This is the probability of getting one green chip given that no red chips are drawn.

$$f_{X_2}(1 \mid 0) = f(x_1 = 0, x_2 = 1)/f_1(x_1 = 0) = \frac{15/45}{28/45} = \frac{15}{28}$$

(b) For $f_{X_1}(2 \mid 0)$, we want to compute the probability that $X_1 = 2$ given that $X_2 = 0$. This is the probability of getting two red chips given that no green chips are drawn.

$$f_{X_1}(2 \mid 0) = f(x_1 = 2, x_2 = 0)/f_2(x_2 = 0) = \frac{1/45}{21/45} = \frac{1}{21}$$

---

When $X$ and $Y$ are continuous random variables, we replace the pf's on the right sides of Definition 12.8 by pdf's and we obtain Definition 12.9.

**Definition 12.9:** Say $f(x, y)$ is the joint pdf of continuous random variables $X$ and $Y$ at the point $(x, y)$, and $f_X(x)$ is the value of the marginal pdf at $X = x$. Then, the equation for $f_Y(y \mid x)$, the conditional pdf of $Y$, given that $X = x$, is

$$f_Y(y \mid x) = f(x, y)/f_X(x), \qquad \text{provided that } f_X(x) > 0 \qquad \textbf{(a)}$$

Suppose that $f_Y(y)$ is the value of the marginal pdf at $Y = y$. Then, the equation for $f_X(x \mid y)$, the conditional pdf of $X$, given that $Y = y$, is

$$f_X(x \mid y) = f(x, y)/f_Y(y), \qquad \text{provided that } f_Y(y) > 0 \qquad \textbf{(b)}$$

*notes*

(i) Part (a) of Definition 12.9 is useful if we want to calculate a conditional probability such as $P(b_1 < Y < b_2$ given that $X = a) = P(b_1 < Y < b_2 \mid X = a)$. The proper equation is given by

$$P(b_1 < Y < b_2 \mid X = a) = \int_{y=b_1}^{b_2} f_Y(y \mid a)\, dy$$

A justification for this equation, by using a limit argument, is presented on pages 149–150 of Larsen and Marx (1986). Similarly, we have

$$P(a_1 < X < a_2 \mid Y = b) = \int_{x=a_1}^{a_2} f_X(x \mid b)\, dx$$

(ii) The above conditional pdf's satisfy all the requirements for a pdf of a single random variable. For example, for an $x$ value such that $f_X(x) > 0$, we have

$$\int_{-\infty}^{\infty} f_Y(y \mid x)\, dy = \frac{\int_{-\infty}^{\infty} f(x, y)\, dy}{f_X(x)} = \frac{f_X(x)}{f_X(x)} = 1$$

Various aspects of Definition 12.9 are illustrated in Problems 12-6 and 12-7.

# 12-4. Extensions to the $n$ Random Variables Case

## A. The joint probability distribution and joint cumulative distribution function

The definitions and theorems of the previous three sections extend in a straightforward way to situations in which there are more than two random variables. Our convention concerning terminology has been to let probability distribution refer to a probability function [abbreviated pf] in the discrete case, and to a probability density function [abbreviated pdf] in the continuous case. We shall continue to follow this convention.

**Definition 12.10:** Given $n$ random variables $X_1, X_2, \ldots, X_n$. If these random variables are discrete, then the symbol for the joint probability function associated with the $n$ random variables is $f(x_1, x_2, \ldots, x_n)$. It is defined by

$$f(x_1, x_2, \ldots, x_n) = P(X_1 = x_1, X_2 = x_2, \ldots, X_n = x_n) \qquad \textbf{(a)}$$

If the $n$ random variables are continuous, then the joint probability density function is that function, also symbolized by $f(x_1, x_2, \ldots, x_n)$, which has the property that for any region $R$ in $n$-dimensional space,

$$P[\text{the point } (x_1, x_2, \ldots, x_n) \text{ is somewhere in } R]$$

$$= \int\int_R \cdots \int f(x_1, x_2, \ldots, x_n) \, dx_1 \, dx_2 \cdot \ldots \cdot dx_n \qquad \textbf{(b)}$$

On the right side, we have multiple integration with respect to the region $R$.

*notes*

(i) Some authors use the term multivariate in place of joint. Thus, for example, for the second part of Definition 12.10, they would refer to the multivariate pdf $f(x_1, x_2, \ldots, x_n)$.

(ii) It is instructive to illustrate the second part of Definition 12.10 for a special case. Suppose we have a three-variable [or trivariate] case, and we wish to find the probability that a point in three-dimensional space is somewhere within the three-dimensional rectangle specified by $a_1 \leq X_1 \leq a_2, b_1 \leq X_2 \leq b_2, c_1 \leq X_3 \leq c_2$. The region $R$ in this case is this rectangle. Then, we would have

$$P(a_1 \leq X_1 \leq a_2, b_1 \leq X_2 \leq b_2, c_1 \leq X_3 \leq c_2)$$

$$= \int_{x_3 = c_1}^{c_2} \int_{x_2 = b_1}^{b_2} \int_{x_1 = a_1}^{a_2} f(x_1, x_2, x_3) \, dx_1 \, dx_2 \, dx_3$$

Here, we have indicated that the order of integration is 123, that is, with respect to $x_1$ first, then with respect to $x_2$, and finally with respect to $x_3$. We assume in this case, and in others like it, that the integrand $f(x_1, x_2, x_3)$ is sufficiently well behaved so that it would be valid to integrate in any of the five other orders, for example, in order 132, or 213, etc.

---

**EXAMPLE 12-12:** The joint pf of $x_1$, $x_2$, and $x_3$ is given by $f(x_1, x_2, x_3) = cx_1 x_2 x_3$ for $x_1 = 1, 2, 3$; $x_2 = 2, 3$; $x_3 = 1, 2$. **(a)** Find the value of $c$. **(b)** Find $P(X_1 = 1, X_2 \leq 3, X_3 = 2)$. **(c)** Find $P(X_1 + X_2 = 4, X_3 = 1)$.

*Solution*

**(a)** It's understood that $f(x_1, x_2, x_3) = 0$ for points other than the 12 points indicated above. From $\Sigma_{x_3} \Sigma_{x_2} \Sigma_{x_1} f(x_1, x_2, x_3) = 1$, we obtain

$$c(1)(2)(1) + c(1)(2)(2) + c(1)(3)(1) + c(1)(3)(2) + c(2)(2)(1) + c(2)(2)(2)$$

$$+ c(2)(3)(1) + c(2)(3)(2) + c(3)(2)(1) + c(3)(2)(2) + c(3)(3)(1)$$

$$+ c(3)(3)(2) = 1 \qquad \textbf{(1)}$$

Thus, $c(90) = 1$, and, hence, $c = \frac{1}{90}$.

**(b)**
$$P(X_1 = 1, X_2 \le 3, X_3 = 2) = f(1,2,2) + f(1,3,2)$$
$$= (\tfrac{1}{90})[(1)(2)(2) + (1)(3)(2)] = \tfrac{10}{90} = \tfrac{1}{9} \quad (2)$$

**(c)**
$$P(X_1 + X_2 = 4, X_3 = 1) = f(1,3,1) + f(2,2,1)$$
$$= (\tfrac{1}{90})[(1)(3)(1) + (2)(2)(1)] = \tfrac{7}{90} \quad (3)$$

Calculations of probabilities pertaining to a joint pdf of three continuous random variables are given in Problem 12-8.

For the joint cdf $F(x_1, x_2, \ldots, x_n)$ of $n$ random variables, we have

$$F(x_1, x_2, \ldots, x_n) = P(X_1 \le x_1, X_2 \le x_2, \ldots, X_n \le x_n) \quad (I)$$

In the continuous random variables case, the joint cdf is given by

$$F(x_1, x_2, \ldots, x_n) = \int_{-\infty}^{x_n} \cdots \int_{-\infty}^{x_2} \int_{-\infty}^{x_1} f(t_1, t_2, \ldots, t_n)\, dt_1\, dt_2 \cdots dt_n \quad (Ia)$$

**EXAMPLE 12-13:** Refer to the probability function of three discrete random variables given in Example 12-12. Determine the following values of the joint cdf $F(x_1, x_2, x_3)$: **(a)** $F(1,3,2)$ and **(b)** $F(4,2,1.5)$.

*Solution*

**(a)** $F(1,3,2) = P(X_1 \le 1, X_2 \le 3, X_3 \le 2)$
$$= f(1,2,1) + f(1,2,2) + f(1,3,1) + f(1,3,2)$$
$$= (\tfrac{1}{90})[(1)(2)(1) + (1)(2)(2) + (1)(3)(1) + (1)(3)(2)] = \tfrac{15}{90} = \tfrac{1}{6} \quad (1)$$

**(b)** $F(4,2,1.5) = P(X_1 \le 4, X_2 \le 2, X_3 \le 1.5) = f(1,2,1) + f(2,2,1) + f(3,2,1)$
$$= (\tfrac{1}{90})[(1)(2)(1) + (2)(2)(1) + (3)(2)(1)] = \tfrac{12}{90} = \tfrac{2}{15} \quad (2)$$

Calculations pertaining to the joint cdf $F(x_1, x_2, x_3)$ of three continuous random variables are given in Problem 12-9.

## B. Joint marginal distributions and the independence concept

**Definition 12.11:** **(A)** Suppose that $f(x_1, x_2, x_3, \ldots, x_n)$ is the joint pf of the $n$ discrete random variables $X_1, X_2, X_3, \ldots, X_n$. Then, the marginal pf of $X_1$ alone, $f_1(x_1)$, is given by

$$f_1(x_1) = \sum_{x_n} \cdots \sum_{x_3} \sum_{x_2} f(x_1, x_2, x_3, \ldots, x_n)$$

Here, the summation is with respect to each of the remaining $(n-1)$ symbols $x_2, x_3, \ldots, x_n$. Then, the marginal pf of $X_2$ alone, $f_2(x_2)$, is given by

$$f_2(x_2) = \sum_{x_n} \cdots \sum_{x_3} \sum_{x_1} f(x_1, x_2, x_3, \ldots, x_n)$$

Here, the summation is with respect to each of the remaining $(n-1)$ symbols $x_1, x_3, \ldots, x_n$. Similar equations apply to the marginal pf's $f_3(x_3), f_4(x_4), \ldots, f_n(x_n)$.

**(B)** Suppose that $f(x_1, x_2, x_3, \ldots, x_n)$ is the joint pdf of the $n$ continuous random variables $X_1, X_2, X_3, \ldots, X_n$. Then the pattern is similar to what is presented in (A) above, except that summation ($\Sigma$) is replaced by integration ($\int$). In the following, $\int_{x_i}$ will stand for $\int_{x_i=-\infty}^{\infty}$. Thus, for example, the marginal pf of $X_2$ alone, $f_2(x_2)$, is given by

$$f_2(x_2) = \int_{x_n} \cdots \int_{x_3} \int_{x_1} f(x_1, x_2, x_3, \ldots, x_n)\, dx_1\, dx_3 \cdots dx_n$$

Here, the integration is with respect to each of the remaining $(n - 1)$ symbols $x_1, x_3, \ldots, x_n$.

---

**EXAMPLE 12-14:** Refer to Example 12-12. Determine the marginal probability functions **(a)** $f_1(x_1)$, **(b)** $f_2(x_2)$, and **(c)** $f_3(x_3)$.

***Solution:*** *Preliminaries:* The joint pf is given by $f(x_1, x_2, x_3) = (\frac{1}{90})x_1 x_2 x_3$ for $x_1 = 1, 2, 3$; $x_2 = 2, 3$; $x_3 = 1, 2$.

**(a)** First of all, note that $f_1(x_1)$ is meaningful only for $x_1 = 1, 2$, and 3. For $f_1(1)$, we have

$$f_1(1) = \sum_{x_3=1}^{2} \sum_{x_2=2}^{3} f(1, x_2, x_3) = f(1, 2, 1) + f(1, 2, 2) + f(1, 3, 1) + f(1, 3, 2)$$

$$= (\tfrac{1}{90})[(1)(2)(1) + (1)(2)(2) + (1)(3)(1) + (1)(3)(2)] = \tfrac{15}{90} = \tfrac{1}{6} \qquad (1)$$

Let us develop a general approach for determining $f_1(x_1)$. First of all, observe that

$$f_1(x_1) = \sum_{x_3=1}^{2} \sum_{x_2=2}^{3} f(x_1, x_2, x_3) = \left(\frac{x_1}{90}\right) \sum_{x_3=1}^{2} \sum_{x_2=2}^{3} x_2 x_3 \qquad (2)$$

Here, we have factored the constant $(x_1/90)$ to the left of the summation symbols. Observe that the summation is with respect to $x_2$ and $x_3$, and thus, $x_1$ may be regarded as a constant. Now,

$$\sum_{x_3=1}^{2} \sum_{x_2=2}^{3} x_2 x_3 = (2)(1) + (3)(1) + (2)(2) + (3)(2) = 15 \qquad (3)$$

Then, substitution from Eq. (3) into Eq. (2) leads to

$$f_1(x_1) = x_1/6 \qquad \text{for } x_1 = 1, 2, 3 \qquad (4)$$

Thus, we see that $f_1(1) = 1/6$, $f_1(2) = 2/6$, and $f_1(3) = 3/6$.

**(b)** The marginal pf $f_2(x_2)$ is defined only for $x_2 = 2$ and 3. We have

$$f_2(x_2) = \sum_{x_3=1}^{2} \sum_{x_1=1}^{3} f(x_1, x_2, x_3) = \left(\frac{x_2}{90}\right) \sum_{x_3=1}^{2} \sum_{x_1=1}^{3} x_1 x_3 \qquad (5)$$

Now,

$$\sum_{x_3=1}^{2} \sum_{x_1=1}^{3} x_1 x_3 = (1)(1) + (2)(1) + (3)(1) + (1)(2) + (2)(2) + (3)(2) = 18 \qquad (6)$$

Then, substitution from Eq. (6) into Eq. (5) leads to

$$f_2(x_2) = x_2/5 \qquad \text{for } x_2 = 2, 3 \qquad (7)$$

Thus, we see that $f_2(2) = 2/5$ and $f_2(3) = 3/5$.

**(c)** The marginal pf $f_3(x_3)$ is defined only for $x_3 = 1$ and 2. We have

$$f_3(x_3) = \sum_{x_2=2}^{3} \sum_{x_1=1}^{3} f(x_1, x_2, x_3) = \left(\frac{x_3}{90}\right) \sum_{x_2=2}^{3} \sum_{x_1=1}^{3} x_1 x_2 \qquad (8)$$

Now,

$$\sum_{x_2=2}^{3} \sum_{x_1=1}^{3} x_1 x_2 = (1)(2) + (2)(2) + (3)(2) + (1)(3) + (2)(3) + (3)(3) = 30 \qquad (9)$$

Then, substitution from Eq. (9) into Eq. (8) leads to

$$f_3(x_3) = x_3/3 \qquad \text{for } x_3 = 1, 2 \qquad (10)$$

Thus, we see that $f_3(1) = 1/3$ and $f_3(2) = 2/3$.

Calculations dealing with determining marginal pdf's for a situation in which there is a joint pdf of three continuous random variables are done in Problem 12-10.

In Section 3-2 [especially Definitions 3.3 and 3.4], we saw that extending the notion of independence from two events to $n$ events proved to be rather difficult: the independence of each subset of the $n$ events had to be checked separately. This is not the situation for the case of $n$ random variables. The appropriate test of independence is provided in Definition 12.12, which is a generalization of Theorem 12.3.

**Definition 12.12:** The $n$ random variables $X_1, X_2, \ldots, X_n$ are said to be independent if and only if

$$f(x_1, x_2, \ldots, x_n) = f_1(x_1) f_2(x_2) \cdot \ldots \cdot f_n(x_n)$$

for all $x_1, x_2, \ldots, x_n$.

This definition applies to both the discrete and continuous random variable cases.

---

**EXAMPLE 12-15:** Determine if the discrete random variables of Example 12-14 are independent.

**Solution:** For $X_1$, $X_2$, and $X_3$ to be independent, $f(x_1, x_2, x_3)$ would have to be equal to $f_1(x_1) f_2(x_2) f_3(x_3)$ for the 12 points for which $f(x_1, x_2, x_3)$ is positive. Now, we know that $f(x_1, x_2, x_3) = (\frac{1}{90}) x_1 x_2 x_3$ for $x_1 = 1, 2, 3$; $x_2 = 2, 3$; $x_3 = 1, 2$. Also, $f_1(x_1) = x_1/6$ for $x_1 = 1, 2, 3$; $f_2(x_2) = x_2/5$ for $x_2 = 2, 3$; and $f_3(x_3) = x_3/3$ for $x_3 = 1, 2$. Thus, we see that the random variables $X_1$, $X_2$, and $X_3$ are independent since $90 = (6)(5)(3)$.

---

A test of independence for three continuous random variables is illustrated in part (d) of Problem 12-10.

## C. Joint conditional probability distributions

Ideas pertaining to joint probability distributions are applicable to situations involving more than two random variables. For example, in the three discrete random variable case, for which $f(x_1, x_2, x_3)$ is the joint pf of discrete random variables $X_1$, $X_2$, and $X_3$, the term

$$f_{X_1, X_2}(x_1, x_2 \mid x_3) = \frac{f(x_1, x_2, x_3)}{f_3(x_3)} \qquad \textbf{(I)}$$

represents the conditional probability that $X_1$ equals $x_1$ and $X_2$ equals $x_2$, given that $X_3$ equals $x_3$. In the expression given by Eq. (I), we think of $x_3$ as being held constant. Also, the quantity on the left side of Eq. (I) is defined only for $x_3$ values such that $f_3(x_3) > 0$.

---

**EXAMPLE 12-16:** Refer to Examples 12-14 and 12-15. Determine the expression for $f_{X_1, X_2}(x_1, x_2 \mid x_3)$.

**Solution:** First, observe that $f_3(x_3) > 0$ for $x_3 = 1$ and 2. Thus, $f_{X_1, X_2}(x_1, x_2 \mid x_3)$ is only defined for those $x_3$ values. Thus, we will suppose that $x_3$ is either equal to 1 or 2. Now, let us focus on values of $x_1$ and $x_2$ for which $x_1 = 1, 2,$ or 3 and $x_2 = 2$ or 3. Then, for those values we have the following expression, after making use of the equations for $f(x_1, x_2, x_3)$ and $f_3(x_3)$ from Example 12-14:

$$f_{X_1, X_2}(x_1, x_2 \mid x_3) = \frac{(\frac{1}{90}) x_1 x_2 x_3}{(\frac{1}{3}) x_3} = (\frac{1}{30}) x_1 x_2 \qquad \textbf{(1)}$$

For values of $x_1$ and $x_2$ other than those mentioned above, we have $f_{X_1,X_2}(x_1, x_2 \mid x_3) = 0$. It is not surprising that $f_{X_1,X_2}(x_1, x_2 \mid x_3) = f_1(x_1)f_2(x_2)$ when we recall that $X_1$, $X_2$, and $X_3$ are independent.

---

Similar ideas apply to joint probability density functions of continuous random variables, where, for example, $f_{X_1,X_2}(x_1, x_2 \mid x_3)$ would again be given by

$$f_{X_1,X_2}(x_1, x_2 \mid x_3) = \frac{f(x_1, x_2, x_3)}{f_3(x_3)}$$

only now the numerator and denominator terms are recognized as probability density functions, and must not be interpreted as probabilities [as in the discrete case].

## 12-5. The Multinomial and Multivariate Hypergeometric Distributions

Two frequently occurring multivariate probability functions are the multinomial and multivariate hypergeometric distributions.

### A. The multinomial distribution

In the binomial distribution [see Section 5-3] situation, there were a sequence of $n$ independent and identical trials in which there were only two mutually exclusive outcomes of interest for each trial. These outcomes were often denoted as "success" and "failure." The binomial distribution [or probability function] was the probability associated with obtaining exactly $x$ successes in $n$ trials [see Definition 5.8 and Theorem 5.5].

Let us consider the situation where there is an experiment which is performed $n$ times in succession. In each of these $n$ independent and identical trials, let us focus on $k$ mutually exclusive outcomes $A_1$, $A_2$, ..., $A_k$ [where $k \geq 2$] for which the probabilities are labeled $p_1$, $p_2$, ..., $p_k$. Clearly, $p_1 + p_2 + \ldots + p_k = 1$; in the binomial case, $k = 2$, $p_1 = p$, and $p_2 = 1 - p = q$.

We shall be interested in calculating the probability of obtaining exactly $x_1$ outcomes of type $A_1$, $x_2$ outcomes of type $A_2$, ..., and $x_k$ outcomes of type $A_k$. Clearly, $x_1 + x_2 + \ldots + x_k = n$. Associated with each $x_i$ is a random variable $X_i$.

We shall denote the probability of obtaining exactly $x_1$ outcomes of type $A_1$, $x_2$ outcomes of type $A_2$, ..., and $x_k$ outcomes of type $A_k$, which is $P(X_1 = x_1, X_2 = x_2, \ldots, X_k = x_k)$, by the symbol

$$f(x_1, x_2, \ldots, x_k; n, p_1, p_2, \ldots, p_k)$$

Further, we say that random variables $X_1$, $X_2$, ..., $X_k$ have a multinomial [probability] distribution.

For the derivation, we proceed as in the derivation of the equation for the binomial probability function. For a particular [or specific] order in which there are $x_1$ outcomes of type $A_1$, $x_2$ outcomes of type $A_2$, ..., and $x_k$ outcomes of type $A_k$, the probability is

$$p_1^{x_1}p_2^{x_2}\cdot\ldots\cdot p_k^{x_k} \tag{I}$$

Now, the number of favorable orders in which there are $x_1$ outcomes of type $A_1$, $x_2$ outcomes of type $A_2$, ..., and $x_k$ outcomes of type $A_k$ is equal to the number of permutations of $n$ things taken all at a time, for which $x_1$ are of one kind, $x_2$ are of a second kind, ..., and $x_k$ are of a $k$th kind. This quantity, symbolized by $P(n; x_1, x_2, \ldots, x_k)$ in accord with Theorem 4.5, is equal to

$$\frac{n!}{x_1! x_2! \cdot \ldots \cdot x_k!} \qquad \textbf{(II)}$$

Now, for the probability $f(x_1, x_2, \ldots, x_k; n, p_1, p_2, \ldots, p_k)$, we multiply the terms from Eqs. (I) and (II) together.

**Theorem 12.4:** Suppose that random variables $X_1, X_2, \ldots, X_k$ have a multinomial [probability] distribution or multinomial probability function. Then, their joint probability function is given by

$$f(x_1, x_2, \ldots, x_k; n, p_1, p_2, \ldots, p_k) = \frac{n!}{x_1! x_2! \cdot \ldots \cdot x_k!} p_1^{x_1} p_2^{x_2} \cdot \ldots \cdot p_k^{x_k}$$

Here, each $x_i$ is a nonnegative integer, and $x_1 + x_2 + \ldots + x_k = n$. Also, $p_1 + p_2 + \ldots + p_k = 1$.

*notes*

(a) The name "multinomial" is used because the expression in Theorem 12.4 represents the general term in the expansion of the multinomial function $(p_1 + p_2 + \ldots + p_k)^n$.

(b) There are actually $(k - 1)$ true variables among the random variables $X_1$, $X_2, \ldots, X_k$. This is so because if the values for $(k - 1)$ of the random variables are specified, then the remaining value can be determined by subtraction, as with $x_k = n - (x_1 + x_2 + \ldots + x_{k-1})$.

(c) An important situation in which the multinomial distribution arises is when there is drawing with replacement from a set of objects, where, on any draw, the focus is on $k$ possible mutually exclusive outcomes.

---

**EXAMPLE 12-17:** A box contains three red, five blue, and two green balls. **(a)** If five balls are drawn with replacement, what is the probability of getting two red, two blue, and one green balls? **(b)** Suppose that two balls are drawn with replacement from the box containing three red, five blue, and two green balls. Let $X_i$ for $i = 1, 2$, and 3 denote the number of red, blue, and green balls obtained. Determine the joint probability function of $X_1$ and $X_2$, and display it in a table. **(c)** By using the table from part (b), determine values for the marginal pf's $f_1(x_1)$, and $f_2(x_2)$.

*Solution*

(a) The equation of Theorem 12.4 applies where $x_1 = 2$, $x_2 = 2$, $x_3 = 1$, $p_1 = 3/10 = .3$, $p_2 = .5$, and $p_3 = .2$. Thus,

$$P(X_1 = 2, X_2 = 2, X_3 = 1) = \frac{5!}{2! 2! 1!}(.3)^2(.5)^2(.2)^1 = .135 \qquad \textbf{(1)}$$

(b) Here, we have

$$f(x_1, x_2, x_3; 2, .3, .5, .2) = \frac{2!}{x_1! x_2!(2 - x_1 - x_2)!}(.3)^{x_1}(.5)^{x_2}(.2)^{(2 - x_1 - x_2)} \qquad \textbf{(2)}$$

Note how we express $x_3$ as $2 - x_1 - x_2$. Observe that $x_1$ and $x_2$ are integers, where $x_1 \geq 0$, $x_2 \geq 0$ and $(2 - x_1 - x_2) \geq 0$, or equivalently, $x_1 + x_2 \leq 2$.

We see that our probability function in Eq. (2) is truly a function of two [and not three] random variables. Here, the random variables are $X_1$ and $X_2$, but we could just as well have used any two of the three original random variables $X_1$, $X_2$, and $X_3$. Let us now denote the pf of Eq. (2) as $\tilde{f}(x_1, x_2)$ to stress the fact that we are actually dealing with a pf of two random variables. The $\tilde{f}(x_1, x_2)$ probabilities, as calculated from Eq. (2), are listed in Table 12-3.

**TABLE 12-3: Table for pf $\tilde{f}(x_1, x_2)$ and Associated Marginal pf's for Example 12-17**

| $\downarrow x_2 \quad \overrightarrow{x_1}$ | 0 | 1 | 2 | $f_2(x_2)$ |
|---|---|---|---|---|
| 0 | .04 | .12 | .09 | .25 |
| 1 | .20 | .30 | 0 | .50 |
| 2 | .25 | 0 | 0 | .25 |
| $f_1(x_1)$ | .49 | .42 | .09 | 1 |

Observe that it is impossible to have $x_1 + x_2 > 2$. That is why $\tilde{f}(x_1, x_2) = 0$ for such cases [for example, $\tilde{f}(1, 2) = 0$]. As an example of a calculation,

$$\tilde{f}(1, 1) = P(1 \text{ red and } 1 \text{ blue}) = 2(.3)(.5) = .30$$

(c) Summing in the columns for $x_1 = 0$, 1, and 2, we obtain the values for $f_1(x_1)$. For example, $f_1(x_1 = 1) = .12 + .30 + 0 = .42$.

   Summing in the rows for $x_2 = 0$, 1, and 2, we obtain the values for $f_2(x_2)$. For example, $f_2(x_2 = 0) = .04 + .12 + .09 = .25$.

---

Observe that the marginal pf $f_1(x_1)$ is the binomial pf with parameters $n = 2$ and $p = p_1 = .3$, and that the marginal pf $f_2(x_2)$ is the binomial pf with parameters $n = 2$ and $p = p_2 = .5$. This is no accident as the following theorem shows. [A proof for the $k = 3$ case is given in Larsen and Marx (1986).]

**Theorem 12.5:** Suppose the random variables $X_1, X_2, \ldots, X_k$ have a multinomial distribution with parameters $n$, $p_1, p_2, \ldots, p_k$. Then, the marginal probability function of each $x_i$, for $i = 1, 2, \ldots, k$, is a binomial distribution with parameters $n$ and $p_i$.

---

**EXAMPLE 12-18:** Determine whether the random variables $X_1$ and $X_2$ of Example 12-17 are independent.

*Solution:* Refer to Table 12-3 of Example 12-17. Consider the pair of values $x_1 = 2, x_2 = 1$, for example. For such a pair, we have $\tilde{f}(2, 1) = 0$, but $f_1(2)f_2(1) = (.09)(.50) > 0$. Thus, $\tilde{f}(2, 1) \neq f_1(2)f_2(1)$, and hence random variables $X_1$ and $X_2$ are not independent.

---

Similar results hold with respect to other multinomial distributions.

## B. The multivariate hypergeometric distribution

Recall how a change was made from a binomial distribution to a hypergeometric distribution when we changed from a drawing with replacement to a drawing without replacement situation, provided everything else stayed the same.

Likewise, a change is made from a multinomial distribution to a multivariate hypergeometric distribution when we change from a drawing with replacement to a drawing without replacement situation, provided everything else stays the same.

Consider an original set of $M$ objects [or elements or things] of which $a_1$ are objects of the first kind, $a_2$ are objects of the second kind, $\ldots$, and $a_k$ are objects of the $k$th kind. Here, $a_1 + a_2 + \ldots + a_k = M$. Now, suppose that

$n$ objects are drawn [or chosen] from the original set. [These $n$ objects are drawn without replacement.] As with the multinomial distribution derivation, we would like to calculate the probability of obtaining $x_1$ objects of the first kind, $x_2$ objects of the second kind, ..., and $x_k$ objects of the $k$th kind. Clearly, $x_1 + x_2 + ... + x_k = n$. Associated with each $x_i$ is a random variable $X_i$.

We shall denote the probability of obtaining exactly $x_1$ objects of the first kind, $x_2$ objects of the second kind, ..., and $x_k$ objects of the $k$th kind, which is $P(X_1 = x_1, X_2 = x_2, ..., X_k = x_k)$, by the symbol

$$f(x_1, x_2, ..., x_k; n, a_1, a_2, ..., a_k)$$

We say the random variables $X_1$, $X_2$, ..., $X_k$ have a multivariate hypergeometric [probability] distribution. The quantities $n$, $a_1$, $a_2$, ..., $a_k$ are called parameters.

Now, our approach for developing a probability equation will be similar to that used in deriving the result in Theorem 5.7 for $h(x; n, a, M)$, the ordinary hypergeometric probability function.

There are $\binom{a_1}{x_1}$ ways of choosing $x_1$ objects from the $a_1$ objects of the first kind, $\binom{a_2}{x_2}$ ways of choosing $x_2$ objects from the $a_2$ objects of the second kind, ..., and $\binom{a_k}{x_k}$ ways of choosing $x_k$ objects from the $a_k$ objects of the $k$th kind. Thus, by the fundamental counting principle [Theorem 3.7], there are

$$\binom{a_1}{x_1}\binom{a_2}{x_2} \cdot \ldots \cdot \binom{a_k}{x_k} \tag{I}$$

different ways of simultaneously choosing the required $x_1 + x_2 + ... + x_k = n$ objects. Now, the total number of different [and equally likely] ways of drawing $n$ objects from a set of $M$ objects is

$$\binom{M}{n} \tag{II}$$

Thus, the probability $P(X_1 = x_1, X_2 = x_2, ..., X_k = x_k)$ is equal to the quotient of the expressions in Eqs. (I) and (II).

**Theorem 12.6:** Suppose that the random variables $X_1$, $X_2$, ..., $X_k$ have a multivariate hypergeometric [probability] distribution. Then, their joint probability function is given by

$$f(x_1, x_2, ..., x_k; n, a_1, a_2, ..., a_k) = \frac{\binom{a_1}{x_1}\binom{a_2}{x_2} \cdot \ldots \cdot \binom{a_k}{x_k}}{\binom{M}{n}}$$

Here, $x_i$ is a nonnegative integer, and $x_i \leq a_i$ for each $i$. Also, $x_1 + x_2 + ... + x_k = n$, and $a_1 + a_2 + ... + a_k = M$.

*notes*

(a) There are actually $(k - 1)$ true variables among the random variables $X_1$, $X_2$, ..., $X_k$. This is so because if the values for $(k - 1)$ of the random variables are specified, then the remaining value can be determined by subtraction, as with $x_k = n - (x_1 + x_2 + ... + x_{k-1})$.

(b) For $k = 2$, the expression in Theorem 12.6 reduces to the expression for a usual hypergeometric probability function, as given in Theorem 5.7. Comparing symbols, we see that $a_1 = a$, $a_2 = M - a$, $x_1 = x$, and $x_2 = n - x$.

Problem 12-11 deals with a multivariate hypergeometric distribution. Also, a table displaying a multivariate hypergeometric distribution is given in Example 12-3. There, the parameters were $a_1 = 2$, $a_2 = 3$, $a_3 = 5$ [thus, $M = 10$], and $n = 2$.

Theorem 12.7, which follows, bears the same relationship to Theorem 12.6 as Theorem 12.5 bears to Theorem 12.4.

**Theorem 12.7:** Suppose the random variables $X_1$, $X_2$, ..., $X_k$ have a multivariate hypergeometric distribution with parameters $a_1$, $a_2$, ..., $a_k$, where $a_1 + a_2 + ... + a_k = M$. Then, the marginal pf of $X_i$, for $i = 1, 2, ..., k$, is a hypergeometric distribution with parameters $n$, $a_i$, and $M$. If we employ the symbolism of Theorem 5.7, we have that the marginal probability function for $X_i$ is given by $h(x_i; n, a_i, M)$. This means that

$$f_i(x_i) = h(x_i; n, a_i, M) = \frac{\binom{a_i}{x_i}\binom{M - a_i}{n - x_i}}{\binom{M}{n}}$$

# SOLVED PROBLEMS

### Introduction to Multivariate Probability Distributions

**PROBLEM 12-1**   A fair coin is tossed three times. Let $X$ denote the number of heads, and let $Y$ denote the absolute value of the difference between the number of heads and the number of tails. (a) Develop a table displaying the joint probability function of $X$ and $Y$. (b) Determine the marginal probability functions of $X$ and $Y$.

*Solution  Preliminaries*: The table immediately below shows the 8 equally likely elements pertaining to the experiment, together with values for $X$ and $Y$ corresponding to each element.

| Element | $HHH$ | $HHT$ | $HTH$ | $HTT$ | $THH$ | $THT$ | $TTH$ | $TTT$ |
|---|---|---|---|---|---|---|---|---|
| $x$ | 3 | 2 | 2 | 1 | 2 | 1 | 1 | 0 |
| $y$ | 3 | 1 | 1 | 1 | 1 | 1 | 1 | 3 |

(a) By careful examination of the rows of $x$ and $y$ values, we can develop the table for the joint pf $f(x, y)$. See Table 12-4. As an example, $f(x = 2, y = 1) = P[\{HHT, HTH, THH\}] = 3/8$.

**TABLE 12-4:  Table for pf $f(x, y)$ and to Associated Marginal pf's**

| $\downarrow y \quad \overrightarrow{x}$ | 0 | 1 | 2 | 3 | $f_Y(y)$ |
|---|---|---|---|---|---|
| 1 | 0 | 3/8 | 3/8 | 0 | 3/4 |
| 3 | 1/8 | 0 | 0 | 1/8 | 1/4 |
| $f_X(x)$ | 1/8 | 3/8 | 3/8 | 1/8 | 1 |

(b) By summing probabilities within columns and rows, we obtain values for $f_X(x)$ and $f_Y(y)$. Observe that $f_X(x)$ is the pf of random variable $X$ alone, which is the binomial distribution for $n = 3$ and $p = 1/2$.

**PROBLEM 12-2** Given $f(x,y) = 2e^{-x-2y}$ for $x > 0$ and $y > 0$, and $f(x,y) = 0$, elsewhere. **(a)** Determine $P(1 < X < 2, 1 < Y < 3)$. **(b)** Interpret the pdf, and the probability calculated in part (a) in terms of a surface in three-dimensional space. **(c)** Determine $f_X(x)$. **(d)** Determine $f_Y(y)$.

**Solution**

**(a)** First, we rewrite $e^{-x-2y}$ as $e^{-x}e^{-2y}$. We have

$$P(1 < X < 2, 1 < Y < 3) = 2 \int_{x=1}^{2} \left( \int_{y=1}^{3} e^{-x}e^{-2y}\, dy \right) dx$$

$$= 2 \int_{x=1}^{2} e^{-x}(e^{-2y}/(-2)]_{y=1}^{3})\, dx = (e^{-2} - e^{-6}) \int_{x=1}^{2} e^{-x}\, dx$$

$$= (e^{-2} - e^{-6})(e^{-1} - e^{-2})$$

$$= e^{-3} - e^{-4} - e^{-7} + e^{-8} = .030895 \tag{1}$$

**(b)** Refer to Figure 12-6. For two random variables, we can interpret the joint pdf as a surface in three-dimensional space. We can do this by plotting $z = f(x,y)$ with respect to the $xy$-plane as is done in Figure 12-6. Here, we have $z = 2e^{-x-2y}$ for $x > 0$ and $y > 0$.

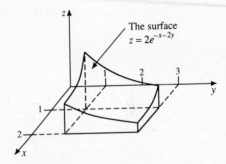

The probability of part (a) is equal to the volume between the surface with equation $z = 2e^{-x-2y}$ and the rectangle $1 < x < 2$, $1 < y < 3$ in the $xy$-plane. See Figure 12-6.

**(c)** In general, $f_X(x) = \int_{-\infty}^{\infty} f(x,y)\, dy$, for a fixed $x$ value. Here, for $x > 0$, this becomes

**Figure 12-6.** Diagram for Problem 12-2.

$$f_X(x) = 2 \int_{0}^{\infty} e^{-x}e^{-2y}\, dy = 2e^{-x} \int_{0}^{\infty} e^{-2y}\, dy = 2e^{-x}(e^{-2y}/(-2)]_{y=0}^{\infty}) = e^{-x} \tag{2}$$

For $x \le 0$, we get

$$f_X(x) = \int_{-\infty}^{\infty} 0\, dy = 0$$

**(d)** In general, $f_Y(y) = \int_{-\infty}^{\infty} f(x,y)\, dx$, for a fixed $y$ value. Here, for $y > 0$, this becomes

$$f_Y(y) = 2 \int_{0}^{\infty} e^{-x}e^{-2y}\, dx = 2e^{-2y} \int_{0}^{\infty} e^{-x}\, dx = 2e^{-2y} \tag{3}$$

For $y \le 0$, we get $f_Y(y) = \int_{-\infty}^{\infty} 0\, dx = 0$.

## The Multivariate Cumulative Distribution Function

**PROBLEM 12-3** Refer to the pf of Problem 12-1. Determine **(a)** $F(2,3)$, **(b)** $F(2.5, 3.2)$, **(c)** $F(0, 3)$, and **(d)** $F(4, 3)$.

**Solution**

**(a)** $F(2,3) = P(X \le 2, Y \le 3) = f(0,1) + f(1,1) + f(2,1) + f(0,3) + f(1,3) + f(2,3) = 0 + 3/8 + 3/8 + 1/8 + 0 + 0 = 7/8$.

**(b)** $F(2.5, 3.2) = P(X \le 2.5, Y \le 3.2) = F(2,3) = 7/8$.

**(c)** $F(0,3) = P(X \le 0, Y \le 3) = f(0,1) + f(0,3) = 0 + 1/8 = 1/8$.

**(d)** $F(4,3) = P(X \le 4, Y \le 3) = 1$.

**PROBLEM 12-4** Given two continuous random variables $X$ and $Y$ with joint pdf $f(x,y)$. Several major properties of the joint cdf $F(x,y)$ are as follows:

$$\frac{\partial^2 F(x,y)}{\partial x\, \partial y} = f(x,y) \tag{I}$$

provided the second partial derivative on the left exists. [The second partial derivative may be replaced by $\partial^2 F(x, y)/(\partial y \, \partial x)$.]

$$P(a_1 < X < a_2, b_1 < Y < b_2) = F(a_2, b_2) - F(a_2, b_1) - F(a_1, b_2) + F(a_1, b_1) \quad \text{(II)}$$

$$F(-\infty, -\infty) = F(-\infty, y_0) = F(x_0, -\infty) = 0 \quad \text{(IIIa)}$$

$$F(\infty, \infty) = 1 \quad \text{(IIIb)}$$

Here, $x_0$ and $y_0$ are any fixed values of $X$ and $Y$. Now, refer to Example 12-7. (a) Verify Eq. (I) above for any point $(x, y)$ such that $0 < x < 2$ and $0 < y < 1$. (b) Use Eq. (II) above to compute $P(1/2 < X < 1, 0 < Y < 1)$.

*Solution*

(a) In Example 12-7, we derived

$$F(x, y) = (\tfrac{1}{12})x^3 y^2 + (\tfrac{1}{6})xy^2 \qquad \text{for } 0 < x < 2 \text{ and } 0 < y < 1 \quad \text{(1)}$$

Thus, we have the following results:

$$\frac{\partial F(x, y)}{\partial y} = \left(\frac{1}{6}\right)x^3 y + \left(\frac{1}{3}\right)xy \qquad \text{for } 0 < x < 2 \text{ and } 0 < y < 1 \quad \text{(2)}$$

$$\frac{\partial^2 F(x, y)}{\partial x \, \partial y} = \frac{\partial}{\partial x}\left(\frac{\partial F(x, y)}{\partial y}\right) = \left(\frac{1}{2}\right)x^2 y + \left(\frac{1}{3}\right)y \qquad \text{for } 0 < x < 2 \text{ and } 0 < y < 1 \quad \text{(3)}$$

The term on the right side of Eq. (3) is none other than $f(x, y)$ for $0 < x < 2$ and $0 < y < 1$. (b) According to Eq. (II) above, we have

$$P(1/2 < X < 1, 0 < Y < 1) = F(1, 1) - F(1, 0) - F(1/2, 1) + F(1/2, 0) \quad \text{(4)}$$

First of all, we have $F(1, 0) = F(1/2, 0) = 0$. From the expression for $F(x, y)$ for $0 < x < 2$, $y \geq 1$ [see solution of Example 12-7], we have $F(x, y) = (\tfrac{1}{12})x^3 + (\tfrac{1}{6})x$. Thus,

$$F(1, 1) = (\tfrac{1}{12})(1) + (\tfrac{1}{6})(1) = \tfrac{3}{12} = \tfrac{1}{4} \quad \text{(5a)}$$

$$F(1/2, 1) = (\tfrac{1}{12})(\tfrac{1}{8}) + (\tfrac{1}{6})(\tfrac{1}{2}) = \tfrac{1}{96} + \tfrac{1}{12} = \tfrac{9}{96} \quad \text{(5b)}$$

After substitution of these four values of $F(x, y)$ into Eq. (4), we get

$$P(1/2 < X < 1, 0 < Y < 1) = \tfrac{24}{96} - 0 - \tfrac{9}{96} + 0 = \tfrac{5}{32} = .15625 \quad \text{(6)}$$

## Independent Random Variables and Conditional Probability Distributions

**PROBLEM 12-5**  Given the five-card deck of Example 12-1 [ace, two, and three of spades and ace and two of hearts]. The experiment is to draw two cards with replacement. Here, $X = 0$ or $1$ depending on whether the first card is heart or spade, and $Y = 0$ or $1$ depending on whether the second card is heart or spade. (a) Determine $f(x, y)$ for $x = 0, 1$ and $y = 0, 1$, and display the results in a table. (b) Determine the marginal pf $f_X(x)$. (c) Determine the marginal pf $f_Y(y)$. (d) Determine whether the random variables $X$ and $Y$ are independent.

*Solution*

(a) Let $H_i$ and $S_i$ indicate heart and spade on card $i$, for $i = 1, 2$.

$$f(0, 0) = P(X = 0, Y = 0) = P(H_1 \text{ and } H_2) = (\tfrac{2}{5})(\tfrac{2}{5}) = \tfrac{4}{25} \quad \text{(1)}$$

$$f(0, 1) = P(H_1 \text{ and } S_2) = (\tfrac{2}{5})(\tfrac{3}{5}) = \tfrac{6}{25} \quad \text{(2)}$$

Likewise, one can show that $f(1, 0) = \tfrac{6}{25}$ and $f(1, 1) = (\tfrac{3}{5})^2 = \tfrac{9}{25}$. The table of values for $f(x, y)$ follows:

| ↓y  →x | 0 | 1 | $f_Y(y)$ |
|--------|------|------|----------|
| 0 | 4/25 | 6/25 | 2/5 |
| 1 | 6/25 | 9/25 | 3/5 |
| $f_X(x)$ | 2/5 | 3/5 | 1 |

**(b)** By summing probabilities in the first and second columns, we obtain the following values:

$$f_X(0) = P(H_1) = f(0,0) + f(0,1) = 2/5$$

$$f_X(1) = P(S_1) = f(1,0) + f(1,1) = 3/5$$

The $f_X(x)$ values are listed in the bottom margin.

**(c)** By summing probabilities in the first and second rows, we obtain the following values:

$$f_Y(0) = P(H_2) = f(0,0) + f(1,0) = 2/5$$

$$f_Y(1) = P(S_2) = f(0,1) + f(1,1) = 3/5$$

The $f_Y(y)$ values are listed in the right margin.

**(d)** We see that $f(x,y) = f_X(x)f_Y(y)$ for all possible $(x,y)$ pairs listed in the table. For example, $f(0,1) = 6/25$ while $f_X(0) = 2/5$ and $f_Y(1) = 3/5$. Thus, random variables $X$ and $Y$ are independent. This result should be intuitively plausible since the result on the first draw [value of $X$] is independent of the result on the second draw [value of $Y$] when the cards are drawn with replacement.

---

**PROBLEM 12-6**  Given that $f(x,y) = 2(x + y)$ for $0 < y < x$, $0 < x < 1$, and $f(x,y) = 0$, elsewhere. [The region of interest can be described by writing $0 < y < x < 1$.] **(a)** Determine $f_X(x)$. **(b)** Determine $f_Y(y)$. **(c)** Compute the probability $P(0 < X < \frac{1}{2})$. **(d)** Determine the conditional pdf $f_X(x \mid y)$. **(e)** Compute the probability $P(0 < X < \frac{1}{2} \mid Y = \frac{1}{4})$.

*Solution*

**(a)** For $0 < x < 1$, we have

$$f_X(x) = \int_{-\infty}^{\infty} f(x,y)\,dy = 2\int_0^x (x + y)\,dy = 2(xy + y^2/2)]_{y=0}^x = 3x^2 \quad \textbf{(1)}$$

The portion of a typical [vertical] line $x = $ constant on which integration is done for $y$ between $y = 0$ and $y = x$ is shown in Figure 12-7a.

For $x \le 0$ and $x \ge 1$, we have $f_X(x) = \int_{-\infty}^{\infty} f(x,y)\,dy = \int_{-\infty}^{\infty} 0\,dy = 0$.

**(b)** For $0 < y < 1$, we have

$$f_Y(y) = \int_{-\infty}^{\infty} f(x,y)\,dx = 2\int_{x=y}^{x=1} (x + y)\,dx$$

$$= 2(x^2/2 + yx)]_{x=y}^1 = 1 + 2y - 3y^2 \quad \textbf{(2)}$$

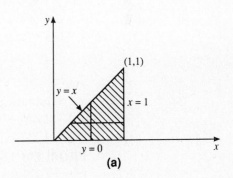

**(a)**

The portion of a typical [horizontal] line $y = $ constant on which integration is done for $x$ between $x = y$ and $x = 1$ is shown in Figure 12-7a.

For $y \le 0$ and $y \ge 1$, we have $f_Y(y) = \int_{-\infty}^{\infty} f(x,y)\,dx = \int_{-\infty}^{\infty} 0\,dx = 0$.

**(c)** Since we have two random variables here, the proper interpretation of the interval $0 < x < \frac{1}{2}$ is as the region $0 < x < \frac{1}{2}$, $-\infty < y < \infty$. Geometrically, this region is the vertical band of infinite extent between the lines $x = 0$ and $x = \frac{1}{2}$. Now, proceeding in general, we have

$$P\left(0 < X < \frac{1}{2}\right) = P\left(0 < X < \frac{1}{2}, -\infty < Y < \infty\right)$$

$$= \int_0^{1/2}\left(\int_{-\infty}^{\infty} f(x,y)\,dy\right)dx \quad \textbf{(3)}$$

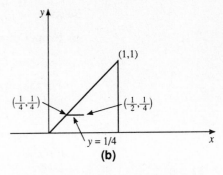

**(b)**

**Figure 12-7.** Diagrams for Problem 12-6. **(a)** Region of interest. **(b)** Interval for probability calculation for part (e).

But, the inner integral is none other than the marginal pdf $f_X(x)$. Thus, we can rewrite Eq. (3) as

$$P\left(0 < X < \frac{1}{2}\right) = \int_0^{1/2} f_X(x)\,dx \tag{4}$$

That is, we obtain an expression similar to that which applies in the single random variable case. [See Definition 8.1 in Section 8-1.]

Now, using the results for $f_X(x)$ from part (a), we obtain

$$P\left(0 < X < \frac{1}{2}\right) = 3\int_0^{1/2} x^2\,dx = \frac{1}{8} \tag{5}$$

**(d)** For a value of $y$ such that $f_Y(y) > 0$, we have

$$f_X(x \mid y) = f(x, y)/f_Y(y) \tag{6}$$

For the current situation, $f_X(x \mid y)$ is defined only for $0 < y < 1$ since $f_Y(y) = 0$ for $y \le 0$ and $y \ge 1$, and $f_Y(y) = 1 + 2y - 3y^2 > 0$ for $0 < y < 1$. [See part (b).] Thus, for a $y$ value such that $0 < y < 1$, we have

$$f_X(x \mid y) = 2(x + y)/f_Y(y) \qquad \text{for } y < x < 1 \tag{7a}$$

and

$$f_X(x \mid y) = 0 \qquad \text{for } x \le y \text{ and for } x \ge 1 \tag{7b}$$

**(e)** For $P(0 < X < \frac{1}{2} \mid Y = \frac{1}{4})$, we have

$$P\left(0 < X < \frac{1}{2} \,\middle|\, Y = \frac{1}{4}\right) = \int_0^{1/2} f_X\left(x \,\middle|\, y = \frac{1}{4}\right) dx$$

$$= \int_0^{1/4} 0\,dx + \left[1/f_Y\left(y = \frac{1}{4}\right)\right] \int_{1/4}^{1/2} f\left(x, \frac{1}{4}\right) dx$$

$$= 0 + \left[1/f_Y\left(y = \frac{1}{4}\right)\right] \int_{1/4}^{1/2} f\left(x, \frac{1}{4}\right) dx \tag{8}$$

Refer to Figure 12-7b for a diagram of the interval of integration indicated on the right side of Eq. (8), namely the line segment $y = \frac{1}{4}$, for $x$ between $\frac{1}{4}$ and $\frac{1}{2}$. Now, from part (b), $f_Y(y = \frac{1}{4}) = \frac{21}{16}$. For the integral on the right side of Eq. (8), we have

$$\int_{1/4}^{1/2} f\left(x, \frac{1}{4}\right) dx = 2\int_{1/4}^{1/2} \left(x + \frac{1}{4}\right) dx = \frac{5}{16} \tag{9}$$

Substituting the results from Eq. (9) and for $f_Y(y = \frac{1}{4})$ into Eq. (8), we get

$$P(0 < X < \tfrac{1}{2} \mid Y = \tfrac{1}{4}) = (\tfrac{5}{16})(\tfrac{16}{21}) = \tfrac{5}{21} \tag{10}$$

**PROBLEM 12-7**   Refer to the pdf's of Example 12-10 and Problem 12-6. **(a)** Find the conditional pdf of $Y$ given $x$. **(b)** Use the result from part (a) to find $P(Y \le .25 \mid X = .5)$.

*Solution*

**(a)** From Definition 12.9, we have

$$f_Y(y \mid x) = f(x, y)/f_X(x) \qquad \text{for } x \text{ such that } f_X(x) > 0 \tag{1}$$

For the current situation, $f_Y(y \mid x)$ is defined for $0 < x < 1$ since $f_X(x) = 0$ for $x \le 0$ and $x \ge 1$, and $f_X(x) = 3x^2$ for $0 < x < 1$. [See part (a).] Thus, for an $x$ value such that $0 < x < 1$, we have

$$f_Y(y \mid x) = 2(x + y)/f_X(x) \qquad \text{for } 0 < y < x \tag{2a}$$

and

$$f_Y(y \mid x) = 0 \qquad \text{for } y \le 0 \text{ and for } y \ge x \tag{2b}$$

**(b)** For $P(Y \le \frac{1}{4} \mid X = \frac{1}{2})$, we have

$$P\left(Y \le \frac{1}{4} \middle| X = \frac{1}{2}\right) = \int_{y=-\infty}^{1/4} f_Y\left(y \middle| x = \frac{1}{2}\right) dy$$

$$= \int_{y=-\infty}^{0} 0 \, dy + \left[1/f_X\left(x = \frac{1}{2}\right)\right] \int_{y=0}^{1/4} f\left(\frac{1}{2}, y\right) dy$$

$$= 0 + \left[1/f_X\left(x = \frac{1}{2}\right)\right] \int_{y=0}^{1/4} f\left(\frac{1}{2}, y\right) dy \qquad (3)$$

Refer to Figure 12-8 for a diagram of the interval of integration indicated on the right side of Eq. (3), namely the line segment $x = \frac{1}{2}$, for $y$ between 0 and $\frac{1}{4}$. Now, from part (a), $f_X(x = \frac{1}{2}) = 3(\frac{1}{2})^2 = \frac{3}{4}$. For the integral on the right side of Eq. (3), we have

$$\int_{y=0}^{1/4} f\left(\frac{1}{2}, y\right) dy = 2 \int_{y=0}^{1/4} \left(\frac{1}{2} + y\right) dy = \frac{5}{16} \qquad (4)$$

Substituting the results from Eq. (4) and for $f_X(x = \frac{1}{2})$ into Eq. (3), we get

$$P(Y \le \frac{1}{4} \mid X = \frac{1}{2}) = (\tfrac{4}{3})(\tfrac{5}{16}) = \tfrac{5}{12} \qquad (5)$$

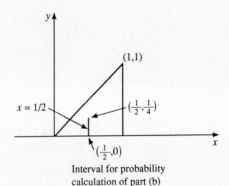

Interval for probability calculation of part (b)

**Figure 12-8.** Diagram for Problem 12-7.

## Extensions to the *n* Random Variables Case

**PROBLEM 12-8** Given that the joint pdf of continuous random variables $X_1$, $X_2$, and $X_3$ is given by $f(x_1, x_2, x_3) = k(x_1 + 2x_2 + 3x_3)$ for $0 < x_1 < 1$, $0 < x_2 < 1$, $0 < x_3 < 1$, and $f(x_1, x_2, x_3) = 0$ elsewhere. **(a)** Show that $k = 1/3$. **(b)** Determine the probability $P(0 < X_1 < 1/2, 0 < X_2 < 1/2, 1/2 < X_3 < 1)$.

*Solution*

**(a)** We know that

$$\int_{-\infty}^{\infty} \int_{-\infty}^{\infty} \int_{-\infty}^{\infty} f(x_1, x_2, x_3) \, dx_1 \, dx_2 \, dx_3 = 1 \qquad (1)$$

Here, this becomes

$$k \int_0^1 \int_0^1 \int_0^1 (x_1 + 2x_2 + 3x_3) \, dx_1 \, dx_2 \, dx_3 = 1 \qquad (2)$$

Denoting the triple integral in Eq. (2) by $I$, we have $k = 1/I$. Solving for $I$, we have

$$I = \int_0^1 \int_0^1 (x_1^2/2 + 2x_2 x_1 + 3x_3 x_1)]_{x_1=0}^{1} \, dx_2 \, dx_3$$

$$= \int_0^1 \int_0^1 (1/2 + 2x_2 + 3x_3) \, dx_2 \, dx_3 = \int_0^1 \left[\left(\frac{1}{2}\right)x_2 + x_2^2 + 3x_3 x_2\right]\Bigg]_{x_2=0}^{1} \, dx_3$$

$$= \int_0^1 \left[\left(\frac{1}{2}\right) + 1 + 3x_3\right] dx_3 = \left[\left(\frac{3}{2}\right)x_3 + \left(\frac{3}{2}\right)x_3^2\right]\Bigg]_{x_3=0}^{1} = 3 \qquad (3)$$

Thus

$$k = 1/I = \tfrac{1}{3} \qquad (4)$$

**(b)** Now,

$$P(0 < X_1 < 1/2, 0 < X_2 < 1/2, 1/2 < X_3 < 1)$$

$$= \left(\frac{1}{3}\right) \int_{1/2}^{1} \int_0^{1/2} \int_0^{1/2} (x_1 + 2x_2 + 3x_3) \, dx_1 \, dx_2 \, dx_3$$

$$= \left(\frac{1}{3}\right) \int_{1/2}^{1} \int_{0}^{1/2} \left[\frac{1}{8} + x_2 + \left(\frac{3}{2}\right)x_3\right] dx_2 \, dx_3$$

$$= \left(\frac{1}{3}\right) \int_{1/2}^{1} \left[\frac{3}{16} + \left(\frac{3}{4}\right)x_3\right] dx_3 = \left(\frac{1}{3}\right)\left(\frac{6}{16}\right) = \frac{1}{8} \qquad \textbf{(5)}$$

Here, several intermediate steps have been omitted.

**PROBLEM 12-9** Refer to the joint pdf of Problem 12-8. (a) Determine the joint cdf $F(x_1, x_2, x_3)$ for a point $(x_1, x_2, x_3)$ in the region $0 < x_1 < 1, 0 < x_2 < 1, 0 < x_3 < 1$. (b) Determine $F(\frac{1}{2}, \frac{1}{2}, \frac{1}{4})$.

*Solution*

(a) Using $r$, $s$, and $t$ as the integration variables corresponding to $x_1$, $x_2$, and $x_3$, we have

$$F(x_1, x_2, x_3) = \left(\frac{1}{3}\right) \int_{0}^{x_3} \int_{0}^{x_2} \int_{0}^{x_1} (r + 2s + 3t) \, dr \, ds \, dt$$

$$= \left(\frac{1}{3}\right) \int_{0}^{x_3} \int_{0}^{x_2} \left[\left(\frac{1}{2}\right)x_1^2 + 2sx_1 + 3tx_1\right] ds \, dt$$

$$= \left(\frac{1}{3}\right) \int_{0}^{x_3} \left[\left(\frac{1}{2}\right)x_1^2 x_2 + x_2^2 x_1 + 3tx_1 x_2\right] dt$$

$$= \left(\frac{1}{3}\right)\left[\left(\frac{1}{2}\right)x_1^2 x_2 x_3 + x_1 x_2^2 x_3 + \left(\frac{3}{2}\right)x_1 x_2 x_3^2\right]$$

Several intermediate steps were omitted in the above development.

(b) From the general expression for $F(x_1, x_2, x_3)$ from part (a), we have

$$F(\tfrac{1}{2}, \tfrac{1}{2}, \tfrac{1}{4}) = (\tfrac{1}{3})[(\tfrac{1}{2})(\tfrac{1}{4})(\tfrac{1}{2})(\tfrac{1}{4}) + (\tfrac{1}{2})(\tfrac{1}{4})(\tfrac{1}{4}) + (\tfrac{3}{2})(\tfrac{1}{2})(\tfrac{1}{2})(\tfrac{1}{16})] = \tfrac{3}{128} = .0234375$$

**PROBLEM 12-10** Refer to the joint pdf of Problems 12-8 and 12-9. Determine the marginal pdf's (a) $f_1(x_1)$, (b) $f_2(x_2)$, and (c) $f_3(x_3)$. (d) Determine whether random variables $X_1$, $X_2$, and $X_3$ are independent.

*Solution*

(a) For $0 < x_1 < 0$, we have

$$f_1(x_1) = \int_{-\infty}^{\infty} \int_{-\infty}^{\infty} f(x_1, x_2, x_3) \, dx_2 \, dx_3 = \left(\frac{1}{3}\right) \int_{0}^{1} \int_{0}^{1} (x_1 + 2x_2 + 3x_3) \, dx_2 \, dx_3 \qquad \textbf{(1)}$$

where $x_1$ is considered to be constant. Now, for the inner integral, we have

$$\int_{0}^{1} (x_1 + 2x_2 + 3x_3) \, dx_2 = (x_1 x_2 + x_2^2 + 3x_3 x_2)\Big]_{x_2=0}^{1} = (x_1 + 1 + 3x_3) \qquad \textbf{(2)}$$

Now, substituting from Eq. (2) into Eq. (1), and then integrating with respect to $x_1$, we get

$$f_1(x_1) = \left(\frac{1}{3}\right)\left[x_1 x_3 + x_3 + \left(\frac{3}{2}\right)x_3^2\right]\Big]_{x_3=0}^{1} = \left(\frac{1}{3}\right)\left(x_1 + \frac{5}{2}\right) \qquad \text{for } 0 < x_1 < 1 \qquad \textbf{(3)}$$

Also, $f_1(x_1) = 0$, elsewhere.

(b) Proceeding in similar fashion for $f_2(x_2)$, only now eliminating some intermediate steps, we have

$$f_2(x_2) = \left(\frac{1}{3}\right) \int_{0}^{1} \int_{0}^{1} (x_1 + 2x_2 + 3x_3) \, dx_1 \, dx_3$$

$$= \left(\frac{1}{3}\right) \int_{0}^{1} \left(\frac{1}{2} + 2x_2 + 3x_3\right) dx_3 = \left(\frac{2}{3}\right)(1 + x_2) \qquad \text{for } 0 < x_2 < 1 \qquad \textbf{(4)}$$

Also, $f_2(x_2) = 0$, elsewhere.

**(c)** For $f_3(x_3)$, we follow a similar type of presentation as in part (b), and obtain

$$f_3(x_3) = \left(\frac{1}{3}\right) \int_0^1 \int_0^1 (x_1 + 2x_2 + 3x_3)\, dx_1\, dx_2$$

$$= \left(\frac{1}{3}\right) \int_0^1 \left(\frac{1}{2} + 2x_2 + 3x_3\right) dx_2 = \left(\frac{1}{2} + x_3\right) \qquad \text{for } 0 < x_3 < 1 \qquad (5)$$

Also, $f_3(x_3) = 0$, elsewhere.

**(d)** First of all, we see that $f(x_1, x_2, x_3)$ doesn't factor into the product $f_1(x_1)f_2(x_2)f_3(x_3)$ for $0 < x_1 < 1$, $0 < x_2 < 1$, $0 < x_3 < 1$. Thus, the random variables $X_1$, $X_2$, and $X_3$ are not independent.

For another approach to show non-independence, we show that $f(x_1, x_2, x_3)$ is not equal to $f_1(x_1)f_2(x_2)f_3(x_3)$ for a particular point $(x_1, x_2, x_3) = (a, b, c)$. For the point $(\frac{1}{4}, \frac{1}{4}, \frac{1}{4})$, we have $f(\frac{1}{4}, \frac{1}{4}, \frac{1}{4}) = \frac{1}{2}$, while $f_1(\frac{1}{4}) = \frac{11}{12}$, $f_2(\frac{1}{4}) = \frac{5}{6}$, $f_3(\frac{1}{4}) = \frac{3}{4}$, and thus $f_1(\frac{1}{4})f_2(\frac{1}{4})f_3(\frac{1}{4}) = \frac{55}{96}$. Thus, since $f(\frac{1}{4}, \frac{1}{4}, \frac{1}{4}) \neq f_1(\frac{1}{4})f_2(\frac{1}{4})f_3(\frac{1}{4})$, we see that random variables $X_1$, $X_2$, and $X_3$ are not independent.

Note that it may be true that $f(x_1, x_2, x_3)$ is equal to $f_1(x_1)f_2(x_2)f_3(x_3)$ for certain points when the random variables $X_1$, $X_2$, and $X_3$ are not independent. For example, in the current problem, we have that $f(\frac{1}{2}, \frac{1}{2}, \frac{1}{4}) = f_1(\frac{1}{2})f_2(\frac{1}{2})f_3(\frac{1}{4}) = \frac{3}{4}$. But, for the random variables $X_1$, $X_2$, and $X_3$ to be independent, the equality $f(x_1, x_2, x_3) = f_1(x_1)f_2(x_2)f_3(x_3)$ must hold for *all* points.

## The Multinomial and Multivariate Hypergeometric Distributions

**PROBLEM 12-11** Consider again a box containing three red, five blue, and two green balls, as in Example 12-17. If five balls are drawn without replacement, what is the probability of getting two red, two blue, and one green balls?

**Solution** Let $x_i$, for $i = 1$, 2, and 3, denote the number of red, blue, and green balls obtained. The equation of Theorem 12.6 applies, with $x_1 = 2$, $x_2 = 2$, $x_3 = 1$, $a_1 = 3$, $a_2 = 5$, and $a_3 = 2$. Thus, $n = 5$ and $M = 10$. We have

$$P(X_1 = 2, X_2 = 2, X_3 = 1) = \frac{\binom{3}{2}\binom{5}{2}\binom{2}{1}}{\binom{10}{5}} = \frac{(3)(10)(2)}{252} = \frac{5}{21} = .2381$$

# Supplementary Problems

**PROBLEM 12-12** Given a fair four-sided [pyramidal] die whose sides [faces] are labeled 1, 2, 3, and 4. A side is said to occur if it is facing downward. The die is tossed two times. Let $X$ denote the number of times a 1 occurs, and let $Y$ denote the number of times a 3 or 4 occurs. **(a)** Determine the probability function $f(x, y)$. **(b)** Determine the marginal pf $f_X(x)$. **(c)** Determine the marginal pf $f_Y(y)$. **(d)** Determine the conditional probability $P(Y = 1 \mid X = 1)$.

**Answer** **(a)** $f(0,0) = 1/16$, $f(1,0) = 2/16$, $f(2,0) = 1/16$, $f(0,1) = f(1,1) = f(0,2) = 4/16$
**(b)** $f_X(0) = 9/16$, $f_X(1) = 6/16$, $f_X(2) = 1/16$ **(c)** $f_Y(0) = f_Y(2) = 1/4$, $f_Y(1) = 2/4$
**(d)** $P(Y = 1 \mid X = 1) = f(1,1)/f_X(1) = 2/3$

**PROBLEM 12-13** Here, $P$, the price per unit of a certain item [in dollars], and $S$, the total sales [in 10,000 units], are random variables whose joint pdf is given by $f(p, s) = 2.5pe^{-ps}$ for $.40 < p < .80$ and $s > 0$, and $f(p, s) = 0$, elsewhere. **(a)** Find the probability the price per unit will be less than 50 cents and sales will be less than 20,000 units. **(b)** Determine the cdf $F(p, s)$ for $.40 < p < .80$ and $s > 0$. **(c)** Determine the marginal pdf $f_P(p)$. **(d)** Find the probability that the price per unit will be less than 50 cents. **(e)** Find the probability that sales will be less than 20,000 units, given that $p = .6$ dollars.

*Answer* **(a)** .1482     **(b)** $F(p, s) = (2.5)[(p - .4) + (e^{-ps} - e^{-.4s})/s]$     **(c)** $f_P(p) = 2.5$ for $.40 < p < .80$, and $f_P(p) = 0$ elsewhere     **(d)** .25     **(e)** $1 - e^{-(.6)(2)} = .6988$

**PROBLEM 12-14**  Given that $f(x, y) = kxy$ for $0 < x < y, 0 < y < 1$, and $f(x, y) = 0$ elsewhere. [The region of interest can also be described by writing $0 < x < y < 1$.] **(a)** Determine $k$. **(b)** Determine the marginal pdf $f_X(x)$. **(c)** Determine the marginal pdf $f_Y(y)$. **(d)** Determine the probability $P(0 < X < \frac{1}{4}, 0 < Y < \frac{1}{2})$. [*Hint:* This can be expressed as $\int_{x=0}^{1/4} \int_{y=x}^{1/2} f(x, y) \, dy \, dx$.] **(e)** Compute $P(0 < X < \frac{1}{4})$.

*Answer* **(a)** $k = 8$     **(b)** $f_X(x) = 4x - 4x^3$ for $0 < x < 1$, and $f_X(x) = 0$ elsewhere     **(c)** $f_Y(y) = 4y^3$ for $0 < y < 1$, and $f_Y(y) = 0$ elsewhere     **(d)** 7/256     **(e)** 31/256

**PROBLEM 12-15**  Suppose that two continuous random variables $X$ and $Y$ with pdf $f(x, y)$ are independent. Prove that $P(a_1 < X < a_2, b_1 < Y < b_2) = P(a_1 < X < a_2)P(b_1 < Y < b_2)$. [*Hint:* Try to express the double integral for the left side as the product of $\int_{a_1}^{a_2} f_X(x) \, dx$ times $\int_{b_1}^{b_2} f_Y(y) \, dy$.]

**PROBLEM 12-16**  Refer to Problem 12-2. **(a)** Show that random variables $X$ and $Y$ are independent **(b)** Use the result from Problem 12-15 to calculate $P(0 < X < 1, 1 < Y < 2)$.

*Answer* **(a)** $f(x, y) = e^{-x} 2e^{-2y} = f_X(x) f_Y(y)$ for $x > 0, y > 0$     **(b)** $(1 - e^{-1})(e^{-2} - e^{-4}) = .07397$

**PROBLEM 12-17**  The useful life, in hours, of a certain computer component is a random variable having the following pdf: $f(x) = 200/(x + 200)^2$ for $x > 0$, and $f(x) = 0$ elsewhere. Suppose that three of these components operate independently. **(a)** Find the joint pdf of $X_1$, $X_2$, and $X_3$, where $X_i$ represents the useful life of component $i$, for $i = 1, 2$, and 3. **(b)** Find the probability $P(X_1 < 200, X_2 < 400, X_3 > 400)$.

*Answer* **(a)** $f(x_1, x_2, x_3) = \dfrac{(200)^3}{(x_1 + 200)^2 (x_2 + 200)^2 (x_3 + 200)^2}$ for $x_1 > 0, x_2 > 0, x_3 > 0$, and

$f(x_1, x_2, x_3) = 0$ elsewhere     **(b)** $\left(1 - \dfrac{200}{400}\right)\left(1 - \dfrac{200}{600}\right)\left(\dfrac{200}{600}\right) = \dfrac{1}{9}$

**PROBLEM 12-18**  In a test of a set of 20 fuses, 14 are labeled "good," four are labeled "passable," and two are labeled as "defectives." Suppose that two fuses are drawn at random [meaning without replacement]. Let $X$ be the number of "passable" fuses and let $Y$ be the number of "defective" fuses in this drawing. **(a)** Construct a table showing the joint pf of $X$ and $Y$. **(b)** Find the marginal probability function of $X$, namely $f_X(x)$. **(c)** Find the marginal probability function of $Y$, namely $f_Y(y)$. **(d)** Find the conditional probability distribution of $Y$ given that $X = 1$.

*Answer* **(a), (b),** and **(c)** Here, $f(x, y)$ values are tabulated in the body of the table below. For example, $f(1, 1) = 8/190$. Also, $f_X(x)$ values are tabulated in the bottom margin, and $f_Y(y)$ values are tabulated in the right margin.     **(d)** $f_Y(y \mid x = 1) = f(1, y)/f_X(1) = f(1, y)/[64/190]$. In particular, $f_Y(y \mid x = 1) = 7/8, 1/8$, and 0 for $y = 0, 1$, and 2, respectively.

The probability distribution for $f(x, y)$ is a multivariate hypergeometric distribution.

| $\downarrow y \quad \overrightarrow{x}$ | 0 | 1 | 2 | $f_Y(y)$ |
|---|---|---|---|---|
| 0 | 91/190 | 56/190 | 6/190 | 153/190 |
| 1 | 28/190 | 8/190 | 0 | 36/190 |
| 2 | 1/190 | 0 | 0 | 1/190 |
| $f_X(x)$ | 120/190 | 64/190 | 6/190 | 1 |

**PROBLEM 12-19** Given the same situation as in Problem 12-18, except that now the fuses are drawn with replacement. Repeat parts (a), (b), and (c) of Problem 12-18. (d) Find the conditional probability distribution of $X$ given that $Y = 0$.

*Answer* (a), (b), and (c) Here, $f(x, y)$ values are tabulated in the body of the table below. Also, $f_X(x)$ and $f_Y(y)$ values are tabulated in the bottom and right margins, respectively. (d) $f_X(x \mid Y = 0) = f(x, 0)/f_Y(0) = f(x, 0)/(.81)$. In particular, $f_X(x \mid Y = 0) = 49/81, 28/81$, and $4/81$ for $x = 0, 1$, and 2, respectively. The probability distribution for $f(x, y)$ is a multinomial distribution.

| $\downarrow y \quad \overrightarrow{x}$ | 0 | 1 | 2 | $f_Y(y)$ |
|---|---|---|---|---|
| 0 | .49 | .28 | .04 | .81 |
| 1 | .14 | .04 | 0 | .18 |
| 2 | .01 | 0 | 0 | .01 |
| $f_X(x)$ | .64 | .32 | .04 | 1.00 |

# FINAL EXAM (Chapters 7–12)

1. A manufacturer found that the lifetime $X$, in months, of a certain type of battery has an exponential pdf with $\theta = 5$ months. (a) Determine the cdf, $F(x)$, for all values of $x$. (b) Compute the probability that a battery of this type will have a lifetime between 3 and 5 months, inclusive. (c) Compute the mean lifetime of this type of battery. (d) Determine the lifetime $x_0$ such that $P(X \geq x_0) = .2865 \ [\approx e^{-1.25}]$.

2. The number of telephone calls arriving at a certain switchboard is a Poisson random variable. The mean number of calls that arrive in 1 minute is 6. (a) What is the probability of 12 or fewer calls arriving in a 2 minute interval? (b) What is the mean number of calls that arrive in a 2 minute interval?

3. In a certain city, the daily consumption of electrical energy, in millions of kilowatt-hours, is approximately a gamma random variable with $\alpha = 2$ and $\beta = 5$. The energy sources in this city can provide a maximum of 30 million kilowatt-hours for a day. (a) What is the probability that the energy sources will be adequate on any given day? (b) What is the probability that the energy sources will be adequate for each of 5 days in succession?

4. Refer to the probability function of Problem 9 on the Midterm Exam. For the table of values for $f(x)$ in the solution of Problem 9, (a) compute the mean and (b) the variance of random variable $X$.

5. Suppose that the moment generating function for a random variable $X$ is given by $M_X(t) = (1 - 3t)^{-4}$. Determine $E(X)$ and $\text{Var}(X)$.

6. In a gambling game, five fair coins are tossed. For a bet of \$5, a gambler will win \$10 if three heads occur. Otherwise, the gambler loses the \$5 bet. (a) What is the expected gain for a typical bet of \$5? (b) What is the house percentage for a bet of \$5? (c) If the above game were a "fair game," how much would the gambler be awarded if three heads occurred?

7. In a certain country, the heights for adult males are normally distributed with a mean of 68 inches and a standard deviation of 4 inches. Let $X$ symbolize the height. (a) Determine the probability $P(66 < X < 73)$. (b) Determine the height $x_0$, which is the 90th percentile.

8. The average daily temperature [in degrees Fahrenheit] in January in the town of Koolberg is a normal random variable $T$ with mean $\mu_T = 20$ and $\sigma_T = 8$. The daily heating cost, $Q$, in dollars, for an office complex is related to $T$ by $Q = -100T + 12,000$. Compute the probability that the daily heating cost for the office complex on a typical January day will exceed \$11,072.

9. A fair die is tossed 180 times. Compute the approximate probability of obtaining between 24 and 32 aces, inclusive. Use the continuity correction.

10. Given a community of about 50,000 voters, in which the true proportion of Republicans is $p_{\text{Rep}} = .45$. How large a sample should be taken so that the probability is .9282 that the sample proportion of Republicans is within .030 of $p_{\text{Rep}}$?

11. A frequency distribution for the maximum loads, in tons, that can be sustained by 80 cables produced by the Alpha Corporation is given in the following table [$N = \Sigma f = 80$]:

| Class mark. tons | 10.7 | 11.2 | 11.7 | 12.2 | 12.7 | 13.2 | 13.7 | Sum |
|---|---|---|---|---|---|---|---|---|
| Class freq., $f$ | 4 | 10 | 16 | 25 | 12 | 8 | 5 | 80 |

(a) Develop a table listing values of the cumulative relative frequency at the class boundaries. (b) Determine the median, the first and third quartiles, and the interquartile range. (c) What percentile is a maximum load of 13.15 tons?

12. For the frequency distribution of Problem 11, compute (a) the sample mean $\bar{x}$, (b) the sample variance $\tilde{s}^2$, and sample standard deviation $\tilde{s}$, and (c) the sample variance $\hat{s}^2$, and sample standard deviation $\hat{s}$.

13. The joint probability function [pf] for discrete random variables $X$ and $Y$ is given by $f(x, y) = k(2x + y^2)$, where $x$ and $y$ take on all integer values such that $0 \leq x \leq 3$ and $0 \leq y \leq 2$. (a) Determine the value of $k$. (b) Determine the marginal pf's $f_X(x)$ and $f_Y(y)$. (c) Determine the probability $P(0 \leq X < 2.3, 0 \leq Y \leq 1.4)$. (d) Determine the value of $F(1.5, 2.6)$, where $F(x, y)$ denotes the cumulative distribution function of $X$ and $Y$.

14. Refer to Problem 13. (a) Determine the conditional probability $P(X = 2 \mid Y = 1)$. (b) Determine the conditional pf $f_X(x \mid y)$ for the value $y = 1$. (c) Determine the probability $P(.5 < X < 2.3)$. (d) Determine the conditional probability $P(.8 < Y \leq 2.2 \mid X = 1)$.

15. Consider the joint pdf of $X$ and $Y$ given by $f(x, y) = (\frac{3}{11})(x^2 + xy)$ for $0 < x < 2$, $0 < y < 1$, and $f(x, y) = 0$, elsewhere. (a) Determine the marginal pdf's $f_X(x)$ and $f_Y(y)$. (b) Determine the probability $P(0 < X < 1, 0 < Y < 3)$. (c) Determine the probability $P(X + Y < 1)$. (d) Determine $P(0 < Y < 1/2 \mid X = 1)$.

# Solutions to Final Exam

1. (a) Here, $f(x) = (\frac{1}{5})e^{-x/5}$ for $x > 0$, and $f(x) = 0$ for $x \leq 0$, where $x$ is the lifetime in months. Now, $F(x_0) = P(X \leq x_0)$. Thus, for $x_0 \geq 0$,

$$F(x_0) = \left(\frac{1}{5}\right) \int_0^{x_0} e^{-x/5} \, dx = 1 - e^{-x_0/5} \tag{1}$$

and $F(x_0) = 0$ for $x_0 \leq 0$.

(b) In general, for a continuous random variable,

$$P(a \leq X \leq b) = F(b) - F(a) \tag{2}$$

Thus,

$$P(3 \leq X \leq 5) = F(5) - F(3) = e^{-.6} - e^{-1} = .18093 \tag{3}$$

after making use of the equation for $F(x_0)$ from Eq. (1) above.

(c) For an exponential pdf, $E(X) = \theta$, and so $E(X) = 5$ months here.

(d)

$$P(X \geq x_0) = .2865 \tag{4}$$

But, in general, for a continuous random variable $X$,

$$P(X \geq x_0) = 1 - F(x_0) \tag{5}$$

Then, using the result from Eq. (1), we have

$$e^{-x_0/5} = .2865 \tag{6}$$

Taking logarithms to the base $e$ in Eq. (6), we have

$$x_0 = -5 \ln(.2865) = (-5)(-1.25002) \approx 6.25 \text{ months} \tag{7}$$

2. (a) The coefficient $\alpha$ is given by 6 calls/minute since $\alpha$ is the mean number of Poisson occurrences in 1 minute. For a 2 minute period, $\lambda = \alpha t = (6)(2) = 12$ calls, and

$$f_2(x) = e^{-12}(12)^x/(x!) \qquad \text{for } x = 0, 1, 2, \ldots$$

Now, $P(X \leq 12) = f_2(0) + f_2(1) + \ldots + f_2(12)$, and this is given to be

$$P(X \leq 12) = .5760$$

from Table IV, or from computer/calculator software.

(b) The mean number of calls that arrive in 2 minutes equals $\alpha t = (6)(2) = 12$ calls.

3. (a) See Section 8-3B for properties of the gamma random variable. For $\alpha = 2$ and $\beta = 5$, the value of $k$ for a gamma random variable is given by $k = 1/[\beta^\alpha \Gamma(\alpha)]$. Thus, here,

$$k = 1/[5^2 \Gamma(2)] = 1/25 \tag{1}$$

Thus, for a gamma random variable $X$, where $x$ refers to the daily consumption of power, in millions of kilowatt-hours, we have

$$f(x) = kx^{\alpha-1}e^{-x/\beta} = (\tfrac{1}{25})xe^{-x/5} \qquad \text{for } x > 0 \tag{2}$$

Thus, if we let $E$ stand for the event "energy sources are adequate on a given day," $E$ is equivalent to the inequality $X \leq 30$. Thus,

$$P(E) = P(X \leq 30) = \left(\frac{1}{25}\right)\int_{x=0}^{30} xe^{-x/5}\, dx$$

$$= \left(\frac{1}{25}\right)\left[25e^{-x/5}\left(-\frac{x}{5}-1\right)\right]_{x=0}^{30} = 1 - 7e^{-6} = .98265 \tag{3}$$

(b) Here, we have a binomial situation for which the binomial random variable $\hat{X}$ stands for the number of successes in $n = 5$ trials [or days]. Here, "success" is a day on which the energy sources are adequate. Thus, the binomial $p$ value is given by $p = P(\text{"Success"}) = .98265$, and

$$P(\text{"Energy adequate for each of 5 successive days"}) = p^5 = (.98265)^5 = .91620 \tag{4}$$

4. *Method 1 for parts (a) and (b):* From the table of the probability function [see Midterm Exam, Problem 9], we first calculate the $xf(x)$ and $x^2f(x)$ columns.

| $x$ | $f(x)$ | $xf(x)$ | $x^2f(x)$ |
|---|---|---|---|
| 0 | 4/120 | 0 | 0 |
| 1 | 36/120 | 36/120 | 36/120 |
| 2 | 60/120 | 120/120 | 240/120 |
| 3 | 20/120 | 60/120 | 180/120 |
| $\Sigma$ | 1.0 | 216/120 | 456/120 |

From the table above, we have

$$E(X) = \sum xf(x) = 216/120 = 1.8 \tag{1}$$

$$E(X^2) = \sum x^2f(x) = 456/120 = 3.8 \tag{2}$$

$$\text{Var}(X) = E(X^2) - [E(X)]^2 = 3.8 - (1.8)^2 = .56 \tag{3}$$

*Method 2 for parts (a) and (b):* From Theorem 9.14, we see that

$$E(X) = np \qquad \text{and} \qquad \text{Var}(X) = npq\left(\frac{M-n}{M-1}\right) \tag{4}, (5)$$

for the hypergeometric pf. Here, we have a hypergeometric pf for which $n = 3$, $a = 6$, and $M = 10$. Thus, $p = a/M = .6$ and $q = 1 - p = .4$. Thus, substituting into Eqs. (4) and (5), we have

$$E(X) = (3)(.6) = 1.8, \qquad \text{Var}(X) = (3)(.6)(.4)(\tfrac{7}{9}) = .56 \tag{6}, (7)$$

5. *Method 1:* Recall that

$$E(X^r) = M_X^{(r)}(0) \tag{1}$$

Thus, here, we have

$$M_X'(t) = (-4)(1 - 3t)^{-5}(-3) \tag{2}$$

$$E(X) = M_X'(0) = (4)(1)(3) = 12 \tag{3}$$

$$M_X^{(2)}(t) = (-5)(-4)(1 - 3t)^{-6}(-3)^2 \tag{4}$$

$$E(X^2) = M_X^{(2)}(0) = (5)(4)(-3)^2 = 180 \tag{5}$$

$$\text{Var}(X) = E(X^2) - [E(X)]^2 = 180 - 144 = 36 \tag{6}$$

*Method 2:* From Theorem 9.20 of Section 9-5D, we have $M_X(t) = (1 - \beta t)^{-\alpha}$ for $t < 1/\beta$, for a gamma random variable. Thus, here we have a gamma random variable with $\alpha = 4$ and $\beta = 3$. Referring now to Theorem 9.17 of Section 9-5C, we have $\mu = E(X) = \alpha\beta = (4)(3) = 12$, and $\text{Var}(X) = \alpha\beta^2 = (4)(9) = 36$.

**6. (a)** Let $G$ be the gain in dollars for a play of the game. Thus,

$$P(G = 10) = b(3; 5, .5) = 10/32 \tag{1}$$

$$P(G = -5) = 1 - b(3; 5, .5) = 22/32 \tag{2}$$

$$E(G) = \sum gP(G = g) = (-5)(22/32) + (10)(10/32) = -10/32 = -.3125 \tag{3}$$

Thus on the average, the gambler will lose $.3125 for each $5 bet.

**(b)** Here, $H = -100E(G)/B = 100(.3125)/5 = 6.25\%$.

**(c)** Let $g^*$ be the amount the gambler wins when 3 heads occur in a fair game. For a fair game, we have $E(G) = 0$. In general, one has $E(G) = \Sigma gP(G = g)$. Thus, here we have

$$0 = (-5)(22/32) + (g^*)(10/32) \tag{4}$$

which leads to $g^* = (5)(22)/10 = \$11$.

**7. (a)** In general, $z = (x - \mu)/\sigma = (x - 68)/4$, where $z$ is a value of the standard normal random variable. For $x_1 = 66$, $z_1 = (66 - 68)/4 = -.5$, and for $x_2 = 73$, $z_2 = (73 - 68)/4 = 1.25$. Thus,

$$P(66 < X < 73) = P(-.5 < Z < 1.25) = P(-.5 < Z < 0) + P(0 < Z < 1.25)$$

$$= .1915 + .3944 = .5859 \tag{1}$$

after making use of Table VI.

**(b)** Here, we seek the value $x_0$ such that $P(X \le x_0) = .90$. Thus, for the corresponding standard random variable value $z_0$, we have

$$P(Z \le z_0) = .90 \tag{2}$$

where

$$z_0 = (x_0 - 68)/4 \quad \text{and thus} \quad x_0 = 4z_0 + 68 \tag{3a), (3b}$$

From Eq. (2) and interpolation within Table VI, we obtain

$$z_0 = 1.282 \tag{4}$$

Thus, from Eq. (3b), we have

$$x_0 = 4(1.282) + 68 = 73.128 \approx 73.13 \text{ in.} \tag{5}$$

as the height which is the 90th percentile.

**8.** Here, $Q$ is related to $T$ by $Q = aT + b$, where $a = -100$ and $b = 12,000$. Thus, since $T$ is a normal random variable, it follows from Theorem 10.2 [Section 10-1B] that $Q$ is a normal random variable with mean $\mu_Q = a\mu_T + b$, and variance $\sigma_Q^2 = a^2\sigma_T^2$. Thus, here,

$$\mu_Q = (-100)(20) + 12,000 = \$10,000 \tag{1}$$

and

$$\sigma_Q = (100)\sigma_T = (100)(8) = 800 \tag{2}$$

Here, we wish to compute $P(Q > 11,072)$. Now,

$$P(Q > 11,072) = P(Z > z_1) \tag{3}$$

where $z_1$ is the standard normal random variable value corresponding to $q_1 = 11,072$. Thus,

$$z_1 = (q_1 - \mu_Q)/\sigma_Q = (11,072 - 10,000)/800 = 1.34 \tag{4}$$

From Table VI, we see that our answer is

$$P(Q > 11,072) = P(Z > 1.34) = .0901 \tag{5}$$

**9.** Here, let $X$ stand for the number of aces obtained in 180 tosses. Thus, $X$ is a binomial random variable for which $n = 180$ and $p = 1/6$, and hence $E(X) = np = 30$ and $nq = 150$. Thus, the normal approximation should be accurate. [See Sections 10-4A and 10-4B.] We have

$$P_B(24 \le X \le 32) = \sum_{x=24}^{32} b(x; 180, 1/6) \approx P_N(23.5 \le X \le 32.5) \tag{1}$$

where subscripts $B$ and $N$ refer to binomial and normal, respectively. Here, we have used the continuity correction. Introducing $z = (x - 30)/5$, where $z$ is the value of the standard normal random variable corresponding to $x$ [recall that $\sigma = (npq)^{1/2} = 5$], we have

$$P_B(24 \le X \le 32) \approx P_N(23.5 \le X \le 32.5) = P_N(-1.3 \le Z \le .5)$$

$$= .4032 + .1915 = .5947 \text{ [Table VI]} \tag{2}$$

*notes*

**(a)** The exact value [from calculator/computer software] is .6035.

**(b)** An approximate estimate, obtained without using a continuity correction, is

$$P_N(24 \le X \le 32) = P_N(-1.2 \le Z \le .4) = .5403$$

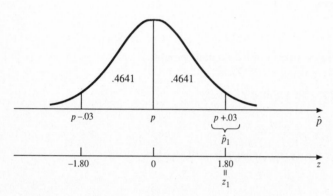

**Figure F-1.** Normal curve approximation for Problem 10 [$p = .45$].

**10.** First of all, we assume that $n$ is large enough so that the normal approximation to the binomial holds. Also, we'll work with the binomial proportion random variable $\hat{P}$, where $\hat{P} = X/n$. Also, we shall neglect the continuity correction here. Refer to Figure F-1. We have

$$P(p - .03 \le \hat{P} \le p + .03) = .9281 \tag{1}$$

Thus,

$$P(p \le \hat{P} \le p + .03) = .4641 \tag{2}$$

Now, $P(0 \le \hat{P} \le z_1) = .4641$ for $z_1 = 1.80$ [Table VI]. This means that $z_1 = 1.80$ corresponds to $\hat{p}_1 = p + .03 = .48$. [See Figure F-1.] Now, the correspondence equation here is

$$z_1 = [\hat{p}_1 - E(\hat{P})]/\sigma_{\hat{P}} \tag{3}$$

where $E(\hat{P}) = p$ and $\sigma_{\hat{P}} = [p(1 - p)/n]^{1/2}$. Substituting the values for $z_1$, $\hat{p}_1$, and $p = .45$ into (3), we get

$$1.80 = \frac{(.03)(n^{1/2})}{[(.45)(.55)]^{1/2}} \tag{4}$$

or

$$n = (.45)(.55)\left(\frac{1.8}{.03}\right)^2 = 891 \tag{5}$$

Thus, the sample size should be 891. Note that $np \approx 401$ and $nq \approx 490$ then, so that the normal approximation should be very accurate.

**11. (a)** From the given frequency table, we can construct the following table for cumulative frequency and cumulative relative frequency $[\hat{F}(x)]$ in terms of class boundary, $x$.

**TABLE F-1: Cumulative Relative Frequency Table for Problem 11**

| Class boundary, $x$ | 10.45 | 10.95 | 11.45 | 11.95 | 12.45 | 12.95 | 13.45 | 13.95 |
|---|---|---|---|---|---|---|---|---|
| Cumulative frequency | 0 | 4 | 14 | 30 | 55 | 67 | 75 | 80 |
| Cumulative relative frequency | .0000 | .0500 | .1750 | .3750 | .6875 | .8375 | .9375 | 1.0 |

For example, the constant class width is $w = .5$, and thus the lowest class boundary is $10.70 - w/2 = 10.45$. Each successive class boundary is .5 greater than the prior class boundary. The cumulative relative frequency corresponding to a class boundary equals the corresponding cumulative frequency, divided by the total frequency, 80.

(b) We indicate the calculation for the median $Md$ [recall that $Md = p_{50}$, the 50th percentile] by using the following subtable:

$$
.50 \quad \Delta \begin{bmatrix} 11.95 \\ p_{50} \\ 12.45 \end{bmatrix} \begin{array}{c} .3750 \\ .5000 \\ .6875 \end{array} \Bigg\rbrack .125 \quad .3125
$$

$$
\begin{array}{c|c}
x & \hat{F}(x) \\
\hline
11.95 & .3750 \\
p_{50} & .5000 \\
12.45 & .6875 \\
\end{array}
$$

By interpolating, we have

$$\Delta = (p_{50} - 11.95) = \left(\frac{.125}{.3125}\right)(.5) = .20 \qquad (1)$$

Thus,

$$Md = p_{50} = 11.95 + .20 = 12.15 \qquad (2)$$

For the first quartile $Q_1 = p_{25}$, we do interpolation with respect to $\hat{F}(x) = 25/100 = .25$. Thus, proceeding as in the above calculations [see Table F-1 for relevant values], we have

$$Q_1 = 11.45 + \left(\frac{.25 - .1750}{.375 - .175}\right)(.5) = 11.6375 \approx 11.64 \qquad (3)$$

For the third quartile $Q_3 = p_{75}$, we do interpolation with respect to $\hat{F}(x) = 75/100 = .75$. Thus, proceeding as in the above calculations, we have

$$Q_3 = 12.45 + \left(\frac{.75 - .6875}{.8375 - .6875}\right)(.5) \approx 12.66 \qquad (4)$$

Thus,

$$\text{Interquartile Range} = (Q_3 - Q_1) = 1.02 \qquad (5)$$

(c) First, we interpolate within Table F-1 to find the cumulative relative frequency $\hat{F}(x)$ that corresponds to the score $x = 13.15$. Thus,

$$\hat{F}(x) = .8375 + \left(\frac{13.15 - 12.95}{13.45 - 12.95}\right)(.9375 - .8375) = .8775 \qquad (6)$$

Thus, $x = 13.15$ is the 87.75th percentile. That is, $p_{87.75} = 13.15$.

12. First, we form a table similar to that of Example 11-17 [Section 11-6C].

| $i$ | Class mark, $x_i$ | Class frequency, $f_i$ | $x_i f_i$ | $x_i^2 f_i$ |
|---|---|---|---|---|
| 1 | 10.7 | 4 | 42.8 | 457.96 |
| 2 | 11.2 | 10 | 112.0 | 1254.4 |
| 3 | 11.7 | 16 | 187.2 | 2190.20 |
| 4 | 12.2 | 25 | 305.0 | 3721.0 |
| 5 | 12.7 | 12 | 152.4 | 1935.48 |
| 6 | 13.2 | 8 | 105.6 | 1393.92 |
| 7 (= k) | 13.7 | 5 | 68.5 | 938.45 |
| $\sum_{i=1}^{7}$ | | 80 | 973.5 | 11891.45 |

Then, referring to Section 11-6C for similar work, we have

$$n = \sum f = 80; \qquad \bar{x} = \left(\frac{1}{n}\right)\sum xf = 973.5/80 = 12.16875 \approx 12.17 \qquad \text{(1a), (1b)}$$

$$m_2' = \left(\frac{1}{n}\right)\sum x^2 f = 11891.45/80 = 148.6431 \qquad (2)$$

$$\tilde{s}^2 = m_2' - (\bar{x})^2 = .564648; \qquad \tilde{s} = (.564648)^{1/2} \approx .7514 \qquad \text{(3a), (3b)}$$

$$\hat{s}^2 = \left(\frac{n}{n-1}\right)\tilde{s}^2 = .571796; \qquad \hat{s} = (.571796)^{1/2} \approx .7561 \qquad \textbf{(4a), (4b)}$$

These sample statistics can also be easily calculated using computer/calculator software.

**13. (a)** Refer to Figure F-2a for the display of the table of the joint probability function $f(x, y)$. From

$$\sum_x \sum_y f(x, y) = 0 + k + 4k + 2k + 3k + \ldots + 7k + 10k = 56k = 1 \qquad \textbf{(1)}$$

we have $k = 1/56$.

**(b)** Now, $f_X(1)$ is $f_X$ for $x = 1$. Thus, $f_X(1)$ is obtained by summing $f(x, y)$ values in the $x = 1$ column of Figure F-2a.

$$f_X(1) = \sum_y f(1, y) = f(1, 0) + f(1, 1) + f(1, 2) = 11k = 11/56 \qquad \textbf{(2)}$$

The values for $f_X(0)$, $f_X(2)$, and $f_X(3)$ are computed in similar fashion with respect to the $x = 0$, 2, and 3 columns. The values for $f_X(x)$ are listed in the bottom margin of Figure F-2a.

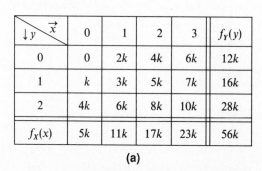
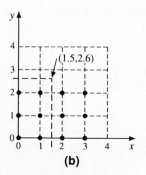

| $\downarrow y \quad \overrightarrow{x}$ | 0 | 1 | 2 | 3 | $f_Y(y)$ |
|---|---|---|---|---|---|
| 0 | 0 | 2k | 4k | 6k | 12k |
| 1 | k | 3k | 5k | 7k | 16k |
| 2 | 4k | 6k | 8k | 10k | 28k |
| $f_X(x)$ | 5k | 11k | 17k | 23k | 56k |

**(a)**                  **(b)**

**Figure F-2.** Diagrams for Problem 13. **(a)** Table of joint probability function $f(x, y)$. **(b)** Geometrical sample space diagram.

Now, $f_Y(2)$ is $f_Y$ for $y = 2$. Thus, $f_Y(2)$ is obtained by summing $f(x, y)$ values in the $y = 2$ row of Figure F-2a.

$$f_Y(2) = \sum_x f(x, 2) = f(0, 2) + f(1, 2) + f(2, 2) + f(3, 2) = 28k = 28/56 \qquad \textbf{(3)}$$

The values for $f_Y(0)$ and $f_Y(1)$ are computed in similar fashion with respect to the $y = 0$ and 1 rows. The values for $f_Y(y)$ are listed in the right-hand margin of Figure F-2a.

**(c)** Refer to the geometrical "sample space" diagram of possible $(x, y)$ points. [See accentuated dots in Figure F-2b.] We locate those accentuated dots of the sample space which are within the rectangle described by $0 \leq x < 2.3$, $0 \leq y \leq 1.4$. Thus,

$$P(0 \leq X < 2.3, 0 \leq Y \leq 1.4) = f(0, 0) + f(0, 1) + f(1, 0) + f(1, 1) + f(2, 0) + f(2, 1) = 15/56 \qquad \textbf{(4)}$$

**(d)** Refer to the "wedge" of Figure F-2b, which lies below and to the left of the point $(1.5, 2.6)$. (See dashed lines.) Summing probabilities for all the dots within the wedge-shaped region, we get

$$F(1.5, 2.6) = P(X \leq 1.5, Y \leq 2.6) = f(0, 0) + f(0, 1) + f(0, 2) + f(1, 0) + f(1, 1) + f(1, 2)$$

$$= [0 + 1 + 4 + 2 + 3 + 6]/56 = 16/56 = 2/7 \qquad \textbf{(5)}$$

**14. (a)**
$$P(X = 2 \mid Y = 1) = f(2, 1)/f_Y(y = 1) = \frac{(5/56)}{(16/56)} = \frac{5}{16} \qquad \textbf{(1)}$$

**(b)** The conditional pf $f_X(x \mid y = 1) = f_X(x \mid 1)$ is given by

$$f_X(x \mid 1) = f(x, 1)/f_Y(y = 1) = f(x, 1)/[16/56] = (\tfrac{7}{2})f(x, 1) \qquad \textbf{(2)}$$

for $x = 0, 1, 2, 3$. Thus, reading off the $f(x, 1)$ values from the $y = 1$ row of Figure F-2a, and recalling that $k = 1/56$, we have $f_X(x = 0 \mid 1) = (\tfrac{7}{2})(\tfrac{1}{56}) = \tfrac{1}{16}$; $f_X(x = 1 \mid 1) = (\tfrac{7}{2})(\tfrac{3}{56}) = \tfrac{3}{16}$; $f_X(x = 2 \mid 1) = \tfrac{5}{16}$; and $f_X(x = 3 \mid 1) = \tfrac{7}{16}$.
Note that $\Sigma_x f_X(x \mid y = 1) = 1$, as expected.

(c) Refer to Figure F-2b. Here, we sum over all possible points for which $x = 1$ and $2$. Thus,

$$P(.5 < X < 2.3) = f(1, 0) + f(1, 1) + f(1, 2) + f(2, 0) + f(2, 1) + f(2, 2) \tag{3}$$

The sum of the first three terms on the right is equal to $f_X(1)$, and the sum of the second three terms on the right is equal to $f_X(2)$. Thus, we get

$$P(.5 < X < 2.3) = f_X(1) + f_X(2) = (11 + 17)/56 = 1/2 \tag{4}$$

Thus, the situation is as if we had just the single random variable $X$.

(d) $$P(.8 \leq Y \leq 2.2 \mid X = 1) = f_Y(y = 1 \mid x = 1) + f_Y(y = 2 \mid x = 1) \tag{5}$$

Thus, we are summing two conditional probabilities along the vertical line $x = 1$ of Figure F-2b. Thus, since $f_Y(y = 1 \mid x = 1) = f(1, 1)/f_X(x = 1)$, and similarly for $f_Y(y = 2 \mid x = 1)$, we have

$$P(.8 \leq Y \leq 2.2 \mid X = 1) = [f(1, 1) + f(1, 2)]/f_X(x = 1) = [3/56 + 6/56]/(11/56) = 9/11 \tag{6}$$

Observe that $y = 1$ and $y = 2$ are the only two $y$ values on the interval $.8 \leq y \leq 2.2$ for which $f(1, y)$ is positive.

15. (a) Refer to Figure F-3, which shows the region of interest [shown cross-hatched] for $f(x, y)$. In general, $f_X(x) = \int_y f(x, y) \, dy$. For $0 < x < 2$, this becomes

$$f_X(x) = \left(\frac{3}{11}\right) \int_{y=0}^{1} (x^2 + xy) \, dy = \left(\frac{3}{11}\right)(x^2 y + xy^2/2) \Big]_{y=0}^{1} = \left(\frac{3}{11}\right)(x^2 + x/2) \tag{1}$$

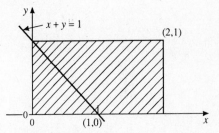

**Figure F-3.** Region of interest for Problem 15.

Clearly, $f_X(x) = 0$ for $x \leq 0$ and $x \geq 2$.

In general, $f_Y(y) = \int_x f(x, y) \, dx$. For $0 < y < 1$, this becomes

$$f_Y(y) = \left(\frac{3}{11}\right) \int_{x=0}^{2} (x^2 + xy) \, dx = \left(\frac{3}{11}\right)(x^3/3 + x^2 y/2) \Big]_{x=0}^{2} = \left(\frac{3}{11}\right)(8/3 + 2y) \tag{2}$$

Clearly, $f_Y(y) = 0$ for $y \leq 0$ and $y \geq 1$.

(b) $$P(0 < X < 1, 0 < Y < 3) = P(0 < X < 1, 0 < Y < 1) + P(0 < X < 1, 1 < Y < 3) \tag{3}$$

Now, the second probability term on the right of Eq. (3) equals zero since $f(x, y) = 0$ for points $(x, y)$ such that $0 < x < 1$ and $1 < y < 3$. Thus,

$$P(0 < X < 1, 0 < Y < 3) = \left(\frac{3}{11}\right) \int_{x=0}^{1} \int_{y=0}^{1} (x^2 + xy) \, dy \, dx = \left(\frac{3}{11}\right) \int_{x=0}^{1} (x^2 + x/2) \, dx$$

$$= \left(\frac{3}{11}\right)(x^3/3 + x^2/4) \Big]_{x=0}^{1} = 7/44 \tag{4}$$

(c) Refer to Figure F-3. We wish to find the probability that a point $(x, y)$ lies in the cross-hatched region below and to the left of the straight line with equation $x + y = 1$. Thus,

$$P(X + Y < 1) = \left(\frac{3}{11}\right) \int_{x=0}^{1} \int_{y=0}^{1-x} (x^2 + xy) \, dy \, dx \tag{5}$$

Now,

$$\int_{y=0}^{1-x} (x^2 + xy) \, dy = (x^2 y + xy^2/2) \Big]_{y=0}^{1-x} = (x - x^3)/2 \tag{6}$$

Substituting from Eq. (6) into Eq. (5), we obtain

$$P(X + Y < 1) = \left(\frac{3}{11}\right)\left(\frac{1}{2}\right) \int_{x=0}^{1} (x - x^3) \, dx = \frac{3}{88} \tag{7}$$

(d) $$P(0 < Y < \tfrac{1}{2} \mid X = 1) = \int_{y=0}^{1/2} f_Y(y \mid x = 1) \, dy \tag{8}$$

Here, we will integrate the conditional pdf $f_Y(y \mid x = 1)$ along the vertical line segment on which $0 < y < \frac{1}{2}$ and $x = 1$. Now,

$$f_Y(y \mid x = 1) = f(1, y)/f_X(x = 1) = \left(\tfrac{22}{9}\right) f(1, y) \tag{9}$$

since from part (a) we have $f_X(x = 1) = \frac{9}{22}$. Now, $f(1, y) = (\frac{3}{11})(1 + y)$ for $0 < y < 1$, and thus,

$$f_Y(y \mid x = 1) = (\tfrac{2}{3})(1 + y) \qquad \text{for } 0 < y < 1 \tag{10}$$

Substituting from Eq. (10) into Eq. (8), we get

$$P(0 < Y < \tfrac{1}{2} \mid X = 1) = \left(\frac{2}{3}\right) \int_{y=0}^{1/2} (1 + y)\, dy = \frac{5}{12} \tag{11}$$

# APPENDIX A   STATISTICAL TABLES

## TABLE I:  Binomial Probabilities, $b(x;n,p) = \dfrac{n!}{x!(n-x)!}\,p^x(1-p)^{n-x}$

| | | | | | | $p$ | | | | | |
|---|---|---|---|---|---|---|---|---|---|---|---|
| $n$ | $x$ | .05 | .10 | .15 | .20 | .25 | .30 | .35 | .40 | .45 | .50 |
| 1 | 0 | .9500 | .9000 | .8500 | .8000 | .7500 | .7000 | .6500 | .6000 | .5500 | .5000 |
|   | 1 | .0500 | .1000 | .1500 | .2000 | .2500 | .3000 | .3500 | .4000 | .4500 | .5000 |
| 2 | 0 | .9025 | .8100 | .7225 | .6400 | .5625 | .4900 | .4225 | .3600 | .3025 | .2500 |
|   | 1 | .0950 | .1800 | .2550 | .3200 | .3750 | .4200 | .4550 | .4800 | .4950 | .5000 |
|   | 2 | .0025 | .0100 | .0225 | .0400 | .0625 | .0900 | .1225 | .1600 | .2025 | .2500 |
| 3 | 0 | .8574 | .7290 | .6141 | .5120 | .4219 | .3430 | .2746 | .2160 | .1664 | .1250 |
|   | 1 | .1354 | .2430 | .3251 | .3840 | .4219 | .4410 | .4436 | .4320 | .4084 | .3750 |
|   | 2 | .0071 | .0270 | .0574 | .0960 | .1406 | .1890 | .2389 | .2880 | .3341 | .3750 |
|   | 3 | .0001 | .0010 | .0034 | .0080 | .0156 | .0270 | .0429 | .0640 | .0911 | .1250 |
| 4 | 0 | .8145 | .6561 | .5220 | .4096 | .3164 | .2401 | .1785 | .1296 | .0915 | .0625 |
|   | 1 | .1715 | .2916 | .3685 | .4096 | .4219 | .4116 | .3845 | .3456 | .2995 | .2500 |
|   | 2 | .0135 | .0486 | .0975 | .1536 | .2109 | .2646 | .3105 | .3456 | .3675 | .3750 |
|   | 3 | .0005 | .0036 | .0115 | .0256 | .0469 | .0756 | .1115 | .1536 | .2005 | .2500 |
|   | 4 | .0000 | .0001 | .0005 | .0016 | .0039 | .0081 | .0150 | .0256 | .0410 | .0625 |
| 5 | 0 | .7738 | .5905 | .4437 | .3277 | .2373 | .1681 | .1160 | .0778 | .0503 | .0312 |
|   | 1 | .2036 | .3280 | .3915 | .4096 | .3955 | .3602 | .3124 | .2592 | .2059 | .1562 |
|   | 2 | .0214 | .0729 | .1382 | .2048 | .2637 | .3087 | .3364 | .3456 | .3369 | .3125 |
|   | 3 | .0011 | .0081 | .0244 | .0512 | .0879 | .1323 | .1811 | .2304 | .2757 | .3125 |
|   | 4 | .0000 | .0004 | .0022 | .0064 | .0146 | .0284 | .0488 | .0768 | .1128 | .1562 |
|   | 5 | .0000 | .0000 | .0001 | .0003 | .0010 | .0024 | .0053 | .0102 | .0185 | .0312 |
| 6 | 0 | .7351 | .5314 | .3771 | .2621 | .1780 | .1176 | .0754 | .0467 | .0277 | .0156 |
|   | 1 | .2321 | .3543 | .3993 | .3932 | .3560 | .3025 | .2437 | .1866 | .1359 | .0938 |
|   | 2 | .0305 | .0984 | .1762 | .2458 | .2966 | .3241 | .3280 | .3110 | .2780 | .2344 |
|   | 3 | .0021 | .0146 | .0145 | .0819 | .1318 | .1852 | .2355 | .2765 | .3032 | .3125 |
|   | 4 | .0001 | .0012 | .0055 | .0154 | .0330 | .0595 | .0951 | .1382 | .1861 | .2344 |
|   | 5 | .0000 | .0001 | .0004 | .0015 | .0044 | .0102 | .0205 | .0369 | .0609 | .0938 |
|   | 6 | .0000 | .0000 | .0000 | .0001 | .0002 | .0007 | .0018 | .0041 | .0083 | .0156 |
| 7 | 0 | .6983 | .4783 | .3206 | .2097 | .1335 | .0824 | .0490 | .0280 | .0152 | .0078 |
|   | 1 | .2573 | .3720 | .3960 | .3670 | .3115 | .2471 | .1848 | .1306 | .0872 | .0547 |
|   | 2 | .0406 | .1240 | .2097 | .2753 | .3115 | .3177 | .2985 | .2613 | .2140 | .1641 |
|   | 3 | .0036 | .0230 | .0617 | .1147 | .1730 | .2269 | .2679 | .2903 | .2918 | .2734 |
|   | 4 | .0002 | .0026 | .0109 | .0287 | .0577 | .0972 | .1442 | .1935 | .2388 | .2734 |
|   | 5 | .0000 | .0002 | .0012 | .0043 | .0115 | .0250 | .0466 | .0774 | .1172 | .1641 |
|   | 6 | .0000 | .0000 | .0001 | .0004 | .0013 | .0036 | .0084 | .0172 | .0320 | .0547 |
|   | 7 | .0000 | .0000 | .0000 | .0000 | .0001 | .0002 | .0006 | .0016 | .0037 | .0078 |
| 8 | 0 | .6634 | .4305 | .2725 | .1678 | .1001 | .0576 | .0319 | .0168 | .0084 | .0039 |
|   | 1 | .2793 | .3826 | .3847 | .3355 | .2670 | .1977 | .1373 | .0896 | .0548 | .0312 |
|   | 2 | .0515 | .1488 | .2376 | .2936 | .3115 | .2965 | .2587 | .2090 | .1569 | .1094 |
|   | 3 | .0054 | .0331 | .0839 | .1468 | .2076 | .2541 | .2786 | .2787 | .2568 | .2188 |
|   | 4 | .0004 | .0046 | .0185 | .0459 | .0865 | .1361 | .1875 | .2322 | .2627 | .2734 |
|   | 5 | .0000 | .0004 | .0026 | .0092 | .0231 | .0467 | .0808 | .1239 | .1719 | .2188 |
|   | 6 | .0000 | .0000 | .0002 | .0011 | .0038 | .0100 | .0217 | .0413 | .0703 | .1094 |
|   | 7 | .0000 | .0000 | .0000 | .0001 | .0004 | .0012 | .0033 | .0079 | .0164 | .0312 |
|   | 8 | .0000 | .0000 | .0000 | .0000 | .0000 | .0001 | .0002 | .0007 | .0017 | .0039 |
| 9 | 0 | .6302 | .3874 | .2316 | .1342 | .0751 | .0404 | .0207 | .0101 | .0046 | .0020 |
|   | 1 | .2985 | .3874 | .3679 | .3020 | .2253 | .1556 | .1004 | .0605 | .0339 | .0176 |
|   | 2 | .0629 | .1722 | .2597 | .3020 | .3003 | .2668 | .2162 | .1612 | .1110 | .0703 |
|   | 3 | .0077 | .0446 | .1069 | .1762 | .2336 | .2668 | .2716 | .2508 | .2119 | .1641 |
|   | 4 | .0006 | .0074 | .0283 | .0661 | .1168 | .1715 | .2194 | .2508 | .2600 | .2461 |
|   | 5 | .0000 | .0008 | .0050 | .0165 | .0389 | .0735 | .1181 | .1672 | .2128 | .2461 |
|   | 6 | .0000 | .0001 | .0006 | .0028 | .0087 | .0210 | .0424 | .0743 | .1160 | .1641 |
|   | 7 | .0000 | .0000 | .0000 | .0003 | .0012 | .0039 | .0098 | .0212 | .0407 | .0703 |
|   | 8 | .0000 | .0000 | .0000 | .0000 | .0001 | .0004 | .0013 | .0035 | .0083 | .0176 |
|   | 9 | .0000 | .0000 | .0000 | .0000 | .0000 | .0000 | .0001 | .0003 | .0008 | .0020 |
| 10 | 0 | .5987 | .3487 | .1969 | .1074 | .0563 | .0282 | .0135 | .0060 | .0025 | .0010 |
|   | 1 | .3151 | .3874 | .3474 | .2684 | .1877 | .1211 | .0725 | .0403 | .0207 | .0098 |
|   | 2 | .0746 | .1937 | .2759 | .3020 | .2816 | .2335 | .1757 | .1209 | .0763 | .0439 |

## TABLE I: Binomial Probabilities (continued)

| n | x | .05 | .10 | .15 | .20 | .25 | .30 | .35 | .40 | .45 | .50 |
|---|---|-----|-----|-----|-----|-----|-----|-----|-----|-----|-----|
|   | 3 | .0105 | .0574 | .1298 | .2013 | .2503 | .2668 | .2522 | .2150 | .1665 | .1172 |
|   | 4 | .0010 | .0112 | .0401 | .0881 | .1460 | .2001 | .2377 | .2508 | .2384 | .2051 |
|   | 5 | .0001 | .0015 | .0085 | .0264 | .0584 | .1092 | .1536 | .2007 | .2340 | .2461 |
|   | 6 | .0000 | .0001 | .0012 | .0055 | .0162 | .0368 | .0689 | .1115 | .1596 | .2051 |
|   | 7 | .0000 | .0000 | .0001 | .0008 | .0031 | .0090 | .0212 | .0425 | .0746 | .1172 |
|   | 8 | .0000 | .0000 | .0000 | .0001 | .0004 | .0014 | .0043 | .0106 | .0229 | .0439 |
|   | 9 | .0000 | .0000 | .0000 | .0000 | .0000 | .0001 | .0005 | .0016 | .0042 | .0098 |
|   | 10 | .0000 | .0000 | .0000 | .0000 | .0000 | .0000 | .0000 | .0001 | .0003 | .0010 |
| 11 | 0 | .5688 | .3138 | .1673 | .0859 | .0422 | .0198 | .0088 | .0036 | .0014 | .0005 |
|   | 1 | .3293 | .3835 | .3248 | .2362 | .1549 | .0932 | .0518 | .0266 | .0125 | .0054 |
|   | 2 | .0867 | .2131 | .2866 | .2953 | .2581 | .1998 | .1395 | .0887 | .0513 | .0269 |
|   | 3 | .0137 | .0710 | .1517 | .2215 | .2581 | .2568 | .2254 | .1774 | .1259 | .0806 |
|   | 4 | .0014 | .0158 | .0536 | .1107 | .1721 | .2201 | .2428 | .2365 | .2060 | .1611 |
|   | 5 | .0001 | .0025 | .0132 | .0388 | .0803 | .1321 | .1830 | .2207 | .2360 | .2256 |
|   | 6 | .0000 | .0003 | .0023 | .0097 | .0268 | .0566 | .0985 | .1471 | .1931 | .2255 |
|   | 7 | .0000 | .0000 | .0003 | .0017 | .0064 | .0173 | .0379 | .0701 | .1128 | .1611 |
|   | 8 | .0000 | .0000 | .0000 | .0002 | .0011 | .0037 | .0102 | .0234 | .0462 | .0806 |
|   | 9 | .0000 | .0000 | .0000 | .0000 | .0001 | .0005 | .0018 | .0052 | .0126 | .0269 |
|   | 10 | .0000 | .0000 | .0000 | .0000 | .0000 | .0000 | .0002 | .0007 | .0021 | .0054 |
|   | 11 | .0000 | .0000 | .0000 | .0000 | .0000 | .0000 | .0000 | .0000 | .0002 | .0005 |
| 12 | 0 | .5404 | .2824 | .1422 | .0687 | .0317 | .0138 | .0057 | .0022 | .0008 | .0002 |
|   | 1 | .3413 | .3766 | .3012 | .2062 | .1267 | .0712 | .0368 | .0174 | .0075 | .0029 |
|   | 2 | .0988 | .2301 | .2924 | .2835 | .2323 | .1678 | .1088 | .0639 | .0339 | .0121 |
|   | 3 | .0173 | .0852 | .1720 | .2362 | .2581 | .2397 | .1954 | .1419 | .0923 | .0537 |
|   | 4 | .0021 | .0213 | .0683 | .1329 | .1936 | .2311 | .2367 | .2128 | .1700 | .1208 |
|   | 5 | .0002 | .0038 | .0193 | .0532 | .1032 | .1585 | .2039 | .2270 | .2225 | .1934 |
|   | 6 | .0000 | .0005 | .0040 | .0155 | .0401 | .0792 | .1281 | .1766 | .2124 | .2256 |
|   | 7 | .0000 | .0000 | .0006 | .0033 | .0115 | .0291 | .0591 | .1009 | .1489 | .1934 |
|   | 8 | .0000 | .0000 | .0001 | .0005 | .0024 | .0078 | .0199 | .0420 | .0762 | .1208 |
|   | 9 | .0000 | .0000 | .0000 | .0001 | .0004 | .0015 | .0048 | .0125 | .0277 | .0537 |
|   | 10 | .0000 | .0000 | .0000 | .0000 | .0000 | .0002 | .0008 | .0025 | .0068 | .0161 |
|   | 11 | .0000 | .0000 | .0000 | .0000 | .0000 | .0000 | .0001 | .0003 | .0010 | .0029 |
|   | 12 | .0000 | .0000 | .0000 | .0000 | .0000 | .0000 | .0000 | .0000 | .0001 | .0002 |
| 13 | 0 | .5133 | .2542 | .1209 | .0550 | .0238 | .0097 | .0037 | .0013 | .0004 | .0001 |
|   | 1 | .3512 | .3672 | .2774 | .1787 | .1029 | .0540 | .0259 | .0113 | .0045 | .0016 |
|   | 2 | .1109 | .2448 | .2937 | .2680 | .2059 | .1388 | .0836 | .0453 | .0220 | .0095 |
|   | 3 | .0214 | .0997 | .1900 | .2457 | .2517 | .2181 | .1651 | .1107 | .0660 | .0349 |
|   | 4 | .0028 | .0277 | .0838 | .1535 | .2097 | .2337 | .2222 | .1845 | .1350 | .0873 |
|   | 5 | .0003 | .0055 | .0266 | .0691 | .1258 | .1803 | .2154 | .2214 | .1989 | .1571 |
|   | 6 | .0000 | .0008 | .0063 | .0230 | .0559 | .1030 | .1546 | .1968 | .2169 | .2095 |
|   | 7 | .0000 | .0001 | .0011 | .0058 | .0186 | .0442 | .0833 | .1312 | .1775 | .2095 |
|   | 8 | .0000 | .0000 | .0001 | .0011 | .0047 | .0142 | .0336 | .0656 | .1089 | .1571 |
|   | 9 | .0000 | .0000 | .0000 | .0001 | .0009 | .0034 | .0101 | .0243 | .0495 | .0873 |
|   | 10 | .0000 | .0000 | .0000 | .0000 | .0001 | .0006 | .0022 | .0065 | .0162 | .0349 |
|   | 11 | .0000 | .0000 | .0000 | .0000 | .0000 | .0001 | .0003 | .0012 | .0036 | .0095 |
|   | 12 | .0000 | .0000 | .0000 | .0000 | .0000 | .0000 | .0000 | .0001 | .0005 | .0016 |
|   | 13 | .0000 | .0000 | .0000 | .0000 | .0000 | .0000 | .0000 | .0000 | .0000 | .0001 |
| 14 | 0 | .4877 | .2288 | .1028 | .0440 | .0178 | .0068 | .0024 | .0008 | .0002 | .0001 |
|   | 1 | .3593 | .3559 | .2539 | .1539 | .0832 | .0407 | .0181 | .0073 | .0027 | .0009 |
|   | 2 | .1229 | .2570 | .2912 | .2501 | .1802 | .1134 | .0634 | .0317 | .0141 | .0056 |
|   | 3 | .0259 | .1142 | .2056 | .2501 | .2402 | .1943 | .1366 | .0845 | .0462 | .0222 |
|   | 4 | .0037 | .0349 | .0998 | .1720 | .2202 | .2290 | .2022 | .1549 | .1040 | .0611 |
|   | 5 | .0004 | .0078 | .0352 | .0860 | .1468 | .1963 | .2178 | .2066 | .1701 | .1222 |
|   | 6 | .0000 | .0013 | .0093 | .0322 | .0734 | .1262 | .1759 | .2066 | .2088 | .1833 |
|   | 7 | .0000 | .0002 | .0019 | .0092 | .0280 | .0618 | .1082 | .1574 | .1952 | .2095 |
|   | 8 | .0000 | .0000 | .0003 | .0020 | .0082 | .0232 | .0510 | .0918 | .1398 | .1833 |
|   | 9 | .0000 | .0000 | .0000 | .0003 | .0018 | .0066 | .0183 | .0408 | .0762 | .1222 |
|   | 10 | .0000 | .0000 | .0000 | .0000 | .0003 | .0014 | .0049 | .0136 | .0312 | .0611 |

## TABLE I:  Binomial Probabilities (continued)

| n | x | .05 | .10 | .15 | .20 | .25 | .30 | .35 | .40 | .45 | .50 |
|---|---|-----|-----|-----|-----|-----|-----|-----|-----|-----|-----|
|    | 11 | .0000 | .0000 | .0000 | .0000 | .0000 | .0002 | .0010 | .0033 | .0093 | .0222 |
|    | 12 | .0000 | .0000 | .0000 | .0000 | .0000 | .0000 | .0001 | .0005 | .0019 | .0056 |
|    | 13 | .0000 | .0000 | .0000 | .0000 | .0000 | .0000 | .0000 | .0001 | .0002 | .0009 |
|    | 14 | .0000 | .0000 | .0000 | .0000 | .0000 | .0000 | .0000 | .0000 | .0000 | .0001 |
| 15 | 0 | .4633 | .2059 | .0874 | .0352 | .0134 | .0047 | .0016 | .0005 | .0001 | .0000 |
|    | 1 | .3658 | .3432 | .2312 | .1319 | .0668 | .0305 | .0126 | .0047 | .0016 | .0005 |
|    | 2 | .1348 | .2669 | .2856 | .2309 | .1559 | .0916 | .0476 | .0219 | .0090 | .0032 |
|    | 3 | .0307 | .1285 | .2184 | .2501 | .2252 | .1700 | .1110 | .0634 | .0318 | .0139 |
|    | 4 | .0049 | .0428 | .1156 | .1876 | .2252 | .2186 | .1792 | .1268 | .0780 | .0417 |
|    | 5 | .0006 | .0105 | .0449 | .1032 | .1651 | .2061 | .2123 | .1859 | .1404 | .0916 |
|    | 6 | .0000 | .0019 | .0132 | .0430 | .0917 | .1472 | .1906 | .2066 | .1914 | .1527 |
|    | 7 | .0000 | .0003 | .0030 | .0138 | .0393 | .0811 | .1319 | .1771 | .2013 | .1964 |
|    | 8 | .0000 | .0000 | .0005 | .0035 | .0131 | .0348 | .0710 | .1181 | .1647 | .1964 |
|    | 9 | .0000 | .0000 | .0001 | .0007 | .0034 | .0116 | .0298 | .0612 | .1048 | .1527 |
|    | 10 | .0000 | .0000 | .0000 | .0001 | .0007 | .0030 | .0096 | .0245 | .0515 | .0916 |
|    | 11 | .0000 | .0000 | .0000 | .0000 | .0001 | .0006 | .0024 | .0074 | .0191 | .0417 |
|    | 12 | .0000 | .0000 | .0000 | .0000 | .0000 | .0001 | .0004 | .0016 | .0052 | .0139 |
|    | 13 | .0000 | .0000 | .0000 | .0000 | .0000 | .0000 | .0001 | .0003 | .0010 | .0032 |
|    | 14 | .0000 | .0000 | .0000 | .0000 | .0000 | .0000 | .0000 | .0000 | .0001 | .0005 |
|    | 15 | .0000 | .0000 | .0000 | .0000 | .0000 | .0000 | .0000 | .0000 | .0000 | .0000 |
| 16 | 0 | .4401 | .1853 | .0743 | .0281 | .0100 | .0033 | .0010 | .0003 | .0001 | .0000 |
|    | 1 | .3706 | .3204 | .2097 | .1126 | .0535 | .0228 | .0087 | .0030 | .0009 | .0002 |
|    | 2 | .1463 | .2745 | .2775 | .2111 | .1336 | .0732 | .0353 | .0150 | .0056 | .0018 |
|    | 3 | .0359 | .1423 | .2285 | .2463 | .2079 | .1465 | .0888 | .0468 | .0215 | .0085 |
|    | 4 | .0061 | .0514 | .1311 | .2001 | .2252 | .2040 | .1553 | .1014 | .0572 | .0278 |
|    | 5 | .0008 | .0137 | .0555 | .1201 | .1802 | .2099 | .2008 | .1623 | .1123 | .0667 |
|    | 6 | .0001 | .0028 | .0180 | .0550 | .1101 | .1649 | .1982 | .1983 | .1684 | .1222 |
|    | 7 | .0000 | .0004 | .0045 | .0197 | .0524 | .1010 | .1524 | .1889 | .1969 | .1746 |
|    | 8 | .0000 | .0001 | .0009 | .0055 | .0197 | .0487 | .0923 | .1417 | .1812 | .1964 |
|    | 9 | .0000 | .0000 | .0001 | .0012 | .0058 | .0185 | .0442 | .0840 | .1318 | .1746 |
|    | 10 | .0000 | .0000 | .0000 | .0002 | .0014 | .0056 | .0167 | .0392 | .0755 | .1222 |
|    | 11 | .0000 | .0000 | .0000 | .0000 | .0002 | .0013 | .0049 | .0142 | .0337 | .0667 |
|    | 12 | .0000 | .0000 | .0000 | .0000 | .0000 | .0002 | .0011 | .0040 | .0115 | .0278 |
|    | 13 | .0000 | .0000 | .0000 | .0000 | .0000 | .0000 | .0002 | .0008 | .0029 | .0085 |
|    | 14 | .0000 | .0000 | .0000 | .0000 | .0000 | .0000 | .0000 | .0001 | .0005 | .0018 |
|    | 15 | .0000 | .0000 | .0000 | .0000 | .0000 | .0000 | .0000 | .0000 | .0001 | .0002 |
|    | 16 | .0000 | .0000 | .0000 | .0000 | .0000 | .0000 | .0000 | .0000 | .0000 | .0000 |
| 17 | 0 | .4181 | .1668 | .0631 | .0225 | .0075 | .0023 | .0007 | .0002 | .0000 | .0000 |
|    | 1 | .3741 | .3150 | .1893 | .0957 | .0426 | .0169 | .0060 | .0019 | .0005 | .0001 |
|    | 2 | .1575 | .2800 | .2673 | .1914 | .1136 | .0581 | .0260 | .0102 | .0035 | .0010 |
|    | 3 | .0415 | .1556 | .2359 | .2393 | .1893 | .1245 | .0701 | .0341 | .0144 | .0052 |
|    | 4 | .0076 | .0605 | .1457 | .2093 | .2209 | .1868 | .1320 | .0796 | .0411 | .0182 |
|    | 5 | .0010 | .0175 | .0668 | .1361 | .1914 | .2081 | .1849 | .1379 | .0875 | .0472 |
|    | 6 | .0001 | .0039 | .0236 | .0680 | .1276 | .1784 | .1991 | .1839 | .1432 | .0944 |
|    | 7 | .0000 | .0007 | .0065 | .0267 | .0668 | .1201 | .1685 | .1927 | .1841 | .1484 |
|    | 8 | .0000 | .0001 | .0014 | .0084 | .0279 | .0644 | .1134 | .1606 | .1883 | .1855 |
|    | 9 | .0000 | .0000 | .0003 | .0021 | .0093 | .0276 | .0611 | .1070 | .1540 | .1855 |
|    | 10 | .0000 | .0000 | .0000 | .0004 | .0025 | .0095 | .0263 | .0571 | .1008 | .1484 |
|    | 11 | .0000 | .0000 | .0000 | .0001 | .0005 | .0026 | .0090 | .0242 | .0525 | .0944 |
|    | 12 | .0000 | .0000 | .0000 | .0000 | .0001 | .0006 | .0024 | .0081 | .0215 | .0472 |
|    | 13 | .0000 | .0000 | .0000 | .0000 | .0000 | .0001 | .0005 | .0021 | .0068 | .0182 |
|    | 14 | .0000 | .0000 | .0000 | .0000 | .0000 | .0000 | .0001 | .0004 | .0016 | .0052 |
|    | 15 | .0000 | .0000 | .0000 | .0000 | .0000 | .0000 | .0000 | .0001 | .0003 | .0010 |
|    | 16 | .0000 | .0000 | .0000 | .0000 | .0000 | .0000 | .0000 | .0000 | .0000 | .0001 |
|    | 17 | .0000 | .0000 | .0000 | .0000 | .0000 | .0000 | .0000 | .0000 | .0000 | .0000 |
| 18 | 0 | .3972 | .1501 | .0536 | .0180 | .0056 | .0016 | .0004 | .0001 | .0000 | .0000 |
|    | 1 | .3763 | .3002 | .1704 | .0811 | .0338 | .0126 | .0042 | .0012 | .0003 | .0001 |
|    | 2 | .1683 | .2835 | .2556 | .1723 | .0958 | .0458 | .0190 | .0069 | .0022 | .0006 |

## TABLE III:  Random Numbers Table

| | | | | | | | | | |
|---|---|---|---|---|---|---|---|---|---|
| 04839 | 96423 | 24878 | 82651 | 66566 | 14778 | 76797 | 14780 | 13300 | 87074 |
| 68086 | 26432 | 46901 | 29849 | 89768 | 81536 | 86645 | 12659 | 92259 | 57102 |
| 39064 | 66432 | 84673 | 40027 | 32832 | 61362 | 98947 | 96067 | 64760 | 64584 |
| 25669 | 26422 | 44407 | 44048 | 37937 | 63904 | 45766 | 66134 | 75470 | 66520 |
| 64117 | 94305 | 26766 | 25940 | 39972 | 22209 | 71500 | 64568 | 91402 | 42416 |
| 87917 | 77341 | 42206 | 35126 | 74087 | 99547 | 81817 | 42607 | 43808 | 76655 |
| 62797 | 56170 | 86324 | 88072 | 76222 | 36086 | 84637 | 93161 | 76038 | 65855 |
| 95876 | 55293 | 18988 | 27354 | 26575 | 08625 | 40801 | 59920 | 29841 | 80140 |
| 29888 | 88604 | 67917 | 48708 | 18912 | 82271 | 65424 | 69774 | 33611 | 54262 |
| 73577 | 12908 | 30883 | 18317 | 28290 | 35797 | 05998 | 41688 | 34952 | 37888 |
| 27958 | 30134 | 04024 | 86385 | 29880 | 99730 | 55536 | 84855 | 29080 | 09250 |
| 90999 | 49127 | 20044 | 59931 | 06115 | 20542 | 18059 | 02008 | 73708 | 83517 |
| 18845 | 49618 | 02304 | 51038 | 20655 | 58727 | 28168 | 15475 | 56942 | 53389 |
| 94824 | 78171 | 84610 | 82834 | 09922 | 25417 | 44137 | 48413 | 25555 | 21246 |
| 35615 | 81263 | 39667 | 47358 | 56873 | 56307 | 61607 | 49518 | 89656 | 20103 |
| 33362 | 64270 | 01638 | 92477 | 66969 | 98420 | 04880 | 45585 | 46565 | 04102 |
| 88720 | 82765 | 34476 | 17032 | 87589 | 40836 | 32427 | 70002 | 70663 | 88863 |
| 39475 | 46473 | 23219 | 53416 | 94970 | 25832 | 69975 | 94884 | 19661 | 72828 |
| 06990 | 67245 | 68350 | 82948 | 11398 | 42878 | 80287 | 88267 | 47363 | 46634 |
| 40980 | 07391 | 58745 | 25774 | 22987 | 80059 | 39911 | 96189 | 41151 | 14222 |
| 83974 | 29992 | 65831 | 38857 | 50490 | 83765 | 55657 | 14361 | 31720 | 57375 |
| 33339 | 31926 | 14883 | 24413 | 59744 | 92351 | 97473 | 89286 | 35931 | 04110 |
| 31662 | 25388 | 61642 | 34072 | 81249 | 35648 | 56891 | 69352 | 48373 | 45578 |
| 93526 | 70765 | 10592 | 04542 | 76463 | 54328 | 02349 | 17247 | 28865 | 14777 |
| 20492 | 38391 | 91132 | 21999 | 59516 | 81652 | 27195 | 48223 | 46751 | 22923 |
| 04153 | 53381 | 79401 | 21438 | 83035 | 92350 | 36693 | 31238 | 59649 | 91754 |
| 05520 | 91962 | 04739 | 13092 | 97662 | 24882 | 94730 | 06496 | 35090 | 04822 |
| 47498 | 87637 | 99016 | 71060 | 88824 | 71013 | 18735 | 20286 | 23153 | 72924 |
| 23167 | 49323 | 45021 | 33132 | 12544 | 41035 | 80780 | 45393 | 44812 | 12515 |
| 23792 | 14422 | 15059 | 45799 | 22716 | 19792 | 09983 | 74353 | 68668 | 30429 |
| 85900 | 98275 | 32388 | 52390 | 16815 | 69298 | 82732 | 38480 | 73817 | 32523 |
| 42559 | 78985 | 05300 | 22164 | 24369 | 54224 | 35083 | 19687 | 11052 | 91491 |
| 14349 | 82674 | 66523 | 44133 | 00697 | 35552 | 35970 | 19124 | 63318 | 29686 |
| 17403 | 53363 | 44167 | 64486 | 64758 | 75366 | 76554 | 31601 | 12614 | 33072 |
| 23632 | 27889 | 47914 | 02584 | 37680 | 20801 | 72152 | 39339 | 34806 | 08930 |

Source: Reprinted by permission from *First Course in Probability* by Sheldon Ross. New York: Macmillan Publishing Co., Inc. (Copyright © 1976 by Sheldon Ross) Page 290.

## TABLE IV: Poisson Probabilities, $p(x; \lambda) = \dfrac{\lambda^x e^{-\lambda}}{x!}$

| $x$ | 0.1 | 0.2 | 0.3 | 0.4 | 0.5 | 0.6 | 0.7 | 0.8 | 0.9 | 1.0 |
|---|---|---|---|---|---|---|---|---|---|---|
| 0 | .9048 | .8187 | .7408 | .6703 | .6065 | .5488 | .4966 | .4493 | .4066 | .3679 |
| 1 | .0905 | .1637 | .2222 | .2681 | .3033 | .3293 | .3476 | .3595 | .3659 | .3679 |
| 2 | .0045 | .0164 | .0333 | .0536 | .0758 | .0988 | .1217 | .1438 | .1647 | .1839 |
| 3 | .0002 | .0011 | .0033 | .0072 | .0126 | .0198 | .0284 | .0383 | .0494 | .0613 |
| 4 | .0000 | .0001 | .0002 | .0007 | .0016 | .0030 | .0050 | .0077 | .0111 | .0153 |
| 5 | .0000 | .0000 | .0000 | .0001 | .0002 | .0004 | .0007 | .0012 | .0020 | .0031 |
| 6 | .0000 | .0000 | .0000 | .0000 | .0000 | .0000 | .0000 | .0002 | .0003 | .0005 |
| 7 | .0000 | .0000 | .0000 | .0000 | .0000 | .0000 | .0000 | .0000 | .0000 | .0001 |

| $x$ | 1.1 | 1.2 | 1.3 | 1.4 | 1.5 | 1.6 | 1.7 | 1.8 | 1.9 | 2.0 |
|---|---|---|---|---|---|---|---|---|---|---|
| 0 | .3329 | .3012 | .2725 | .2466 | .2231 | .2019 | .1827 | .1653 | .1496 | .1353 |
| 1 | .3662 | .3614 | .3543 | .3452 | .3347 | .3230 | .3106 | .2975 | .2842 | .2707 |
| 2 | .2014 | .2169 | .2303 | .2417 | .2510 | .2584 | .2640 | .2678 | .2700 | .2707 |
| 3 | .0738 | .0867 | .0998 | .1128 | .1255 | .1378 | .1496 | .1607 | .1710 | .1804 |
| 4 | .0203 | .0260 | .0324 | .0395 | .0471 | .0551 | .0636 | .0723 | .0812 | .0902 |
| 5 | .0045 | .0062 | .0084 | .0111 | .0141 | .0176 | .0216 | .0260 | .0309 | .0361 |
| 6 | .0008 | .0012 | .0018 | .0026 | .0035 | .0047 | .0061 | .0078 | .0098 | .0120 |
| 7 | .0001 | .0002 | .0003 | .0005 | .0008 | .0011 | .0015 | .0020 | .0027 | .0034 |
| 8 | .0000 | .0000 | .0001 | .0001 | .0001 | .0002 | .0003 | .0005 | .0006 | .0009 |
| 9 | .0000 | .0000 | .0000 | .0000 | .0000 | .0000 | .0001 | .0001 | .0001 | .0002 |

| $x$ | 2.1 | 2.2 | 2.3 | 2.4 | 2.5 | 2.6 | 2.7 | 2.8 | 2.9 | 3.0 |
|---|---|---|---|---|---|---|---|---|---|---|
| 0 | .1225 | .1108 | .1003 | .0907 | .0821 | .0743 | .0672 | .0608 | .0550 | .0498 |
| 1 | .2572 | .2438 | .2306 | .2177 | .2052 | .1931 | .1815 | .1703 | .1596 | .1494 |
| 2 | .2700 | .2681 | .2652 | .2613 | .2565 | .2510 | .2450 | .2384 | .2314 | .2240 |
| 3 | .1890 | .1966 | .2033 | .2090 | .2138 | .2176 | .2205 | .2225 | .2237 | .2240 |
| 4 | .0992 | .1082 | .1169 | .1254 | .1336 | .1414 | .1488 | .1557 | .1622 | .1680 |
| 5 | .0417 | .0476 | .0538 | .0602 | .0668 | .0735 | .0804 | .0872 | .0940 | .1008 |
| 6 | .0146 | .0174 | .0206 | .0241 | .0278 | .0319 | .0362 | .0407 | .0455 | .0504 |
| 7 | .0044 | .0055 | .0068 | .0083 | .0099 | .0118 | .0139 | .0163 | .0188 | .0216 |
| 8 | .0011 | .0015 | .0019 | .0025 | .0031 | .0038 | .0047 | .0057 | .0068 | .0081 |
| 9 | .0003 | .0004 | .0005 | .0007 | .0009 | .0011 | .0014 | .0018 | .0022 | .0027 |
| 10 | .0001 | .0001 | .0001 | .0002 | .0002 | .0003 | .0004 | .0005 | .0006 | .0008 |
| 11 | .0000 | .0000 | .0000 | .0000 | .0000 | .0001 | .0001 | .0001 | .0002 | .0002 |
| 12 | .0000 | .0000 | .0000 | .0000 | .0000 | .0000 | .0000 | .0000 | .0000 | .0001 |

**TABLE IV: Poisson Probabilities (continued)**

| | | | | | λ | | | | | |
|---|---|---|---|---|---|---|---|---|---|---|
| x | 3.1 | 3.2 | 3.3 | 3.4 | 3.5 | 3.6 | 3.7 | 3.8 | 3.9 | 4.0 |
| 0 | .0450 | .0408 | .0369 | .0334 | .0302 | .0273 | .0247 | .0224 | .0202 | .0183 |
| 1 | .1397 | .1304 | .1217 | .1135 | .1057 | .0984 | .0915 | .0850 | .0789 | .0733 |
| 2 | .2165 | .2087 | .2008 | .1929 | .1850 | .1771 | .1692 | .1615 | .1539 | .1465 |
| 3 | .2237 | .2226 | .2209 | .2186 | .2158 | .2125 | .2087 | .2046 | .2001 | .1954 |
| 4 | .1734 | .1781 | .1823 | .1858 | .1888 | .1912 | .1931 | .1944 | .1951 | .1954 |
| 5 | .1075 | .1140 | .1203 | .1264 | .1322 | .1377 | .1429 | .1477 | .1522 | .1563 |
| 6 | .0555 | .0608 | .0662 | .0716 | .0771 | .0826 | .0881 | .0936 | .0989 | .1042 |
| 7 | .0246 | .0278 | .0312 | .0348 | .0385 | .0425 | .0466 | .0508 | .0551 | .0595 |
| 8 | .0095 | .0111 | .0129 | .0148 | .0169 | .0191 | .0215 | .0241 | .0269 | .0298 |
| 9 | .0033 | .0040 | .0047 | .0056 | .0066 | .0076 | .0089 | .0102 | .0116 | .0132 |
| 10 | .0010 | .0013 | .0016 | .0019 | .0023 | .0028 | .0033 | .0039 | .0045 | .0053 |
| 11 | .0003 | .0004 | .0005 | .0006 | .0007 | .0009 | .0011 | .0013 | .0016 | .0019 |
| 12 | .0001 | .0001 | .0001 | .0002 | .0002 | .0003 | .0003 | .0004 | .0005 | .0006 |
| 13 | .0000 | .0000 | .0000 | .0000 | .0001 | .0001 | .0001 | .0001 | .0002 | .0002 |
| 14 | .0000 | .0000 | .0000 | .0000 | .0000 | .0000 | .0000 | .0000 | .0000 | .0001 |

| | | | | | λ | | | | | |
|---|---|---|---|---|---|---|---|---|---|---|
| x | 4.1 | 4.2 | 4.3 | 4.4 | 4.5 | 4.6 | 4.7 | 4.8 | 4.9 | 5.0 |
| 0 | .0166 | .0150 | .0136 | .0123 | .0111 | .0101 | .0091 | .0082 | .0074 | .0067 |
| 1 | .0679 | .0630 | .0583 | .0540 | .0500 | .0462 | .0427 | .0395 | .0365 | .0337 |
| 2 | .1393 | .1323 | .1254 | .1188 | .1125 | .1063 | .1005 | .0948 | .0894 | .0842 |
| 3 | .1904 | .1852 | .1798 | .1743 | .1687 | .1631 | .1574 | .1517 | .1460 | .1404 |
| 4 | .1951 | .1944 | .1933 | .1917 | .1898 | .1875 | .1849 | .1820 | .1789 | .1755 |
| 5 | .1600 | .1633 | .1662 | .1687 | .1708 | .1725 | .1738 | .1747 | .1753 | .1755 |
| 6 | .1093 | .1143 | .1191 | .1237 | .1281 | .1323 | .1362 | .1398 | .1432 | .1462 |
| 7 | .0640 | .0686 | .0732 | .0778 | .0824 | .0869 | .0914 | .0959 | .1002 | .1044 |
| 8 | .0328 | .0360 | .0393 | .0428 | .0463 | .0500 | .0537 | .0575 | .0614 | .0653 |
| 9 | .0150 | .0168 | .0188 | .0209 | .0232 | .0255 | .0280 | .0307 | .0334 | .0363 |
| 10 | .0061 | .0071 | .0081 | .0092 | .0104 | .0118 | .0132 | .0147 | .0164 | .0181 |
| 11 | .0023 | .0027 | .0032 | .0037 | .0043 | .0049 | .0056 | .0064 | .0073 | .0082 |
| 12 | .0008 | .0009 | .0011 | .0014 | .0016 | .0019 | .0022 | .0026 | .0030 | .0034 |
| 13 | .0002 | .0003 | .0004 | .0005 | .0006 | .0007 | .0008 | .0009 | .0011 | .0013 |
| 14 | .0001 | .0001 | .0001 | .0001 | .0002 | .0002 | .0003 | .0003 | .0004 | .0005 |
| 15 | .0000 | .0000 | .0000 | .0000 | .0001 | .0001 | .0001 | .0001 | .0001 | .0002 |

| | | | | | λ | | | | | |
|---|---|---|---|---|---|---|---|---|---|---|
| x | 5.1 | 5.2 | 5.3 | 5.4 | 5.5 | 5.6 | 5.7 | 5.8 | 5.9 | 6.0 |
| 0 | .0061 | .0055 | .0050 | .0045 | .0041 | .0037 | .0033 | .0030 | .0027 | .0025 |
| 1 | .0311 | .0287 | .0265 | .0244 | .0225 | .0207 | .0191 | .0176 | .0162 | .0149 |
| 2 | .0793 | .0746 | .0701 | .0659 | .0618 | .0580 | .0544 | .0509 | .0477 | .0446 |
| 3 | .1348 | .1293 | .1239 | .1185 | .1133 | .1082 | .1033 | .0985 | .0938 | .0892 |
| 4 | .1719 | .1681 | .1641 | .1600 | .1558 | .1515 | .1472 | .1428 | .1383 | .1339 |
| 5 | .1753 | .1748 | .1740 | .1728 | .1714 | .1697 | .1678 | .1656 | .1632 | .1606 |
| 6 | .1490 | .1515 | .1537 | .1555 | .1571 | .1584 | .1594 | .1601 | .1605 | .1606 |
| 7 | .1086 | .1125 | .1163 | .1200 | .1234 | .1267 | .1298 | .1326 | .1353 | .1377 |
| 8 | .0692 | .0731 | .0771 | .0810 | .0849 | .0887 | .0925 | .0962 | .0998 | .1033 |
| 9 | .0392 | .0423 | .0454 | .0486 | .0519 | .0552 | .0586 | .0620 | .0654 | .0688 |
| 10 | .0200 | .0220 | .0241 | .0262 | .0285 | .0309 | .0334 | .0359 | .0386 | .0413 |
| 11 | .0093 | .0104 | .0116 | .0129 | .0143 | .0157 | .0173 | .0190 | .0207 | .0225 |
| 12 | .0039 | .0045 | .0051 | .0058 | .0065 | .0073 | .0082 | .0092 | .0102 | .0113 |
| 13 | .0015 | .0018 | .0021 | .0024 | .0028 | .0032 | .0036 | .0041 | .0046 | .0052 |
| 14 | .0006 | .0007 | .0008 | .0009 | .0011 | .0013 | .0015 | .0017 | .0019 | .0022 |
| 15 | .0002 | .0002 | .0003 | .0003 | .0004 | .0005 | .0006 | .0007 | .0008 | .0009 |
| 16 | .0001 | .0001 | .0001 | .0001 | .0001 | .0002 | .0002 | .0002 | .0003 | .0003 |
| 17 | .0000 | .0000 | .0000 | .0000 | .0000 | .0001 | .0001 | .0001 | .0001 | .0001 |

**TABLE IV:  Poisson Probabilities (continued)**

λ

| x | 6.1 | 6.2 | 6.3 | 6.4 | 6.5 | 6.6 | 6.7 | 6.8 | 6.9 | 7.0 |
|---|-----|-----|-----|-----|-----|-----|-----|-----|-----|-----|
| 0 | .0022 | .0020 | .0018 | .0017 | .0015 | .0014 | .0012 | .0011 | .0010 | .0009 |
| 1 | .0137 | .0126 | .0116 | .0106 | .0098 | .0090 | .0082 | .0076 | .0070 | .0064 |
| 2 | .0417 | .0390 | .0364 | .0340 | .0318 | .0296 | .0276 | .0258 | .0240 | .0223 |
| 3 | .0848 | .0806 | .0765 | .0726 | .0688 | .0652 | .0617 | .0584 | .0552 | .0521 |
| 4 | .1294 | .1249 | .1205 | .1162 | .1118 | .1076 | .1034 | .0992 | .0952 | .0912 |
| 5 | .1579 | .1549 | .1519 | .1487 | .1454 | .1420 | .1385 | .1349 | .1314 | .1277 |
| 6 | .1605 | .1601 | .1595 | .1586 | .1575 | .1562 | .1546 | .1529 | .1511 | .1490 |
| 7 | .1399 | .1418 | .1435 | .1450 | .1462 | .1472 | .1480 | .1486 | .1489 | .1490 |
| 8 | .1066 | .1099 | .1130 | .1160 | .1188 | .1215 | .1240 | .1263 | .1284 | .1304 |
| 9 | .0723 | .0757 | .0791 | .0825 | .0858 | .0891 | .0923 | .0954 | .0985 | .1014 |
| 10 | .0441 | .0469 | .0498 | .0528 | .0558 | .0588 | .0618 | .0649 | .0679 | .0710 |
| 11 | .0245 | .0265 | .0285 | .0307 | .0330 | .0353 | .0377 | .0401 | .0426 | .0452 |
| 12 | .0124 | .0137 | .0150 | .0164 | .0179 | .0194 | .0210 | .0227 | .0245 | .0264 |
| 13 | .0058 | .0065 | .0073 | .0081 | .0089 | .0098 | .0108 | .0119 | .0130 | .0142 |
| 14 | .0025 | .0029 | .0033 | .0037 | .0041 | .0046 | .0052 | .0058 | .0064 | .0071 |
| 15 | .0010 | .0012 | .0014 | .0016 | .0018 | .0020 | .0023 | .0026 | .0029 | .0033 |
| 16 | .0004 | .0005 | .0005 | .0006 | .0007 | .0008 | .0010 | .0011 | .0013 | .0014 |
| 17 | .0001 | .0002 | .0002 | .0002 | .0003 | .0003 | .0004 | .0004 | .0005 | .0006 |
| 18 | .0000 | .0001 | .0001 | .0001 | .0001 | .0001 | .0001 | .0002 | .0002 | .0002 |
| 19 | .0000 | .0000 | .0000 | .0000 | .0000 | .0000 | .0000 | .0001 | .0001 | .0001 |

λ

| x | 7.1 | 7.2 | 7.3 | 7.4 | 7.5 | 7.6 | 7.7 | 7.8 | 7.9 | 8.0 |
|---|-----|-----|-----|-----|-----|-----|-----|-----|-----|-----|
| 0 | .0008 | .0007 | .0007 | .0006 | .0006 | .0005 | .0005 | .0004 | .0004 | .0003 |
| 1 | .0059 | .0054 | .0049 | .0045 | .0041 | .0038 | .0035 | .0032 | .0029 | .0027 |
| 2 | .0208 | .0198 | .0180 | .0167 | .0156 | .0145 | .0134 | .0125 | .0116 | .0107 |
| 3 | .0492 | .0464 | .0438 | .0413 | .0389 | .0366 | .0345 | .0324 | .0305 | .0286 |
| 4 | .0874 | .0836 | .0799 | .0764 | .0729 | .0696 | .0663 | .0632 | .0602 | .0573 |
| 5 | .1241 | .1204 | .1167 | .1130 | .1094 | .1057 | .1021 | .0986 | .0951 | .0916 |
| 6 | .1468 | .1445 | .1420 | .1394 | .1367 | .1339 | .1311 | .1282 | .1252 | .1221 |
| 7 | .1489 | .1486 | .1481 | .1474 | .1465 | .1454 | .1442 | .1428 | .1413 | .1396 |
| 8 | .1321 | .1337 | .1351 | .1363 | .1373 | .1382 | .1388 | .1392 | .1395 | .1396 |
| 9 | .1042 | .1070 | .1096 | .1121 | .1144 | .1167 | .1187 | .1207 | .1224 | .1241 |
| 10 | .0740 | .0770 | .0800 | .0829 | .0858 | .0887 | .0914 | .0941 | .0967 | .0993 |
| 11 | .0478 | .0504 | .0531 | .0558 | .0585 | .0613 | .0640 | .0667 | .0695 | .0722 |
| 12 | .0283 | .0303 | .0323 | .0344 | .0366 | .0388 | .0411 | .0434 | .0457 | .0481 |
| 13 | .0154 | .0168 | .0181 | .0196 | .0211 | .0227 | .0243 | .0260 | .0278 | .0296 |
| 14 | .0078 | .0086 | .0095 | .0104 | .0113 | .0123 | .0134 | .0145 | .0157 | .0169 |
| 15 | .0037 | .0041 | .0046 | .0051 | .0057 | .0062 | .0069 | .0075 | .0083 | .0090 |
| 16 | .0016 | .0019 | .0021 | .0024 | .0026 | .0030 | .0033 | .0037 | .0041 | .0045 |
| 17 | .0007 | .0008 | .0009 | .0010 | .0012 | .0013 | .0015 | .0017 | .0019 | .0021 |
| 18 | .0003 | .0003 | .0004 | .0004 | .0005 | .0006 | .0006 | .0007 | .0008 | .0009 |
| 19 | .0001 | .0001 | .0001 | .0002 | .0002 | .0002 | .0003 | .0003 | .0003 | .0004 |
| 20 | .0000 | .0000 | .0001 | .0001 | .0001 | .0001 | .0001 | .0001 | .0001 | .0002 |
| 21 | .0000 | .0000 | .0000 | .0000 | .0000 | .0000 | .0000 | .0000 | .0001 | .0001 |

**TABLE IV:  Poisson Probabilities (continued)**

$\lambda$

| x | 8.1 | 8.2 | 8.3 | 8.4 | 8.5 | 8.6 | 8.7 | 8.8 | 8.9 | 9.0 |
|---|-----|-----|-----|-----|-----|-----|-----|-----|-----|-----|
| 0 | .0003 | .0003 | .0002 | .0002 | .0002 | .0002 | .0002 | .0002 | .0001 | .0001 |
| 1 | .0025 | .0023 | .0021 | .0019 | .0017 | .0016 | .0014 | .0013 | .0012 | .0011 |
| 2 | .0100 | .0092 | .0086 | .0079 | .0074 | .0068 | .0063 | .0058 | .0054 | .0050 |
| 3 | .0269 | .0252 | .0237 | .0222 | .0208 | .0195 | .0183 | .0171 | .0160 | .0150 |
| 4 | .0544 | .0517 | .0419 | .0466 | .0443 | .0420 | .0398 | .0377 | .0357 | .0337 |
| 5 | .0882 | .0849 | .0816 | .0784 | .0752 | .0722 | .0692 | .0663 | .0635 | .0607 |
| 6 | .1191 | .1160 | .1128 | .1097 | .1066 | .1034 | .1003 | .0972 | .0941 | .0911 |
| 7 | .1378 | .1358 | .1338 | .1317 | .1294 | .1271 | .1247 | .1222 | .1197 | .1171 |
| 8 | .1395 | .1392 | .1388 | .1382 | .1375 | .1366 | .1356 | .1344 | .1332 | .1318 |
| 9 | .1256 | .1269 | .1280 | .1290 | .1299 | .1306 | .1311 | .1315 | .1317 | .1318 |
| 10 | .1017 | .1040 | .1063 | .1084 | .1104 | .1123 | .1140 | .1157 | .1172 | .1186 |
| 11 | .0749 | .0776 | .0802 | .0828 | .0853 | .0878 | .0902 | .0925 | .0948 | .0970 |
| 12 | .0505 | .0530 | .0555 | .0579 | .0604 | .0629 | .0654 | .0679 | .0703 | .0728 |
| 13 | .0315 | .0334 | .0354 | .0374 | .0395 | .0416 | .0438 | .0459 | .0481 | .0504 |
| 14 | .0182 | .0196 | .0210 | .0225 | .0240 | .0256 | .0272 | .0289 | .0306 | .0324 |
| 15 | .0098 | .0107 | .0116 | .0126 | .0136 | .0147 | .0158 | .0169 | .0182 | .0194 |
| 16 | .0050 | .0055 | .0060 | .0066 | .0072 | .0079 | .0086 | .0093 | .0101 | .0109 |
| 17 | .0024 | .0026 | .0029 | .0033 | .0036 | .0040 | .0044 | .0048 | .0053 | .0058 |
| 18 | .0011 | .0012 | .0014 | .0015 | .0017 | .0019 | .0021 | .0024 | .0026 | .0029 |
| 19 | .0005 | .0005 | .0006 | .0007 | .0008 | .0009 | .0010 | .0011 | .0012 | .0014 |
| 20 | .0002 | .0002 | .0002 | .0003 | .0003 | .0004 | .0004 | .0005 | .0005 | .0006 |
| 21 | .0001 | .0001 | .0001 | .0001 | .0001 | .0002 | .0002 | .0002 | .0002 | .0003 |
| 22 | .0000 | .0000 | .0000 | .0000 | .0001 | .0001 | .0001 | .0001 | .0001 | .0001 |

$\lambda$

| x | 9.1 | 9.2 | 9.3 | 9.4 | 9.5 | 9.6 | 9.7 | 9.8 | 9.9 | 10 |
|---|-----|-----|-----|-----|-----|-----|-----|-----|-----|-----|
| 0 | .0001 | .0001 | .0001 | .0001 | .0001 | .0001 | .0001 | .0001 | .0001 | .0000 |
| 1 | .0010 | .0009 | .0009 | .0008 | .0007 | .0007 | .0006 | .0005 | .0005 | .0005 |
| 2 | .0046 | .0043 | .0040 | .0037 | .0034 | .0031 | .0029 | .0027 | .0025 | .0023 |
| 3 | .0140 | .0131 | .0123 | .0115 | .0107 | .0100 | .0093 | .0087 | .0081 | .0076 |
| 4 | .0319 | .0302 | .0285 | .0269 | .0254 | .0240 | .0226 | .0213 | .0201 | .0189 |
| 5 | .0581 | .0555 | .0530 | .0506 | .0483 | .0460 | .0439 | .0418 | .0398 | .0378 |
| 6 | .0881 | .0851 | .0822 | .0793 | .0764 | .0736 | .0709 | .0682 | .0656 | .0631 |
| 7 | .1145 | .1118 | .1091 | .1064 | .1037 | .1010 | .0982 | .0955 | .0928 | .0901 |
| 8 | .1302 | .1286 | .1269 | .1251 | .1232 | .1212 | .1191 | .1170 | .1148 | .1126 |
| 9 | .1317 | .1315 | .1311 | .1306 | .1300 | .1293 | .1284 | .1274 | .1263 | .1251 |
| 10 | .1198 | .1210 | .1219 | .1228 | .1235 | .1241 | .1245 | .1249 | .1250 | .1251 |
| 11 | .0091 | .1012 | .1031 | .1049 | .1067 | .1083 | .1098 | .1112 | .1125 | .1137 |
| 12 | .0752 | .0776 | .0799 | .0822 | .0844 | .0866 | .0888 | .0908 | .0928 | .0948 |
| 13 | .0526 | .0549 | .0572 | .0594 | .0617 | .0640 | .0662 | .0685 | .0707 | .0729 |
| 14 | .0342 | .0361 | .0380 | .0399 | .0419 | .0439 | .0459 | .0479 | .0500 | .0521 |
| 15 | .0208 | .0221 | .0235 | .0250 | .0265 | .0281 | .0297 | .0313 | .0330 | .0347 |
| 16 | .0118 | .0127 | .0137 | .0147 | .0157 | .0168 | .0180 | .0192 | .0204 | .0217 |
| 17 | .0063 | .0069 | .0075 | .0081 | .0088 | .0095 | .0103 | .0111 | .0119 | .0128 |
| 18 | .0032 | .0035 | .0039 | .0042 | .0046 | .0051 | .0055 | .0060 | .0065 | .0071 |
| 19 | .0015 | .0017 | .0019 | .0021 | .0023 | .0026 | .0028 | .0031 | .0034 | .0037 |
| 20 | .0007 | .0008 | .0009 | .0010 | .0011 | .0012 | .0014 | .0015 | .0017 | .0019 |
| 21 | .0003 | .0003 | .0004 | .0004 | .0005 | .0006 | .0006 | .0007 | .0008 | .0009 |
| 22 | .0001 | .0001 | .0002 | .0002 | .0002 | .0002 | .0003 | .0003 | .0004 | .0004 |
| 23 | .0000 | .0001 | .0001 | .0001 | .0001 | .0001 | .0001 | .0001 | .0002 | .0002 |
| 24 | .0000 | .0000 | .0000 | .0000 | .0000 | .0000 | .0000 | .0001 | .0001 | .0001 |

## TABLE IV:  Poisson Probabilities (continued)

| $x$ | 11 | 12 | 13 | 14 | 15 | 16 | 17 | 18 | 19 | 20 |
|---|---|---|---|---|---|---|---|---|---|---|
| 0 | .0000 | .0000 | .0000 | .0000 | .0000 | .0000 | .0000 | .0000 | .0000 | .0000 |
| 1 | .0002 | .0001 | .0000 | .0000 | .0000 | .0000 | .0000 | .0000 | .0000 | .0000 |
| 2 | .0010 | .0004 | .0002 | .0001 | .0000 | .0000 | .0000 | .0000 | .0000 | .0000 |
| 3 | .0037 | .0018 | .0008 | .0004 | .0002 | .0001 | .0000 | .0000 | .0000 | .0000 |
| 4 | .0102 | .0053 | .0027 | .0013 | .0006 | .0003 | .0001 | .0001 | .0000 | .0000 |
| 5 | .0224 | .0127 | .0070 | .0037 | .0019 | .0010 | .0005 | .0002 | .0001 | .0001 |
| 6 | .0411 | .0255 | .0152 | .0087 | .0048 | .0026 | .0014 | .0007 | .0004 | .0002 |
| 7 | .0646 | .0437 | .0281 | .0174 | .0104 | .0060 | .0034 | .0018 | .0010 | .0005 |
| 8 | .0888 | .0655 | .0457 | .0304 | .0194 | .0120 | .0072 | .0042 | .0024 | .0013 |
| 9 | .1085 | .0874 | .0661 | .0473 | .0324 | .0213 | .0135 | .0083 | .0050 | .0029 |
| 10 | .1194 | .1048 | .0859 | .0663 | .0486 | .0341 | .0230 | .0150 | .0095 | .0058 |
| 11 | .1194 | .1144 | .1015 | .0844 | .0663 | .0496 | .0355 | .0245 | .0164 | .0106 |
| 12 | .1094 | .1144 | .1099 | .0984 | .0829 | .0661 | .0504 | .0368 | .0259 | .0176 |
| 13 | .0926 | .1056 | .1099 | .1060 | .0956 | .0814 | .0658 | .0509 | .0378 | .0271 |
| 14 | .0728 | .0905 | .1021 | .1060 | .1024 | .0930 | .0800 | .0655 | .0514 | .0387 |
| 15 | .0534 | .0724 | .0885 | .0989 | .1024 | .0992 | .0906 | .0786 | .0650 | .0516 |
| 16 | .0367 | .0543 | .0719 | .0866 | .0960 | .0992 | .0963 | .0884 | .0772 | .0646 |
| 17 | .0237 | .0383 | .0550 | .0713 | .0847 | .0934 | .0963 | .0936 | .0863 | .0760 |
| 18 | .0145 | .0256 | .0397 | .0554 | .0706 | .0830 | .0909 | .0936 | .0911 | .0844 |
| 19 | .0084 | .0161 | .0272 | .0409 | .0557 | .0699 | .0814 | .0887 | .0911 | .0888 |
| 20 | .0046 | .0097 | .0177 | .0286 | .0418 | .0559 | .0692 | .0798 | .0866 | .0888 |
| 21 | .0024 | .0055 | .0109 | .0191 | .0299 | .0426 | .0560 | .0684 | .0783 | .0846 |
| 22 | .0012 | .0030 | .0065 | .0121 | .0204 | .0310 | .0433 | .0560 | .0676 | .0769 |
| 23 | .0006 | .0016 | .0037 | .0074 | .0133 | .0216 | .0320 | .0438 | .0559 | .0669 |
| 24 | .0003 | .0008 | .0020 | .0043 | .0083 | .0144 | .0226 | .0328 | .0442 | .0557 |
| 25 | .0001 | .0004 | .0010 | .0024 | .0050 | .0092 | .0154 | .0237 | .0336 | .0446 |
| 26 | .0000 | .0002 | .0005 | .0013 | .0029 | .0057 | .0101 | .0164 | .0246 | .0343 |
| 27 | .0000 | .0001 | .0002 | .0007 | .0016 | .0034 | .0063 | .0109 | .0173 | .0254 |
| 28 | .0000 | .0000 | .0001 | .0003 | .0009 | .0019 | .0038 | .0070 | .0117 | .0181 |
| 29 | .0000 | .0000 | .0001 | .0002 | .0004 | .0011 | .0023 | .0044 | .0077 | .0125 |
| 30 | .0000 | .0000 | .0000 | .0001 | .0002 | .0006 | .0013 | .0026 | .0049 | .0083 |
| 31 | .0000 | .0000 | .0000 | .0000 | .0001 | .0003 | .0007 | .0015 | .0030 | .0054 |
| 32 | .0000 | .0000 | .0000 | .0000 | .0001 | .0001 | .0004 | .0009 | .0018 | .0034 |
| 33 | .0000 | .0000 | .0000 | .0000 | .0000 | .0001 | .0002 | .0005 | .0010 | .0020 |
| 34 | .0000 | .0000 | .0000 | .0000 | .0000 | .0000 | .0001 | .0002 | .0006 | .0012 |
| 35 | .0000 | .0000 | .0000 | .0000 | .0000 | .0000 | .0000 | .0001 | .0003 | .0007 |
| 36 | .0000 | .0000 | .0000 | .0000 | .0000 | .0000 | .0000 | .0001 | .0002 | .0004 |
| 37 | .0000 | .0000 | .0000 | .0000 | .0000 | .0000 | .0000 | .0000 | .0001 | .0002 |
| 38 | .0000 | .0000 | .0000 | .0000 | .0000 | .0000 | .0000 | .0000 | .0000 | .0001 |
| 39 | .0000 | .0000 | .0000 | .0000 | .0000 | .0000 | .0000 | .0000 | .0000 | .0001 |

Source: Reprinted by permission from *Handbook of Probability and Statistics with Tables*, by R. S. Burington and D. C. May, Jr. New York: McGraw-Hill Book Company, 1953.

## TABLE V:  Chi-Square Critical Values

| $v$ | $\alpha = .995$ | $\alpha = .99$ | $\alpha = .975$ | $\alpha = .95$ | $\alpha = .90$ | $\alpha = .10$ | $\alpha = .05$ | $\alpha = .025$ | $\alpha = .01$ | $\alpha = .005$ | $v$ |
|---|---|---|---|---|---|---|---|---|---|---|---|
| 1 | .0000393 | .000157 | .000982 | .00393 | .0158 | 2.706 | 3.841 | 5.024 | 6.635 | 7.879 | 1 |
| 2 | .0100 | .0201 | .0506 | .103 | .2107 | 4.605 | 5.991 | 7.378 | 9.210 | 10.597 | 2 |
| 3 | .0717 | .115 | .216 | .352 | .5844 | 6.251 | 7.815 | 9.348 | 11.345 | 12.838 | 3 |
| 4 | .207 | .297 | .484 | .711 | 1.064 | 7.779 | 9.488 | 11.143 | 13.277 | 14.860 | 4 |
| 5 | .412 | .554 | .831 | 1.145 | 1.610 | 9.236 | 11.070 | 12.832 | 15.086 | 16.750 | 5 |
| 6 | .676 | .872 | 1.237 | 1.635 | 2.204 | 10.64 | 12.592 | 14.449 | 16.812 | 18.548 | 6 |
| 7 | .989 | 1.239 | 1.690 | 2.167 | 2.833 | 12.02 | 14.067 | 16.013 | 18.475 | 20.278 | 7 |
| 8 | 1.344 | 1.646 | 2.180 | 2.733 | 3.490 | 13.36 | 15.507 | 17.535 | 20.090 | 21.955 | 8 |
| 9 | 1.735 | 2.088 | 2.700 | 3.325 | 4.168 | 14.68 | 16.919 | 19.023 | 21.666 | 23.589 | 9 |
| 10 | 2.156 | 2.558 | 3.247 | 3.940 | 4.865 | 15.99 | 18.307 | 20.483 | 23.209 | 25.188 | 10 |
| 11 | 2.603 | 3.053 | 3.816 | 4.575 | 5.578 | 17.27 | 19.675 | 21.920 | 24.725 | 26.757 | 11 |
| 12 | 3.074 | 3.571 | 4.404 | 5.226 | 6.304 | 18.55 | 21.026 | 23.337 | 26.217 | 28.300 | 12 |
| 13 | 3.565 | 4.107 | 5.009 | 5.892 | 7.042 | 19.81 | 22.362 | 24.736 | 27.688 | 29.819 | 13 |
| 14 | 4.075 | 4.660 | 5.629 | 6.571 | 7.790 | 21.06 | 23.685 | 26.119 | 29.141 | 31.319 | 14 |
| 15 | 4.601 | 5.229 | 6.262 | 7.261 | 8.547 | 22.31 | 24.996 | 27.488 | 30.578 | 32.801 | 15 |
| 16 | 5.142 | 5.812 | 6.908 | 7.962 | 9.312 | 23.54 | 26.296 | 28.845 | 32.000 | 34.267 | 16 |
| 17 | 5.697 | 6.408 | 7.564 | 8.672 | 10.09 | 24.77 | 27.587 | 30.191 | 33.409 | 35.718 | 17 |
| 18 | 6.265 | 7.015 | 8.231 | 9.390 | 10.86 | 25.99 | 28.869 | 31.526 | 34.805 | 37.156 | 18 |
| 19 | 6.844 | 7.633 | 8.907 | 10.117 | 11.65 | 27.20 | 30.144 | 32.852 | 36.191 | 38.582 | 19 |
| 20 | 7.434 | 8.260 | 9.591 | 10.851 | 12.44 | 28.41 | 31.410 | 34.170 | 37.566 | 39.997 | 20 |
| 21 | 8.034 | 8.897 | 10.283 | 11.591 | 13.24 | 29.62 | 32.671 | 35.479 | 38.932 | 41.401 | 21 |
| 22 | 8.643 | 9.542 | 10.982 | 12.338 | 14.04 | 30.81 | 33.924 | 36.781 | 40.289 | 42.796 | 22 |
| 23 | 9.260 | 10.196 | 11.689 | 13.091 | 14.85 | 32.01 | 35.172 | 38.076 | 41.638 | 44.181 | 23 |
| 24 | 9.886 | 10.856 | 12.401 | 13.848 | 15.66 | 33.20 | 36.415 | 39.364 | 42.980 | 45.558 | 24 |
| 25 | 10.520 | 11.524 | 13.120 | 14.611 | 16.47 | 34.38 | 37.652 | 40.646 | 44.314 | 46.928 | 25 |
| 30 | 13.787 | 14.953 | 16.791 | 18.493 | 20.60 | 40.26 | 43.773 | 46.979 | 50.892 | 53.672 | 30 |
| 40 | 20.71 | 22.16 | 24.43 | 26.51 | 29.05 | 51.80 | 55.76 | 59.34 | 63.69 | 66.77 | 40 |
| 50 | 27.99 | 29.71 | 32.36 | 34.76 | 37.69 | 63.17 | 67.51 | 71.42 | 76.15 | 79.49 | 50 |
| 60 | 35.53 | 37.48 | 40.48 | 43.19 | 46.46 | 74.40 | 79.08 | 83.30 | 88.38 | 91.95 | 60 |
| 70 | 43.27 | 45.44 | 48.76 | 51.74 | 55.33 | 85.53 | 90.53 | 95.02 | 100.4 | 104.2 | 70 |
| 80 | 51.17 | 53.54 | 57.15 | 60.39 | 64.28 | 96.58 | 101.9 | 106.6 | 112.3 | 116.3 | 80 |
| 90 | 59.20 | 61.75 | 65.65 | 69.13 | 73.29 | 107.6 | 113.1 | 118.1 | 124.1 | 128.3 | 90 |
| 100 | 67.33 | 70.06 | 74.22 | 77.93 | 82.86 | 118.5 | 124.3 | 129.6 | 135.8 | 140.2 | 100 |

Adapted from *Biometrika Tables for Statisticians*, Table 8, Vol. I, Cambridge University Press, 1954, by permission of the *Biometrika* trustees.

*notes*

(a) For chi-square random variable $X$, $P(X > x_0) = \alpha$. Values for $\alpha$ are at headings of columns, and values for $x_0$ are in the body of the table. Degrees of freedom $v$ are listed in the extreme left and right margin columns.

(b) Other names for $x_0$ are $x_{\alpha,v}$ and $\chi^2_{\alpha,v}$. For example, suppose that $v = 3$ and $\alpha = .05$. Then, from the body of the table, we have $x_0 = 7.815$, and, thus $\chi^2_{.05,3} = 7.815$.

## TABLE VI: Standard Normal Distribution Probabilities

| $z_1$ | .00 | .01 | .02 | .03 | .04 | .05 | .06 | .07 | .08 | .09 |
|---|---|---|---|---|---|---|---|---|---|---|
| .0 | .0000 | .0040 | .0080 | .0120 | .0160 | .0199 | .0239 | .0279 | .0319 | .0359 |
| .1 | .0398 | .0438 | .0478 | .0517 | .0557 | .0596 | .0636 | .0675 | .0714 | .0753 |
| .2 | .0793 | .0832 | .0871 | .0910 | .0948 | .0987 | .1026 | .1064 | .1103 | .1141 |
| .3 | .1179 | .1217 | .1255 | .1293 | .1331 | .1368 | .1406 | .1443 | .1480 | .1517 |
| .4 | .1554 | .1591 | .1628 | .1664 | .1700 | .1736 | .1772 | .1808 | .1844 | .1879 |
| .5 | .1915 | .1950 | .1985 | .2019 | .2054 | .2088 | .2123 | .2157 | .2190 | .2224 |
| .6 | .2257 | .2291 | .2324 | .2357 | .2389 | .2422 | .2454 | .2486 | .2517 | .2549 |
| .7 | .2580 | .2611 | .2642 | .2673 | .2704 | .2734 | .2764 | .2794 | .2823 | .2852 |
| .8 | .2881 | .2910 | .2939 | .2967 | .2995 | .3023 | .3051 | .3078 | .3106 | .3133 |
| .9 | .3159 | .3186 | .3212 | .3238 | .3264 | .3289 | .3315 | .3340 | .3365 | .3389 |
| 1.0 | .3413 | .3438 | .3461 | .3485 | .3508 | .3531 | .3554 | .3577 | .3599 | .3621 |
| 1.1 | .3643 | .3665 | .3686 | .3708 | .3729 | .3749 | .3770 | .3790 | .3810 | .3830 |
| 1.2 | .3849 | .3869 | .3888 | .3907 | .3925 | .3944 | .3962 | .3980 | .3997 | .4015 |
| 1.3 | .4032 | .4049 | .4066 | .4082 | .4099 | .4115 | .4131 | .4147 | .4162 | .4177 |
| 1.4 | .4192 | .4207 | .4222 | .4236 | .4251 | .4265 | .4279 | .4292 | .4306 | .4319 |
| 1.5 | .4332 | .4345 | .4357 | .4370 | .4382 | .4394 | .4406 | .4418 | .4429 | .4441 |
| 1.6 | .4452 | .4463 | .4474 | .4484 | .4495 | .4505 | .4515 | .4525 | .4535 | .4545 |
| 1.7 | .4554 | .4564 | .4573 | .4582 | .4591 | .4599 | .4608 | .4616 | .4625 | .4633 |
| 1.8 | .4641 | .4649 | .4656 | .4664 | .4671 | .4678 | .4686 | .4693 | .4699 | .4706 |
| 1.9 | .4713 | .4719 | .4726 | .4732 | .4738 | .4744 | .4750 | .4756 | .4761 | .4767 |
| 2.0 | .4772 | .4778 | .4783 | .4788 | .4793 | .4798 | .4803 | .4808 | .4812 | .4817 |
| 2.1 | .4821 | .4826 | .4830 | .4834 | .4838 | .4842 | .4846 | .4850 | .4854 | .4857 |
| 2.2 | .4861 | .4864 | .4868 | .4871 | .4875 | .4878 | .4881 | .4884 | .4887 | .4890 |
| 2.3 | .4893 | .4896 | .4898 | .4901 | .4904 | .4906 | .4909 | .4911 | .4913 | .4916 |
| 2.4 | .4918 | .4920 | .4922 | .4925 | .4927 | .4929 | .4931 | .4932 | .4934 | .4936 |
| 2.5 | .4938 | .4940 | .4941 | .4943 | .4945 | .4946 | .4948 | .4949 | .4951 | .4952 |
| 2.6 | .4953 | .4955 | .4956 | .4957 | .4959 | .4960 | .4961 | .4962 | .4963 | .4964 |
| 2.7 | .4965 | .4966 | .4967 | .4968 | .4969 | .4970 | .4971 | .4972 | .4973 | .4974 |
| 2.8 | .4974 | .4975 | .4976 | .4977 | .4977 | .4978 | .4979 | .4979 | .4980 | .4981 |
| 2.9 | .4981 | .4982 | .4982 | .4983 | .4984 | .4984 | .4985 | .4985 | .4986 | .4986 |
| 3.0 | .4987 | .4987 | .4987 | .4988 | .4988 | .4989 | .4989 | .4989 | .4990 | .4990 |

*notes*

(a) Here, $P(0 < Z < z_1) = (1/\sqrt{2\pi}) \int_0^{z_1} e^{-z^2/2} \, dz$.

(b) As an example of table usage, $P(0 < Z < 1.96) = .4750$. Go down in the extreme left column to 1.9. Go across in the extreme top row to .06. The row for 1.9 and the column for .06 intersect at .4750.

(c) For $z_1 = 4.0$, 5.0, and 6.0, the probabilities are .49997, .4999997, and .499999999.

# APPENDIX B
# USEFUL MATHEMATICAL RESULTS

## 1-1. Algebraic Results

**(A)** Given the quadratic equation $ax^2 + bx + c = 0$, where $a \neq 0$. Let $\Delta = b^2 - 4ac$. The two solutions of the quadratic equation are

$$x_1 = (-b + \sqrt{\Delta})/(2a) \qquad \text{and} \qquad x_2 = (-b - \sqrt{\Delta})/(2a)$$

**(B)**
$$\sum_{i=1}^{n} i = 1 + 2 + 3 + \ldots + n = \frac{n(n+1)}{2}$$

**(C)**
$$\sum_{i=1}^{n} i^2 = 1^2 + 2^2 + 3^2 + \ldots + n^2 = \frac{n(n+1)(2n+1)}{6}$$

**(D)** *Sum of Geometric Progression:*

$$\sum_{i=1}^{n} ar^{(i-1)} = a + ar + ar^2 + \ldots + ar^{n-1} = \frac{a(1-r^n)}{1-r} \qquad \text{provided that } r \neq 1$$

**(E)** *Binomial Sum:*

$$\sum_{r=0}^{n} \frac{n!}{r!(n-r)!} a^{n-r}b^r = a^n + na^{n-1}b + [n(n-1)/2]a^{n-2}b^2$$

$$+ \ldots + \frac{n!}{r!(n-r)!} a^{n-r}b^r + \ldots + [n(n-1)/2]a^2b^{n-2}$$

$$+ nab^{n-1} + b^n = (a+b)^n$$

## 1-2. Basic Results from Calculus

Results (A) through (E) are frequently occurring anti-derivative [integral] formulas.

**(A)**
$$\int x^n \, dx = x^{n+1}/(n+1) + C \qquad \text{for } n \neq -1$$

**(B)**
$$\int \frac{dx}{x} = \ln|x| + C$$

**(C)**
$$\int e^{ax} \, dx = \frac{e^{ax}}{a} + C \qquad \text{for } a \neq 0$$

**(D)**
$$\int xe^{ax} \, dx = \left(\frac{e^{ax}}{a^2}\right)(ax - 1) + C \qquad \text{for } a \neq 0$$

**(E)**
$$\int x^2 e^{ax} \, dx = \left(\frac{e^{ax}}{a^3}\right)(a^2x^2 - 2ax + 2) + C \qquad \text{for } a \neq 0$$

**(F)** *Fundamental Theorem of Calculus:* If $f(x)$ is continuous and $F'(x) = f(x)$ [equivalently, $\int f(x)\,dx = F(x) + C$] on $a \leq x \leq b$, then $\int_a^b f(x)\,dx = F(b) - F(a)$. *Example:* Since $\int e^{3x}\,dx = (\frac{1}{3})e^{3x} + C$, then $\int_0^2 e^{3x}\,dx = (\frac{1}{3})e^{3x}]_0^2 = (\frac{1}{3})[e^6 - 1]$.

**(G)** Taylor infinite series for $e^z$ in powers of $z$ [series below converges to $e^z$ for $-\infty < z < \infty$]:

$$e^z = 1 + z + z^2/(2!) + \ldots + z^x/(x!) + \ldots = \sum_{x=0}^{\infty} z^x/(x!)$$

**(H)** Geometric infinite series. [The series on the right converges to $a/(1 - z)$ for $-1 < z < 1$.]

$$\frac{a}{1 - z} = a + az + az^2 + \ldots + az^x + \ldots = \sum_{x=0}^{\infty} az^x$$

**(I)**    $$\ln(1 + z) = z - z^2/2 + z^3/3 - z^4/4 + \ldots = \sum_{x=1}^{\infty} [(-1)^{(x+1)}z^x]/x$$

$$\text{for } -1 < z \leq 1$$

## 1-3.  Other Useful Results from Calculus

**(A)** *L'Hôpital's Rule; 0/0 Case:* If $\lim_{x \to a} f(x) = 0$ and $\lim_{x \to a} g(x) = 0$, then $\lim_{x \to a} f(x)/g(x)$ is called an *indeterminate form of the 0/0 type*. Then, if $\lim_{x \to a} f'(x)/g'(x)$ has a finite value $L$, or if this limit is $+\infty$ or $-\infty$, then

$$\lim_{x \to a} f(x)/g(x) = \lim_{x \to a} f'(x)/g'(x)$$

Here, $\lim_{x \to a}$ can be replaced by $\lim_{x \to a^+}$, $\lim_{x \to a^-}$, $\lim_{x \to \infty}$, or $\lim_{x \to -\infty}$.

**(B)** *L'Hôpital's Rule; $\pm\infty/\pm\infty$ Case:* If $\lim_{x \to a} f(x) = \pm\infty$ and $\lim_{x \to a} g(x) = \pm\infty$, then $\lim_{x \to a} f(x)/g(x)$ is called an *indeterminate form of the $\pm\infty/\pm\infty$ type*. Then, if $\lim_{x \to a} f'(x)/g'(x)$ has a finite value $L$ or if this limit is $+\infty$ or $-\infty$, then

$$\lim_{x \to a} f(x)/g(x) = \lim_{x \to a} f'(x)/g'(x)$$

Here, too, $\lim_{x \to a}$ can be replaced by $\lim_{x \to a^+}$, $\lim_{x \to a^-}$, $\lim_{x \to \infty}$, or $\lim_{x \to -\infty}$.

**(C)** *Extensions of L'Hôpital's Rule:* Refer to cases (A) and (B) above. If $\lim_{x \to a} f'(x)/g'(x)$ is itself an indeterminate form of the 0/0 or $\pm\infty/\pm\infty$ type, then the process indicated in case (A) or case (B) can be repeated. For example, if $\lim_{x \to a} f''(x)/g''(x)$ has a finite value $L$ or if this limit is $+\infty$ or $-\infty$, then

$$\lim_{x \to a} f'(x)/g'(x) = \lim_{x \to a} f''(x)/g''(x)$$

and thus

$$\lim_{x \to a} f(x)/g(x) = \lim_{x \to a} f''(x)/g''(x)$$

**(D)** *Leibnitz's Rule for Differentiation Under the Integral Sign:* Given that $H(x) = \int_{a(x)}^{b(x)} f(t, x) \, dt$, where $a(x)$ and $b(x)$ are functions of $x$ alone and $f(t, x)$ is a function of both $t$ and $x$. Then,

$$\frac{dH}{dx} = \int_{a(x)}^{b(x)} \frac{\partial f}{\partial x} \, dt + f\big(b(x), x\big)b'(x) - f\big(a(x), x\big)a'(x)$$

provided that $f(t, x)$, $\partial f/\partial x$, $a(x)$, $a'(x)$, $b(x)$, and $b'(x)$ satisfy certain continuity conditions with respect to their independent variables.

**(E)** *Leibnitz's Rule—Special Cases:* Refer to case (D). Suppose that $f(t, x)$ is merely $f(t)$, a function of $t$ alone. Then, $H(x) = \int_{a(x)}^{b(x)} f(t) \, dt$, and the equation for $dH/dx$ in case (D) reduces to

$$\frac{dH}{dx} = f\big(b(x)\big)b'(x) - f\big(a(x)\big)a'(x) \tag{E-1}$$

If $a(x)$ is a constant in Eq. (E-1), which means we can write $a(x) = a$, then Eq. (E-1) reduces to

$$\frac{dH}{dx} = f\big(b(x)\big)b'(x) \tag{E-2}$$

If $b(x)$ is a constant in Eq. (E-1), which means we can write $b(x) = b$, then Eq. (E-1) reduces to

$$\frac{dH}{dx} = -f(a(x))a'(x) \qquad \textbf{(E-3)}$$

If $a(x)$ is constant [say, $a(x) = a$] and $b(x) = x$ in Eq. (E-1) or (E-2), then it follows that

$$\frac{dH}{dx} = f(x) \qquad \text{or, equivalently} \qquad \frac{d}{dx}\left(\int_a^x f(t)\,dt\right) = f(x) \qquad \textbf{(E-4)}$$

**(F)** *Gamma Function Results:* The gamma function, with argument $\alpha$, denoted by $\Gamma(\alpha)$, is defined as follows:

$$\Gamma(\alpha) = \int_{t=0}^{\infty} t^{(\alpha-1)}e^{-t}\,dt \qquad \text{for } \alpha > 0 \qquad \textbf{(F-1)}$$

It is true that $\Gamma(1) = 1$, and

$$\Gamma(\alpha + 1) = \alpha\Gamma(\alpha) \qquad \text{for } \alpha > 0 \qquad \textbf{(F-2)}$$

Also, the following are true:

$$\Gamma(\alpha + 1) = \alpha! \qquad \text{if } \alpha \text{ is a nonnegative integer} \qquad \textbf{(F-3)}$$

$$\Gamma(\tfrac{1}{2}) = \sqrt{\pi} \qquad \textbf{(F-4)}$$

**(G)** *Miscellaneous Improper Integrals of the $\int_0^\infty$ Type:*

$$\int_0^{\infty} \exp(-at^2)\,dt = \left(\frac{1}{2}\right)\sqrt{\pi/a} \qquad \text{for } a > 0 \qquad \textbf{(G-1)}$$

$$\int_0^{\infty} t^m \exp(-at^2)\,dt = \left(\frac{1}{2a^{(m+1)/2}}\right)\Gamma[(m+1)/2] \qquad \text{for } a > 0,\, m > -1 \qquad \textbf{(G-2)}$$

$$\int_0^{\infty} t^{m-1} \exp(-at)\,dt = \Gamma(m)/[a^m] \qquad \text{for } a > 0,\, m > 0 \qquad \textbf{(G-3)}$$

# 1-4. Results for Key Discrete Probability Distributions

**(A)** *Binomial* $[b(x; n, p)]$:

$$f(x) = \left(\frac{n!}{x!(n-x)!}\right)p^x(1-p)^{n-x} \qquad \text{for } x = 0, 1, 2, \ldots, n$$

$$\mu = np; \qquad \sigma^2 = np(1-p); \qquad M_X(t) = [1 + p(e^t - 1)]^n$$

(Bernoulli distribution is binomial distribution for $n = 1$.)

**(B)** *Geometric* $[g(x; p)]$:

$$f(x) = (1-p)^{x-1}p \qquad \text{for } x = 1, 2, 3, \ldots$$

$$\mu = 1/p; \qquad \sigma^2 = (1-p)/[p^2]; \qquad M_X(t) = [pe^t]/[1 - (1-p)e^t]$$
$$\text{for } t < -\ln(1-p)$$

**(C)** *Negative Binomial* $[b^*(x; k, p)]$:

$$f(x) = \binom{x-1}{k-1}p^k(1-p)^{x-k} \qquad \text{for } x = k, k+1, k+2, \ldots$$

$$\mu = k/p; \qquad \sigma^2 = [k(1-p)]/[p^2]; \qquad M_X(t) = \{[pe^t]/[1 - (1-p)e^t]\}^k$$
$$\text{for } t < -\ln(1-p)$$

**(D)** *Hypergeometric* $[h(x; n, a, M)]$:

$$f(x) = \left[\binom{a}{x}\binom{M-a}{n-x}\right]\Big/\binom{M}{n} \qquad \text{for } x = 0, 1, \ldots, n$$

Here, $n$ objects are drawn without replacement from a set of $M$ objects containing $a$ successes. The random variable $X$ stands for the number of successes in the $n$ objects drawn. In the following, let $p = a/M$. Then,

$$\mu = np; \qquad \sigma^2 = np(1-p)\left(\frac{M-n}{M-1}\right)$$

A simple expression for $M_X(t)$ is not available.

**(E)** *Poisson* $[p(x; \lambda)]$:

$$f(x) = \frac{\lambda^x e^{-\lambda}}{x!} \qquad \text{for } x = 0, 1, 2, \ldots$$

$$\mu = \lambda; \qquad \sigma^2 = \lambda; \qquad M_X(t) = \exp[\lambda(e^t - 1)]$$

## 1-5.  Results for Key Continuous Probability Distributions

**(A)** *Uniform:*

$$f(x) = 1/(b-a) \qquad \text{for } a < x < b, f(x) = 0 \text{ elsewhere}$$

$$\mu = (a+b)/2; \qquad \sigma^2 = (b-a)^2/12; \qquad M_X(t) = (e^{bt} - e^{at})/[t(b-a)]$$
$$\text{for } t \neq 0$$

**(B)** *Normal:*

$$f(x) = \frac{1}{\sigma\sqrt{2\pi}}\exp\left[-\frac{1}{2}\left(\frac{x-\mu}{\sigma}\right)^2\right] \qquad \text{for } -\infty < x < \infty$$

The mean is $\mu$ and the standard deviation is $\sigma$ in the above equation.

$$M_X(t) = \exp[\mu t]\exp[\sigma^2 t^2/2]$$

**(C)** *Exponential:*

$$f(x) = \left(\frac{1}{\theta}\right)e^{-x/\theta} \qquad \text{for } x > 0, f(x) = 0 \text{ elsewhere}$$

Also, $\theta > 0$.

$$\mu = \theta; \qquad \sigma^2 = \theta^2; \qquad M_X(t) = (1 - \theta t)^{-1} \qquad \text{for } t < 1/\theta$$

**(D)** *Gamma:*

$$f(x) = kx^{\alpha-1}e^{-x/\beta} \qquad \text{for } x > 0, f(x) = 0 \text{ elsewhere}$$

Here, $k = 1/[\beta^\alpha \Gamma(\alpha)]$, and $\alpha > 0, \beta > 0; \mu = \alpha\beta, \sigma^2 = \alpha\beta^2, M_X(t) = (1 - \beta t)^{-\alpha}$ for $t < 1/\beta$.

**(E)** *Chi-Square* [This is gamma distribution for $\beta = 2$ and $\alpha = v/2$]:

$$f(x) = kx^{(v-2)/2}e^{-x/2} \qquad \text{for } x > 0, f(x) = 0 \text{ elsewhere}$$

Here, $k = 1/[2^{v/2}\Gamma(v/2)]$, and $v > 0; \mu = v, \sigma^2 = 2v, M_X(t) = (1 - 2t)^{-v/2}$ for $t < 1/2$.

**(F)** *Beta:*

$$f(x) = kx^{\alpha-1}(1-x)^{\beta-1} \qquad \text{for } 0 < x < 1, f(x) = 0 \text{ elsewhere}$$

Here, $k = \Gamma(\alpha + \beta)/[\Gamma(\alpha)\Gamma(\beta)]$, and $\alpha > 0, \beta > 0; \mu = \alpha/(\alpha + \beta), \sigma^2 = [\alpha\beta]/[(\alpha + \beta)^2(\alpha + \beta + 1)]$. A simple expression for $M_X(t)$ is not available.

# REFERENCES

Anton, H. *Calculus*, 3rd ed. (New York, N.Y.: Wiley, 1988).

Blum, J. R. and Rosenblatt, J. I. *Probability and Statistics* (Phila., Pa.: Saunders, 1972).

DeGroot, M. H. *Probability and Statistics*, 2nd ed. (Reading, Ma.: Addison-Wesley, 1986).

Feller, W. *An Introduction to Probability Theory and Its Applications*, Vol. 1, 3rd ed. (New York, N.Y.: Wiley, 1968).

Freund, J. E. *Modern Elementary Statistics*, 7th ed. (Englewood Cliffs, N.J.: Prentice-Hall, 1988).

Freund, J. E. and Walpole, R. *Mathematical Statistics*, 4th ed. (Englewood Cliffs, N.J.: Prentice-Hall, 1987).

Goldberg, S. *Probability, an Introduction* (Englewood Cliffs, N.J.: Prentice-Hall, 1960).

Goodman, R. *Introduction to Stochastic Models* (Menlo Park, Ca.: Benjamin Cummings, 1988).

Hoel, P. G. *Introduction to Mathematical Statistics*, 5th ed. (New York, N.Y.: Wiley, 1984).

Hogg, R. V. and Craig, A. T. *Introduction to Mathematical Statistics*, 4th ed. (New York, N.Y.: Macmillan, 1978).

Hogg, R. V. and Tanis, E. A. *Probability and Statistical Inference*, 2nd ed. (New York, N.Y.: Macmillan, 1983).

Kreysig, E. *Introductory Mathematical Statistics* (New York, N.Y.: Wiley, 1970).

Lapin, L. L. *Business Statistics* (San Diego, Ca.: Harcourt Brace Jovanovich—College Outline Series, 1984).

Larsen, R. J. and Marx, M. L. *An Introduction to Mathematical Statistics and Its Applications*, 2nd ed. (Englewood Cliffs, N.J.: Prentice-Hall, 1986).

Larson, H. J. *Introduction to Probability Theory and Statistical Inference*, 2nd ed. (New York, N.Y.: Wiley, 1974).

Mendenhall, W., Scheaffer, R. L., and Wackerly, D. D. *Mathematical Statistics with Applications*, 3rd ed. (Boston, Ma.: Duxbury Press, 1986).

Meyer, P. L. *Introductory Probability and Statistical Applications*, 2nd ed. (Reading, Ma.: Addison-Wesley, 1970).

Miller, I. and Freund, J. E. *Probability and Statistics for Engineers*, 3rd ed. (Englewood Cliffs, N.J.: Prentice-Hall, 1985).

Mood, A. M., Graybill, F. A., and Boes, D. C. *Introduction to the Theory of Statistics*, 3rd ed. (New York, N.Y.: McGraw-Hill, 1974).

Niven, I. *Mathematics of Choice* (Washington, D.C.: Mathematical Association of America-New Mathematical Library, 1965).

Ross, S. M. *A First Course in Probability*, 2nd ed. (New York, N.Y.: Macmillan, 1984).

Snell, J. L. *Introduction to Probability* (New York, N.Y.: Random House, 1988).

Strait, P. T. *A First Course in Probability and Statistics with Applications*, 2nd ed. (San Diego, Ca.: Harcourt Brace Jovanovich, 1989).

Tanis, E. A. *Statistics I: Descriptive Statistics and Probability* (San Diego, Ca.: Harcourt Brace Jovanovich—College Outline Series, 1987a).

Tanis, E. A. *Statistics II: Estimation and Tests of Hypotheses* (San Diego, Ca.: Harcourt Brace Jovanovich—College Outline Series, 1987b).

Walpole, R. E. and Myers, R. H. *Probability and Statistics for Engineers and Scientists*, 3rd ed. (New York, N.Y.: Macmillan, 1985).

# INDEX